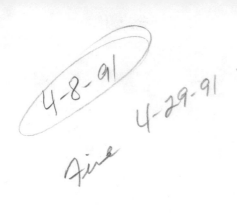

4-8-91

Fine 4-29-91

W9-BVT-635

ELEMENTARY
LINEAR ALGEBRA

FIFTH EDITION

HOWARD ANTON

Drexel University

JOHN WILEY & SONS
New York · Chichester · Brisbane · Toronto · Singapore

Copyright © 1973, 1977, 1981, 1984, and 1987, by Anton Textbooks, Inc.

All rights reserved. Published simultaneously in Canada.

Reproduction or translation of any part of
this work beyond that permitted by Sections
107 and 108 of the 1976 United States Copyright
Act without the permission of the copyright
owner is unlawful. Requests for permission
or further information should be addressed to
the Permissions Department, John Wiley & Sons.

Library of Congress Cataloging in Publication Data
Anton, Howard.
    Elementary linear algebra.

    Includes index.
    1. Algebra, Linear. I. Title.
QA184.A57 1987   512'.5   86-11087
ISBN 0-471-84819-0

Printed in the United States of America

10 9 8

To Pat and my children:
Brian, David, and Lauren

Preface

This textbook provides an elementary treatment of linear algebra that is suitable for students in their freshman or sophomore year. Calculus is *not* a prerequisite. I have, however, included a number of exercises for students with a calculus background; these are clearly marked: "For students who have studied calculus."

My aim in writing this book is to present the fundamentals of linear algebra in the clearest possible way. Pedagogy is the main consideration; formalism is secondary. Where possible, basic ideas are studied by means of computational examples (over 200 of them) and geometrical interpretation.

My treatment of proofs varies. Those proofs that are elementary and have significant pedagogical content are presented precisely, in a style tailored for beginners. A few proofs that are more difficult, but pedagogically valuable, are placed at the ends of the sections and marked "Optional." Still other proofs are omitted completely, with emphasis placed on applying the theorem. Whenever a proof is omitted, I try to motivate the result, often with a discussion about its interpretation in 2-space or 3-space.

It is my experience that Σ-notation is more of a hindrance than a help for beginners in linear algebra. Therefore, I have generally avoided its use.

It is a pedagogical axiom that a teacher should proceed from the familiar to the unfamiliar and from the concrete to the abstract. The ordering of the chapters reflects my adherence to this tenet.

Chapter 1 deals with systems of linear equations, how to solve them, and some of their properties. It also contains the basic material on matrices and their arithmetic properties.

Chapter 2 introduces determinants. I have used the classical permutation approach. In my opinion, it is less abstract than the approach through n-linear alternating forms, gives the student a better intuitive grasp of the subject than an inductive development, and provides the best foundation for a future study of advanced topics in linear algebra.

Chapter 3 introduces vectors in 2-space and 3-space as arrows and develops the analytic geometry of lines and planes in 3-space. Depending on the background of the students, this chapter can be omitted without a loss of continuity (see the guide for the instructor that follows this preface).

Chapter 4 and **Chapter 5** develop the basic results about real finite-dimensional vector spaces and linear transformations. I begin with a study of R^n and proceed slowly to the general concept of a vector.

Chapter 6 deals with the eigenvalue problem and diagonalization.

Chapter 7 gives some applications of linear algebra to problems of approximation, systems of differential equations, Fourier series, and identifying conic sections and quadric surfaces. Applications to business, biology, engineering, economics, the social sciences, and the physical sciences are included in an optional paperback supplement to this text, *Applications of Linear Algebra*, by Chris Rorres and the author. Included in the supplement are such topics as curves of best fit to empirical data, population dynamics, Markov processes, optimal harvesting, Leontief models in economics, engineering applications, graph theory, and an introduction to linear programming. There is also an alternate version of this text, called *Elementary Linear Algebra with Applications*, that combines this text and most of the material in *Applications of Linear Algebra*.

Chapter 8 introduces numerical methods of linear algebra; it does not require access to computing facilities since the exercises can be solved by hand computation or with the use of a pocket calculator. This chapter gives the student a basic understanding of how certain linear algebra problems are solved practically. Too many students complete their linear algebra studies with the naive belief that eigenvalues are found in practice by solving the characteristic equation. Some instructors may wish to use this section in conjunction with the *LINEAR-KIT*^{TM} software package that is available as a supplement to this text.

Chapter 9 gives a brief introduction to complex numbers and then proceeds to the development of complex vector spaces and inner product spaces. Unitary, Hermitian, and normal matrices are studied, and the chapter concludes with the proof that symmetric matrices have real eigenvalues.

I have included a large number of exercises. Each exercise set begins with routine drill problems and progresses towards more theoretical problems. Answers to all the computational problems are given at the end of the text.

Since there is more material in this book than can be covered in a one-semester or one-quarter course, the instructor will have to make a selection of topics. To help in this selection, I have provided a guide to the instructor following this preface.

NEW FEATURES IN THE FIFTH EDITION

The wide acceptance of the first four editions has been most gratifying and I am grateful for the many favorable comments and constructive suggestions received from users. In response to these suggestions I have made the following changes:

- During the 14 years that have passed since this text was first published, the widespread use of computers has significantly increased the importance of numerical methods in linear algebra. In recognition of this I have rewritten Chapter 8, which deals with this topic. There is a new section on *LU*-decompositions and a new section in which the various methods for solving linear systems are compared in terms of the number of operations they require. The inverse power method for finding "smallest" eigenvalues has been introduced, and the student is given more information on the advantages and disadvantages of various numerical methods. However, this chapter is still intended only as a brief introduction to the topic.

- The Supplementary Exercises have been greatly expanded to provide a richer variety of problems.

- Examples have been added and exposition improved in many parts of the text.

- To reduce the level of abstraction, the material on length and angle in inner product spaces has been rewritten, and more examples of inner products have been added.

- As a general rule, I prefer to explain concepts in words rather than symbols. However, so many readers asked that I introduce the notation $[T]_B$ for the matrix of a linear operator with respect to a basis B and $[T]_{B,B'}$ for the matrix of a linear transformation with respect to bases B and B' that I have done so.

- New material on positive definite matrices and quadratic forms has been added.

SUPPLEMENTARY MATERIALS

- *Student Solutions Manual for Elementary Linear Algebra* by Elizabeth M. Grobe and Charles A. Grobe, Jr. contains detailed solutions to most theoretical exercises and many computational exercises.

- *Applications of Linear Algebra* by Chris Rorres and Howard Anton contains applications to business, biology, engineering, economics, the social sciences, and the physical sciences. This supplement also contains a "minicourse" in linear programming which can be covered in about six lectures.

- *Solutions Manual for Applications of Linear Algebra* contains detailed solutions to all exercises.

- *Elementary Linear Algebra with Applications* by Howard Anton and Chris Rorres is an expanded version of this text that includes most of the material from *Applications of Linear Algebra*.

- *LINEAR-KIT*™—A new linear algebra software package that can perform most of the basic linear algebra computations using either fractions or decimals. It is available for IBM compatible computers, the Apple IIe, and the Apple IIc.

- *LINEAR-KIT*™ *Problem Book*—A book of linear algebra problems designed to be solved using the *LINEAR-KIT*™ software package.

If any of these supplements is not in stock, ask your bookstore manager to order a copy for you.

HOWARD ANTON

A Guide
for the Instructor

STANDARD COURSE

Chapter 3 can be omitted without loss of continuity if the students have previously studied lines, planes, and geometric vectors in 2-space and 3-space. Depending on the available time and the background of the students, the instructor may wish to add all or part of this chapter to the following suggested core material:

Chapter 1	7 lectures
Chapter 2	5 lectures
Chapter 4	14 lectures
Chapter 5	6 lectures
Chapter 6	5 lectures

This schedule is rather liberal; it allows a fair amount of classroom time for discussion of homework problems but assumes that little classroom time is devoted to the material marked "optional." The instructor can build on this core as time permits by including lectures on optional material, Chapter 3, Chapter 7, Chapter 8, and Chapter 9.

APPLICATIONS-ORIENTED COURSE

Instructors who want to emphasize applications or numerical methods may prefer the following program, which reaches eigenvalues and eigenvectors more quickly:

Chapter 1	7 lectures
Chapter 2	5 lectures
Sections 4.1–4.6	8 lectures
Sections 5.1–5.2	3 lectures
Sections 6.1–6.2	3 lectures

Whenever Sections 5.4 and 5.5 are deleted, the instructor should omit the optional material at the end of 6.1, begin Section 6.2 with the matrix form of the diagonalization problem, and delete Example 9 in that section.

Once the above material is covered, the instructor can choose topics from Chapter 7, Chapter 8, or Chapter 9 in the text or else choose topics from the supplement, *Applications of Linear Algebra.* Depending on which topics are selected, it may be necessary to cover some of the material in Sections 4.7–4.10 or 5.3–5.5.

COMPUTER-ORIENTED COURSE

Either the standard course or the application-oriented course can be taught in conjunction with the *LINEAR-KIT*™ software together with the *LINEAR-KIT*™ *Problem Book* to supplement the exercises.

Because problems can be solved so quickly using *LINEAR-KIT*™, students will need much less time to solve homework problems, which should allow them to cover more material or the same material in more depth.

Acknowledgments

I express my appreciation for the helpful guidance provided by the following people:

Joseph Buckley, *Western Michigan University*
Harold S. Engelsohn, *Kingsborough Community College*
Lawrence D. Kugler, *University of Michigan*
Robert W. Negus, *Rio Hondo Junior College*
Hal G. Moore, *Brigham Young University*
William A. Brown, *University of Maine at Portland-Gorham*
Ralph P. Grimaldi, *Rose-Hulman Institute of Technology*
Robert M. McConnel, *University of Tennessee*
James R. Wall, *Auburn University*
Roger H. Marty, *Cleveland State University*
Donald R. Sherbert, *University of Illinois*
Joseph L. Ullman, *University of Michigan*
Arthur G. Wasserman, *University of Michigan*
Collin J. Hightower, *University of Colorado*
Marjorie E. Fitting, *San Jose State University*
Bruce Edwards, *University of Florida*
Garret Etgen, *University of Houston*
Donald P. Minassian, *Butler University*
David E. Flesner, *Gettysburg College*
Arlene Kleinstein
Bart S. Ng, *Purdue University*
Mathew Gould, *Vanderbilt University*
C. S. Ballantine, *Oregon State University*
Douglas McLeod, *Drexel University*
Craig Miller, *University of Pennsylvania*
William Scott, *University of Utah*
F. P. J. Rimrott, *University of Toronto*

I am also grateful to William F. Trench of Trinity University, whose suggestions immeasurably improved both the style and content of the text and Steven C. Althoen of the University of Michigan-Flint who pointed out that, contrary to my statement in previous editions, the Gauss-Jordan reduction procedure is named after the German engineer Wilhelm Jordan and not the famous mathematician, Camille Jordan. Thanks are also due to Dale Lick who encouraged my work in its early stages, to Frederick C. Corey who helped make the first edition a reality, and to my former editor, Gary W. Ostedt, who guided this book through three successful editions. Finally, I am indebted to the entire production staff of Wiley and especially to my editor, Robert Pirtle, for his invaluable help during the development of this new edition.

H. A.

Contents

---

* Additional applications to business, economics, and the physical and social sciences are available in the supplement to this text, *Applications of Linear Algebra* or in the expanded version of this text, *Elementary Linear Algebra with Applications.*

CHAPTER ONE

Systems of Linear Equations and Matrices

1.1 INTRODUCTION TO SYSTEMS OF LINEAR EQUATIONS

In this section we introduce basic terminology and discuss a method for solving systems of linear equations.

A line in the xy-plane can be represented algebraically by an equation of the form

$$a_1 x + a_2 y = b$$

An equation of this kind is called a linear equation in the variables x and y. More generally, we define a *linear equation* in the n variables $x_1, x_2, \ldots, x_n$ to be one that can be expressed in the form

$$a_1 x_1 + a_2 x_2 + \cdots + a_n x_n = b$$

where $a_1, a_2, \ldots, a_n$ and b are real constants.

Example 1

The following are linear equations:

$$x + 3y = 7 \qquad\qquad x_1 - 2x_2 - 3x_3 + x_4 = 7$$
$$y = \tfrac{1}{2}x + 3z + 1 \qquad\qquad x_1 + x_2 + \cdots + x_n = 1$$

Observe that a linear equation does not involve any products or roots of variables. All variables occur only to the first power and do not appear as arguments for trigonometric, logarithmic, or exponential functions. The following are *not* linear equations:

$$x + 3y^2 = 7 \qquad\qquad 3x + 2y - z + xz = 4$$
$$y - \sin x = 0 \qquad\qquad \sqrt{x_1} + 2x_2 + x_3 = 1$$

A *solution* of a linear equation $a_1x_1 + a_2x_2 + \cdots + a_nx_n = b$ is a sequence of n numbers $s_1, s_2, \ldots, s_n$ such that the equation is satisfied when we substitute $x_1 = s_1, x_2 = s_2, \ldots, x_n = s_n$. The set of all solutions of the equation is called its *solution set*.

Example 2

Find the solution set of each of the following:

(i) $4x - 2y = 1$ (ii) $x_1 - 4x_2 + 7x_3 = 5$

To find solutions of (i), we can assign an arbitrary value to x and solve for y, or choose an arbitrary value for y and solve for x. If we follow the first approach and assign x an arbitrary value t, we obtain

$$x = t, \qquad y = 2t - \tfrac{1}{2}$$

These formulas describe the solution set in terms of the arbitrary parameter t. Particular numerical solutions can be obtained by substituting specific values for t. For example, $t = 3$ yields the solution $x = 3$, $y = 11/2$ and $t = -1/2$ yields the solution $x = -1/2$, $y = -3/2$.

If we follow the second approach and assign y the arbitrary value t, we obtain

$$x = \tfrac{1}{2}t + \tfrac{1}{4}, \qquad y = t$$

Although these formulas are different from those obtained above, they yield the same solution set as t varies over all possible real numbers. For example, the previous formulas gave the solution $x = 3$, $y = 11/2$ when $t = 3$, while these formulas yield this solution when $t = 11/2$.

To find the solution set of (ii) we can assign arbitrary values to any two variables and solve for the third variable. In particular, if we assign arbitrary values s and t to x_2 and x_3, respectively, and solve for x_1, we obtain

$$x_1 = 5 + 4s - 7t, \qquad x_2 = s, \qquad x_3 = t$$

A finite set of linear equations in the variables $x_1, x_2, \ldots, x_n$ is called a *system of linear equations* or a *linear system*. A sequence of numbers $s_1, s_2, \ldots, s_n$ is called a *solution* of the system if $x_1 = s_1, x_2 = s_2, \ldots, x_n = s_n$ is a solution of every equation in the system. For example, the system

$$4x_1 - x_2 + 3x_3 = -1$$
$$3x_1 + x_2 + 9x_3 = -4$$

has the solution $x_1 = 1, x_2 = 2, x_3 = -1$ since these values satisfy both equations. However, $x_1 = 1, x_2 = 8, x_3 = 1$ is not a solution since these values satisfy only the first of the two equations in the system.

Not all systems of linear equations have solutions. For example, if we multiply the second equation of the system

$$x + \ y = 4$$
$$2x + 2y = 6$$

by 1/2, it becomes evident that there are no solutions since the resulting system

$$x + y = 4$$
$$x + y = 3$$

has contradictory equations.

A system of equations that has no solutions is said to be ***inconsistent***. If there is at least one solution, it is called ***consistent***. To illustrate the possibilities that can occur in solving systems of linear equations, consider a general system of two linear equations in the unknowns x and y:

$$a_1 x + b_1 y = c_1 \qquad (a_1, b_1 \text{ both not zero})$$
$$a_2 x + b_2 y = c_2 \qquad (a_2, b_2 \text{ both not zero})$$

The graphs of these equations are lines; call them l_1 and l_2. Since a point (x, y) lies on a line if and only if the numbers x and y satisfy the equation of the line, the solutions of the system of equations will correspond to points of intersection of l_1 and l_2. There are three possibilitiees (Figure 1.1):

(a) the lines l_1 and l_2 may be parallel, in which case there is no intersection and consequently no solution to the system;
(b) the lines l_1 and l_2 may intersect at only one point, in which case the system has exactly one solution;
(c) the lines l_1 and l_2 may coincide, in which case there are infinitely many points of intersection and consequently infinitely many solutions to the system.

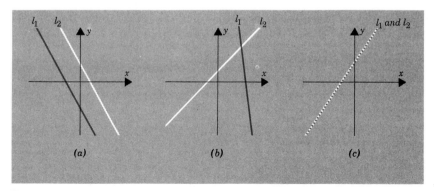

Figure 1.1 (*a*) No solution. (*b*) One solution. (*c*) Infinitely many solutions.

Although we have considered only two equations with two unknowns here, we will show later that this same result holds for arbitrary systems; that is,

every system of linear equations has either no solutions, exactly one solution, or infinitely many solutions.

An arbitrary system of m linear equations in n unknowns will be written

$$a_{11}x_1 + a_{12}x_2 + \cdots + a_{1n}x_n = b_1$$
$$a_{21}x_1 + a_{22}x_2 + \cdots + a_{2n}x_n = b_2$$
$$\vdots \qquad \vdots \qquad \qquad \vdots \qquad \vdots$$
$$a_{m1}x_1 + a_{m2}x_2 + \cdots + a_{mn}x_n = b_m$$

where $x_1, x_2, \ldots, x_n$ are the unknowns and the subscripted a's and b's denote constants.

For example, a general system of three linear equations in four unknowns will be written

$$a_{11}x_1 + a_{12}x_2 + a_{13}x_3 + a_{14}x_4 = b_1$$
$$a_{21}x_1 + a_{22}x_2 + a_{23}x_3 + a_{24}x_4 = b_2$$
$$a_{31}x_1 + a_{32}x_2 + a_{33}x_3 + a_{34}x_4 = b_3$$

The double subscripting on the coefficients of the unknowns is a useful device that is used to specify the location of the coefficient in the system. The first subscript on the coefficient a_{ij} indicates the equation in which the coefficient occurs, and the second subscript indicates which unknown it multiplies. Thus, a_{12} is in the first equation and multiplies unknown x_2.

If we mentally keep track of the location of the $+$'s, the x's, and the $=$'s, a system of m linear equations in n unknowns can be abbreviated by writing only the rectangular array of numbers:

$$\begin{bmatrix} a_{11} & a_{12} & \cdots & a_{1n} & b_1 \\ a_{21} & a_{22} & \cdots & a_{2n} & b_2 \\ \vdots & \vdots & & \vdots & \vdots \\ a_{m1} & a_{m2} & \cdots & a_{mn} & b_m \end{bmatrix}$$

This is called the **augmented matrix** for the system. (The term *matrix* is used in mathematics to denote a rectangular array of numbers. Matrices arise in many contexts; we shall study them in more detail in later sections.) For example, the augmented matrix for the system of equations

$$x_1 + x_2 + 2x_3 = 9$$
$$2x_1 + 4x_2 - 3x_3 = 1$$
$$3x_1 + 6x_2 - 5x_3 = 0$$

is

$$\begin{bmatrix} 1 & 1 & 2 & 9 \\ 2 & 4 & -3 & 1 \\ 3 & 6 & -5 & 0 \end{bmatrix}$$

REMARK. When constructing an augmented matrix, the unknowns must be written in the same order in each equation.

The basic method for solving a system of linear equations is to replace the given system by a new system that has the same solution set but which is easier to solve. This new system is generally obtained in a series of steps by applying the following three types of operations to eliminate unknowns systematically.

1. Multiply an equation through by a nonzero constant.
2. Interchange two equations.
3. Add a multiple of one equation to another.

Since the rows (horizontal lines) of an augmented matrix correspond to the equations in the associated system, these three operations correspond to the following operations on the rows of the augmented matrix.

1. Multiply a row through by a nonzero constant.
2. Interchange two rows.
3. Add a multiple of one row to another row.

These are called ***elementary row operations***. The following example illustrates how these operations can be used to solve systems of linear equations. Since a systematic procedure for finding solutions will be derived in the next section, it is not necessary to worry about how the steps in this example were selected. The main effort at this time should be devoted to understanding the computations and the discussion.

Example 3

In the left column below we solve a system of linear equations by operating on the equations in the system, and in the right column we solve the same system by operating on the rows of the augmented matrix.

$$\begin{array}{r} x + y + 2z = 9 \\ 2x + 4y - 3z = 1 \\ 3x + 6y - 5z = 0 \end{array} \qquad \begin{bmatrix} 1 & 1 & 2 & 9 \\ 2 & 4 & -3 & 1 \\ 3 & 6 & -5 & 0 \end{bmatrix}$$

Add -2 times the first equation to the second to obtain

Add -2 times the first row to the second to obtain

$$\begin{array}{r} x + y + 2z = 9 \\ 2y - 7z = -17 \\ 3x + 6y - 5z = 0 \end{array} \qquad \begin{bmatrix} 1 & 1 & 2 & 9 \\ 0 & 2 & -7 & -17 \\ 3 & 6 & -5 & 0 \end{bmatrix}$$

Add -3 times the first equation to the third to obtain

$$\begin{aligned} x + y + 2z &= 9 \\ 2y - 7z &= -17 \\ 3y - 11z &= -27 \end{aligned}$$

Add -3 times the first row to the third to obtain

$$\begin{bmatrix} 1 & 1 & 2 & 9 \\ 0 & 2 & -7 & -17 \\ 0 & 3 & -11 & -27 \end{bmatrix}$$

Multiply the second equation by 1/2 to obtain

$$\begin{aligned} x + y + 2z &= 9 \\ y - \tfrac{7}{2}z &= -\tfrac{17}{2} \\ 3y - 11z &= -27 \end{aligned}$$

Multiply the second row by 1/2 to obtain

$$\begin{bmatrix} 1 & 1 & 2 & 9 \\ 0 & 1 & -\tfrac{7}{2} & -\tfrac{17}{2} \\ 0 & 3 & -11 & -27 \end{bmatrix}$$

Add -3 times the second equation to the third to obtain

$$\begin{aligned} x + y + 2z &= 9 \\ y - \tfrac{7}{2}z &= -\tfrac{17}{2} \\ -\tfrac{1}{2}z &= -\tfrac{3}{2} \end{aligned}$$

Add -3 times the second row to the third to obtain

$$\begin{bmatrix} 1 & 1 & 2 & 9 \\ 0 & 1 & -\tfrac{7}{2} & -\tfrac{17}{2} \\ 0 & 0 & -\tfrac{1}{2} & -\tfrac{3}{2} \end{bmatrix}$$

Multiply the third equation by -2 to obtain

$$\begin{aligned} x + y + 2z &= 9 \\ y - \tfrac{7}{2}z &= -\tfrac{17}{2} \\ z &= 3 \end{aligned}$$

Multiply the third row by -2 to obtain

$$\begin{bmatrix} 1 & 1 & 2 & 9 \\ 0 & 1 & -\tfrac{7}{2} & -\tfrac{17}{2} \\ 0 & 0 & 1 & 3 \end{bmatrix}$$

Add -1 times the second equation to the first to obtain

$$\begin{aligned} x \quad + \tfrac{11}{2}z &= \tfrac{35}{2} \\ y - \tfrac{7}{2}z &= -\tfrac{17}{2} \\ z &= 3 \end{aligned}$$

Add -1 times the second row to the first to obtain

$$\begin{bmatrix} 1 & 0 & \tfrac{11}{2} & \tfrac{35}{2} \\ 0 & 1 & -\tfrac{7}{2} & -\tfrac{17}{2} \\ 0 & 0 & 1 & 3 \end{bmatrix}$$

Add $-\tfrac{11}{2}$ times the third equation to the first and $\tfrac{7}{2}$ times the third equation the second to obtain

$$\begin{aligned} x &= 1 \\ y &= 2 \\ z &= 3 \end{aligned}$$

Add $-\tfrac{11}{2}$ times the third row to the first and $\tfrac{7}{2}$ times the third row to the second to obtain

$$\begin{bmatrix} 1 & 0 & 0 & 1 \\ 0 & 1 & 0 & 2 \\ 0 & 0 & 1 & 3 \end{bmatrix}$$

The solution

$$x = 1, \qquad y = 2, \qquad z = 3$$

is now evident. ▲

EXERCISE SET 1.1

1. Which of the following are linear equations in x_1, x_2, and x_3?
   (a) $x_1 + 2x_1x_2 + x_3 = 2$ (b) $x_1 + x_2 + x_3 = \sin k$ (k is a constant)
   (c) $x_1 - 3x_2 + 2x_3^{1/2} = 4$ (d) $x_1 = \sqrt{2}x_3 - x_2 + 7$
   (e) $x_1 + x_2^{-1} - 3x_3 = 5$ (f) $x_1 = x_3$

2. Find the solution set of each of the following:
   (a) $6x - 7y = 3$ (b) $2x_1 + 4x_2 - 7x_3 = 8$
   (c) $-3x_1 + 4x_2 - 7x_3 + 8x_4 = 5$ (d) $2v - w + 3x + y - 4z = 0$

3. Find the augmented matrix for each of the following systems of linear equations.
   (a) $\begin{aligned} x_1 - 2x_2 &= 0 \\ 3x_1 + 4x_2 &= -1 \\ 2x_1 - x_2 &= 3 \end{aligned}$ (b) $\begin{aligned} x_1 \qquad + x_3 &= 1 \\ -x_1 + 2x_2 - x_3 &= 3 \end{aligned}$

   (c) $\begin{aligned} x_1 \qquad + x_3 \qquad &= 1 \\ 2x_2 - x_3 \quad + x_5 &= 2 \\ 2x_3 + x_4 \qquad &= 3 \end{aligned}$ (d) $\begin{aligned} x_1 &= 1 \\ x_2 &= 2 \end{aligned}$

4. Find a system of linear equations corresponding to each of the following augmented matrices.

   (a) $\begin{bmatrix} 1 & 0 & -1 & 2 \\ 2 & 1 & 1 & 3 \\ 0 & -1 & 2 & 4 \end{bmatrix}$ (b) $\begin{bmatrix} 1 & 0 & 0 \\ 0 & 1 & 0 \\ 1 & -1 & 1 \end{bmatrix}$

   (c) $\begin{bmatrix} 1 & 2 & 3 & 4 & 5 \\ 5 & 4 & 3 & 2 & 1 \end{bmatrix}$ (d) $\begin{bmatrix} 1 & 0 & 0 & 0 & 1 \\ 0 & 1 & 0 & 0 & 2 \\ 0 & 0 & 1 & 0 & 3 \\ 0 & 0 & 0 & 1 & 4 \end{bmatrix}$

5. For which value(s) of the constant k does the following system of linear equations have no solutions? Exactly one solution? Infinitely many solutions?

$$x - y = 3$$
$$2x - 2y = k$$

6. Consider the system of equations

$$ax + by = k$$
$$cx + dy = l$$
$$ex + fy = m$$

Discuss the relative positions of the lines $ax + by = k$, $cx + dy = l$, and $ex + fy = m$ when:
   (a) the system has no solutions
   (b) the system has exactly one solution
   (c) the system has infinitely many solutions

7. Show that if the system of equations in Exercise 6 is consistent, then at least one equation can be discarded from the system without altering the solution set.

8. Let $k = l = m = 0$ in Exercise 6; show that the system must be consistent. What can be said about the point of intersection of the three lines if the system has exactly one solution?

9. Consider the system of equations

$$x + y + 2z = a$$
$$x \quad\;\; + z = b$$
$$2x + y + 3z = c$$

Show that in order for this system to be consistent, a, b, and c must satisfy $c = a + b$.

10. Prove: If the linear equations $x_1 + kx_2 = c$ and $x_1 + lx_2 = d$ have the same solution set, then the equations are identical.

1.2 GAUSSIAN ELIMINATION

In this section we give a systematic procedure for solving systems of linear equations; it is based on the idea of reducing the augmented matrix to a form that is simple enough so that the system of equations can be solved by inspection.

In the last step of Example 3 we obtained the augmented matrix

$$\begin{bmatrix} 1 & 0 & 0 & 1 \\ 0 & 1 & 0 & 2 \\ 0 & 0 & 1 & 3 \end{bmatrix} \tag{1.1}$$

from which the solution of the system was evident.

Matrix (1.1) is an example of a matrix that is in *reduced row-echelon form*. To be of this form, a matrix must have the following properties.

1. *If a row does not consist entirely of zeros, then the first nonzero number in the row is a 1. (We call this a **leading 1**.)*

2. *If there are any rows that consist entirely of zeros, then they are grouped together at the bottom of the matrix.*

3. *In any two successive rows that do not consist entirely of zeros, the leading 1 in the lower row occurs farther to the right than the leading 1 in the higher row.*

4. *Each column that contains a leading 1 has zeros everywhere else.*

A matrix having properties 1, 2, and 3 is said to be in *row-echelon form*.

Example 4

The following matrices are in reduced row-echelon form.

$$\begin{bmatrix} 1 & 0 & 0 & 4 \\ 0 & 1 & 0 & 7 \\ 0 & 0 & 1 & -1 \end{bmatrix}, \begin{bmatrix} 1 & 0 & 0 \\ 0 & 1 & 0 \\ 0 & 0 & 1 \end{bmatrix}, \begin{bmatrix} 0 & 1 & -2 & 0 & 1 \\ 0 & 0 & 0 & 1 & 3 \\ 0 & 0 & 0 & 0 & 0 \\ 0 & 0 & 0 & 0 & 0 \end{bmatrix}, \begin{bmatrix} 0 & 0 \\ 0 & 0 \end{bmatrix}$$

The following matrices are in row-echelon form.

$$\begin{bmatrix} 1 & 4 & 3 & 7 \\ 0 & 1 & 6 & 2 \\ 0 & 0 & 1 & 5 \end{bmatrix}, \begin{bmatrix} 1 & 1 & 0 \\ 0 & 1 & 0 \\ 0 & 0 & 0 \end{bmatrix}, \begin{bmatrix} 0 & 1 & 2 & 6 & 0 \\ 0 & 0 & 1 & -1 & 0 \\ 0 & 0 & 0 & 0 & 1 \end{bmatrix}$$

The reader should check to see that each of the above matrices satisfies all the necessary requirements. ▲

REMARK. It is not difficult to see that a matrix in row-echelon form must have zeros below each leading 1 (see Example 4). In contrast, a matrix in reduced row-echelon form must have zeros above and below each leading 1.

If, by a sequence of elementary row operations, the augmented matrix for a system of linear equations is put in reduced row-echelon form, then the solution set for the system can be obtained by inspection or, at worst, after a few simple steps. The next example illustrates this point.

Example 5

Suppose that the augmented matrix for a system of linear equations has been reduced by row operations to the given reduced row-echelon form. Solve the system.

(a) $\begin{bmatrix} 1 & 0 & 0 & 5 \\ 0 & 1 & 0 & -2 \\ 0 & 0 & 1 & 4 \end{bmatrix}$
(b) $\begin{bmatrix} 1 & 0 & 0 & 4 & -1 \\ 0 & 1 & 0 & 2 & 6 \\ 0 & 0 & 1 & 3 & 2 \end{bmatrix}$

(c) $\begin{bmatrix} 1 & 6 & 0 & 0 & 4 & -2 \\ 0 & 0 & 1 & 0 & 3 & 1 \\ 0 & 0 & 0 & 1 & 5 & 2 \\ 0 & 0 & 0 & 0 & 0 & 0 \end{bmatrix}$
(d) $\begin{bmatrix} 1 & 0 & 0 & 0 \\ 0 & 1 & 2 & 0 \\ 0 & 0 & 0 & 1 \end{bmatrix}$

Solution to (a). The corresponding system of equations is

$$\begin{aligned} x_1 \qquad\qquad &= 5 \\ x_2 \qquad &= -2 \\ x_3 &= 4 \end{aligned}$$

By inspection, $x_1 = 5$, $x_2 = -2$, $x_3 = 4$.

Solution to (b). The corresponding system of equations is

$$\begin{aligned} x_1 \qquad\qquad + 4x_4 &= -1 \\ x_2 \quad + 2x_4 &= 6 \\ x_3 + 3x_4 &= 2 \end{aligned}$$

Since x_1, x_2, and x_3 correspond to leading 1's in the augmented matrix, we call them **leading variables**. Solving for the leading variables in terms of x_4 gives

$$\begin{aligned} x_1 &= -1 - 4x_4 \\ x_2 &= 6 - 2x_4 \\ x_3 &= 2 - 3x_4 \end{aligned}$$

Since x_4 can be assigned an arbitrary value, say t, we have infinitely many solutions. The solution set is given by the formulas

$$x_1 = -1 - 4t, \qquad x_2 = 6 - 2t, \qquad x_3 = 2 - 3t, \qquad x_4 = t$$

Solution to (c). The corresponding system of equations is

$$\begin{aligned} x_1 + 6x_2 \qquad\quad + 4x_5 &= -2 \\ x_3 \quad + 3x_5 &= 1 \\ x_4 + 5x_5 &= 2 \end{aligned}$$

Here the leading variables are x_1, x_3, and x_4. Solving for the leading variables in terms of the remaining variables gives

$$\begin{aligned} x_1 &= -2 - 4x_5 - 6x_2 \\ x_3 &= 1 - 3x_5 \\ x_4 &= 2 - 5x_5 \end{aligned}$$

Since x_5 can be assigned an arbitrary value, t, and x_2 can be assigned an arbitrary value, s, there are infinitely many solutions. The solution set is given by the formulas

$$x_1 = -2 - 4t - 6s, \qquad x_2 = s, \qquad x_3 = 1 - 3t, \qquad x_4 = 2 - 5t, \qquad x_5 = t$$

Solution to (d). The last equation in the corresponding system of equations is

$$0x_1 + 0x_2 + 0x_3 = 1$$

Since this equation can never be satisfied, there is no solution to the system.

We have just seen how easy it is to solve a system of linear equations, once its augmented matrix is in reduced row-echelon form. Now we shall give a step-by-step procedure that can be used to reduce any matrix to reduced row-echelon form. As we state each step in the procedure, we shall illustrate the idea by reducing the following matrix to reduced row-echelon form.

$$\begin{bmatrix} 0 & 0 & -2 & 0 & 7 & 12 \\ 2 & 4 & -10 & 6 & 12 & 28 \\ 2 & 4 & -5 & 6 & -5 & -1 \end{bmatrix}$$

Step 1. Locate the leftmost column that does not consist entirely of zeros.

$$\begin{bmatrix} 0 & 0 & -2 & 0 & 7 & 12 \\ 2 & 4 & -10 & 6 & 12 & 28 \\ 2 & 4 & -5 & 6 & -5 & -1 \end{bmatrix}$$

└─**Leftmost nonzero column**

Step 2. Interchange the top row with another row, if necessary, to bring a nonzero entry to the top of the column found in Step 1.

$$\begin{bmatrix} 2 & 4 & -10 & 6 & 12 & 28 \\ 0 & 0 & -2 & 0 & 7 & 12 \\ 2 & 4 & -5 & 6 & -5 & -1 \end{bmatrix}$$

The first and second rows in the previous matrix were interchanged.

Step 3. If the entry that is now at the top of the column found in Step 1 is a, multiply the first row by $1/a$ in order to introduce a leading 1.

$$\begin{bmatrix} 1 & 2 & -5 & 3 & 6 & 14 \\ 0 & 0 & -2 & 0 & 7 & 12 \\ 2 & 4 & -5 & 6 & -5 & -1 \end{bmatrix}$$

The first row of the previous matrix was multiplied by 1/2.

Step 4. Add suitable multiples of the top row to the rows below so that all entries below leading 1 become zeros.

$$\begin{bmatrix} 1 & 2 & -5 & 3 & 6 & 14 \\ 0 & 0 & -2 & 0 & 7 & 12 \\ 0 & 0 & 5 & 0 & -17 & -29 \end{bmatrix}$$

-2 times the first row of the previous matrix was added to the third row.

Step 5. Now cover the top row in the matrix and begin again with Step 1 applied to the submatrix that remains. Continue in this way until the *entire* matrix is in row-echelon form.

$$\begin{bmatrix} 1 & 2 & -5 & 3 & 6 & 14 \\ 0 & 0 & -2 & 0 & 7 & 12 \\ 0 & 0 & 5 & 0 & -17 & -29 \end{bmatrix}$$

↑
└── Leftmost nonzero column in
the submatrix

$$\begin{bmatrix} 1 & 2 & -5 & 3 & 6 & 14 \\ 0 & 0 & 1 & 0 & -\frac{7}{2} & -6 \\ 0 & 0 & 5 & 0 & -17 & -29 \end{bmatrix}$$

> The first row in the submatrix was multiplied by $-1/2$ to introduce a leading 1.

$$\begin{bmatrix} 1 & 2 & -5 & 3 & 6 & 14 \\ 0 & 0 & 1 & 0 & -\frac{7}{2} & -6 \\ 0 & 0 & 0 & 0 & \frac{1}{2} & 1 \end{bmatrix}$$

> -5 times the first row of the submatrix was added to the second row of the submatrix to introduce a zero below the leading 1.

$$\begin{bmatrix} 1 & 2 & -5 & 3 & 6 & 14 \\ 0 & 0 & 1 & 0 & -\frac{7}{2} & -6 \\ 0 & 0 & 0 & 0 & \frac{1}{2} & 1 \end{bmatrix}$$

> The top row in the submatrix was covered, and we returned again to Step 1.

↑
└── Leftmost nonzero column
in the new submatrix

$$\begin{bmatrix} 1 & 2 & -5 & 3 & 6 & 14 \\ 0 & 0 & 1 & 0 & -\frac{7}{2} & -6 \\ 0 & 0 & 0 & 0 & 1 & 2 \end{bmatrix}$$

> The first (and only) row in the new submatrix was multiplied by 2 to introduce a leading 1.

The *entire* matrix is now in row-echelon form. To find the reduced row-echelon form we need the following additional step.

Step 6. Beginning with the last nonzero row and working upward, add suitable multiples of each row to the rows above to introduce zeros above the leading 1's.

$$\begin{bmatrix} 1 & 2 & -5 & 3 & 6 & 14 \\ 0 & 0 & 1 & 0 & 0 & 1 \\ 0 & 0 & 0 & 0 & 1 & 2 \end{bmatrix}$$

> 7/2 times the third row of the previous matrix was added to the second row.

$$\begin{bmatrix} 1 & 2 & -5 & 3 & 0 & 2 \\ 0 & 0 & 1 & 0 & 0 & 1 \\ 0 & 0 & 0 & 0 & 1 & 2 \end{bmatrix}$$

-6 times the third row was added to the first row.

$$\begin{bmatrix} 1 & 2 & 0 & 3 & 0 & 7 \\ 0 & 0 & 1 & 0 & 0 & 1 \\ 0 & 0 & 0 & 0 & 1 & 2 \end{bmatrix}$$

5 times the second row was added to the first row.

The last matrix is in reduced row-echelon form.

The above procedure for reducing a matrix to reduced row-echelon form is called **Gauss-Jordan elimination.*** If we use only the first five steps, the procedure produces a row-echelon form and is called **Gaussian elimination**.

Example 6

Solve by Gauss-Jordan elimination.

$$\begin{aligned} x_1 + 3x_2 - 2x_3 \quad\quad + 2x_5 \quad\quad\quad &= 0 \\ 2x_1 + 6x_2 - 5x_3 - 2x_4 + 4x_5 - 3x_6 &= -1 \\ 5x_3 + 10x_4 \quad\quad + 15x_6 &= 5 \\ 2x_1 + 6x_2 \quad\quad + 8x_4 + 4x_5 + 18x_6 &= 6 \end{aligned}$$

The augmented matrix for the system is

$$\begin{bmatrix} 1 & 3 & -2 & 0 & 2 & 0 & 0 \\ 2 & 6 & -5 & -2 & 4 & -3 & -1 \\ 0 & 0 & 5 & 10 & 0 & 15 & 5 \\ 2 & 6 & 0 & 8 & 4 & 18 & 6 \end{bmatrix}$$

Adding -2 times the first row to the second and fourth rows gives

$$\begin{bmatrix} 1 & 3 & -2 & 0 & 2 & 0 & 0 \\ 0 & 0 & -1 & -2 & 0 & -3 & -1 \\ 0 & 0 & 5 & 10 & 0 & 15 & 5 \\ 0 & 0 & 4 & 8 & 0 & 18 & 6 \end{bmatrix}$$

---

* *Carl Friedrich Gauss (1777–1855)*, sometimes called the "prince of mathematicians," made profound contributions to number theory, theory of functions, probability, and statistics. He discovered a way to calculate the orbits of asteroids, made basic discoveries in electromagnetic theory, and invented a telegraph.

Wilhelm Jordan (1842–1899) was a German engineer who specialized in geodesy. His contribution to solving linear systems appeared in his popular book, *Handbuch der Vermessungskunde* (*Handbook of Geodesy*), in 1888.

Multiplying the second row by -1 and then adding -5 times the second row to the third row and -4 times the second row to the fourth row gives

$$\begin{bmatrix} 1 & 3 & -2 & 0 & 2 & 0 & 0 \\ 0 & 0 & 1 & 2 & 0 & 3 & 1 \\ 0 & 0 & 0 & 0 & 0 & 0 & 0 \\ 0 & 0 & 0 & 0 & 0 & 6 & 2 \end{bmatrix}$$

Interchanging the third and fourth rows and then multiplying the third row of the resulting matrix by $1/6$ gives the row-echelon form

$$\begin{bmatrix} 1 & 3 & -2 & 0 & 2 & 0 & 0 \\ 0 & 0 & 1 & 2 & 0 & 3 & 1 \\ 0 & 0 & 0 & 0 & 0 & 1 & \frac{1}{3} \\ 0 & 0 & 0 & 0 & 0 & 0 & 0 \end{bmatrix}$$

Adding -3 times the third row to the second row and then adding 2 times the second row of the resulting matrix to the first row yields the reduced row-echelon form

$$\begin{bmatrix} 1 & 3 & 0 & 4 & 2 & 0 & 0 \\ 0 & 0 & 1 & 2 & 0 & 0 & 0 \\ 0 & 0 & 0 & 0 & 0 & 1 & \frac{1}{3} \\ 0 & 0 & 0 & 0 & 0 & 0 & 0 \end{bmatrix}$$

The corresponding system of equations is

$$\begin{aligned} x_1 + 3x_2 \quad + 4x_4 + 2x_5 \quad &= 0 \\ x_3 + 2x_4 \quad &= 0 \\ x_6 &= \tfrac{1}{3} \end{aligned}$$

(We have discarded the last equation, $0x_1 + 0x_2 + 0x_3 + 0x_4 + 0x_5 + 0x_6 = 0$, since it will be satisfied automatically by the solutions of the remaining equations.) Solving for the leading variables, we obtain

$$\begin{aligned} x_1 &= -3x_2 - 4x_4 - 2x_5 \\ x_3 &= -2x_4 \\ x_6 &= \tfrac{1}{3} \end{aligned}$$

If we assign x_2, x_4, and x_5 the arbitrary values r, s, and t, respectively, the solution set is given by the formulas

$$x_1 = -3r - 4s - 2t, \quad x_2 = r, \quad x_3 = -2s, \quad x_4 = s, \quad x_5 = t, \quad x_6 = \tfrac{1}{3}$$

Example 7

It is often more convenient to solve a system of linear equations by using Gaussian elimination to bring the augmented matrix into row-echelon form without continuing all the way to the reduced row-echelon form. When this is done, the correspond-

ing system of equations can be solved by a technique called ***back-substitution***. We
shall illustrate this method using the system of equations in Example 6.

From the computations in Example 6, a row-echelon form of the augmented
matrix is

$$\begin{bmatrix} 1 & 3 & -2 & 0 & 2 & 0 & 0 \\ 0 & 0 & 1 & 2 & 0 & 3 & 1 \\ 0 & 0 & 0 & 0 & 0 & 1 & \frac{1}{3} \\ 0 & 0 & 0 & 0 & 0 & 0 & 0 \end{bmatrix}$$

To solve the corresponding system of equations

$$\begin{aligned} x_1 + 3x_2 - 2x_3 \quad\quad + 2x_5 \quad\quad &= 0 \\ x_3 + 2x_4 \quad\quad + 3x_6 &= 1 \\ x_6 &= \tfrac{1}{3} \end{aligned}$$

we proceed as follows:

Step 1. Solve the equations for the leading variables.

$$\begin{aligned} x_1 &= -3x_2 + 2x_3 - 2x_5 \\ x_3 &= 1 - 2x_4 - 3x_6 \\ x_6 &= \tfrac{1}{3} \end{aligned}$$

Step 2. Beginning with the bottom equation and working upward, successively substitute each equation into all the equations above it.

Substituting $x_6 = 1/3$ into the second equation yields

$$\begin{aligned} x_1 &= -3x_2 + 2x_3 - 2x_5 \\ x_3 &= -2x_4 \\ x_6 &= \tfrac{1}{3} \end{aligned}$$

Substituting $x_3 = -2x_4$ into the first equation yields

$$\begin{aligned} x_1 &= -3x_2 - 4x_4 - 2x_5 \\ x_3 &= -2x_4 \\ x_6 &= \tfrac{1}{3} \end{aligned}$$

Step 3. Assign arbitrary values to any nonleading variables.

If we assign x_2, x_4, and x_5 the arbitrary values r, s, and t, respectively, the
solution set is given by the formulas

$$x_1 = -3r - 4s - 2t, \quad x_2 = r, \quad x_3 = -2s, \quad x_4 = s, \quad x_5 = t, \quad x_6 = \tfrac{1}{3}$$

This agrees with the solution obtained in Example 6. ▲

Example 8

Solve

$$
\begin{aligned}
x + y + 2z &= 9 \\
2x + 4y - 3z &= 1 \\
3x + 6y - 5z &= 0
\end{aligned}
$$

by Gaussian elimination and back-substitution.

Solution. This is the system in Example 3. In that example we converted the augmented matrix

$$
\begin{bmatrix}
1 & 1 & 2 & 9 \\
2 & 4 & -3 & 1 \\
3 & 6 & -5 & 0
\end{bmatrix}
$$

to the row-echelon form

$$
\begin{bmatrix}
1 & 1 & 2 & 9 \\
0 & 1 & -\frac{7}{2} & -\frac{17}{2} \\
0 & 0 & 1 & 3
\end{bmatrix}
$$

The system corresponding to this matrix is

$$
\begin{aligned}
x + y + 2z &= 9 \\
y - \tfrac{7}{2}z &= -\tfrac{17}{2} \\
z &= 3
\end{aligned}
$$

Solving for the leading variables yields

$$
\begin{aligned}
x &= 9 - y - 2z \\
y &= -\tfrac{17}{2} + \tfrac{7}{2}z \\
z &= 3
\end{aligned}
$$

Substituting the bottom equation into those above yields

$$
\begin{aligned}
x &= 3 - y \\
y &= 2 \\
z &= 3
\end{aligned}
$$

and substituting the second equation into the top yields

$$
\begin{aligned}
x &= 1 \\
y &= 2 \\
z &= 3
\end{aligned}
$$

This agrees with the result found by Gauss-Jordan elimination in Example 3. ▲

REMARK. The procedures we have given for reducing a matrix to row-echelon form or reduced row-echelon form are well suited for computer computation because they are systematic. However, these procedures sometimes introduce fractions, which might otherwise be avoided by varying the steps in the right way. Thus, once the basic procedure has been mastered, the reader may wish to vary the steps in specific problems to avoid fractions (see Exercise 13). It can be proved, although we shall not do it, that no matter how the elementary row operations are varied, one will always arrive at the same reduced row-echelon form; that is, *the reduced row-echelon form of a matrix is unique*. However, *a row-echelon form is not unique;* by changing the sequence of elementary row operations it is possible to arrive at different row-echelon forms (see Exercise 14).

EXERCISE SET 1.2

1. Which of the following are in reduced row-echelon form?

(a) $\begin{bmatrix} 1 & 0 & 0 \\ 0 & 0 & 0 \\ 0 & 0 & 1 \end{bmatrix}$
(b) $\begin{bmatrix} 0 & 1 & 0 \\ 1 & 0 & 0 \\ 0 & 0 & 0 \end{bmatrix}$
(c) $\begin{bmatrix} 1 & 1 & 0 \\ 0 & 1 & 0 \\ 0 & 0 & 0 \end{bmatrix}$

(d) $\begin{bmatrix} 1 & 2 & 0 & 3 & 0 \\ 0 & 0 & 1 & 1 & 0 \\ 0 & 0 & 0 & 0 & 1 \\ 0 & 0 & 0 & 0 & 0 \end{bmatrix}$
(e) $\begin{bmatrix} 1 & 0 & 0 & 5 \\ 0 & 0 & 1 & 3 \\ 0 & 1 & 0 & 4 \end{bmatrix}$
(f) $\begin{bmatrix} 1 & 0 & 3 & 1 \\ 0 & 1 & 2 & 4 \end{bmatrix}$

2. Which of the following are in row-echelon form?

(a) $\begin{bmatrix} 1 & 2 & 3 \\ 0 & 0 & 0 \\ 0 & 0 & 1 \end{bmatrix}$
(b) $\begin{bmatrix} 1 & -7 & 5 & 5 \\ 0 & 1 & 3 & 2 \end{bmatrix}$
(c) $\begin{bmatrix} 1 & 1 & 0 \\ 0 & 1 & 0 \\ 0 & 0 & 0 \end{bmatrix}$

(d) $\begin{bmatrix} 1 & 3 & 0 & 2 & 0 \\ 1 & 0 & 2 & 2 & 0 \\ 0 & 0 & 0 & 0 & 1 \\ 0 & 0 & 0 & 0 & 0 \end{bmatrix}$
(e) $\begin{bmatrix} 2 & 3 & 4 \\ 0 & 1 & 2 \\ 0 & 0 & 3 \end{bmatrix}$
(f) $\begin{bmatrix} 0 & 0 & 0 \\ 0 & 0 & 0 \\ 0 & 0 & 0 \end{bmatrix}$

3. In each part suppose that the augmented matrix for a system of linear equations has been reduced by row operations to the given reduced row-echelon form. Solve the system.

(a) $\begin{bmatrix} 1 & 0 & 0 & 4 \\ 0 & 1 & 0 & 3 \\ 0 & 0 & 1 & 2 \end{bmatrix}$
(b) $\begin{bmatrix} 1 & 0 & 0 & 3 & 2 \\ 0 & 1 & 0 & -1 & 4 \\ 0 & 0 & 1 & 1 & 2 \end{bmatrix}$

(c) $\begin{bmatrix} 1 & 5 & 0 & 0 & 5 & -1 \\ 0 & 0 & 1 & 0 & 3 & 1 \\ 0 & 0 & 0 & 1 & 4 & 2 \\ 0 & 0 & 0 & 0 & 0 & 0 \end{bmatrix}$
(d) $\begin{bmatrix} 1 & 2 & 0 & 0 \\ 0 & 0 & 1 & 0 \\ 0 & 0 & 0 & 1 \end{bmatrix}$

4. In each part suppose that the augmented matrix for a system of linear equations has been reduced by row operations to the given row-echelon form. Solve the system.

(a) $\begin{bmatrix} 1 & 2 & -4 & 2 \\ 0 & 1 & -2 & -1 \\ 0 & 0 & 1 & 2 \end{bmatrix}$
(b) $\begin{bmatrix} 1 & 0 & 4 & 7 & 10 \\ 0 & 1 & -3 & -4 & -2 \\ 0 & 0 & 1 & 1 & 2 \end{bmatrix}$

(c) $\begin{bmatrix} 1 & 5 & -4 & 0 & -7 & -5 \\ 0 & 0 & 1 & 1 & 7 & 3 \\ 0 & 0 & 0 & 1 & 4 & 2 \\ 0 & 0 & 0 & 0 & 0 & 0 \end{bmatrix}$
(d) $\begin{bmatrix} 1 & 2 & 2 & 2 \\ 0 & 1 & 3 & 3 \\ 0 & 0 & 0 & 1 \end{bmatrix}$

5. Solve each of the following systems by Gauss-Jordan elimination.

(a) $\begin{aligned} x_1 + x_2 + 2x_3 &= 8 \\ -x_1 - 2x_2 + 3x_3 &= 1 \\ 3x_1 - 7x_2 + 4x_3 &= 10 \end{aligned}$
(b) $\begin{aligned} 2x_1 + 2x_2 + 2x_3 &= 0 \\ -2x_1 + 5x_2 + 2x_3 &= 0 \\ -7x_1 + 7x_2 + x_3 &= 0 \end{aligned}$

(c) $\begin{aligned} x - y + 2z - w &= -1 \\ 2x + y - 2z - 2w &= -2 \\ -x + 2y - 4z + w &= 1 \\ 3x \qquad\quad - 3w &= -3 \end{aligned}$

6. Solve each of the systems in Exercise 5 by Gaussian elimination.

7. Solve each of the following systems by Gauss-Jordan elimination.

(a) $\begin{aligned} 2x_1 - 3x_2 &= -2 \\ 2x_1 + x_2 &= 1 \\ 3x_1 + 2x_2 &= 1 \end{aligned}$
(b) $\begin{aligned} 3x_1 + 2x_2 - x_3 &= -15 \\ 5x_1 + 3x_2 + 2x_3 &= 0 \\ 3x_1 + x_2 + 3x_3 &= 11 \\ 11x_1 + 7x_2 &= -30 \end{aligned}$
(c) $\begin{aligned} 4x_1 - 8x_2 &= 12 \\ 3x_1 - 6x_2 &= 9 \\ -2x_1 + 4x_2 &= -6 \end{aligned}$

8. Solve each of the systems in Exercise 7 by Gaussian elimination.

9. Solve each of the following systems by Gauss-Jordan elimination.

(a) $\begin{aligned} 5x_1 + 2x_2 + 6x_3 &= 0 \\ -2x_1 + x_2 + 3x_3 &= 0 \end{aligned}$
(b) $\begin{aligned} x_1 - 2x_2 + x_3 - 4x_4 &= 1 \\ x_1 + 3x_2 + 7x_3 + 2x_4 &= 2 \\ x_1 - 12x_2 - 11x_3 - 16x_4 &= 5 \end{aligned}$

10. Solve each of the systems in Exercise 9 by Gaussian elimination.

11. Solve the following systems, where a, b, and c are constants.

(a) $\begin{aligned} 2x + y &= a \\ 3x + 6y &= b \end{aligned}$
(b) $\begin{aligned} x_1 + x_2 + x_3 &= a \\ 2x_1 \qquad\;\; + 2x_3 &= b \\ 3x_2 + 3x_3 &= c \end{aligned}$

12. For which values of a will the following system have no solutions? Exactly one solution? Infinitely many solutions?

$$\begin{aligned} x + 2y - 3z &= 4 \\ 3x - y + 5z &= 2 \\ 4x + y + (a^2 - 14)z &= a + 2 \end{aligned}$$

13. Reduce

$$\begin{bmatrix} 2 & 1 & 3 \\ 0 & -2 & 7 \\ 3 & 4 & 5 \end{bmatrix}$$

to reduced row-echelon form without introducing any fractions.

14. Find two different row-echelon forms of

$$\begin{bmatrix} 1 & 3 \\ 2 & 7 \end{bmatrix}$$

15. Solve the following system of nonlinear equations for the unknown angles α, β, and γ, where $0 \le \alpha \le 2\pi, 0 \le \beta \le 2\pi$, and $0 \le \gamma < \pi$.

$$2 \sin \alpha - \quad \cos \beta + 3 \tan \gamma = 3$$
$$4 \sin \alpha + 2 \cos \beta - 2 \tan \gamma = 2$$
$$6 \sin \alpha - 3 \cos \beta + \quad \tan \gamma = 9$$

16. Describe the possible reduced row-echelon forms of

$$\begin{bmatrix} a & b & c \\ d & e & f \\ g & h & i \end{bmatrix}$$

17. Show that if $ad - bc \ne 0$, then the reduced row-echelon form of

$$\begin{bmatrix} a & b \\ c & d \end{bmatrix} \quad \text{is} \quad \begin{bmatrix} 1 & 0 \\ 0 & 1 \end{bmatrix}$$

18. Use Exercise 17 to show that if $ad - bc \ne 0$, then the system

$$ax + by = k$$
$$cx + dy = l$$

has exactly one solution.

1.3   HOMOGENEOUS SYSTEMS OF LINEAR EQUATIONS

As we have already pointed out, every system of linear equations has either one solution, infinitely many solutions, or no solutions at all. As we progress, there will be situations in which we will not be interested in finding solutions to a given system, but instead will be concerned with deciding how many solutions the system has. In this section we consider several cases in which it is possible to make statements about the number of solutions by inspection.

A system of linear equations is said to be **homogeneous** if all the constant terms are zero; that is, the system has the form

$$a_{11}x_1 + a_{12}x_2 + \cdots + a_{1n}x_n = 0$$
$$a_{21}x_1 + a_{22}x_2 + \cdots + a_{2n}x_n = 0$$
$$\vdots \qquad \vdots \qquad \qquad \vdots \qquad \vdots$$
$$a_{m1}x_1 + a_{m2}x_2 + \cdots + a_{mn}x_n = 0$$

Every homogeneous system of linear equations is consistent, since $x_1 = 0$, $x_2 = 0, \ldots, x_n = 0$ is always a solution. This solution is called the **trivial solution**; if there are other solutions, they are called **nontrivial solutions**.

Since a homogeneous system of linear equations must be consistent, there is either one solution or infinitely many solutions. Since one of these solutions is the trivial solution, we can make the following statement.

For a homogeneous system of linear equations, exactly one of the following is true:

1. *The system has only the trivial solution.*
2. *The system has infinitely many nontrivial solutions in addition to the trivial solution.*

There is one case in which a homogeneous system is assured of having nontrivial solutions; namely, whenever the system involves more unknowns than equations. To see why, consider the following example of four equations in five unknowns.

Example 9

Solve the following homogeneous system of linear equations by Gauss-Jordan elimination.

$$
\begin{aligned}
2x_1 + 2x_2 - x_3 + x_5 &= 0 \\
-x_1 - x_2 + 2x_3 - 3x_4 + x_5 &= 0 \\
x_1 + x_2 - 2x_3 - x_5 &= 0 \\
x_3 + x_4 + x_5 &= 0
\end{aligned}
\qquad (1.2)
$$

The augmented matrix for the system is

$$
\begin{bmatrix}
2 & 2 & -1 & 0 & 1 & 0 \\
-1 & -1 & 2 & -3 & 1 & 0 \\
1 & 1 & -2 & 0 & -1 & 0 \\
0 & 0 & 1 & 1 & 1 & 0
\end{bmatrix}
$$

Reducing this matrix to reduced row-echelon form, we obtain

$$
\begin{bmatrix}
1 & 1 & 0 & 0 & 1 & 0 \\
0 & 0 & 1 & 0 & 1 & 0 \\
0 & 0 & 0 & 1 & 0 & 0 \\
0 & 0 & 0 & 0 & 0 & 0
\end{bmatrix}
$$

The corresponding system of equations is

$$
\begin{aligned}
x_1 + x_2 \quad\quad\;\; + x_5 &= 0 \\
x_3 \quad + x_5 &= 0 \\
x_4 \quad\quad\;\; &= 0
\end{aligned}
\tag{1.3}
$$

Solving for the leading variables yields

$$
\begin{aligned}
x_1 &= -x_2 - x_5 \\
x_3 &= -x_5 \\
x_4 &= \quad 0
\end{aligned}
$$

The solution set is therefore given by

$$
x_1 = -s - t, \qquad x_2 = s, \qquad x_3 = -t, \qquad x_4 = 0, \qquad x_5 = t
$$

Note that the trivial solution is obtained when $s = t = 0$. ▲

    Example 9 illustrates two important points about solving homogeneous systems of linear equations. First, none of the three elementary row operations can alter the final column of zeros in the augmented matrix, so that the system of equations corresponding to the reduced row-echelon form of the augmented matrix must also be a homogeneous system (see system 1.3 in Example 9). Second, depending on whether the reduced row-echelon form of the augmented matrix has any zero rows, the number of equations in the reduced system is the same or less than the number of equations in the original system (compare systems 1.2 and 1.3 in Example 9). Thus, if the given homogeneous system has m equations in n unknowns with $m < n$, and if there are r nonzero rows in the reduced row-echelon form of the augmented matrix, we will have $r < n$. It follows that the system of equations corresponding to the reduced row-echelon form of the augmented matrix will have the form

$$
\begin{aligned}
\cdots x_{k_1} \quad\quad\quad\quad\quad\quad + \Sigma(\;) &= 0 \\
\cdots x_{k_2} \quad\quad\quad\quad + \Sigma(\;) &= 0 \\
\cdots \quad\quad \ddots \quad\quad \vdots \\
x_{k_r} + \Sigma(\;) &= 0
\end{aligned}
\tag{1.4}
$$

where $x_{k_1}, x_{k_2}, \ldots, x_{k_r}$ are the leading variables and $\Sigma(\;)$ denotes sums (possibly all different) that involve the $n - r$ remaining variables. Solving for the leading variables gives

$$
\begin{aligned}
x_{k_1} &= -\Sigma(\;) \\
x_{k_2} &= -\Sigma(\;) \\
&\vdots \\
x_{k_r} &= -\Sigma(\;)
\end{aligned}
$$

As in Example 9, we can assign arbitrary values to the variables on the right-hand side and thus obtain infinitely many solutions to the system.

In summary, we have the following important theorem.

Theorem 1. *A homogeneous system of linear equations with more unknowns than equations has infinitely many solutions.*

REMARK. Note that Theorem 1 applies only to homogeneous systems. A non-homogeneous system with more unknowns than equations need not be consistent (Exercise 11); however, if the system is consistent, it will have infinitely many solutions. We omit the proof.

EXERCISE SET 1.3

1. Without using pencil and paper, determine which of the following homogeneous systems have nontrivial solutions.

(a) $\begin{aligned} x_1 + 3x_2 + 5x_3 + x_4 &= 0 \\ 4x_1 - 7x_2 - 3x_3 - x_4 &= 0 \\ 3x_1 + 2x_2 + 7x_3 + 8x_4 &= 0 \end{aligned}$
(b) $\begin{aligned} x_1 + 2x_2 + 3x_3 &= 0 \\ x_2 + 4x_3 &= 0 \\ 5x_3 &= 0 \end{aligned}$

(c) $\begin{aligned} a_{11}x_1 + a_{12}x_2 + a_{13}x_3 &= 0 \\ a_{21}x_1 + a_{22}x_2 + a_{23}x_3 &= 0 \end{aligned}$
(d) $\begin{aligned} x_1 + x_2 &= 0 \\ 2x_1 + 2x_2 &= 0 \end{aligned}$

In Exercises 2–5 solve the given homogeneous system of linear equations.

2. $\begin{aligned} 2x_1 + x_2 + 3x_3 &= 0 \\ x_1 + 2x_2 &= 0 \\ x_2 + x_3 &= 0 \end{aligned}$
3. $\begin{aligned} 3x_1 + x_2 + x_3 + x_4 &= 0 \\ 5x_1 - x_2 + x_3 - x_4 &= 0 \end{aligned}$

4. $\begin{aligned} 2x_1 - 4x_2 + x_3 + x_4 &= 0 \\ x_1 - 5x_2 + 2x_3 &= 0 \\ - 2x_2 - 2x_3 - x_4 &= 0 \\ x_1 + 3x_2 + x_4 &= 0 \\ x_1 - 2x_2 - x_3 + x_4 &= 0 \end{aligned}$
5. $\begin{aligned} x + 6y - 2z &= 0 \\ 2x - 4y + z &= 0 \end{aligned}$

6. For which value(s) of λ does the following system of equations have nontrivial solutions?

$$(\lambda - 3)x + y = 0$$
$$x + (\lambda - 3)y = 0$$

7. Consider the system of equations

$$ax + by = 0$$
$$cx + dy = 0$$
$$ex + fy = 0$$

Discuss the relative positions of the lines $ax + by = 0$, $cx + dy = 0$, and $ex + fy = 0$ when:
(a) the system has only the trivial solution
(b) the system has nontrivial solutions

8. Consider the system of equations $a(x_0 + x_1) + b(y_0 + y_1) =$

$$ax + by = 0$$
$$cx + dy = 0$$

   (a) Show that if $x = x_0$, $y = y_0$ is any solution and k is any constant, then $x = kx_0$, $y = ky_0$ is also a solution.
   (b) Show that if $x = x_0$, $y = y_0$ and $x = x_1$, $y = y_1$ are any two solutions, then $x = x_0 + x_1$, $y = y_0 + y_1$ is also a solution.

9. Consider the systems of equations

   $$(I)\ ax + by = k \qquad (II)\ ax + by = 0$$
   $$cx + dy = l \qquad cx + dy = 0$$

   (a) Show that if $x = x_1$, $y = y_1$ and $x = x_2$, $y = y_2$ are both solutions of I, then $x = x_1 - x_2$, $y = y_1 - y_2$ is a solution of II.
   (b) Show that if $x = x_1$, $y = y_1$ is a solution of I and $x = x_0$, $y = y_0$ is a solution of II, then $x = x_1 + x_0$, $y = y_1 + y_0$ is a solution of I.

10. (a) In the system of equations numbered (1.4), explain why it would be incorrect to denote the leading variables by $x_1, x_2, \ldots, x_r$ rather than $x_{k_1}, x_{k_2}, \ldots, x_{k_r}$ as we have done.
    (b) The system of equations numbered (1.3) is a specific case of (1.4). What value does r have in this case? What are $x_{k_1}, x_{k_2}, \ldots, x_{k_r}$ in this case? Write out the sums denoted by $\Sigma(\ )$ in (1.4).

11. Find an inconsistent linear system that has more unknowns than equations.

1.4 MATRICES AND MATRIX OPERATIONS

Rectangular arrays of real numbers arise in many contexts other than as augmented matrices for systems of linear equations. In this section we consider such arrays as objects in their own right and develop some of their properties for use in our later work.

Definition. A *matrix* is a rectangular array of numbers. The numbers in the array are called the *entries* in the matrix.

Example 10

The following are matrices.

$$\begin{bmatrix} 1 & 2 \\ 3 & 0 \\ -1 & 4 \end{bmatrix} \quad \begin{bmatrix} 2 & 1 & 0 & -3 \end{bmatrix} \quad \begin{bmatrix} -\sqrt{2} & \pi & e \\ 3 & \frac{1}{2} & 0 \\ 0 & 0 & 0 \end{bmatrix} \quad \begin{bmatrix} 1 \\ 3 \end{bmatrix} \quad [4]$$

As these examples indicate, matrices vary in size. The *size* of a matrix is described by specifying the number of *rows* (horizontal lines) and *columns* (vertical lines) that occur in the matrix. The first matrix in Example 10 has 3 rows and 2 columns so that its size is 3 by 2 (written 3 × 2). The first number always indicates the number of rows and the second indicates the number of columns. The remaining matrices in Example 10 thus have sizes 1 × 4, 3 × 3, 2 × 1, and 1 × 1, respectively. ▲

REMARK. It is common practice to omit the brackets on a 1 × 1 matrix. Thus, we might write 4 rather than [4]. Although this makes it impossible to tell whether 4 denotes the number "four" or the 1 × 1 matrix whose entry is "four," this rarely causes problems, since it is usually possible to tell which is meant from the context in which the symbol appears.

We shall use capital letters to denote matrices and lowercase letters to denote numerical quantities; thus, we might write

$$A = \begin{bmatrix} 2 & 1 & 7 \\ 3 & 4 & 2 \end{bmatrix} \quad \text{or} \quad C = \begin{bmatrix} a & b & c \\ d & e & f \end{bmatrix}$$

When discussing matrices, it is common to refer to numerical quantities as *scalars*. Until Chapter 9 *all our scalars will be real numbers*.

If A is a matrix, we will use a_{ij} to denote the entry that occurs in row i and column j of A. Thus, a general 3 × 4 matrix can be written

$$A = \begin{bmatrix} a_{11} & a_{12} & a_{13} & a_{14} \\ a_{21} & a_{22} & a_{23} & a_{24} \\ a_{31} & a_{32} & a_{33} & a_{34} \end{bmatrix}$$

Usually, we shall match the letter used to denote a matrix and the letter used for its entries. Thus, for a matrix B we would use b_{ij} for the entry in row i and column j. A general $m \times n$ matrix might be written

$$B = \begin{bmatrix} b_{11} & b_{12} & \cdots & b_{1n} \\ b_{21} & b_{22} & \cdots & b_{2n} \\ \vdots & \vdots & & \vdots \\ b_{m1} & b_{m2} & \cdots & b_{mn} \end{bmatrix} \quad \text{or} \quad [b_{ij}]_{m \times n}$$

A matrix A with n rows and n columns is called a *square matrix of order n*, and the entries $a_{11}, a_{22}, \ldots, a_{nn}$ are said to be on the *main diagonal* of A (see Figure 1.2).

$$\begin{bmatrix} a_{11} & a_{12} & \cdots & a_{1n} \\ a_{21} & a_{22} & \cdots & a_{2n} \\ \vdots & \vdots & & \vdots \\ a_{n1} & a_{n2} & \cdots & a_{nn} \end{bmatrix} \quad \textbf{Figure 1.2}$$

So far we have used matrices to abbreviate the work in solving systems of linear equations. For other applications, however, it is desirable to develop an "arithmetic of matrices" in which matrices can be added and multiplied in a useful way. The remainder of this section will be devoted to developing this arithmetic.

Two matrices are said to be *equal* if they have the same size and the corresponding entries in the two matrices are equal.

Example 11

Consider the matrices

$$A = \begin{bmatrix} 2 & 1 \\ 3 & 4 \end{bmatrix} \qquad B = \begin{bmatrix} 2 & 1 \\ 3 & 5 \end{bmatrix} \qquad C = \begin{bmatrix} 2 & 1 & 0 \\ 3 & 4 & 0 \end{bmatrix}$$

Here $A \neq C$ since A and C do not have the same size. For the same reason $B \neq C$. Also, $A \neq B$ since not all the corresponding entries are equal. ▲

Definition. If A and B are any two matrices of the same size, then the *sum* $A + B$ is the matrix obtained by adding together the corresponding entries in the two matrices. Matrices of different sizes cannot be added.

Example 12

Consider the matrices

$$A = \begin{bmatrix} 2 & 1 & 0 & 3 \\ -1 & 0 & 2 & 4 \\ 4 & -2 & 7 & 0 \end{bmatrix} \qquad B = \begin{bmatrix} -4 & 3 & 5 & 1 \\ 2 & 2 & 0 & -1 \\ 3 & 2 & -4 & 5 \end{bmatrix} \qquad C = \begin{bmatrix} 1 & 1 \\ 2 & 2 \end{bmatrix}$$

Then

$$A + B = \begin{bmatrix} -2 & 4 & 5 & 4 \\ 1 & 2 & 2 & 3 \\ 7 & 0 & 3 & 5 \end{bmatrix}$$

while $A + C$ and $B + C$ are undefined. ▲

Definition. If A is any matrix and c is any scalar, then the *product* cA is the matrix obtained by multiplying each entry of A by c.

Example 13

If A is the matrix

$$A = \begin{bmatrix} 4 & 2 \\ 1 & 3 \\ -1 & 0 \end{bmatrix}$$

then

$$2A = \begin{bmatrix} 8 & 4 \\ 2 & 6 \\ -2 & 0 \end{bmatrix} \qquad \text{and} \qquad (-1)A = \begin{bmatrix} -4 & -2 \\ -1 & -3 \\ 1 & 0 \end{bmatrix}$$

If B is any matrix, then $-B$ will denote the product $(-1)B$. If A and B are two matrices of the same size, then $A - B$ is defined to be the sum $A + (-B) = A + (-1)B$. ▲

Example 14

Consider the matrices

$$A = \begin{bmatrix} 2 & 3 & 4 \\ 1 & 2 & 1 \end{bmatrix} \quad \text{and} \quad B = \begin{bmatrix} 0 & 2 & 7 \\ 1 & -3 & 5 \end{bmatrix}$$

From the above definitions

$$-B = \begin{bmatrix} 0 & -2 & -7 \\ -1 & 3 & -5 \end{bmatrix}$$

and

$$A - B = \begin{bmatrix} 2 & 3 & 4 \\ 1 & 2 & 1 \end{bmatrix} + \begin{bmatrix} 0 & -2 & -7 \\ -1 & 3 & -5 \end{bmatrix} = \begin{bmatrix} 2 & 1 & -3 \\ 0 & 5 & -4 \end{bmatrix}$$

Observe that $A - B$ can be obtained directly by subtracting the entries of B from the corresponding entries of A. ▲

Above, we defined the multiplication of a matrix by a scalar. The next question, then, is how do we multiply two matrices together? Although the most natural definition of matrix multiplication would seem to be "multiply corresponding entries together," this definition would not be very useful for most problems. Experience has led mathematicians to the following less intuitive but more useful definition of matrix multiplication.

Definition. If A is an $m \times r$ matrix and B is an $r \times n$ matrix, then the **product** AB is the $m \times n$ matrix whose entries are determined as follows. To find the entry in row i and column j of AB, single out row i from the matrix A and column j from the matrix B. Multiply the corresponding entries from the row and column together and then add up the resulting products.

Example 15

Consider the matrices

$$A = \begin{bmatrix} 1 & 2 & 4 \\ 2 & 6 & 0 \end{bmatrix} \quad B = \begin{bmatrix} 4 & 1 & 4 & 3 \\ 0 & -1 & 3 & 1 \\ 2 & 7 & 5 & 2 \end{bmatrix}$$

Since A is a 2×3 matrix and B is a 3×4 matrix, the product AB is a 2×4 matrix. To determine, for example, the entry in row 2 and column 3 of AB, we single out row 2 from A and column 3 from B. Then, as illustrated below, we multiply corresponding entries together and add up these products.

$$\begin{bmatrix} 1 & 2 & 4 \\ 2 & 6 & 0 \end{bmatrix} \begin{bmatrix} 4 & 1 & 4 & 3 \\ 0 & -1 & 3 & 1 \\ 2 & 7 & 5 & 2 \end{bmatrix} = \begin{bmatrix} \square & \square & \square & \square \\ \square & \square & 26 & \square \end{bmatrix}$$

$$(2 \cdot 4) + (6 \cdot 3) + (0 \cdot 5) = 26$$

The entry in row 1 and column 4 of AB is computed as follows.

$$\begin{bmatrix} 1 & 2 & 4 \\ 2 & 6 & 0 \end{bmatrix} \begin{bmatrix} 4 & 1 & 4 & 3 \\ 0 & -1 & 3 & 1 \\ 2 & 7 & 5 & 2 \end{bmatrix} = \begin{bmatrix} \square & \square & \square & 13 \\ \square & \square & 26 & \square \end{bmatrix}$$

$$(1 \cdot 3) + (2 \cdot 1) + (4 \cdot 2) = 13$$

The computations for the remaining products are

$$(1 \cdot 4) + (2 \cdot 0) + (4 \cdot 2) = 12$$
$$(1 \cdot 1) - (2 \cdot 1) + (4 \cdot 7) = 27$$
$$(1 \cdot 4) + (2 \cdot 3) + (4 \cdot 5) = 30$$
$$(2 \cdot 4) + (6 \cdot 0) + (0 \cdot 2) = 8$$
$$(2 \cdot 1) - (6 \cdot 1) + (0 \cdot 7) = -4$$
$$(2 \cdot 3) + (6 \cdot 1) + (0 \cdot 2) = 12$$

$$AB = \begin{bmatrix} 12 & 27 & 30 & 13 \\ 8 & -4 & 26 & 12 \end{bmatrix}$$

The definition of matrix multiplication requires that the number of columns of the first factor A be the same as the number of rows of the second factor B in order to form the product AB. If this condition is not satisfied, the product is undefined. A convenient way to determine whether a product of two matrices is defined is to write down the size of the first factor and, to the right of it, write down the size of the second factor. If, as in Figure 1.3, the inside numbers are the same, then the product is defined. The outside numbers then give the size of the product.

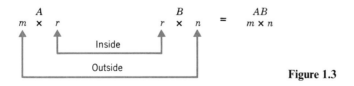

Figure 1.3

Example 16

Suppose that A is a 3×4 matrix, B is a 4×7 matrix, and C is a 7×3 matrix. Then AB is defined and is a 3×7 matrix; CA is defined and is a 7×4 matrix; BC is defined and is a 4×3 matrix. The products AC, CB, and BA are all undefined.

Example 17

If A is a general $m \times r$ matrix and B is a general $r \times n$ matrix, then, as suggested by the shading below, the entry in row i and column j of AB is given by the formula

$$a_{i1}b_{1j} + a_{i2}b_{2j} + a_{i3}b_{3j} + \cdots + a_{ir}b_{rj}$$

$$AB = \begin{bmatrix} a_{11} & a_{12} & \cdots & a_{1r} \\ a_{21} & a_{22} & \cdots & a_{2r} \\ \vdots & \vdots & & \vdots \\ a_{i1} & a_{i2} & \cdots & a_{ir} \\ \vdots & \vdots & & \vdots \\ a_{m1} & a_{m2} & \cdots & a_{mr} \end{bmatrix} \begin{bmatrix} b_{11} & b_{12} & \cdots & b_{1j} & \cdots & b_{1n} \\ b_{21} & b_{22} & \cdots & b_{2j} & \cdots & b_{2n} \\ \vdots & \vdots & & \vdots & & \vdots \\ b_{r1} & b_{r2} & \cdots & b_{rj} & \cdots & b_{rn} \end{bmatrix}$$

Matrix multiplication has an important application to systems of linear equations. Consider any system of m linear equations in n unknowns.

$$\begin{aligned} a_{11}x_1 + a_{12}x_2 + \cdots + a_{1n}x_n &= b_1 \\ a_{21}x_1 + a_{22}x_2 + \cdots + a_{2n}x_n &= b_2 \\ \vdots \qquad\qquad \vdots \qquad\qquad \vdots \qquad & \vdots \\ a_{m1}x_1 + a_{m2}x_2 + \cdots + a_{mn}x_n &= b_m \end{aligned}$$

Since two matrices are equal if and only if their corresponding entries are equal, we can replace the m equations in this system by the single matrix equation

$$\begin{bmatrix} a_{11}x_1 + a_{12}x_2 + \cdots + a_{1n}x_n \\ a_{21}x_1 + a_{22}x_2 + \cdots + a_{2n}x_n \\ \vdots \qquad\qquad \vdots \qquad\qquad \vdots \\ a_{m1}x_1 + a_{m2}x_2 + \cdots + a_{mn}x_n \end{bmatrix} = \begin{bmatrix} b_1 \\ b_2 \\ \vdots \\ b_m \end{bmatrix}$$

The $m \times 1$ matrix on the left side of this equation can be written as a product to give

$$\begin{bmatrix} a_{11} & a_{12} & \cdots & a_{1n} \\ a_{21} & a_{22} & \cdots & a_{2n} \\ \vdots & \vdots & & \vdots \\ a_{m1} & a_{m2} & \cdots & a_{mn} \end{bmatrix} \begin{bmatrix} x_1 \\ x_2 \\ \vdots \\ x_n \end{bmatrix} = \begin{bmatrix} b_1 \\ b_2 \\ \vdots \\ b_m \end{bmatrix}$$

If we designate these matrices by A, X, and B, respectively, the original system of m equations in n unknowns has been replaced by the single matrix equation

$$AX = B \tag{1.5}$$

Some of our later work will be devoted to solving matrix equations like this for the unknown matrix X. As a consequence of this matrix approach, we will obtain effective new methods for solving systems of linear equations. Matrix A in (1.5) is called the *coefficient matrix* for the system.

Example 18

At times it is helpful to be able to find a particular row or column in a product AB without computing the entire product. We leave it as an exercise to show that the entries in the jth column of AB are the entries in the product AB_j, where B_j is the matrix formed using only the jth column of B. Thus, if A and B are the matrices in Example 15, the second column of AB can be obtained by the computation

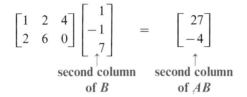

second column   second column
of B of AB

Similarly, the entries in the ith row of AB are the entries in the product A_iB, where A_i is the matrix formed by using only the ith row of A. Thus the first row in the product AB of Example 15 can be obtained by the computation

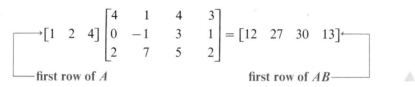

first row of A first row of AB

We conclude this section by defining a matrix operation that has no analogue in the real numbers.

Definition. If A is any $m \times n$ matrix, then the **transpose of A** is denoted by A^t and is defined to be the $n \times m$ matrix whose first column is the first row of A, whose second column is the second row of A, whose third column is the third row of A, and so on.

Example 19

The following are some examples of matrices and their transposes.

$$A = \begin{bmatrix} a_{11} & a_{12} & a_{13} & a_{14} \\ a_{21} & a_{22} & a_{23} & a_{24} \\ a_{31} & a_{32} & a_{33} & a_{34} \end{bmatrix} \qquad A^t = \begin{bmatrix} a_{11} & a_{21} & a_{31} \\ a_{12} & a_{22} & a_{32} \\ a_{13} & a_{23} & a_{33} \\ a_{14} & a_{24} & a_{34} \end{bmatrix}$$

$$B = \begin{bmatrix} 2 & 3 \\ 1 & 4 \\ 5 & 6 \end{bmatrix} \qquad B^t = \begin{bmatrix} 2 & 1 & 5 \\ 3 & 4 & 6 \end{bmatrix}$$

$$C = \begin{bmatrix} 1 & 3 & 5 \end{bmatrix} \qquad C^t = \begin{bmatrix} 1 \\ 3 \\ 5 \end{bmatrix}$$

$$D = \begin{bmatrix} 3 & 5 & -2 \\ 5 & 4 & 1 \\ -2 & 1 & 7 \end{bmatrix} \qquad D^t = \begin{bmatrix} 3 & 5 & -2 \\ 5 & 4 & 1 \\ -2 & 1 & 7 \end{bmatrix}$$

$$E = \begin{bmatrix} 4 \end{bmatrix} \qquad E^t = \begin{bmatrix} 4 \end{bmatrix} \ \triangle$$

REMARK. Observe that not only are the columns of A^t the rows of A, but the rows of A^t are the columns of A.

EXERCISE SET 1.4

1. Let A and B be 4×5 matrices and let C, D, and E be 5×2, 4×2, and 5×4 matrices, respectively. Determine which of the following matrix expressions are defined. For those which are defined, give the size of the resulting matrix.
   (a) BA (b) $AC + D$ (c) $AE + B$ (d) $AB + B$
   (e) $E(A + B)$ (f) $E(AC)$ (g) E^tA (h) $(A^t + E)D$

2. (a) Show that if AB and BA are both defined, then AB and BA are square matrices.
   (b) Show that if A is an $m \times n$ matrix and $A(BA)$ is defined, then B is an $n \times m$ matrix.

3. Solve the following matrix equation for a, b, c, and d.
   $$\begin{bmatrix} a - b & b + c \\ 3d + c & 2a - 4d \end{bmatrix} = \begin{bmatrix} 8 & 1 \\ 7 & 6 \end{bmatrix}$$

4. Consider the matrices
   $$A = \begin{bmatrix} 3 & 0 \\ -1 & 2 \\ 1 & 1 \end{bmatrix} \quad B = \begin{bmatrix} 4 & -1 \\ 0 & 2 \end{bmatrix} \quad C = \begin{bmatrix} 1 & 4 & 2 \\ 3 & 1 & 5 \end{bmatrix}$$
   $$D = \begin{bmatrix} 1 & 5 & 2 \\ -1 & 0 & 1 \\ 3 & 2 & 4 \end{bmatrix} \quad E = \begin{bmatrix} 6 & 1 & 3 \\ -1 & 1 & 2 \\ 4 & 1 & 3 \end{bmatrix}$$
   Compute
   (a) AB (b) $D + E$ (c) $D - E$
   (d) DE (e) ED (f) $-7B$

5. Using the matrices from Exercise 4, perform the indicated operations (where possible).
   (a) $3C - D$ (b) $(3E)D$
   (c) $(AB)C$ (d) $A(BC)$
   (e) $(4B)C + 2B$ (f) $D + E^2$ (where $E^2 = EE$)

6. Using the matrices from Exercise 4, perform the indicated operations (where possible).
   (a) A^t (b) D^t (c) $3B^t$
   (d) $BA^t - C^t$ (e) $(BB^t)C$ (f) $D^tE^t - (ED)^t$

7. Let

$$A = \begin{bmatrix} 3 & -2 & 7 \\ 6 & 5 & 4 \\ 0 & 4 & 9 \end{bmatrix} \quad \text{and} \quad B = \begin{bmatrix} 6 & -2 & 4 \\ 0 & 1 & 3 \\ 7 & 7 & 5 \end{bmatrix}$$

   Use the method of Example 18 to find

   (a) the first row of AB (b) the third row of AB
   (c) the second column of AB (d) the first column of BA
   (e) the third row of AA (f) the third column of AA

8. Let C, D, and E be the matrices in Exercise 4. Using as few computations as possible, determine the entry in row 2 and column 3 of $C(DE)$.

9. (a) Show that if A has a row of zeros and B is any matrix for which AB is defined, then AB also has a row of zeros.
   (b) Find a similar result involving a column of zeros.

10. If a_{ij} is an entry in row i and column j of a matrix A, then in what row and column will a_{ij} appear in A^t?

11. Let A be any $m \times n$ matrix and let 0 be the $m \times n$ matrix each of whose entries is zero. Show that if $kA = 0$, then $k = 0$ or $A = 0$.

12. Let I be the $n \times n$ matrix whose entry in row i and column j is

$$\begin{cases} 1 & \text{if} & i = j \\ 0 & \text{if} & i \neq j \end{cases}$$

   Show that $AI = IA = A$ for every $n \times n$ matrix A.

13. A square matrix is called a ***diagonal matrix*** if all entries off the main diagonal are zeros. Show that the product of diagonal matrices is again a diagonal matrix. State a rule for multiplying diagonal matrices.

14. (a) Show that the entries in the jth column of AB are the entries in the product AB_j, where B_j is the matrix formed from the jth column of B.
    (b) Show that the entries in the ith row of AB are the entries in the product A_iB, where A_i is the matrix formed from the ith row of A.

1.5 RULES OF MATRIX ARITHMETIC

Although many of the rules of arithmetic for real numbers also hold for matrices, there are some exceptions. For real numbers a and b, we always have $ab = ba$, which is called the *commutative law for multiplication*. For matrices, however, AB and BA need not be equal. Equality can fail to hold for three reasons. It can happen, for example, that AB is defined but BA is undefined. This is the case if A is a 2×3 matrix and B is a 3×4 matrix. Also, it can happen that AB and BA are both defined but have different sizes. This is the situation if A is a 2×3 matrix and B is a 3×2 matrix. Finally, as our next example shows, it is possible to have $AB \neq BA$ even if both AB and BA are defined and have the same size.

Example 20

Consider the matrices

$$A = \begin{bmatrix} -1 & 0 \\ 2 & 3 \end{bmatrix} \qquad B = \begin{bmatrix} 1 & 2 \\ 3 & 0 \end{bmatrix}$$

Multiplying gives

$$AB = \begin{bmatrix} -1 & -2 \\ 11 & 4 \end{bmatrix} \qquad BA = \begin{bmatrix} 3 & 6 \\ -3 & 0 \end{bmatrix}$$

Thus, $AB \neq BA$. ▲

Although the commutative law for multiplication is not valid in matrix arithmetic, many familiar laws of arithmetic are valid for matrices. Some of the most important ones and their names are summarized in the following theorem.

Theorem 2. *Assuming that the sizes of the matrices are such that the indicated operations can be performed, the following rules of matrix arithmetic are valid.*

(a) $A + B = B + A$	(*Commutative law for addition*)
(b) $A + (B + C) = (A + B) + C$	(*Associative law for addition*)
(c) $A(BC) = (AB)C$	(*Associative law for multiplication*)
(d) $A(B + C) = AB + AC$	(*Distributive law*)
(e) $(B + C)A = BA + CA$	(*Distributive law*)
(f) $A(B - C) = AB - AC$	
(g) $(B - C)A = BA - CA$	
(h) $a(B + C) = aB + aC$	
(i) $a(B - C) = aB - aC$	
(j) $(a + b)C = aC + bC$	
(k) $(a - b)C = aC - bC$	
(l) $(ab)C = a(bC)$	
(m) $a(BC) = (aB)C = B(aC)$	

Each of the equations in this theorem asserts an equality between matrices. To prove one of these equalities, it is necessary to show that the matrix on the left side has the same size as the matrix on the right side, and that corresponding entries on the two sides are equal. As an illustration, we shall prove (h). Some of the remaining proofs are given as exercises.

Proof of (h). Since the left side involves the operation $B + C$, B and C must have the same size, say $m \times n$. It follows that $a(B + C)$ and $aB + aC$ are also $m \times n$ matrices, and thus have the same size.

Let l_{ij} be any entry on the left side, and let r_{ij} be the corresponding entry on the right side. To complete the proof, we must show $l_{ij} = r_{ij}$. If we let a_{ij}, b_{ij}, and c_{ij} be the entries in the ith row and jth column of A, B, and C, respectively, then from the definitions of the matrix operations

$$l_{ij} = a(b_{ij} + c_{ij}) \qquad \text{and} \qquad r_{ij} = ab_{ij} + ac_{ij}$$

Since $a(b_{ij} + c_{ij}) = ab_{ij} + ac_{ij}$, we have $l_{ij} = r_{ij}$, which completes the proof.

Although the operations of matrix addition and matrix multiplication were defined for pairs of matrices, associative laws (b) and (c) enable us to denote sums and products of three matrices as $A + B + C$ and ABC without inserting any parentheses. This is justified by the fact that no matter how parentheses are inserted, the associative laws guarantee that the same end result will be obtained. Without going into details we observe that similar results are valid for sums or products involving four or more matrices. In general, *given any sum or any product of matrices, pairs of parentheses can be inserted or deleted anywhere within the expression without affecting the end result.*

Example 21

As an illustration of the associative law for matrix multiplication, consider

$$A = \begin{bmatrix} 1 & 2 \\ 3 & 4 \\ 0 & 1 \end{bmatrix} \quad B = \begin{bmatrix} 4 & 3 \\ 2 & 1 \end{bmatrix} \quad C = \begin{bmatrix} 1 & 0 \\ 2 & 3 \end{bmatrix}$$

Then

$$AB = \begin{bmatrix} 1 & 2 \\ 3 & 4 \\ 0 & 1 \end{bmatrix}\begin{bmatrix} 4 & 3 \\ 2 & 1 \end{bmatrix} = \begin{bmatrix} 8 & 5 \\ 20 & 13 \\ 2 & 1 \end{bmatrix}$$

so that

$$(AB)C = \begin{bmatrix} 8 & 5 \\ 20 & 13 \\ 2 & 1 \end{bmatrix}\begin{bmatrix} 1 & 0 \\ 2 & 3 \end{bmatrix} = \begin{bmatrix} 18 & 15 \\ 46 & 39 \\ 4 & 3 \end{bmatrix}$$

On the other hand

$$BC = \begin{bmatrix} 4 & 3 \\ 2 & 1 \end{bmatrix}\begin{bmatrix} 1 & 0 \\ 2 & 3 \end{bmatrix} = \begin{bmatrix} 10 & 9 \\ 4 & 3 \end{bmatrix}$$

so that

$$A(BC) = \begin{bmatrix} 1 & 2 \\ 3 & 4 \\ 0 & 1 \end{bmatrix} \begin{bmatrix} 10 & 9 \\ 4 & 3 \end{bmatrix} = \begin{bmatrix} 18 & 15 \\ 46 & 39 \\ 4 & 3 \end{bmatrix}$$

Thus, $(AB)C = A(BC)$, as guaranteed by Theorem 2(c). ▲

A matrix, all of whose entries are zero, such as

$$\begin{bmatrix} 0 & 0 \\ 0 & 0 \end{bmatrix} \quad \begin{bmatrix} 0 & 0 & 0 \\ 0 & 0 & 0 \\ 0 & 0 & 0 \end{bmatrix} \quad \begin{bmatrix} 0 & 0 & 0 & 0 \\ 0 & 0 & 0 & 0 \end{bmatrix} \quad \begin{bmatrix} 0 \\ 0 \\ 0 \\ 0 \end{bmatrix} \quad [0]$$

is called a *zero matrix*. A zero matrix will be denoted by *0*; if it is important to emphasize the size, we shall write $0_{m \times n}$ for the $m \times n$ zero matrix.

If A is any matrix and *0* is the zero matrix with the same size, it is obvious that $A + 0 = 0 + A = A$. The matrix *0* plays much the same role in these matrix equations as the number 0 plays in the numerical equations $a + 0 = 0 + a = a$.

Since we already know that some of the rules of arithmetic for real numbers do not carry over to matrix arithmetic, it would be foolhardy to assume that all the properties of the real number zero carry over to zero matrices. For example, consider the following two standard results in the arithmetic of real numbers.

(i) If $ab = ac$ and $a \neq 0$, then $b = c$. (This is called the *cancellation law*.)
(ii) If $ad = 0$, then at least one of the factors on the left is 0.

As the next example shows, the corresponding results are not generally true in matrix arithmetic.

Example 22

Consider the matrices

$$A = \begin{bmatrix} 0 & 1 \\ 0 & 2 \end{bmatrix} \quad B = \begin{bmatrix} 1 & 1 \\ 3 & 4 \end{bmatrix} \quad C = \begin{bmatrix} 2 & 5 \\ 3 & 4 \end{bmatrix} \quad D = \begin{bmatrix} 3 & 7 \\ 0 & 0 \end{bmatrix}$$

Here

$$AB = AC = \begin{bmatrix} 3 & 4 \\ 6 & 8 \end{bmatrix}$$

Although $A \neq 0$, it is *incorrect* to cancel the A from both sides of the equation $AB = AC$ and write $B = C$. Thus, the cancellation law fails to hold for matrices.

Also, $AD = 0$, yet $A \neq 0$ and $D \neq 0$, so that result (ii) listed above does not carry over to matrix arithmetic. ▲

In spite of the above examples, there are a number of familiar properties of the real number 0 that do carry over to zero matrices. Some of the more important ones are summarized in the next theorem. The proofs are left as exercises.

Theorem 3. *Assuming that the sizes of the matrices are such that the indicated operations can be performed, the following rules of matrix arithmetic are valid.*

(a) $A + 0 = 0 + A = A$
(b) $A - A = 0$
(c) $0 - A = -A$
(d) $A0 = 0; \quad 0A = 0$

As an application of our results on matrix arithmetic, we shall prove the following theorem, which was anticipated earlier in the text.

Theorem 4. *Every system of linear equations has either no solutions, exactly one solution, or infinitely many solutions.*

Proof. If $AX = B$ is a system of linear equations, exactly one of the following is true: (a) the system has no solutions, (b) the system has exactly one solution, or (c) the system has more than one solution. The proof will be complete if we can show that the system has infinitely many solutions in case (c).

Assume that $AX = B$ has more than one solution and let X_1 and X_2 be two different solutions. Thus, $AX_1 = B$ and $AX_2 = B$. Subtracting these equations gives $AX_1 - AX_2 = 0$ or $A(X_1 - X_2) = 0$. If we let $X_0 = X_1 - X_2$ and let k be any scalar, then

$$A(X_1 + kX_0) = AX_1 + A(kX_0)$$
$$= AX_1 + k(AX_0)$$
$$= B + k0$$
$$= B + 0$$
$$= B$$

But this says that $X_1 + kX_0$ is a solution of $AX = B$. Since there are infinitely many choices for k, $AX = B$ has infinitely many solutions. ▮

Of special interest are square matrices with 1's on the main diagonal and 0's off the main diagonal, such as

$$\begin{bmatrix} 1 & 0 \\ 0 & 1 \end{bmatrix}, \quad \begin{bmatrix} 1 & 0 & 0 \\ 0 & 1 & 0 \\ 0 & 0 & 1 \end{bmatrix}, \quad \begin{bmatrix} 1 & 0 & 0 & 0 \\ 0 & 1 & 0 & 0 \\ 0 & 0 & 1 & 0 \\ 0 & 0 & 0 & 1 \end{bmatrix}, \quad \text{and so on.}$$

A matrix of this form is called an ***identity matrix*** and is denoted by I. If it is important to emphasize the size, we shall write I_n for the $n \times n$ identity matrix.

If A is an $m \times n$ matrix, then, as illustrated in the next example, $AI_n = A$ and $I_m A = A$. Thus, an identity matrix plays much the same role in matrix arithmetic as the number 1 plays in the numerical relationships $a \cdot 1 = 1 \cdot a = a$.

Example 23

Consider the matrix

$$A = \begin{bmatrix} a_{11} & a_{12} & a_{13} \\ a_{21} & a_{22} & a_{23} \end{bmatrix}$$

Then

$$I_2 A = \begin{bmatrix} 1 & 0 \\ 0 & 1 \end{bmatrix} \begin{bmatrix} a_{11} & a_{12} & a_{13} \\ a_{21} & a_{22} & a_{23} \end{bmatrix} = \begin{bmatrix} a_{11} & a_{12} & a_{13} \\ a_{21} & a_{22} & a_{23} \end{bmatrix} = A$$

and

$$AI_3 = \begin{bmatrix} a_{11} & a_{12} & a_{13} \\ a_{21} & a_{22} & a_{23} \end{bmatrix} \begin{bmatrix} 1 & 0 & 0 \\ 0 & 1 & 0 \\ 0 & 0 & 1 \end{bmatrix} = \begin{bmatrix} a_{11} & a_{12} & a_{13} \\ a_{21} & a_{22} & a_{23} \end{bmatrix} = A$$

Definition. If A is a square matrix, and if a matrix B can be found such that $AB = BA = I$, then A is said to be ***invertible*** and B is called an ***inverse*** of A.

Example 24

The matrix

$$B = \begin{bmatrix} 3 & 5 \\ 1 & 2 \end{bmatrix} \quad \text{is an inverse of} \quad A = \begin{bmatrix} 2 & -5 \\ -1 & 3 \end{bmatrix}$$

since

$$AB = \begin{bmatrix} 2 & -5 \\ -1 & 3 \end{bmatrix} \begin{bmatrix} 3 & 5 \\ 1 & 2 \end{bmatrix} = \begin{bmatrix} 1 & 0 \\ 0 & 1 \end{bmatrix} = I$$

and

$$BA = \begin{bmatrix} 3 & 5 \\ 1 & 2 \end{bmatrix} \begin{bmatrix} 2 & -5 \\ -1 & 3 \end{bmatrix} = \begin{bmatrix} 1 & 0 \\ 0 & 1 \end{bmatrix} = I$$

Example 25

The matrix

$$A = \begin{bmatrix} 1 & 4 & 0 \\ 2 & 5 & 0 \\ 3 & 6 & 0 \end{bmatrix}$$

is not invertible. To see why, let

$$B = \begin{bmatrix} b_{11} & b_{12} & b_{13} \\ b_{21} & b_{22} & b_{23} \\ b_{31} & b_{32} & b_{33} \end{bmatrix}$$

be any 3×3 matrix. From Example 18 the third column of BA is

$$\begin{bmatrix} b_{11} & b_{12} & b_{13} \\ b_{21} & b_{22} & b_{23} \\ b_{31} & b_{32} & b_{33} \end{bmatrix} \begin{bmatrix} 0 \\ 0 \\ 0 \end{bmatrix} = \begin{bmatrix} 0 \\ 0 \\ 0 \end{bmatrix}$$

Thus,

$$BA \neq I = \begin{bmatrix} 1 & 0 & 0 \\ 0 & 1 & 0 \\ 0 & 0 & 1 \end{bmatrix}$$

It is reasonable to ask whether an invertible matrix can have more than one inverse. The next theorem shows the answer is no—an invertible matrix has exactly one inverse.

Theorem 5. *If B and C are both inverses of the matrix A, then B = C.*

Proof. Since B is an inverse of A, $BA = I$. Multiplying both sides on the right by C gives $(BA)C = IC = C$. But $(BA)C = B(AC) = BI = B$, so that $B = C$.

As a consequence of this important result, we can now speak of "the" inverse of an invertible matrix. If A is invertible, then its inverse will be denoted by the symbol A^{-1}. Thus,

$$AA^{-1} = I \quad \text{and} \quad A^{-1}A = I$$

The inverse of A plays much the same role in matrix arithmetic that reciprocal a^{-1} plays in the numerical relationships $aa^{-1} = 1$ and $a^{-1}a = 1$.

Example 26

Consider the 2×2 matrix

$$A = \begin{bmatrix} a & b \\ c & d \end{bmatrix}$$

If $ad - bc \neq 0$, then

$$A^{-1} = \frac{1}{ad - bc} \begin{bmatrix} d & -b \\ -c & a \end{bmatrix} = \begin{bmatrix} \dfrac{d}{ad - bc} & -\dfrac{b}{ad - bc} \\ -\dfrac{c}{ad - bc} & \dfrac{a}{ad - bc} \end{bmatrix}$$

since $AA^{-1} = I_2$ and $A^{-1}A = I_2$ (verify). In the next section we shall show how to find inverses of invertible matrices whose sizes are greater than 2×2. ▲

Theorem 6. *If A and B are invertible matrices of the same size, then*

(a) *AB is invertible*
(b) $(AB)^{-1} = B^{-1}A^{-1}$

Proof. If we can show that $(AB)(B^{-1}A^{-1}) = (B^{-1}A^{-1})(AB) = I$, then we will have simultaneously established that AB is invertible and that $(AB)^{-1} = B^{-1}A^{-1}$. But $(AB)(B^{-1}A^{-1}) = A(BB^{-1})A^{-1} = AIA^{-1} = AA^{-1} = I$. Similarly $(B^{-1}A^{-1})(AB) = I$. ▌

Although we will not prove it, this result can be extended to include three or more factors. Thus, we can state the following general rule.

A product of invertible matrices is always invertible, and the inverse of the product is the product of the inverses in the reverse order.

Example 27

Consider the matrices

$$A = \begin{bmatrix} 1 & 2 \\ 1 & 3 \end{bmatrix} \quad B = \begin{bmatrix} 3 & 2 \\ 2 & 2 \end{bmatrix} \quad AB = \begin{bmatrix} 7 & 6 \\ 9 & 8 \end{bmatrix}$$

Applying the formula given in Example 26, we obtain

$$A^{-1} = \begin{bmatrix} 3 & -2 \\ -1 & 1 \end{bmatrix} \quad B^{-1} = \begin{bmatrix} 1 & -1 \\ -1 & \frac{3}{2} \end{bmatrix} \quad (AB)^{-1} = \begin{bmatrix} 4 & -3 \\ -\frac{9}{2} & \frac{7}{2} \end{bmatrix}$$

Also

$$B^{-1}A^{-1} = \begin{bmatrix} 1 & -1 \\ -1 & \frac{3}{2} \end{bmatrix} \begin{bmatrix} 3 & -2 \\ -1 & 1 \end{bmatrix} = \begin{bmatrix} 4 & -3 \\ -\frac{9}{2} & \frac{7}{2} \end{bmatrix}$$

Therefore, $(AB)^{-1} = B^{-1}A^{-1}$ as guaranteed by Theorem 6. ▲

Next, we shall define powers of a square matrix and discuss their properties.

Definition. If A is a square matrix, then we define the nonnegative integer powers of A to be

$$A^0 = I \qquad A^n = \underbrace{AA \cdots A}_{n \text{ factors}} \qquad (n > 0)$$

Moreover, if A is invertible, then we define the negative integer powers to be

$$A^{-n} = (A^{-1})^n = \underbrace{A^{-1}A^{-1} \cdots A^{-1}}_{n \text{ factors}}$$

The following theorem, which we state without proof, shows that the familiar laws of exponents are valid.

Theorem 7. *If A is a square matrix and r and s are integers, then*

$$A^r A^s = A^{r+s} \qquad (A^r)^s = A^{rs}$$

The next theorem provides some additional useful properties of matrix exponents.

Theorem 8. *If A is an invertible matrix, then:*

(a) A^{-1} is invertible and $(A^{-1})^{-1} = A$.
(b) A^n is invertible and $(A^n)^{-1} = (A^{-1})^n$ $\qquad n = 0, 1, 2, \ldots$.
(c) For any nonzero scalar k, kA is invertible and $(kA)^{-1} = \dfrac{1}{k} A^{-1}$.

Proof.
(a) Since $AA^{-1} = A^{-1}A = I$, A^{-1} is invertible and $(A^{-1})^{-1} = A$.
(b) This part is left as an exercise.
(c) If k is any nonzero scalar, results (l) and (m) of Theorem 2 enable us to write

$$(kA)\left(\frac{1}{k}A^{-1}\right) = \frac{1}{k}(kA)A^{-1} = \left(\frac{1}{k}k\right)AA^{-1} = (1)I = I$$

Similarly $\left(\dfrac{1}{k}A^{-1}\right)(kA) = I$ so that kA is invertible and $(kA)^{-1} = \dfrac{1}{k}A^{-1}$. ∎

We conclude this section with a theorem that lists the main properties of the transpose operation. The proofs are discussed in the exercises.

Theorem 9. *If the sizes of the matrices are such that the given operations can be performed, then*

(a) $(A^t)^t = A$
(b) $(A + B)^t = A^t + B^t$
(c) $(kA)^t = kA^t$, *where k is any scalar*
(d) $(AB)^t = B^t A^t$

Although we shall not prove it, part (d) of this theorem can be extended to include three or more factors. More precisely, we can state the following general rule.

The transpose of a product of matrices is equal to the product of their transposes in the reverse order.

REMARK. Note the similarity between this result and the result following Theorem 6 about the inverse of a product of matrices.

EXERCISE SET 1.5

1. Let

$$A = \begin{bmatrix} 3 & 2 \\ -1 & 3 \end{bmatrix} \quad B = \begin{bmatrix} 4 & 0 \\ 1 & 5 \end{bmatrix} \quad C = \begin{bmatrix} 0 & -1 \\ 4 & 6 \end{bmatrix} \quad a = -3 \quad b = 2$$

Show
(a) $A + (B + C) = (A + B) + C$ (b) $(AB)C = A(BC)$
(c) $(a + b)C = aC + bC$ (d) $a(B - C) = aB - aC$

2. Using the matrices and scalars in Exercise 1, show
(a) $a(BC) = (aB)C = B(aC)$ (b) $A(B - C) = AB - AC$

3. Using the matrices and scalars in Exercise 1, show
(a) $(A^t)^t = A$ (b) $(A + B)^t = A^t + B^t$
(c) $(aC)^t = aC^t$ (d) $(AB)^t = B^t A^t$

4. Use the formula given in Example 26 to compute the inverses of the following matrices.

$$A = \begin{bmatrix} 3 & 1 \\ 5 & 2 \end{bmatrix} \quad B = \begin{bmatrix} 2 & -3 \\ 4 & 4 \end{bmatrix} \quad C = \begin{bmatrix} 2 & 0 \\ 0 & 3 \end{bmatrix}$$

5. Verify that the matrices A and B in Exercise 4 satisfy the relationship $(AB)^{-1} = B^{-1}A^{-1}$.

6. Let A and B be square matrices of the same size. Is $(AB)^2 = A^2 B^2$ a valid matrix identity? Justify your answer.

7. Let A be an invertible matrix whose inverse is

$$\begin{bmatrix} 3 & 4 \\ 5 & 6 \end{bmatrix}$$

Find the matrix A.

8. Let A be an invertible matrix, and suppose that the inverse of $7A$ is

$$\begin{bmatrix} -1 & 2 \\ 4 & -7 \end{bmatrix}$$

Find the matrix A.

9. Let A be the matrix

$$\begin{bmatrix} 1 & 0 \\ 2 & 3 \end{bmatrix}$$

Compute A^3, A^{-3}, and $A^2 - 2A + I$.

10. Let A be the matrix

$$\begin{bmatrix} 1 & 1 & 0 \\ 0 & 1 & 1 \\ 1 & 0 & 1 \end{bmatrix}$$

Determine whether A is invertible, and if so, find its inverse. (**Hint.** Solve $AX = I$ by equating corresponding entries on the two sides.)

11. Find the inverse of

$$\begin{bmatrix} \cos\theta & \sin\theta \\ -\sin\theta & \cos\theta \end{bmatrix}$$

12. (a) Find 2×2 matrices A and B such that

$$(A + B)^2 \neq A^2 + 2AB + B^2$$

(b) Show that, if A and B are square matrices such that $AB = BA$, then

$$(A + B)^2 = A^2 + 2AB + B^2$$

(c) Find an expansion of $(A + B)^2$ that is valid for all square matrices A and B having the same size.

13. Consider the matrix

$$A = \begin{bmatrix} a_{11} & 0 & 0 & \cdots & 0 \\ 0 & a_{22} & 0 & \cdots & 0 \\ \vdots & \vdots & \vdots & & \vdots \\ 0 & 0 & 0 & \cdots & a_{nn} \end{bmatrix}$$

where $a_{11} a_{22} \cdots a_{nn} \neq 0$. Show that A is invertible and find its inverse.

14. Assume that A is a square matrix that satisfies $A^2 - 3A + I = 0$. Show that $A^{-1} = 3I - A$.

15. (a) Show that a matrix with a row of zeros cannot have an inverse.
(b) Show that a matrix with a column of zeros cannot have an inverse.

16. Is the sum of two invertible matrices necessarily invertible?

17. Let A and B be square matrices such that $AB = 0$. Show that if A is invertible then $B = 0$.

18. In Theorem 3 why didn't we write part (d) as $A0 = 0 = 0A$?

19. The real equation $a^2 = 1$ has exactly two solutions. Find at least eight different 3×3 matrices that satisfy the matrix equation $A^2 = I_3$. (**Hint.** Look for solutions in which all the entries off the main diagonal are zero.)

20. Let $AX = B$ be any consistent system of linear equations, and let X_1 be a fixed solution. Show that every solution to the system can be written in the form $X = X_1 + X_0$ where X_0 is a solution to $AX = 0$. Show also that every matrix of this form is a solution.

21. Let

$$A = \begin{bmatrix} a_{11} & a_{12} & a_{13} \\ a_{21} & a_{22} & a_{23} \\ a_{31} & a_{32} & a_{33} \end{bmatrix} \quad \text{and} \quad B = \begin{bmatrix} b_{11} & b_{12} & b_{13} \\ b_{21} & b_{22} & b_{23} \\ b_{31} & b_{32} & b_{33} \end{bmatrix}$$

Show that
(a) $(A^t)^t = A$ (b) $(A + B)^t = A^t + B^t$ (c) $(AB)^t = B^t A^t$ (d) $(kA)^t = kA^t$

22. (a) Verify that if the circled entries are interchanged as indicated in the following figure, then the resulting matrix is the transpose of the original matrix.

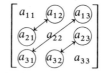

(b) Use the result in (a) to help find a nonzero 3×3 matrix A such that $A^t = A$.
(c) Use the result in (a) to help find a nonzero 3×3 matrix A such that $A^t = -A$.
(d) In general, if A is a square matrix, then A^t results when the entries that are symmetrically placed on the two sides of the main diagonal are interchanged. Use this fact to help find a 4×4 nonzero matrix such that $A = A^t$.

23. A square matrix A is called *symmetric* if $A^t = A$ and *skew-symmetric* if $A^t = -A$. Show that if B is a square matrix, then
(a) BB^t and $B + B^t$ are symmetric   (b) $B - B^t$ is skew-symmetric

24. If A is a square matrix and n is a positive integer, is it true that $(A^n)^t = (A^t)^n$? Justify your answer.

25. Apply parts (d) and (m) of Theorem 2 to the matrices A, B, and $(-1)C$ to derive the result in part (f).

26. Use parts (b) and (c) of Theorem 9 to prove that $(A - B)^t = A^t - B^t$.

27. Prove: If A is invertible, then A^t is invertible and $(A^t)^{-1} = (A^{-1})^t$.

28. Prove part (b) of Theorem 2.

29. Prove Theorem 3.

30. Prove part (c) of Theorem 2.

31. Prove part (b) of Theorem 8.

32. Consider the laws of exponents $A^r A^s = A^{r+s}$ and $(A^r)^s = A^{rs}$.
   (a) Show that if A is any square matrix, these laws are valid for all nonnegative integer values of r and s.
   (b) Show that if A is invertible, these laws hold for all negative integer values of r and s.

33. Show that if A is invertible and k is any nonzero scalar, then $(kA)^n = k^n A^n$ for all integer values of n.

34. (a) Show that if A is invertible and $AB = AC$, then $B = C$.
   (b) Explain why part (a) and Example 21 do not contradict one another.

35. Prove: If R is a square matrix in reduced row-echelon form, and if R has no zero rows, then $R = I$.

1.6   ELEMENTARY MATRICES AND A METHOD FOR FINDING A^{-1}

In this section we shall develop a simple scheme or algorithm for finding the inverse of an invertible matrix.

> **Definition.** An $n \times n$ matrix is called an ***elementary matrix*** if it can be obtained from the $n \times n$ identity matrix I_n by performing a single elementary row operation.

Example 28

Listed below are four elementary matrices and the operations that produce them.

(i) $\begin{bmatrix} 1 & 0 \\ 0 & -3 \end{bmatrix}$ (ii) $\begin{bmatrix} 1 & 0 & 0 & 0 \\ 0 & 0 & 0 & 1 \\ 0 & 0 & 1 & 0 \\ 0 & 1 & 0 & 0 \end{bmatrix}$ (iii) $\begin{bmatrix} 1 & 0 & 3 \\ 0 & 1 & 0 \\ 0 & 0 & 1 \end{bmatrix}$ (iv) $\begin{bmatrix} 1 & 0 & 0 \\ 0 & 1 & 0 \\ 0 & 0 & 1 \end{bmatrix}$

| Multiply the second row of I_2 by -3 | Interchange the second and fourth rows of I_4 | Add 3 times the third row of I_3 to the first row | Multiply the first row of I_3 by 1 |

When a matrix A is multiplied on the *left* by an elementary matrix E, the effect is to perform an elementary row operation on A. This is the content of the following theorem, which we state without proof.

Theorem 10. *If the elementary matrix E results from performing a certain row operation on I_m and if A is an $m \times n$ matrix, then the product EA is the matrix that results when this same row operation is performed on A.*

The following example illustrates this idea.

Example 29

Consider the matrix

$$A = \begin{bmatrix} 1 & 0 & 2 & 3 \\ 2 & -1 & 3 & 6 \\ 1 & 4 & 4 & 0 \end{bmatrix}$$

and consider the elementary matrix

$$E = \begin{bmatrix} 1 & 0 & 0 \\ 0 & 1 & 0 \\ 3 & 0 & 1 \end{bmatrix}$$

which results from adding 3 times the first row of I_3 to the third row. The product EA is

$$EA = \begin{bmatrix} 1 & 0 & 2 & 3 \\ 2 & -1 & 3 & 6 \\ 4 & 4 & 10 & 9 \end{bmatrix}$$

which is precisely the same matrix that results when we add 3 times the first row of A to the third row. ▲

REMARK. Theorem 10 is primarily of theoretical interest and will be used for developing some results about matrices and systems of linear equations. From a computational viewpoint, it is preferable to perform row operations directly rather than multiply on the left by an elementary matrix.

If an elementary row operation is applied to an identity matrix I to produce an elementary matrix E, then there is a second row operation that, when applied to E, produces I back again. For example, if E is obtained by multiplying the ith row of I by a nonzero constant c, then I can be recovered if the ith row of E is multiplied by $1/c$. The various possibilities are listed in Figure 1.4.

Row Operation on I that Produces E	Row Operation on E that Reproduces I
Multiply row i by $c \neq 0$.	Multiply row i by $1/c$
Interchange rows i and j.	Interchange rows i and j.
Add c times row i to row j.	Add $-c$ times row i to row j

Figure 1.4

The operations on the right side of this table are called the ***inverse operations*** of the corresponding operations on the left.

Example 30

Using the results in Figure 1.4, the first three elementary matrices given in Example 28 can be restored to identity matrices by applying the following row operations: multiply the second row in (i) by $-1/3$; interchange the second and fourth rows in (ii); add -3 times the third row of (iii) to the first row. ▲

The next theorem gives an important property of elementary matrices.

Theorem 11. *Every elementary matrix is invertible, and the inverse is also an elementary matrix.*

Proof. If E is an elementary matrix, then E results from performing some row operation on I. Let E_0 be the matrix that results when the inverse of this operation is performed on I. Applying Theorem 10 and using the fact that inverse row operations cancel the effect of each other, it follows that

$$E_0 E = I \quad \text{and} \quad E E_0 = I$$

Thus, the elementary matrix E_0 is the inverse of E. ▮

If a matrix B can be obtained from a matrix A by performing a finite sequence of elementary row operations, then obviously we can get from B back to A by performing the inverses of these elementary row operations in reverse order. Matrices that can be obtained from one another by a finite sequence of elementary row operations are said to be ***row equivalent.***

The next theorem establishes some fundamental relationships between $n \times n$ matrices and systems of n linear equations in n unknowns. These results are extremely important and will be used many times in later sections.

Theorem 12. *If A is an $n \times n$ matrix, then the following statements are equivalent, that is, all true or all false.*

(a) *A is invertible.*
(b) *$AX = 0$ has only the trivial solution.*
(c) *A is row equivalent to I_n.*

Proof. We shall prove the equivalence by establishing the chain of implications: $(a) \Rightarrow (b) \Rightarrow (c) \Rightarrow (a)$.

$(a) \Rightarrow (b)$: Assume A is invertible and let X_0 be any solution to $AX = 0$. Thus, $AX_0 = 0$. Multiplying both sides of this equation by A^{-1} gives $A^{-1}(AX_0) = A^{-1}0$, or $(A^{-1}A)X_0 = 0$, or $IX_0 = 0$, or $X_0 = 0$. Thus, $AX = 0$ has only the trivial solution.

$(b) \Rightarrow (c)$: Let $AX = 0$ be the matrix form of the system

$$
\begin{aligned}
a_{11}x_1 + a_{12}x_2 + \cdots + a_{1n}x_n &= 0 \\
a_{21}x_1 + a_{22}x_2 + \cdots + a_{2n}x_n &= 0 \\
\vdots \qquad \vdots \qquad\qquad \vdots \qquad \vdots \\
a_{n1}x_1 + a_{n2}x_2 + \cdots + a_{nn}x_n &= 0
\end{aligned}
\tag{1.6}
$$

and assume the system has only the trivial solution. If we solve by Gauss-Jordan elimination, then the system of equations corresponding to the reduced row-echelon form of the augmented matrix will be

$$
\begin{aligned}
x_1 \qquad\qquad\quad &= 0 \\
x_2 \qquad\quad &= 0 \\
\ddots \\
x_n &= 0
\end{aligned}
\tag{1.7}
$$

Thus, the augmented matrix

$$
\begin{bmatrix}
a_{11} & a_{12} & \cdots & a_{1n} & 0 \\
a_{21} & a_{22} & \cdots & a_{2n} & 0 \\
\vdots & \vdots & & \vdots & \vdots \\
a_{n1} & a_{n2} & \cdots & a_{nn} & 0
\end{bmatrix}
$$

for (1.6) can be reduced to the augmented matrix

$$
\begin{bmatrix}
1 & 0 & 0 & \cdots & 0 & 0 \\
0 & 1 & 0 & \cdots & 0 & 0 \\
0 & 0 & 1 & \cdots & 0 & 0 \\
\vdots & \vdots & \vdots & & \vdots & \vdots \\
0 & 0 & 0 & \cdots & 1 & 0
\end{bmatrix}
$$

for (1.7) by a sequence of elementary row operations. If we disregard the last column (of zeros) in each of these matrices, we can conclude that A can be reduced to I_n by a sequence of elementary row operations; that is, A is row equivalent to I_n.

$(c) \Rightarrow (a)$: Assume A is row equivalent to I_n, so that A can be reduced to I_n by a finite sequence of elementary row operations. By Theorem 10 each of these operations can be accomplished by multiplying on the left by an appropriate elementary matrix. Thus, we can find elementary matrices $E_1, E_2, \ldots, E_k$ such that

$$E_k \cdots E_2 E_1 A = I_n \tag{1.8}$$

By Theorem 11, $E_1, E_2, \ldots, E_k$ are invertible. Multiplying both sides of equation (1.8) on the left successively by $E_k^{-1}, \ldots, E_2^{-1}, E_1^{-1}$ we obtain

$$A = E_1^{-1} E_2^{-1} \cdots E_k^{-1} I_n = E_1^{-1} E_2^{-1} \cdots E_k^{-1} \tag{1.9}$$

Since (1.9) expresses A as a product of invertible matrices, we can conclude that A is invertible. ▮

REMARK. Because I_n is in reduced row-echelon form and because the reduced row-echelon form of a matrix A is unique, part (c) of Theorem 12 is equivalent to stating that I_n is the reduced row-echelon form of A.

As our first application of this theorem, we shall establish a method for determining the inverse of an invertible matrix.

Inverting the left and right sides of (1.9) yields $A^{-1} = E_k \cdots E_2 E_1$, or equivalently

$$A^{-1} = E_k \cdots E_2 E_1 I_n \tag{1.10}$$

which tells us that A^{-1} can be obtained by multiplying I_n successively on the left by the elementary matrices $E_1, E_2, \ldots, E_k$. Since each multiplication on the left by one of these elementary matrices performs a row operation, it follows, by comparing equations (1.8) and (1.10), that *the sequence of row operations that reduces A to I_n will reduce I_n to A^{-1}.* Thus, to find the inverse of an invertible matrix A, we must find a sequence of elementary row operations that reduces A to the identity and then perform this same sequence of operations on I_n to obtain A^{-1}. A simple method for carrying out this procedure is given in the following example.

Example 31

Find the inverse of

$$A = \begin{bmatrix} 1 & 2 & 3 \\ 2 & 5 & 3 \\ 1 & 0 & 8 \end{bmatrix}$$

We wish to reduce A to the identity matrix by row operations and simultaneously apply these operations to I to produce A^{-1}. This can be accomplished by adjoining the identity matrix to the right of A and applying row operations to both sides until the left side is reduced to I. The final matrix will then have the form $[I \mid A^{-1}]$. The computations can be carried out as follows.

$$\begin{bmatrix} 1 & 2 & 3 & \vdots & 1 & 0 & 0 \\ 2 & 5 & 3 & \vdots & 0 & 1 & 0 \\ 1 & 0 & 8 & \vdots & 0 & 0 & 1 \end{bmatrix}$$

$$\begin{bmatrix} 1 & 2 & 3 & \vdots & 1 & 0 & 0 \\ 0 & 1 & -3 & \vdots & -2 & 1 & 0 \\ 0 & -2 & 5 & \vdots & -1 & 0 & 1 \end{bmatrix}$$

We added -2 times the first row to the second and -1 times the first row to the third.

$$\begin{bmatrix} 1 & 2 & 3 & \vdots & 1 & 0 & 0 \\ 0 & 1 & -3 & \vdots & -2 & 1 & 0 \\ 0 & 0 & -1 & \vdots & -5 & 2 & 1 \end{bmatrix}$$

We added 2 times the second row to the third.

$$\begin{bmatrix} 1 & 2 & 3 & \vdots & 1 & 0 & 0 \\ 0 & 1 & -3 & \vdots & -2 & 1 & 0 \\ 0 & 0 & 1 & \vdots & 5 & -2 & -1 \end{bmatrix}$$

We multiplied the third row by -1.

$$\begin{bmatrix} 1 & 2 & 0 & \vdots & -14 & 6 & 3 \\ 0 & 1 & 0 & \vdots & 13 & -5 & -3 \\ 0 & 0 & 1 & \vdots & 5 & -2 & -1 \end{bmatrix}$$

We added 3 times the third row to the second and -3 times the third row to the first.

$$\begin{bmatrix} 1 & 0 & 0 & \vdots & -40 & 16 & 9 \\ 0 & 1 & 0 & \vdots & 13 & -5 & -3 \\ 0 & 0 & 1 & \vdots & 5 & -2 & -1 \end{bmatrix}$$

We added -2 times the second row to the first.

Thus,

$$A^{-1} = \begin{bmatrix} -40 & 16 & 9 \\ 13 & -5 & -3 \\ 5 & -2 & -1 \end{bmatrix}$$

Often it will not be known in advance whether a given matrix is invertible. If the matrix is not invertible, then by part (*c*) of Theorem 12 it is not row equivalent to I_n; that is, its reduced row-echelon form has at least one row of zeros. Thus, if the procedure in the last example is attempted on a matrix that is not invertible, then at some point in the computations a row of zeros will occur on the *left side*. It can then be concluded that the given matrix is not invertible, and the computations can be stopped.

Example 32

Consider the matrix

$$A = \begin{bmatrix} 1 & 6 & 4 \\ 2 & 4 & -1 \\ -1 & 2 & 5 \end{bmatrix}$$

Applying the procedure of Example 30 yields

$$\begin{bmatrix} 1 & 6 & 4 & \vdots & 1 & 0 & 0 \\ 2 & 4 & -1 & \vdots & 0 & 1 & 0 \\ -1 & 2 & 5 & \vdots & 0 & 0 & 1 \end{bmatrix}$$

$$\begin{bmatrix} 1 & 6 & 4 & \vdots & 1 & 0 & 0 \\ 0 & -8 & -9 & \vdots & -2 & 1 & 0 \\ 0 & 8 & 9 & \vdots & 1 & 0 & 1 \end{bmatrix}$$
> We added −2 times the first row to the second and added the first row to the third.

$$\begin{bmatrix} 1 & 6 & 4 & \vdots & 1 & 0 & 0 \\ 0 & -8 & -9 & \vdots & -2 & 1 & 0 \\ 0 & 0 & 0 & \vdots & -1 & 1 & 1 \end{bmatrix}$$
> We added the second row to the third.

Since we have obtained a row of zeros on the left side, A is not invertible.

Example 33

In Example 31 we showed that

$$A = \begin{bmatrix} 1 & 2 & 3 \\ 2 & 5 & 3 \\ 1 & 0 & 8 \end{bmatrix}$$

is an invertible matrix. From Theorem 12 it follows that the system of equations

$$\begin{aligned} x_1 + 2x_2 + 3x_3 &= 0 \\ 2x_1 + 5x_2 + 3x_3 &= 0 \\ x_1 \qquad\quad + 8x_3 &= 0 \end{aligned}$$

has only the trivial solution.

EXERCISE SET 1.6

1. Which of the following are elementary matrices?

(a) $\begin{bmatrix} 2 & 0 \\ 0 & 1 \end{bmatrix}$

(b) $\begin{bmatrix} 1 & 0 \\ 3 & 1 \end{bmatrix}$

(c) $\begin{bmatrix} 2 & 0 \\ 0 & 2 \end{bmatrix}$

(d) $\begin{bmatrix} 0 & 1 & 0 \\ 1 & 0 & 0 \\ 0 & 0 & 1 \end{bmatrix}$

(e) $\begin{bmatrix} 0 & 1 & 0 \\ 0 & 0 & 1 \\ 0 & 0 & 1 \end{bmatrix}$

(f) $\begin{bmatrix} 1 & 0 & 0 \\ 0 & 1 & -3 \\ 0 & 0 & 1 \end{bmatrix}$

(g) $\begin{bmatrix} 1 & 0 & 0 & 0 \\ 0 & 1 & 0 & 0 \\ 0 & 1 & 1 & 0 \\ 0 & 0 & 0 & 1 \end{bmatrix}$

2. Determine the row operation that will restore the given elementary matrix to an identity matrix.

(a) $\begin{bmatrix} 1 & 0 \\ 5 & 1 \end{bmatrix}$ (b) $\begin{bmatrix} 0 & 0 & 1 \\ 0 & 1 & 0 \\ 1 & 0 & 0 \end{bmatrix}$ (c) $\begin{bmatrix} 1 & 0 & 0 & 0 \\ 0 & 8 & 0 & 0 \\ 0 & 0 & 1 & 0 \\ 0 & 0 & 0 & 1 \end{bmatrix}$

3. Consider the matrices

$$A = \begin{bmatrix} 1 & 2 & 3 \\ 4 & 5 & 6 \\ 7 & 8 & 9 \end{bmatrix} \qquad B = \begin{bmatrix} 7 & 8 & 9 \\ 4 & 5 & 6 \\ 1 & 2 & 3 \end{bmatrix} \qquad C = \begin{bmatrix} 1 & 2 & 3 \\ 4 & 5 & 6 \\ 9 & 12 & 15 \end{bmatrix}$$

Find elementary matrices E_1, E_2, E_3, and E_4 such that
(a) $E_1 A = B$ (b) $E_2 B = A$ (c) $E_3 A = C$ (d) $E_4 C = A$

4. In Exercise 3 is it possible to find an elementary matrix E such that $EB = C$? Justify your answer.

In Exercises 5–7 use the method shown in Examples 31 and 32 to find the inverse of the given matrix if the matrix is invertible.

5. (a) $\begin{bmatrix} 1 & 2 \\ 3 & 5 \end{bmatrix}$ (b) $\begin{bmatrix} -2 & 3 \\ 3 & -5 \end{bmatrix}$ (c) $\begin{bmatrix} 8 & -6 \\ -4 & 3 \end{bmatrix}$

6. (a) $\begin{bmatrix} 3 & 4 & -1 \\ 1 & 0 & 3 \\ 2 & 5 & -4 \end{bmatrix}$ (b) $\begin{bmatrix} 3 & 1 & 5 \\ 2 & 4 & 1 \\ -4 & 2 & -9 \end{bmatrix}$ (c) $\begin{bmatrix} 1 & 0 & 1 \\ 0 & 1 & 1 \\ 1 & 1 & 0 \end{bmatrix}$

(d) $\begin{bmatrix} 2 & 6 & 6 \\ 2 & 7 & 6 \\ 2 & 7 & 7 \end{bmatrix}$ (e) $\begin{bmatrix} 1 & 0 & 1 \\ -1 & 1 & 1 \\ 0 & 1 & 0 \end{bmatrix}$ (f) $\begin{bmatrix} \frac{1}{5} & \frac{1}{5} & \frac{1}{5} \\ \frac{1}{5} & \frac{1}{5} & -\frac{4}{5} \\ -\frac{2}{5} & \frac{1}{10} & \frac{1}{10} \end{bmatrix}$

7. (a) $\begin{bmatrix} \sqrt{\frac{1}{2}} & \sqrt{\frac{1}{2}} & 0 \\ -\sqrt{\frac{1}{2}} & \sqrt{\frac{1}{2}} & 0 \\ 0 & 0 & 1 \end{bmatrix}$ (b) $\begin{bmatrix} 1 & 0 & 0 & 0 \\ 1 & 2 & 0 & 0 \\ 1 & 2 & 4 & 0 \\ 1 & 2 & 4 & 8 \end{bmatrix}$ (c) $\begin{bmatrix} 5 & 11 & 7 & 3 \\ 2 & 1 & 4 & -5 \\ 3 & -2 & 8 & 7 \\ 0 & 0 & 0 & 0 \end{bmatrix}$

8. Show that the matrix

$$A = \begin{bmatrix} \cos\theta & \sin\theta & 0 \\ -\sin\theta & \cos\theta & 0 \\ 0 & 0 & 1 \end{bmatrix}$$

is invertible for all values of θ and find A^{-1}.

9. Consider the matrix

$$A = \begin{bmatrix} 1 & 0 \\ 3 & 4 \end{bmatrix}$$

(a) Find elementary matrices E_1 and E_2 such that $E_2 E_1 A = I$.

(b) Write A^{-1} as a product of two elementary matrices.

(c) Write A as a product of two elementary matrices.

10. Perform the following row operations on

$$A = \begin{bmatrix} 3 & 1 & 0 \\ -2 & 1 & 4 \\ 3 & 5 & 5 \end{bmatrix}$$

by multiplying A on the left by a suitable elementary matrix. Check your answer in each case by performing the row operation directly on A.

(a) Interchange the first and third rows.

(b) Multiply the second row by $1/3$.

(c) Add twice the second row to the first row.

11. Express the matrix

$$A = \begin{bmatrix} 1 & 3 & 3 & 8 \\ -2 & -5 & 1 & -8 \\ 0 & 1 & 7 & 8 \end{bmatrix}$$

in the form $A = EFR$, where E and F are elementary matrices, and R is in row-echelon form.

12. Show that if

$$A = \begin{bmatrix} 1 & 0 & 0 \\ 0 & 1 & 0 \\ a & b & c \end{bmatrix}$$

is an elementary matrix, then at least one entry in the third row must be a zero.

13. Find the inverse of each of the following 4×4 matrices, where k_1, k_2, k_3, k_4, and k are all nonzero.

(a) $\begin{bmatrix} k_1 & 0 & 0 & 0 \\ 0 & k_2 & 0 & 0 \\ 0 & 0 & k_3 & 0 \\ 0 & 0 & 0 & k_4 \end{bmatrix}$ (b) $\begin{bmatrix} 0 & 0 & 0 & k_1 \\ 0 & 0 & k_2 & 0 \\ 0 & k_3 & 0 & 0 \\ k_4 & 0 & 0 & 0 \end{bmatrix}$ (c) $\begin{bmatrix} k & 0 & 0 & 0 \\ 1 & k & 0 & 0 \\ 0 & 1 & k & 0 \\ 0 & 0 & 1 & k \end{bmatrix}$

14. Prove that if A is an $m \times n$ matrix, there is an invertible matrix C such that CA is in reduced row-echelon form.

15. Prove that if A is an invertible matrix and B is row equivalent to A, then B is also invertible.

1.7   FURTHER RESULTS ON SYSTEMS OF EQUATIONS AND INVERTIBILITY

In this section we shall establish more results about systems of linear equations and invertibility of matrices. Our work will lead to a totally new method for solving n equations in n unknowns.

Theorem 13. *If A is an invertible $n \times n$ matrix, then for each $n \times 1$ matrix B, the system of equations $AX = B$ has exactly one solution, namely, $X = A^{-1}B$.*

Proof. Since $A(A^{-1}B) = B$, $X = A^{-1}B$ is a solution of $AX = B$. To show that this is the only solution, we will assume that X_0 is an arbitrary solution and then show that X_0 must be the solution $A^{-1}B$.

If X_0 is any solution, then $AX_0 = B$. Multiplying both sides by A^{-1}, we obtain $X_0 = A^{-1}B$. ∎

Example 34

Consider the system of linear equations

$$\begin{aligned} x_1 + 2x_2 + 3x_3 &= 5 \\ 2x_1 + 5x_2 + 3x_3 &= 3 \\ x_1 \qquad\quad + 8x_3 &= 17 \end{aligned}$$

In matrix form this system can be written as $AX = B$, where

$$A = \begin{bmatrix} 1 & 2 & 3 \\ 2 & 5 & 3 \\ 1 & 0 & 8 \end{bmatrix} \qquad X = \begin{bmatrix} x_1 \\ x_2 \\ x_3 \end{bmatrix} \qquad B = \begin{bmatrix} 5 \\ 3 \\ 17 \end{bmatrix}$$

In Example 31 we showed that A is invertible and

$$A^{-1} = \begin{bmatrix} -40 & 16 & 9 \\ 13 & -5 & -3 \\ 5 & -2 & -1 \end{bmatrix}$$

By Theorem 13 the solution of the system is

$$X = A^{-1}B = \begin{bmatrix} -40 & 16 & 9 \\ 13 & -5 & -3 \\ 5 & -2 & -1 \end{bmatrix} \begin{bmatrix} 5 \\ 3 \\ 17 \end{bmatrix} = \begin{bmatrix} 1 \\ -1 \\ 2 \end{bmatrix}$$

or $x_1 = 1$, $x_2 = -1$, $x_3 = 2$. ▲

REMARK. Note that the method of Example 34 applies only when the system has as many equations as unknowns and the coefficient matrix is invertible.

Frequently, one is concerned with solving a sequence of systems

$$AX = B_1, \quad AX = B_2, \quad AX = B_3, \ldots, \quad AX = B_k$$

each of which has the same square coefficient matrix A. If A is invertible, then the solutions

$$X = A^{-1}B_1, \quad X = A^{-1}B_2, \quad X = A^{-1}B_3, \ldots, \quad X = A^{-1}B_k$$

can be obtained with one matrix inversion and k matrix multiplications. However, a more efficient method is to form the matrix

$$[A|B_1|B_2|\cdots|B_k] \tag{1.11}$$

in which the coefficient matrix A is "augmented" by all k of the matrices $B_1, B_2, \ldots, B_k$. By reducing (1.11) to reduced row-echelon form we can solve all k systems at once by Gauss-Jordan elimination. This method has the added advantage that it applies even when A is not invertible.

Example 35

Solve the systems

(a) $\begin{aligned} x_1 + 2x_2 + 3x_3 &= 4 \\ 2x_1 + 5x_2 + 3x_3 &= 5 \\ x_1 \qquad\quad + 8x_3 &= 9 \end{aligned}$ (b) $\begin{aligned} x_1 + 2x_2 + 3x_3 &= 1 \\ 2x_1 + 5x_2 + 3x_3 &= 6 \\ x_1 \qquad\quad + 8x_3 &= -6 \end{aligned}$

Solution. The two systems have the same coefficient matrix. If we augment this coefficient matrix with the columns of constants on the right sides of these systems, we obtain

$$\begin{bmatrix} 1 & 2 & 3 & | & 4 & | & 1 \\ 2 & 5 & 3 & | & 5 & | & 6 \\ 1 & 0 & 8 & | & 9 & | & -6 \end{bmatrix}$$

Reducing this matrix to reduced row-echelon form yields (verify)

$$\begin{bmatrix} 1 & 0 & 0 & | & 1 & | & 2 \\ 0 & 1 & 0 & | & 0 & | & 1 \\ 0 & 0 & 1 & | & 1 & | & -1 \end{bmatrix}$$

It follows from the last two columns that the solution of system (a) is $x_1 = 1$, $x_2 = 0$, $x_3 = 1$ and of system (b) is $x_1 = 2$, $x_2 = 1$, $x_3 = -1$.

Let us digress for a moment to illustrate a type of application that leads to a set of linear systems, all of which have the same coefficient matrix. In various applied problems physical systems arise that are described as *black boxes*. This term is intended to convey the idea that the internal workings of the system are unknown or unimportant for the problem—one simply imagines that if an input, such as electrical current, force, or heat, is applied to the black box, then a certain output will result (Figure 1.5).

Figure 1.5

Often, the input and output can be described mathematically by matrices with one column. For example, if the black box is a microprocessor, then the input might be an $n \times 1$ matrix whose entries are n voltages read across certain input terminals, and the output might be an $m \times 1$ matrix whose entries are the resulting currents in m output wires. Mathematically speaking, such a system does nothing more than transform an $n \times 1$ input matrix into an $m \times 1$ output matrix.

For many black-box systems the input matrix M_{IN} and the output matrix M_{OUT} are related by a matrix equation

$$AM_{\text{IN}} = M_{\text{OUT}}$$

where the entries in A are physical parameters determined by the system. Such systems are called *linear physical systems.*

In many applications it is important to know what input must be applied to the system to achieve a certain desired output. This amounts to solving the equation $AX = M_{\text{OUT}}$ for the unknown input X, given the desired output M_{OUT}. Thus, if we have a succession of different output matrices $B_1, \ldots, B_k$, and we want to determine the input matrices that produce these given outputs, we must successively solve the k systems of linear equations

$$AX = B_1, \quad AX = B_2, \ldots, \quad AX = B_k$$

each of which has the same coefficient matrix A.

Up to now, to show that an $n \times n$ matrix A is invertible, it has been necessary to find an $n \times n$ matrix B such that

$$AB = I \quad \text{and} \quad BA = I$$

The next theorem shows that if we produce an $n \times n$ matrix B satisfying *either* condition, then the other condition holds automatically.

Theorem 14. *Let A be a square matrix.*

(a) If B is a square matrix satisfying $BA = I$, then $B = A^{-1}$.
(b) If B is a square matrix satisfying $AB = I$, then $B = A^{-1}$.

Proof. We shall prove (a) and leave (b) as an exercise.
(a) Assume $BA = I$. If we can show that A is invertible, the proof can be completed by multiplying $BA = I$ on both sides by A^{-1} to obtain

$$BAA^{-1} = IA^{-1} \quad \text{or} \quad BI = IA^{-1} \quad \text{or} \quad B = A^{-1}$$

To show that A is invertible, it suffices to show that the system $AX = 0$ has only the trivial solution (see Theorem 12). However, if we multiply both sides of $AX = 0$ on the left by B, we obtain $BAX = B0$ or $IX = 0$ or $X = 0$. Thus, the system of equations $AX = 0$ has only the trivial solution. ▌

We are now in a position to add a fourth statement equivalent to the three given in Theorem 12.

Theorem 15. *If A is an n × n matrix, then the following statements are equivalent.*

(a) *A is invertible.*
(b) *AX = 0 has only the trivial solution.*
(c) *A is row equivalent to I_n.*
(d) *AX = B is consistent for every n × 1 matrix B.*

Proof. Since we proved in Theorem 12 that (a), (b), and (c) are equivalent, it will be sufficient to prove (a) $\Rightarrow$ (d) and (d) $\Rightarrow$ (a).

(a) $\Rightarrow$ (d): If A is invertible and B is any $n \times 1$ matrix, then $X = A^{-1}B$ is a solution of $AX = B$ by Theorem 13. Thus, $AX = B$ is consistent.

(d) $\Rightarrow$ (a): If the system $AX = B$ is consistent for every $n \times 1$ matrix B, then in particular, the systems

$$AX = \begin{bmatrix} 1 \\ 0 \\ 0 \\ \vdots \\ 0 \end{bmatrix}, \quad AX = \begin{bmatrix} 0 \\ 1 \\ 0 \\ \vdots \\ 0 \end{bmatrix}, \quad \cdots, \quad AX = \begin{bmatrix} 0 \\ 0 \\ 0 \\ \vdots \\ 1 \end{bmatrix}$$

will be consistent. Let X_1 be a solution of the first system, X_2 a solution of the second system, ..., and X_n a solution of the last system, and let us form an $n \times n$ matrix C having these solutions as columns. Thus, C has the form

$$C = [X_1 \mid X_2 \mid \cdots \mid X_n]$$

As discussed in Example 18, the successive columns of the product AC will be

$$AX_1, AX_2, \ldots, AX_n$$

Thus,

$$AC = [AX_1 \mid AX_2 \mid \cdots \mid AX_n] = \begin{bmatrix} 1 & 0 & \cdots & 0 \\ 0 & 1 & \cdots & 0 \\ 0 & 0 & \cdots & 0 \\ \vdots & \vdots & & \vdots \\ 0 & 0 & \cdots & 1 \end{bmatrix} = I$$

By part (b) of Theorem 14 it follows that $C = A^{-1}$. Thus, A is invertible.

In our later work the following fundamental problem will occur over and over again in various contexts.

A Fundamental Problem. Let A be a fixed $m \times n$ matrix. Find all $m \times 1$ matrices B such that the system of equations $AX = B$ is consistent.

If A is an invertible matrix, Theorem 13 completely solves this problem by asserting that for *every* $m \times 1$ matrix B, $AX = B$ has the unique solution $X = A^{-1}B$. If A is not square, or if A is square but not invertible, then Theorem 13 does not apply. In these cases the matrix B must satisfy certain conditions in order for $AX = B$ to be consistent. The following example illustrates how Gaussian elimination can be used to determine such conditions.

Example 36

What conditions must b_1, b_2, and b_3 satisfy in order for the system of equations

$$x_1 + x_2 + 2x_3 = b_1$$
$$x_1 \quad\quad + x_3 = b_2$$
$$2x_1 + x_2 + 3x_3 = b_3$$

to be consistent?

Solution. The augmented matrix is

$$\begin{bmatrix} 1 & 1 & 2 & b_1 \\ 1 & 0 & 1 & b_2 \\ 2 & 1 & 3 & b_3 \end{bmatrix}$$

which can be reduced to row-echelon form as follows.

$$\begin{bmatrix} 1 & 1 & 2 & b_1 \\ 0 & -1 & -1 & b_2 - b_1 \\ 0 & -1 & -1 & b_3 - 2b_1 \end{bmatrix}$$

-1 times the first row was added to the second and -2 times the first row was added to the third.

$$\begin{bmatrix} 1 & 1 & 2 & b_1 \\ 0 & 1 & 1 & b_1 - b_2 \\ 0 & -1 & -1 & b_3 - 2b_1 \end{bmatrix}$$

The second row was multiplied by -1.

$$\begin{bmatrix} 1 & 1 & 2 & b_1 \\ 0 & 1 & 1 & b_1 - b_2 \\ 0 & 0 & 0 & b_3 - b_2 - b_1 \end{bmatrix}$$

The second row was added to the third.

It is now evident from the third row in the matrix that the system has a solution if and only if b_1, b_2, and b_3 satisfy the condition

$$b_3 - b_2 - b_1 = 0 \qquad \text{or} \qquad b_3 = b_1 + b_2$$

To express this condition another way, $AX = B$ is consistent if and only if B is a matrix of the form

$$B = \begin{bmatrix} b_1 \\ b_2 \\ b_1 + b_2 \end{bmatrix}$$

where b_1 and b_2 are arbitrary.

Example 37

What conditions must b_1, b_2, and b_3 satisfy in order for the system of equations

$$\begin{aligned} x_1 + 2x_2 + 3x_3 &= b_1 \\ 2x_1 + 5x_2 + 3x_3 &= b_2 \\ x_1 \qquad\quad + 8x_3 &= b_3 \end{aligned}$$

to be consistent?

Solution. The augmented matrix is

$$\begin{bmatrix} 1 & 2 & 3 & b_1 \\ 2 & 5 & 3 & b_2 \\ 1 & 0 & 8 & b_3 \end{bmatrix}$$

Reducing this to reduced row-echelon form yields (verify):

$$\begin{bmatrix} 1 & 0 & 0 & -40b_1 + 16b_2 + 9b_3 \\ 0 & 1 & 0 & 13b_1 - 5b_2 - 3b_3 \\ 0 & 0 & 1 & 5b_1 - 2b_2 - b_3 \end{bmatrix} \qquad (1.12)$$

In this case there are no restrictions on b_1, b_2, and b_3; that is, the given system $AX = B$ has the unique solution

$$x_1 = -40b_1 + 16b_2 + 9b_3, \quad x_2 = 13b_1 - 5b_2 - 3b_3, \quad x_3 = 5b_1 - 2b_2 - b_3 \tag{1.13}$$

for all B.

REMARK. Because the system $AX = B$ in Example 37 is consistent for all B, it follows from Theorem 15 that A is invertible. We leave it for the reader to verify that formulas (1.13) can also be obtained by calculating $X = A^{-1}B$.

EXERCISE SET 1.7

In Exercises 1–8 solve the system using the method of Example 34.

1. $x_1 + 2x_2 = 7$
$2x_1 + 5x_2 = -3$

2. $3x_1 - 6x_2 = 8$
$2x_1 + 5x_2 = 1$

3. $x_1 + 2x_2 + 2x_3 = -1$
$x_1 + 3x_2 + x_3 = 4$
$x_1 + 3x_2 + 2x_3 = 3$

4. $2x_1 + x_2 + x_3 = 7$
$3x_1 + 2x_2 + x_3 = -3$
$x_2 + x_3 = 5$

5. $\frac{1}{5}x + \frac{1}{5}y + \frac{1}{5}z = 1$
$\frac{1}{5}x + \frac{1}{5}y - \frac{4}{5}z = 2$
$-\frac{2}{5}x + \frac{1}{10}y + \frac{1}{10}z = 0$

6. $3w + x + 7y + 9z = 4$
$w + x + 4y + 4z = 7$
$-w \quad\quad - 2y - 3z = 0$
$-2w - x - 4y - 6z = 6$

7. $3x_1 + 5x_2 = b_1$
$x_1 + 2x_2 = b_2$

8. $x_1 + 2x_2 + 3x_3 = b_1$
$2x_1 + 5x_2 + 5x_3 = b_2$
$3x_1 + 5x_2 + 8x_3 = b_3$

9. Solve the following general system using the method of Example 34.

$$x_1 + 2x_2 + x_3 = b_1$$
$$x_1 - x_2 + x_3 = b_2$$
$$x_1 + x_2 \quad\quad = b_3$$

and then use the resulting formulas to find the solution if
(a) $b_1 = -1$, $b_2 = 3$, $b_3 = 4$ (b) $b_1 = 5$, $b_2 = 0$, $b_3 = 0$
(c) $b_1 = -1$, $b_2 = -1$, $b_3 = 3$

10. Solve the three systems in Exercise 9 using the method of Example 35.

In Exercises 11–14 use the method of Example 35 to solve the systems in all the parts simultaneously.

11. $x_1 - 3x_2 = b_1$
$4x_1 - 2x_2 = b_2$
(a) $b_1 = 1$, $b_2 = 4$,   (b) $b_1 = -2$, $b_2 = 5$

12. $x_1 - 3x_2 - x_3 = b_1$
$-2x_1 + 7x_2 + 2x_3 = b_2$
$3x_1 + 2x_2 - 4x_3 = b_3$
(a) $b_1 = 0$, $b_2 = 1$, $b_3 = 0$ (b) $b_1 = -3$, $b_2 = 4$, $b_3 = -5$

13. $2x_1 - 5x_2 = b_1$
$x_1 + 3x_2 = b_2$
(a) $b_1 = 0$, $b_2 = 1$ (b) $b_1 = -4$, $b_2 = 6$
(c) $b_1 = -1$, $b_2 = 3$ (d) $b_1 = -5$, $b_2 = 1$

14. $x_1 + 2x_2 - x_3 = b_1$
$2x_1 + 5x_2 + 4x_3 = b_2$
$3x_1 + 7x_2 + 4x_3 = b_3$
(a) $b_1 = 1$, $b_2 = 0$, $b_3 = -1$ (b) $b_1 = 0$, $b_2 = 1$, $b_3 = 1$
(c) $b_1 = -1$, $b_2 = -1$, $b_3 = 0$

15. The method of Example 35 can be used for linear systems with infinitely many solutions. Use that method to solve the systems in both parts together.

(a) $x_1 - 2x_2 + x_3 = -2$
$2x_1 - 5x_2 + x_3 = 1$
$3x_1 - 7x_2 + 2x_3 = -1$

(b) $x_1 - 2x_2 + x_3 = 1$
$2x_1 - 5x_2 + x_3 = -1$
$3x_1 - 7x_2 + 2x_3 = 0$

In Exercises 16—19 find the conditions that the b's must satisfy for the system to be consistent.

16. $4x_1 - 2x_2 = b_1$
$2x_1 - x_2 = b_2$

17. $x_1 - x_2 + 3x_3 = b_1$
$3x_1 - 3x_2 + 9x_3 = b_2$
$-2x_1 + 2x_2 - 6x_3 = b_3$

18. $x_1 + 2x_2 - x_3 = b_1$
$2x_1 + 5x_2 + 4x_3 = b_2$
$3x_1 + 7x_2 + 4x_3 = b_3$

19. $2x_1 + 3x_2 - x_3 + x_4 = b_1$
$x_1 + 5x_2 + x_3 - 2x_4 = b_2$
$-x_1 + 2x_2 + 2x_3 - 3x_4 = b_3$
$3x_1 + x_2 - 3x_3 + 4x_4 = b_4$

20. Consider the matrices

$$A = \begin{bmatrix} 2 & 2 & 3 \\ 1 & 2 & 1 \\ 2 & -2 & 1 \end{bmatrix} \quad \text{and} \quad X = \begin{bmatrix} x_1 \\ x_2 \\ x_3 \end{bmatrix}$$

(a) Show that the equation $AX = X$ can be rewritten as $(A - I)X = 0$ and use this result to solve $AX = X$ for X.

(b) Solve $AX = 4X$.

21. Without using pencil and paper, determine whether the following matrices are invertible.

(a) $\begin{bmatrix} 2 & 1 & -3 & 1 \\ 0 & 5 & 4 & 3 \\ 0 & 0 & 1 & 2 \\ 0 & 0 & 0 & 3 \end{bmatrix}$
(b) $\begin{bmatrix} 5 & 1 & 4 & 1 \\ 0 & 0 & 2 & -1 \\ 0 & 0 & 1 & 1 \\ 0 & 0 & 0 & 7 \end{bmatrix}$

Hint. Consider the associated homogeneous systems

$2x_1 + x_2 - 3x_3 + x_4 = 0$
$5x_2 + 4x_3 + 3x_4 = 0$
$x_3 + 2x_4 = 0$
$3x_4 = 0$

and

$5x_1 + x_2 + 4x_3 + x_4 = 0$
$2x_3 - x_4 = 0$
$x_3 + x_4 = 0$
$7x_4 = 0$

22. Let $AX = 0$ be a homogeneous system of n linear equations in n unknowns that has only the trivial solution. Show that if k is any positive integer, then the system $A^k X = 0$ also has only the trivial solution.

23. Let $AX = 0$ be a homogeneous system of n linear equations in n unknowns, and let Q be an invertible matrix. Show that $AX = 0$ has just the trivial solution if and only if $(QA)X = 0$ has just the trivial solution.

24. Show that an $n \times n$ matrix A is invertible if and only if it can be written as a product of elementary matrices.

25. Use part (a) of Theorem 14 to prove part (b).

SUPPLEMENTARY EXERCISES

1. Use Gauss-Jordan elimination to solve for x' and y' in terms of x and y.

$$x = \tfrac{3}{5}x' - \tfrac{4}{5}y'$$
$$y = \tfrac{4}{5}x' + \tfrac{3}{5}y'$$

2. Use Gauss-Jordan elimination to solve for x' and y' in terms of x and y.

$$x = x' \cos\theta - y' \sin\theta$$
$$y = x' \sin\theta + y' \cos\theta$$

3. Find a homogeneous linear system with two equations that are not multiples of one another and such that

$$x_1 = 1, \quad x_2 = -1, \quad x_3 = 1, \quad x_4 = 2$$

and

$$x_1 = 2, \quad x_2 = 0, \quad x_3 = 3, \quad x_4 = -1$$

are solutions of the system.

4. A box containing pennies, nickels, and dimes has 13 coins with a total value of 83 cents. How many coins of each type are in the box?

5. Find positive integers that satisfy

$$x + y + z = 9$$
$$x + 5y + 10z = 44$$

6. For which value(s) of a does the following system have zero, one, infinitely many solutions?

$$x_1 + x_2 + x_3 = 4$$
$$x_3 = 2$$
$$(a^2 - 4)x_3 = a - 2$$

7. Let

$$\begin{bmatrix} a & 0 & b & 2 \\ a & a & 4 & 4 \\ 0 & a & 2 & b \end{bmatrix}$$

be the augmented matrix for a linear system. For what values of a and b does the system have

(a) a unique solution,
(b) a one-parameter solution,
(c) a two-parameter solution,
(d) no solution?

8. Find a matrix K such that $AKB = C$ given that

$$A = \begin{bmatrix} 1 & 4 \\ -2 & 3 \\ 1 & -2 \end{bmatrix} \quad B = \begin{bmatrix} 2 & 0 & 0 \\ 0 & 1 & -1 \end{bmatrix} \quad C = \begin{bmatrix} 8 & 6 & -6 \\ 6 & -1 & 1 \\ -4 & 0 & 0 \end{bmatrix}$$

9. How should the coefficients a, b, and c be chosen so that the system

$$\begin{aligned} ax + by - 3z &= -3 \\ -2x - by + cz &= -1 \\ ax + 3y - cz &= -3 \end{aligned}$$

has the solution $x = 1$, $y = -1$, and $z = 2$?

10. In each part solve the matrix equation for X.

(a) $X \begin{bmatrix} -1 & 0 & 1 \\ 1 & 1 & 0 \\ 3 & 1 & -1 \end{bmatrix} = \begin{bmatrix} 1 & 2 & 0 \\ -3 & 1 & 5 \end{bmatrix}$

(b) $X \begin{bmatrix} 1 & -1 & 2 \\ 3 & 0 & 1 \end{bmatrix} = \begin{bmatrix} -5 & -1 & 0 \\ 6 & -3 & 7 \end{bmatrix}$

(c) $\begin{bmatrix} 3 & 1 \\ -1 & 2 \end{bmatrix} X - X \begin{bmatrix} 1 & 4 \\ 2 & 0 \end{bmatrix} = \begin{bmatrix} 2 & -2 \\ 5 & 4 \end{bmatrix}$

11. (a) Express the equations

$$\begin{aligned} y_1 &= x_1 - x_2 + x_3 \\ y_2 &= 3x_1 + x_2 - 4x_3 \quad \text{and} \\ y_3 &= -2x_1 - 2x_2 + 3x_3 \end{aligned} \qquad \begin{aligned} z_1 &= 4y_1 - y_2 + y_3 \\ z_2 &= -3y_1 + 5y_2 - y_3 \end{aligned}$$

in the matrix forms $Y = AX$ and $Z = BY$. Then use these to obtain a direct relationship $Z = CX$ between Z and X.

(b) Use the equation $Z = CX$ obtained in (a) to express z_1 and z_2 in terms of x_1, x_2, and x_3.

(c) Check the result in (b) by directly substituting the equations for y_1, y_2, and y_3 into the equations for z_1 and z_2 and then simplifying.

12. If A is $m \times n$ and B is $n \times p$, how many multiplication operations and how many addition operations are needed to calculate the matrix product AB?

13. Let A be a square matrix.
(a) Show that $(I - A)^{-1} = I + A + A^2 + A^3$ if $A^4 = 0$.
(b) Show that $(I - A)^{-1} = I + A + A^2 + \cdots + A^n$ if $A^{n+1} = 0$.

14. Find values of a, b, and c so that the graph of the polynomial $p(x) = ax^2 + bx + c$ passes through the points $(1, 2)$, $(-1, 6)$, and $(2, 3)$.

15. **(For readers who have studied calculus.)** Find values of a, b, and c so that the graph of the polynomial $p(x) = ax^2 + bx + c$ passes through the point $(-1, 0)$ and has a horizontal tangent at $(2, -9)$.

16. Let J_n be the $n \times n$ matrix each of whose entries is 1. Show that

$$(I - J_n)^{-1} = I - \frac{1}{n-1} J_n$$

17. Show that if a square matrix A satisfies

$$A^3 + 4A^2 - 2A + 7I = 0$$

then so does A^t.

18. Prove: If B is invertible, then $AB^{-1} = B^{-1}A$ if and only if $AB = BA$.

19. Prove: If A is invertible, then $A + B$ and $I + BA^{-1}$ are both invertible or both not invertible.

20. The **trace** of a square matrix A is denoted by $\text{tr}(A)$ and is defined to be the sum of the entries on the main diagonal. Prove that if A and B are $n \times n$ matrices, then
(a) $\text{tr}(A + B) = \text{tr}(A) + \text{tr}(B)$
(b) $\text{tr}(kA) = k\,\text{tr}(A)$
(c) $\text{tr}(A^t) = \text{tr}(A)$
(d) $\text{tr}(AB) = \text{tr}(BA)$

21. Use Exercise 20 to show that there are no square matrices A and B such that

$$AB - BA = I.$$

22. Prove: If A is an $m \times n$ matrix and B is the $n \times 1$ matrix each of whose entries is $1/n$, then

$$AB = \begin{bmatrix} \bar{r}_1 \\ \bar{r}_2 \\ \vdots \\ \bar{r}_m \end{bmatrix}$$

where $\bar{r}_i$ is the average of the entries in the ith row of A.

23. (For readers who have studied calculus.) If

$$C = \begin{bmatrix} c_{11}(x) & c_{12}(x) & \cdots & c_{1n}(x) \\ c_{21}(x) & c_{22}(x) & \cdots & c_{2n}(x) \\ \vdots & \vdots & & \vdots \\ c_{m1}(x) & c_{m2}(x) & \cdots & c_{mn}(x) \end{bmatrix}$$

where $c_{ij}(x)$ is a differentiable function of x, then we define

$$\frac{dC}{dx} = \begin{bmatrix} c'_{11}(x) & c'_{12}(x) & \cdots & c'_{1n}(x) \\ c'_{21}(x) & c'_{22}(x) & \cdots & c'_{2n}(x) \\ \vdots & \vdots & & \vdots \\ c'_{m1}(x) & c'_{m2}(x) & \cdots & c'_{mn}(x) \end{bmatrix}$$

Show that if the entries in A and B are differentiable functions of x and the sizes of the matrices are such that the stated operations can be performed, then

(a) $\dfrac{d}{dx}(kA) = k\dfrac{dA}{dx}$

(b) $\dfrac{d}{dx}(A + B) = \dfrac{dA}{dx} + \dfrac{dB}{dx}$

(c) $\dfrac{d}{dx}(AB) = \dfrac{dA}{dx}B + A\dfrac{dB}{dx}$

24. **(For readers who have studied calculus.)** Use part (c) of Exercise 23 to show that

$$\dfrac{dA^{-1}}{dx} = -A^{-1}\dfrac{dA}{dx}A^{-1}$$

State all the assumptions you make in obtaining this formula.

25. Find the values of A, B, and C that will make the equation

$$\dfrac{x^2 + x - 2}{(3x - 1)(x^2 + 1)} = \dfrac{A}{3x - 1} + \dfrac{Bx + C}{x^2 + 1}$$

an identity. [**Hint:** Multiply through by $(3x - 1)(x^2 + 1)$ and equate the corresponding coefficients of the polynomials on each side of the resulting equation.]

$x^2 + x - 2 = A(x^2 + 1) + (Bx + C)(3x - 1)$
$ = Ax^2 + A + 3Bx^2 - Bx + 3Cx - C$
$ = Ax^2 + 3Bx^2 + 3Cx - Bx + A - C$
$ = x^2(A + 3B) + x(3C - B) + A - C$

Determinants

2.1 THE DETERMINANT FUNCTION

We are all familiar with functions such as $f(x) = \sin x$ and $f(x) = x^2$, which associate a real number $f(x)$ with a real value of the variable x. Since both x and $f(x)$ assume only real values, such functions can be described as *real-valued functions of a real variable*. In this section we initiate the study of *real-valued functions of a matrix variable*, that is, functions that associate a real number $f(X)$ with a matrix X. Our primary effort will be devoted to the study of one such function called the determinant function. Our work on the determinant function will have important applications to the theory of systems of linear equations and will also lead us to an explicit formula for the inverse of an invertible matrix.

Before we can define the determinant function, it will be necessary to establish some results concerning permutations.

Definition. A *permutation* of the set of integers $\{1, 2, \ldots, n\}$ is an arrangement of these integers in some order without omissions or repetitions.

Example 1

There are six different permutations of the set of integers $\{1, 2, 3\}$. These are

$$(1, 2, 3) \quad (2, 1, 3) \quad (3, 1, 2)$$
$$(1, 3, 2) \quad (2, 3, 1) \quad (3, 2, 1)$$

One convenient method of systematically listing permutations is to use a *permutation tree*. This method is illustrated in our next example.

Example 2

List all permutations of the set of integers {1, 2, 3, 4}.

Solution. Consider Figure 2.1. The four dots labeled 1, 2, 3, 4 at the top of the figure represent the possible choices for the first number in the permutation. The three branches emanating from these dots represent the possible choices for the second position in the permutation. Thus, if the permutation begins (2, –, –, –), the three possibilities for the second position are 1, 3, and 4. The two branches emanating from each dot in the second position represent the possible choices for the third position. Thus, if the permutation begins (2, 3, –, –), the two possible choices for the third position are 1 and 4. Finally, the single branch emanating from each dot in the third position represents the only possible choice for the fourth position. Thus, if the permutation begins (2, 3, 4, –), the only choice for the fourth position is 1. The different permutations can now be listed by tracing out all the possible paths through the "tree" from the first position to the last position. We obtain the following list by this process.

(1, 2, 3, 4)	(2, 1, 3, 4)	(3, 1, 2, 4)	(4, 1, 2, 3)
(1, 2, 4, 3)	(2, 1, 4, 3)	(3, 1, 4, 2)	(4, 1, 3, 2)
(1, 3, 2, 4)	(2, 3, 1, 4)	(3, 2, 1, 4)	(4, 2, 1, 3)
(1, 3, 4, 2)	(2, 3, 4, 1)	(3, 2, 4, 1)	(4, 2, 3, 1)
(1, 4, 2, 3)	(2, 4, 1, 3)	(3, 4, 1, 2)	(4, 3, 1, 2)
(1, 4, 3, 2)	(2, 4, 3, 1)	(3, 4, 2, 1)	(4, 3, 2, 1)

From this example we see that there are 24 permutations of {1, 2, 3, 4}. This result could have been anticipated without actually listing the permutations by arguing as follows. Since the first position can be filled in four ways and then the second position in three ways, there are 4 · 3 ways of filling the first two positions. Since the third position can then be filled in two ways, there are 4 · 3 · 2 ways of filling the first three positions. Finally, since the last position can then be filled in only one way, there are 4 · 3 · 2 · 1 = 24 ways of filling all four positions. In

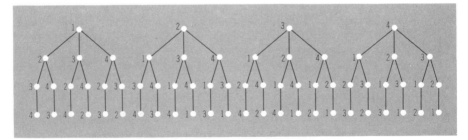

Figure 2.1

general, the set $\{1, 2, \ldots, n\}$ will have $n(n-1)(n-2)\cdots 2\cdot 1 = n!$ different permutations.

To denote a general permutation of the set $\{1, 2, \ldots, n\}$, we shall write $(j_1, j_2, \ldots, j_n)$. Here, j_1 is the first integer in the permutation, j_2 is the second, and so on. An *inversion* is said to occur in a permutation $(j_1, j_2, \ldots, j_n)$ whenever a larger integer precedes a smaller one. The total number of inversions occurring in a permutation can be obtained as follows: (1) find the number of integers that are less than j_1 and that follow j_1 in the permutation; (2) find the number of integers that are less than j_2 and that follow j_2 in the permutation. Continue this counting process for $j_3, \ldots, j_{n-1}$. The sum of these numbers will be the total number of inversions in the permutation.

Example 3

Determine the number of inversions in the following permutations:

(i) $(6, 1, 3, 4, 5, 2)$ (ii) $(2, 4, 1, 3)$ (iii) $(1, 2, 3, 4)$

(i) The number of inversions is $5 + 0 + 1 + 1 + 1 = 8$.
(ii) The number of inversions is $1 + 2 + 0 = 3$.
(iii) There are no inversions in this permutation.

Definition. A permutation is called *even* if the total number of inversions is an even integer and is called *odd* if the total number of inversions is an odd integer.

Example 4

The following table classifies the various permutations of $\{1, 2, 3\}$ as even or odd.

Permutation	Number of Inversions	Classification
(1, 2, 3)	0	even
(1, 3, 2)	1	odd
(2, 1, 3)	1	odd
(2, 3, 1)	2	even
(3, 1, 2)	2	even
(3, 2, 1)	3	odd

Consider the $n \times n$ matrix

$$A = \begin{bmatrix} a_{11} & a_{12} & \cdots & a_{1n} \\ a_{21} & a_{22} & \cdots & a_{2n} \\ \vdots & \vdots & & \vdots \\ a_{n1} & a_{n2} & \cdots & a_{nn} \end{bmatrix}$$

By an ***elementary product from A*** we shall mean any product of n entries from A, no two of which come from the same row or same column.

Example 5

List all elementary products from the matrices

(i) $\begin{bmatrix} a_{11} & a_{12} \\ a_{21} & a_{22} \end{bmatrix}$ (ii) $\begin{bmatrix} a_{11} & a_{12} & a_{13} \\ a_{21} & a_{22} & a_{23} \\ a_{31} & a_{32} & a_{33} \end{bmatrix}$

(i) Since each elementary product has two factors, and since each factor comes from a different row, an elementary product can be written in the form

$$a_{1\_}a_{2\_}$$

where the blanks designate column numbers. Since no two factors in the product come from the same column, the column numbers must be $1\ 2$ or $2\ 1$. Thus, the only elementary products are $a_{11}a_{22}$ and $a_{12}a_{21}$.

(ii) Since each elementary product has three factors, each of which comes from a different row, an elementary product can be written in the form

$$a_{1\_}a_{2\_}a_{3\_}$$

Since no two factors in the product come from the same column, the column numbers have no repetitions; consequently, they must form a permutation of the set $\{1, 2, 3\}$. These $3! = 6$ permutations yield the following list of elementary products.

$$a_{11}a_{22}a_{33} \qquad a_{12}a_{21}a_{33} \qquad a_{13}a_{21}a_{32}$$
$$a_{11}a_{23}a_{32} \qquad a_{12}a_{23}a_{31} \qquad a_{13}a_{22}a_{31}$$

As this example points out, an $n \times n$ matrix A has $n!$ elementary products. They are the products of the form $a_{1j_1}a_{2j_2} \ldots a_{nj_n}$, where $(j_1, j_2, \ldots, j_n)$ is a permutation of the set $\{1, 2, \ldots, n\}$. By a ***signed elementary product from A*** we shall mean an elementary product $a_{1j_1}a_{2j_2} \ldots a_{nj_n}$ multiplied by $+1$ or -1. We use the $+$ if $(j_1, j_2, \ldots, j_n)$ is an even permutation and the $-$ if $(j_1, j_2, \ldots, j_n)$ is an odd permutation.

Example 6

List all signed elementary products from the matrices.

(i) $\begin{bmatrix} a_{11} & a_{12} \\ a_{21} & a_{22} \end{bmatrix}$ (ii) $\begin{bmatrix} a_{11} & a_{12} & a_{13} \\ a_{21} & a_{22} & a_{23} \\ a_{31} & a_{32} & a_{33} \end{bmatrix}$

| (i) | | | Signed |
Elementary Product	Associated Permutation	Even or Odd	Elementary Product
$a_{11}a_{22}$	$(1, 2)$	even	$a_{11}a_{22}$
$a_{12}a_{21}$	$(2, 1)$	odd	$-a_{12}a_{21}$

| (ii) | | | Signed |
Elementary Product	Associated Permutation	Even or Odd	Elementary Product
$a_{11}a_{22}a_{33}$	$(1, 2, 3)$	even	$a_{11}a_{22}a_{33}$
$a_{11}a_{23}a_{32}$	$(1, 3, 2)$	odd	$-a_{11}a_{23}a_{32}$
$a_{12}a_{21}a_{33}$	$(2, 1, 3)$	odd	$-a_{12}a_{21}a_{33}$
$a_{12}a_{23}a_{31}$	$(2, 3, 1)$	even	$a_{12}a_{23}a_{31}$
$a_{13}a_{21}a_{32}$	$(3, 1, 2)$	even	$a_{13}a_{21}a_{32}$
$a_{13}a_{22}a_{31}$	$(3, 2, 1)$	odd	$-a_{13}a_{22}a_{31}$

We are now in a position to define the determinant function.

> **Definition.** Let A be a square matrix. The ***determinant function*** is denoted by ***det***, and we define $\det(A)$ to be the sum of all signed elementary products from A. The number $\det(A)$ is called the ***determinant of A***.

Example 7

Referring to Example 6, we obtain

(i) $\det \begin{bmatrix} a_{11} & a_{12} \\ a_{21} & a_{22} \end{bmatrix} = a_{11}a_{22} - a_{12}a_{21}$

(ii) $\det \begin{bmatrix} a_{11} & a_{12} & a_{13} \\ a_{21} & a_{22} & a_{23} \\ a_{31} & a_{32} & a_{33} \end{bmatrix} = a_{11}a_{22}a_{33} + a_{12}a_{23}a_{31} + a_{13}a_{21}a_{32}$
$$- a_{13}a_{22}a_{31} - a_{12}a_{21}a_{33} - a_{11}a_{23}a_{32}$$

It is useful to have the two formulas in this example available for ready reference. To avoid memorizing these unwieldy expressions, however, we suggest using the mnemonic devices described in Figure 2.2. The first formula in Example 7

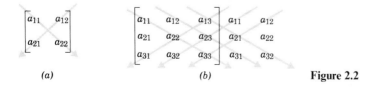

(a)   *(b)*   **Figure 2.2**

is obtained from Figure 2.2*a* by multiplying the entries on the rightward arrow and subtracting the product of the entries on the leftward arrow. The second formula in Example 7 is obtained by recopying the first and second columns as shown in Figure 2.2*b*. The determinant is then computed by summing the products on the rightward arrows and subtracting the products on the leftward arrows.

Example 8

Evaluate the determinants of

$$A = \begin{bmatrix} 3 & 1 \\ 4 & -2 \end{bmatrix} \quad \text{and} \quad B = \begin{bmatrix} 1 & 2 & 3 \\ -4 & 5 & 6 \\ 7 & -8 & 9 \end{bmatrix}$$

Using the method of Figure 2.2*a* gives

$$\det(A) = (3)(-2) - (1)(4) = -10$$

Using the method of Figure 2.2*b* gives

$$\det(B) = (45) + (84) + (96) - (105) - (-48) - (-72) = 240$$

Warning. We emphasize that the methods described in Figure 2.2 *do not* work for determinants of 4 × 4 matrices or higher.

Evaluating determinants directly from the definition leads to computational difficulties. Indeed, evaluating a 4 × 4 determinant directly would involve computing 4! = 24 signed elementary products, and a 10 × 10 determinant would involve 10! = 3,628,800 signed elementary products. Even the fastest of digital computers cannot handle the computation of a 25 × 25 determinant by this method in a practical amount of time. Much of the remainder of this chapter is devoted, therefore, to developing properties of determinants that will simplify their evaluation.

We conclude this section with some comments about terminology and notation. First, we note that the symbol $|A|$ is an alternative notation for $\det(A)$. For example, the determinant of a 3 × 3 matrix can be written as

$$\det \begin{bmatrix} a_{11} & a_{12} & a_{13} \\ a_{21} & a_{22} & a_{23} \\ a_{31} & a_{32} & a_{33} \end{bmatrix} \quad \text{or} \quad \begin{vmatrix} a_{11} & a_{12} & a_{13} \\ a_{21} & a_{22} & a_{23} \\ a_{31} & a_{32} & a_{33} \end{vmatrix}$$

In the latter notation the determinant of the matrix A in Example 8 would be written

$$\begin{vmatrix} 3 & 1 \\ 4 & -2 \end{vmatrix} = -10$$

REMARK. Strictly speaking, the determinant of a matrix is a number. However, it is common practice to "abuse" the terminology slightly and use the term "determinant" to refer to the *matrix* whose determinant is being computed. Thus, we might refer to

$$\begin{vmatrix} 3 & 1 \\ 4 & -2 \end{vmatrix}$$

as a 2 × 2 determinant and call 3 the entry in the first row and first column of the determinant.

Finally, we note that the determinant of A is often written symbolically as

$$\det(A) = \sum \pm a_{1j_1} a_{2j_2} \ldots a_{nj_n}$$

where $\sum$ indicates that the terms are to be summed over all permutations $(j_1, j_2, \ldots, j_n)$ and the $+$ or $-$ is selected in each term according to whether the permutation is even or odd. This notation is useful when the definition of a determinant needs to be emphasized.

EXERCISE SET 2.1

1. Find the number of inversions in each of the following permutations of $\{1, 2, 3, 4, 5\}$.

    (a) (3 4 1 5 2)   (b) (4 2 5 3 1)   (c) (5 4 3 2 1)
    (d) (1 2 3 4 5)   (e) (1 3 5 4 2)   (f) (2 3 5 4 1)

2. Classify each of the permutations in Exercise 1 as even or odd.

In Exercises 3–10 evaluate the determinant.

3. $\begin{vmatrix} 1 & 2 \\ -1 & 3 \end{vmatrix}$
    4. $\begin{vmatrix} 6 & 4 \\ 3 & 2 \end{vmatrix}$
    5. $\begin{vmatrix} -1 & 7 \\ -8 & -3 \end{vmatrix}$
    6. $\begin{vmatrix} k-1 & 2 \\ 4 & k-3 \end{vmatrix}$

7. $\begin{vmatrix} 1 & -2 & 7 \\ 3 & 5 & 1 \\ 4 & 3 & 8 \end{vmatrix}$
    8. $\begin{vmatrix} 8 & 2 & -1 \\ -3 & 4 & -6 \\ 1 & 7 & 2 \end{vmatrix}$
    9. $\begin{vmatrix} 1 & 0 & 3 \\ 4 & 0 & -1 \\ 2 & 8 & 6 \end{vmatrix}$
    10. $\begin{vmatrix} k & -3 & 9 \\ 2 & 4 & k+1 \\ 1 & k^2 & 3 \end{vmatrix}$

11. Find all values of λ for which $\det(A) = 0$.

    (a) $A = \begin{bmatrix} \lambda-1 & -2 \\ 1 & \lambda-4 \end{bmatrix}$
    (b) $A = \begin{bmatrix} \lambda-6 & 0 & 0 \\ 0 & \lambda & -1 \\ 0 & 4 & \lambda-4 \end{bmatrix}$

12. Classify each permutation of $\{1, 2, 3, 4\}$ as even or odd.

13. Use the results in Exercise 12 to construct a formula for the determinant of a 4×4 matrix.

14. Use the formula obtained in Exercise 13 to evaluate

$$\begin{vmatrix} 1 & 4 & -3 & 1 \\ 2 & 0 & 6 & 3 \\ 4 & -1 & 2 & 5 \\ 1 & 0 & -2 & 4 \end{vmatrix}$$

15. Use the determinant definition to evaluate

$$\text{(a)} \begin{vmatrix} 0 & 0 & 0 & 0 & 1 \\ 0 & 0 & 0 & 2 & 0 \\ 0 & 0 & 3 & 0 & 0 \\ 0 & 4 & 0 & 0 & 0 \\ 5 & 0 & 0 & 0 & 0 \end{vmatrix} \quad \text{(b)} \begin{vmatrix} 0 & 4 & 0 & 0 & 0 \\ 0 & 0 & 0 & 2 & 0 \\ 0 & 0 & 3 & 0 & 0 \\ 0 & 0 & 0 & 0 & 1 \\ 5 & 0 & 0 & 0 & 0 \end{vmatrix}$$

16. Prove that if a square matrix A has a column of zeros, then $\det(A) = 0$.

2.2 EVALUATING DETERMINANTS BY ROW REDUCTION

In this section we show that the determinant of a matrix can be evaluated by reducing the matrix to row-echelon form. This method is of importance since it avoids the lengthy computations involved in directly applying the determinant definition.

We first consider two classes of matrices whose determinants can be easily evaluated, regardless of the size of the matrix.

> **Theorem 1.** *If A is any square matrix that contains a row of zeros, then* $\det(A) = 0$.

Proof. Since a signed elementary product from A contains one factor from each row of A, every signed elementary product contains a factor from the row of zeros and consequently has value zero. Since $\det(A)$ is the sum of all signed elementary products, we obtain $\det(A) = 0$.

A square matrix is called *upper triangular* if all the entries below the main diagonal are zeros. Similarly, a square matrix is called *lower triangular* if all the entries above the main diagonal are zeros. A matrix that is either upper or lower triangular is called *triangular*.

Example 9

A general 4 × 4 upper triangular matrix has the form

$$
\begin{bmatrix}
a_{11} & a_{12} & a_{13} & a_{14} \\
0 & a_{22} & a_{23} & a_{24} \\
0 & 0 & a_{33} & a_{34} \\
0 & 0 & 0 & a_{44}
\end{bmatrix}
$$

A general 4 × 4 lower triangular matrix has the form

$$
\begin{bmatrix}
a_{11} & 0 & 0 & 0 \\
a_{21} & a_{22} & 0 & 0 \\
a_{31} & a_{32} & a_{33} & 0 \\
a_{41} & a_{42} & a_{43} & a_{44}
\end{bmatrix}
$$

Example 10

Evaluate det(A), where

$$
A =
\begin{bmatrix}
a_{11} & 0 & 0 & 0 \\
a_{21} & a_{22} & 0 & 0 \\
a_{31} & a_{32} & a_{33} & 0 \\
a_{41} & a_{42} & a_{43} & a_{44}
\end{bmatrix}
$$

The only elementary product from A that can be nonzero is $a_{11}a_{22}a_{33}a_{44}$. To see this, consider a typical elementary product $a_{1j_1}a_{2j_2}a_{3j_3}a_{4j_4}$. Since $a_{12} = a_{13} = a_{14} = 0$, we must have $j_1 = 1$ in order to have a nonzero elementary product. If $j_1 = 1$, we must have $j_2 \neq 1$, since no two factors come from the same column. Further, since $a_{23} = a_{24} = 0$, we must have $j_2 = 2$ in order to have a nonzero product. Continuing in this way, we obtain $j_3 = 3$ and $j_4 = 4$. Since $a_{11}a_{22}a_{33}a_{44}$ is multiplied by $+1$ in forming the signed elementary product, we obtain

$$
\det(A) = a_{11}a_{22}a_{33}a_{44}
$$

An argument similar to the one just presented can be applied to any triangular matrix to yield the following general result.

Theorem 2. *If A is an $n \times n$ triangular matrix, then det(A) is the product of the entries on the main diagonal; that is, $\det(A) = a_{11}a_{22} \cdots a_{nn}$.*

Example 11

$$
\begin{vmatrix}
2 & 7 & -3 & 8 & 3 \\
0 & -3 & 7 & 5 & 1 \\
0 & 0 & 6 & 7 & 6 \\
0 & 0 & 0 & 9 & 8 \\
0 & 0 & 0 & 0 & 4
\end{vmatrix} = (2)(-3)(6)(9)(4) = -1296
$$

The next theorem shows how an elementary row operation on a matrix affects the value of its determinant.

Theorem 3. *Let A be any n × n matrix.*

(a) *If A′ is the matrix that results when a single row of A is multiplied by a constant k, then det(A′) = k det(A).*

(b) *If A′ is the matrix that results when two rows of A are interchanged, then det(A′) = − det(A).*

(c) *If A′ is the matrix that results when a multiple of one row of A is added to another row, then det(A′) = det(A).*

We shall omit the proof (see Exercise 15).

Example 12

Consider the matrices

$$A = \begin{bmatrix} 1 & 2 & 3 \\ 0 & 1 & 4 \\ 1 & 2 & 1 \end{bmatrix} \qquad A_1 = \begin{bmatrix} 4 & 8 & 12 \\ 0 & 1 & 4 \\ 1 & 2 & 1 \end{bmatrix} \qquad A_2 = \begin{bmatrix} 0 & 1 & 4 \\ 1 & 2 & 3 \\ 1 & 2 & 1 \end{bmatrix}$$

$$A_3 = \begin{bmatrix} 1 & 2 & 3 \\ -2 & -3 & 2 \\ 1 & 2 & 1 \end{bmatrix}$$

If we evaluate the determinants of these matrices by the method used in Example 8 we obtain

$$\det(A) = -2, \quad \det(A_1) = -8, \quad \det(A_2) = 2, \quad \det(A_3) = -2$$

Observe that A_1 is obtained by multiplying the first row of A by 4; A_2 by interchanging the first two rows; and A_3 by adding -2 times the third row of A to the second. As predicted by Theorem 3, we have the relationships

$$\det(A_1) = 4 \det(A) \qquad \det(A_2) = -\det(A) \qquad \text{and} \qquad \det(A_3) = \det(A)$$

Example 13

Statement (a) in Theorem 3 has an alternate interpretation that is sometimes useful. This result allows us to bring a "common factor" from any row of a square matrix through the determinant sign. To illustrate, consider the matrices

$$A = \begin{bmatrix} a_{11} & a_{12} & a_{13} \\ a_{21} & a_{22} & a_{23} \\ a_{31} & a_{32} & a_{33} \end{bmatrix} \qquad B = \begin{bmatrix} a_{11} & a_{12} & a_{13} \\ ka_{21} & ka_{22} & ka_{23} \\ a_{31} & a_{32} & a_{33} \end{bmatrix}$$

where the second row of B has a common factor of k. Since B is the matrix that results when the second row of A is multiplied by k, statement (a) in Theorem 3 asserts that $\det(B) = k \det(A)$; that is,

$$\begin{vmatrix} a_{11} & a_{12} & a_{13} \\ ka_{21} & ka_{22} & ka_{23} \\ a_{31} & a_{32} & a_{33} \end{vmatrix} = k \begin{vmatrix} a_{11} & a_{12} & a_{13} \\ a_{21} & a_{22} & a_{23} \\ a_{31} & a_{32} & a_{33} \end{vmatrix}$$

We shall now formulate an alternate method for evaluating determinants that will avoid the large amount of computation involved in directly applying the determinant definition. The basic idea of this method is to apply elementary row operations to reduce the given matrix A to a matrix R that is in row-echelon form. Since a row-echelon form of a square matrix is upper triangular (Exercise 14), $\det(R)$ can be evaluated using Theorem 2. The value of $\det(A)$ can then be obtained by using Theorem 3 to relate the unknown value of $\det(A)$ to the known value of $\det(R)$. The following example illustrates this method.

Example 14

Evaluate $\det(A)$ where

$$A = \begin{bmatrix} 0 & 1 & 5 \\ 3 & -6 & 9 \\ 2 & 6 & 1 \end{bmatrix}$$

Solution. Reducing A to row-echelon form and applying Theorem 3, we obtain

$$\det(A) = \begin{vmatrix} 0 & 1 & 5 \\ 3 & -6 & 9 \\ 2 & 6 & 1 \end{vmatrix} = - \begin{vmatrix} 3 & -6 & 9 \\ 0 & 1 & 5 \\ 2 & 6 & 1 \end{vmatrix}$$

> The first and second rows of A were interchanged.

$$= -3 \begin{vmatrix} 1 & -2 & 3 \\ 0 & 1 & 5 \\ 2 & 6 & 1 \end{vmatrix}$$

> A common factor of 3 from the first row of the preceding matrix was taken through the det sign (see Example 13).

$$= -3 \begin{vmatrix} 1 & -2 & 3 \\ 0 & 1 & 5 \\ 0 & 10 & -5 \end{vmatrix}$$

> -2 times the first row of the preceding matrix was added to the third row.

$$= -3 \begin{vmatrix} 1 & -2 & 3 \\ 0 & 1 & 5 \\ 0 & 0 & -55 \end{vmatrix}$$

> -10 times the second row of the preceding matrix was added to the third row.

$$= (-3)(-55) \begin{vmatrix} 1 & -2 & 3 \\ 0 & 1 & 5 \\ 0 & 0 & 1 \end{vmatrix}$$

> A common factor of -55 from the last row of the preceding matrix was taken through the det sign.

$$= (-3)(-55)(1) = 165$$

REMARK. The method of row reduction is well suited for computer evaluation of determinants because it is systematic and easily programmed. However, for hand computation, the methods we will develop in subsequent sections are often easier.

Example 15

Evaluate det(A), where

$$A = \begin{bmatrix} 1 & 3 & -2 & 4 \\ 2 & 6 & -4 & 8 \\ 3 & 9 & 1 & 5 \\ 1 & 1 & 4 & 8 \end{bmatrix}$$

$$\det(A) = \begin{vmatrix} 1 & 3 & -2 & 4 \\ 0 & 0 & 0 & 0 \\ 3 & 9 & 1 & 5 \\ 1 & 1 & 4 & 8 \end{vmatrix} \qquad \boxed{\begin{array}{l} -2 \text{ times the first} \\ \text{row of } A \text{ was added} \\ \text{to the second row.} \end{array}}$$

No further reduction is needed since it follows from Theorem 1 that det(A) = 0.

It should be evident from this example that whenever a square matrix has two proportional rows (like the first and second rows of A), it is possible to introduce a row of zeros by adding a suitable multiple of one of these rows to the other. Thus, *if a square matrix has two proportional rows, its determinant is zero.*

Example 16

Each of the following matrices has two proportional rows; thus, by inspection, each has a zero determinant.

$$\begin{bmatrix} -1 & 4 \\ -2 & 8 \end{bmatrix} \qquad \begin{bmatrix} 2 & 7 & 8 \\ 3 & 2 & 4 \\ 2 & 7 & 8 \end{bmatrix} \qquad \begin{bmatrix} 3 & -1 & 4 & -5 \\ 6 & -2 & 5 & 2 \\ 5 & 8 & 1 & 4 \\ -9 & 3 & -12 & 15 \end{bmatrix}$$

EXERCISE SET 2.2

1. Evaluate the following by inspection.

(a) $\begin{vmatrix} 2 & -40 & 17 \\ 0 & 1 & 11 \\ 0 & 0 & 3 \end{vmatrix}$
(b) $\begin{vmatrix} 1 & 0 & 0 & 0 \\ -9 & -1 & 0 & 0 \\ 12 & 7 & 8 & 0 \\ 4 & 5 & 7 & 2 \end{vmatrix}$

(c) $\begin{vmatrix} 1 & 2 & 3 \\ 3 & 7 & 6 \\ 1 & 2 & 3 \end{vmatrix}$
(d) $\begin{vmatrix} 3 & -1 & 2 \\ 6 & -2 & 4 \\ 1 & 7 & 3 \end{vmatrix}$

In Exercises 2–9 evaluate the determinants of the given matrices by reducing the matrix to row-echelon form.

2. $\begin{bmatrix} 2 & 3 & 7 \\ 0 & 0 & -3 \\ 1 & -2 & 7 \end{bmatrix}$
3. $\begin{bmatrix} 2 & 1 & 1 \\ 4 & 2 & 3 \\ 1 & 3 & 0 \end{bmatrix}$

4. $\begin{bmatrix} 1 & -2 & 0 \\ -3 & 5 & 1 \\ 4 & -3 & 2 \end{bmatrix}$ **5.** $\begin{bmatrix} 2 & -4 & 8 \\ -2 & 7 & -2 \\ 0 & 1 & 5 \end{bmatrix}$

6. $\begin{bmatrix} 3 & 6 & 9 & 3 \\ -1 & 0 & 1 & 0 \\ 1 & 3 & 2 & -1 \\ -1 & -2 & -2 & 1 \end{bmatrix}$ **7.** $\begin{bmatrix} 2 & 1 & 3 & 1 \\ 1 & 0 & 1 & 1 \\ 0 & 2 & 1 & 0 \\ 0 & 1 & 2 & 3 \end{bmatrix}$

8. $\begin{bmatrix} \frac{1}{2} & \frac{1}{2} & 1 & \frac{1}{2} \\ -\frac{1}{2} & \frac{1}{2} & 0 & \frac{1}{2} \\ \frac{2}{3} & \frac{1}{3} & \frac{1}{3} & 0 \\ \frac{1}{3} & 1 & \frac{1}{3} & 0 \end{bmatrix}$ **9.** $\begin{bmatrix} 1 & 3 & 1 & 5 & 3 \\ -2 & -7 & 0 & -4 & 2 \\ 0 & 0 & 1 & 0 & 1 \\ 0 & 0 & 2 & 1 & 1 \\ 0 & 0 & 0 & 1 & 1 \end{bmatrix}$

10. Assume $\det \begin{bmatrix} a & b & c \\ d & e & f \\ g & h & i \end{bmatrix} = 5$. Find

(a) $\det \begin{bmatrix} d & e & f \\ g & h & i \\ a & b & c \end{bmatrix}$ (b) $\det \begin{bmatrix} -a & -b & -c \\ 2d & 2e & 2f \\ -g & -h & -i \end{bmatrix}$

(c) $\det \begin{bmatrix} a+d & b+e & c+f \\ d & e & f \\ g & h & i \end{bmatrix}$ (d) $\det \begin{bmatrix} a & b & c \\ d-3a & e-3b & f-3c \\ 2g & 2h & 2i \end{bmatrix}$

11. Use row reduction to show that

$$\begin{vmatrix} 1 & 1 & 1 \\ a & b & c \\ a^2 & b^2 & c^2 \end{vmatrix} = (b-a)(c-a)(c-b)$$

12. Use an argument like that given in Example 10 to show

(a) $\det \begin{bmatrix} 0 & 0 & a_{13} \\ 0 & a_{22} & a_{23} \\ a_{31} & a_{32} & a_{33} \end{bmatrix} = -a_{13}a_{22}a_{31}$

(b) $\det \begin{bmatrix} 0 & 0 & 0 & a_{14} \\ 0 & 0 & a_{23} & a_{24} \\ 0 & a_{32} & a_{33} & a_{34} \\ a_{41} & a_{42} & a_{43} & a_{44} \end{bmatrix} = a_{14}a_{23}a_{32}a_{41}$

13. Prove that Theorem 1 is true when the word "row" is replaced by "column."

14. Prove that a row-echelon form of a square matrix is upper triangular.

15. Prove the following special cases of Theorem 3.

(a) $\begin{vmatrix} ka_{11} & ka_{12} & ka_{13} \\ a_{21} & a_{22} & a_{23} \\ a_{31} & a_{32} & a_{33} \end{vmatrix} = k \begin{vmatrix} a_{11} & a_{12} & a_{13} \\ a_{21} & a_{22} & a_{23} \\ a_{31} & a_{32} & a_{33} \end{vmatrix}$

(b) $\begin{vmatrix} a_{21} & a_{22} & a_{23} \\ a_{11} & a_{12} & a_{13} \\ a_{31} & a_{32} & a_{33} \end{vmatrix} = - \begin{vmatrix} a_{11} & a_{12} & a_{13} \\ a_{21} & a_{22} & a_{23} \\ a_{31} & a_{32} & a_{33} \end{vmatrix}$

(c) $\begin{vmatrix} a_{11} + ka_{21} & a_{12} + ka_{22} & a_{13} + ka_{23} \\ a_{21} & a_{22} & a_{23} \\ a_{31} & a_{32} & a_{33} \end{vmatrix} = \begin{vmatrix} a_{11} & a_{12} & a_{13} \\ a_{21} & a_{22} & a_{23} \\ a_{31} & a_{32} & a_{33} \end{vmatrix}$

2.3 PROPERTIES OF THE DETERMINANT FUNCTION

In this section we develop some of the fundamental properties of the determinant function. Our work here will give us some further insight into the relationship between a square matrix and its determinant. One of the immediate consequences of this material will be an important determinant test for the invertibility of a matrix.

Recall that the determinant of an $n \times n$ matrix A is defined to be the sum of all signed elementary products from A. Since an elementary product has one factor from each row and one factor from each column, it is obvious that A and A^t have precisely the same set of elementary products. Although we shall omit the details, it can be shown that A and A^t actually have the same *signed* elementary products; this yields the following theorem.

Theorem 4. *If A is any square matrix, then $det(A) = det(A^t)$.*

REMARK. Because of this result, nearly every theorem about determinants that contains the word "row" in its statement is also true when the word "column" is substituted for "row." To prove a column statement one need only transpose the matrix in question to convert the column statement to a row statement, and then apply the corresponding known result for rows.

Example 17

Use Theorem 4 and the fact that interchanging two rows of a square matrix A changes the sign of its determinant to prove that interchanging two columns of A changes the sign of its determinant.

Solution. Let A' be the matrix that results when column r and column s of A are interchanged. Thus, $(A')^t$ is the matrix that results when *row r* and *row s* of A^t are interchanged. Therefore,

$$\det(A') = \det(A')^t \qquad \text{(Theorem 4)}$$
$$= -\det(A^t) \qquad \text{(Theorem 3}b\text{)}$$
$$= -\det(A) \qquad \text{(Theorem 4)}$$

which proves the result. ▲

The next two examples use column properties of determinants.

Example 18

By inspection, the matrix

$$\begin{bmatrix} 1 & -2 & 7 \\ -4 & 8 & 5 \\ 2 & -4 & 3 \end{bmatrix}$$

has a zero determinant since the first and second columns are proportional. ▲

Example 19

Compute the determinant of

$$A = \begin{bmatrix} 1 & 0 & 0 & 3 \\ 2 & 7 & 0 & 6 \\ 0 & 6 & 3 & 0 \\ 7 & 3 & 1 & -5 \end{bmatrix}$$

This determinant could be computed as before by using elementary row operations to reduce A to row-echelon form. In contrast, we can put A in lower triangular form in one step by adding -3 times the first column to the fourth to obtain

$$\det(A) = \det \begin{bmatrix} 1 & 0 & 0 & 0 \\ 2 & 7 & 0 & 0 \\ 0 & 6 & 3 & 0 \\ 7 & 3 & 1 & -26 \end{bmatrix} = (1)(7)(3)(-26) = -546 \quad ▲$$

This example points out that it is always wise to keep an eye open for column operations that can shorten computations.

Suppose that A and B are $n \times n$ matrices and k is any scalar. We now consider possible relationships between $\det(A)$, $\det(B)$, and

$$\det(kA), \quad \det(A + B), \quad \text{and } \det(AB)$$

Since a common factor of any row of a matrix can be moved through the det sign, and since each of the n rows in kA has a common factor of k, we obtain

$$\det(kA) = k^n \det(A) \qquad (2.1)$$

Example 20

Consider the matrices

$$A = \begin{bmatrix} 3 & 1 \\ 2 & 2 \end{bmatrix} \quad \text{and} \quad 5A = \begin{bmatrix} 15 & 5 \\ 10 & 10 \end{bmatrix}$$

By direct calculation $\det(A) = 4$ and $\det(5A) = 100$. This agrees with relationship (2.1), which asserts that $\det(5A) = 5^2\det(A)$. ▲

Unfortunately, no simple relationship exists between $\det(A)$, $\det(B)$, and $\det(A + B)$ in general. In particular, we emphasize that $\det(A + B)$ is usually *not* equal to $\det(A) + \det(B)$. The following example illustrates this fact.

Example 21

Consider

$$A = \begin{bmatrix} 1 & 2 \\ 2 & 5 \end{bmatrix} \quad B = \begin{bmatrix} 3 & 1 \\ 1 & 3 \end{bmatrix} \quad A + B = \begin{bmatrix} 4 & 3 \\ 3 & 8 \end{bmatrix}$$

We have $\det(A) = 1$, $\det(B) = 8$, and $\det(A + B) = 23$; thus $\det(A + B) \neq \det(A) + \det(B)$.

In spite of this negative result, there is one important relationship concerning sums of determinants that is often useful. To illustrate, consider two 2×2 matrices

$$A = \begin{bmatrix} a_{11} & a_{12} \\ a_{21} & a_{22} \end{bmatrix} \quad \text{and} \quad A' = \begin{bmatrix} a_{11} & a_{12} \\ a'_{21} & a'_{22} \end{bmatrix}$$

that differ only in the second row. From the formula in Example 7, we obtain

$$\det(A) + \det(A') = (a_{11}a_{22} - a_{12}a_{21}) + (a_{11}a'_{22} - a_{12}a'_{21})$$
$$= a_{11}(a_{22} + a'_{22}) - a_{12}(a_{21} + a'_{21})$$
$$= \det \begin{bmatrix} a_{11} & a_{12} \\ a_{21} + a'_{21} & a_{22} + a'_{22} \end{bmatrix}$$

Thus,

$$\det \begin{bmatrix} a_{11} & a_{12} \\ a_{21} & a_{22} \end{bmatrix} + \det \begin{bmatrix} a_{11} & a_{12} \\ a'_{21} & a'_{22} \end{bmatrix} = \det \begin{bmatrix} a_{11} & a_{12} \\ a_{21} + a'_{21} & a_{22} + a'_{22} \end{bmatrix}$$ ▲

This example is a special case of the following general result.

Theorem 5. *Let A, A', and A'' be n × n matrices that differ only in a single row, say the rth, and assume that the rth row of A'' can be obtained by adding corresponding entries in the rth rows of A and A'. Then*

$$\det(A'') = \det(A) + \det(A').$$

The same result holds for columns.

Example 22

By evaluating the determinants, the reader can check that

$$\det \begin{bmatrix} 1 & 7 & 5 \\ 2 & 0 & 3 \\ 1+0 & 4+1 & 7+(-1) \end{bmatrix} = \det \begin{bmatrix} 1 & 7 & 5 \\ 2 & 0 & 3 \\ 1 & 4 & 7 \end{bmatrix} + \det \begin{bmatrix} 1 & 7 & 5 \\ 2 & 0 & 3 \\ 0 & 1 & -1 \end{bmatrix}$$

Theorem 6. *If A and B are square matrices of the same size, then*

$$\det(AB) = \det(A)\det(B).$$

When one considers how complicated the definitions of matrix multiplication and determinants are, the elegant simplicity of this result is surprising. We shall omit the proof.

Example 23

Consider the matrices

$$A = \begin{bmatrix} 3 & 1 \\ 2 & 1 \end{bmatrix} \qquad B = \begin{bmatrix} -1 & 3 \\ 5 & 8 \end{bmatrix} \qquad AB = \begin{bmatrix} 2 & 17 \\ 3 & 14 \end{bmatrix}$$

We have $\det(A)\det(B) = (1)(-23) = -23$. In contrast, by direct computation $\det(AB) = -23$, so that $\det(AB) = \det(A)\det(B)$.

In Theorem 15 of Chapter 1, we listed three important statements that are equivalent to the invertibility of a matrix. The next example will help us to add another result to that list.

Example 24

The purpose of this example is to show that if the reduced row-echelon form R of a *square* matrix has no rows consisting entirely of zeros, then R must be the identity matrix. This can be illustrated by considering the following $n \times n$ matrix:

$$R = \begin{bmatrix} r_{11} & r_{12} & \cdots & r_{1n} \\ r_{21} & r_{22} & \cdots & r_{2n} \\ \vdots & \vdots & & \vdots \\ r_{n1} & r_{n2} & \cdots & r_{nn} \end{bmatrix}$$

Either the last row in this matrix consists entirely of zeros or it does not. If not, the matrix contains no zero rows, and consequently each of the n rows has a leading entry of 1. Since these leading 1's occur progressively further to the right as we move down the matrix, each of these 1's must occur on the main diagonal. Since the other entries in the same column as one of these 1's are zero, R must be I. Thus, either R has a row of zeros or $R = I$.

> **Theorem 7.** *A square matrix A is invertible if and only if* $\det(A) \neq 0$.

Proof. If A is invertible, then $I = AA^{-1}$ so that $1 = \det(I) = \det(A)\det(A^{-1})$. Thus, $\det(A) \neq 0$. Conversely, assume that $\det(A) \neq 0$. We shall show that A is row equivalent to I, and thus conclude from Theorem 12 in Section 1.6 that A is invertible. Let R be the reduced row-echelon form of A. Since R can be obtained from A by a finite sequence of elementary row operations, we can find elementary matrices $E_1, E_2, \ldots, E_k$ such that $E_k \cdots E_2 E_1 A = R$ or $A = E_1^{-1}E_2^{-1} \cdots E_k^{-1}R$. Thus,

$$\det(A) = \det(E_1^{-1})\det(E_2^{-1}) \cdots \det(E_k^{-1})\det(R)$$

Since we are assuming that $\det(A) \neq 0$, it follows from this equation that $\det(R) \neq 0$. Therefore, R does not have any zero rows, so that $R = I$ (see Example 24).

> **Corollary.** If A is invertible, then
>
> $$\det(A^{-1}) = \frac{1}{\det(A)}$$

Proof. Since $A^{-1}A = I$, $\det(A^{-1}A) = \det(I)$; that is, $\det(A^{-1})\det(A) = 1$. Since $\det(A) \neq 0$, the proof can be completed by dividing through by $\det(A)$.

Example 25

Since the first and third rows of

$$A = \begin{bmatrix} 1 & 2 & 3 \\ 1 & 0 & 1 \\ 2 & 4 & 6 \end{bmatrix}$$

are proportional, $\det(A) = 0$. Thus, A is not invertible.

EXERCISE SET 2.3

1. Verify that $\det(A) = \det(A^t)$ for

(a) $A = \begin{bmatrix} 1 & -3 \\ 2 & 5 \end{bmatrix}$ (b) $A = \begin{bmatrix} 1 & 2 & 7 \\ -1 & 0 & 6 \\ 3 & 2 & 8 \end{bmatrix}$

2. Verify that $\det(AB) = \det(A)\det(B)$ when

$$A = \begin{bmatrix} 2 & 1 & 0 \\ 3 & 4 & 0 \\ 0 & 0 & 2 \end{bmatrix} \quad \text{and} \quad B = \begin{bmatrix} 1 & -1 & 3 \\ 7 & 1 & 2 \\ 5 & 0 & 1 \end{bmatrix}$$

3. By inspection, explain why $\det(A) = 0$.

$$A = \begin{bmatrix} -3 & 4 & 7 & -2 \\ 2 & 6 & 1 & -3 \\ 1 & 0 & 0 & 0 \\ 2 & -8 & 3 & 4 \end{bmatrix}$$

4. Use Theorem 7 to determine which of the following matrices are invertible.

inv. (a) $\begin{bmatrix} 1 & 0 & 0 \\ 3 & 6 & 7 \\ 0 & 8 & -1 \end{bmatrix}$ (b) $\begin{bmatrix} -2 & 1 & -4 \\ 1 & 1 & 2 \\ 3 & 1 & 6 \end{bmatrix}$ not

not (c) $\begin{bmatrix} 7 & 2 & 1 \\ 7 & 2 & 1 \\ 3 & 6 & 6 \end{bmatrix}$ (d) $\begin{bmatrix} 0 & 7 & 5 \\ 0 & 1 & -1 \\ 0 & 3 & 2 \end{bmatrix}$ not

5. Assume $\det(A) = 5$, where

$$A = \begin{bmatrix} a & b & c \\ d & e & f \\ g & h & i \end{bmatrix}$$

Find

(a) $\det(3A)$ (b) $\det(2A^{-1})$ (c) $\det((2A)^{-1})$ (d) $\det \begin{bmatrix} a & g & d \\ b & h & e \\ c & i & f \end{bmatrix}$

6. Without directly evaluating, show that $x = 0$ and $x = 2$ satisfy

$$\begin{vmatrix} x^2 & x & 2 \\ 2 & 1 & 1 \\ 0 & 0 & -5 \end{vmatrix} = 0$$

7. Without directly evaluating, show that

$$\det \begin{bmatrix} b+c & c+a & b+a \\ a & b & c \\ 1 & 1 & 1 \end{bmatrix} = 0$$

In Exercises 8–11 prove the identity without evaluating the determinants.

8. $\begin{vmatrix} a_1 & b_1 & a_1 + b_1 + c_1 \\ a_2 & b_2 & a_2 + b_2 + c_2 \\ a_3 & b_3 & a_3 + b_3 + c_3 \end{vmatrix} = \begin{vmatrix} a_1 & b_1 & c_1 \\ a_2 & b_2 & c_2 \\ a_3 & b_3 & c_3 \end{vmatrix}$

9. $\begin{vmatrix} a_1 + b_1 & a_1 - b_1 & c_1 \\ a_2 + b_2 & a_2 - b_2 & c_2 \\ a_3 + b_3 & a_3 - b_3 & c_3 \end{vmatrix} = -2 \begin{vmatrix} a_1 & b_1 & c_1 \\ a_2 & b_2 & c_2 \\ a_3 & b_3 & c_3 \end{vmatrix}$

10. $\begin{vmatrix} a_1 + b_1 t & a_2 + b_2 t & a_3 + b_3 t \\ a_1 t + b_1 & a_2 t + b_2 & a_3 t + b_3 \\ c_1 & c_2 & c_3 \end{vmatrix} = (1 - t^2) \begin{vmatrix} a_1 & a_2 & a_3 \\ b_1 & b_2 & b_3 \\ c_1 & c_2 & c_3 \end{vmatrix}$

11. $\begin{vmatrix} a_1 & b_1 + ta_1 & c_1 + rb_1 + sa_1 \\ a_2 & b_2 + ta_2 & c_2 + rb_2 + sa_2 \\ a_3 & b_3 + ta_3 & c_3 + rb_3 + sa_3 \end{vmatrix} = \begin{vmatrix} a_1 & a_2 & a_3 \\ b_1 & b_2 & b_3 \\ c_1 & c_2 & c_3 \end{vmatrix}$

12. For which value(s) of k does A fail to be invertible?

(a) $A = \begin{bmatrix} k - 3 & -2 \\ -2 & k - 2 \end{bmatrix}$ (b) $A = \begin{bmatrix} 1 & 2 & 4 \\ 3 & 1 & 6 \\ k & 3 & 2 \end{bmatrix}$

13. Use Theorem 7 to show that

$$\begin{bmatrix} \sin^2 \alpha & \sin^2 \beta & \sin^2 \gamma \\ \cos^2 \alpha & \cos^2 \beta & \cos^2 \gamma \\ 1 & 1 & 1 \end{bmatrix}$$

is not invertible for any values of α, β, and γ.

14. Let A and B be $n \times n$ matrices. Show that if A is invertible, then $\det(B) = \det(A^{-1}BA)$.

15. (a) Express

$$\begin{vmatrix} a_1 + b_1 & c_1 + d_1 \\ a_2 + b_2 & c_2 + d_2 \end{vmatrix}$$

as a sum of four determinants whose entries contain no sums.

(b) Express

$$\begin{vmatrix} a_1 + b_1 & c_1 + d_1 & e_1 + f_1 \\ a_2 + b_2 & c_2 + d_2 & e_2 + f_2 \\ a_3 + b_3 & c_3 + d_3 & e_3 + f_3 \end{vmatrix}$$

as a sum of eight determinants whose entries contain no sums.

2.4 COFACTOR EXPANSION; CRAMER'S RULE

In this section we consider a method for evaluating determinants that is useful for hand computations and important theoretically. As a consequence of our work here, we shall obtain a formula for the inverse of an invertible matrix as well as a formula for the solution to certain systems of linear equations in terms of determinants.

Definition. If A is a square matrix, then the *minor of entry a_{ij}* is denoted by M_{ij} and is defined to be the determinant of the submatrix that remains after the ith row and jth column are deleted from A. The number $(-1)^{i+j}M_{ij}$ is denoted by C_{ij} and is called the *cofactor of entry a_{ij}*.

Example 26

Let

$$A = \begin{bmatrix} 3 & 1 & -4 \\ 2 & 5 & 6 \\ 1 & 4 & 8 \end{bmatrix}$$

The minor of entry a_{11} is

$$M_{11} = \begin{vmatrix} 3 & 1 & -4 \\ 2 & 5 & 6 \\ 1 & 4 & 8 \end{vmatrix} = \begin{vmatrix} 5 & 6 \\ 4 & 8 \end{vmatrix} = 16$$

The cofactor of a_{11} is

$$C_{11} = (-1)^{1+1}M_{11} = M_{11} = 16$$

Similarly, the minor of entry a_{32} is

$$M_{32} = \begin{vmatrix} 3 & 1 & -4 \\ 2 & 5 & 6 \\ 1 & 4 & 8 \end{vmatrix} = \begin{vmatrix} 3 & -4 \\ 2 & 6 \end{vmatrix} = 26$$

The cofactor of a_{32} is

$$C_{32} = (-1)^{3+2}M_{32} = -M_{32} = -26$$

Notice that the cofactor and the minor of an element a_{ij} differ only in sign, that is, $C_{ij} = \pm M_{ij}$. A quick way for determining whether to use the $+$ or $-$ is to use the fact that the sign relating C_{ij} and M_{ij} is in the ith row and jth column of the array

$$\begin{bmatrix} + & - & + & - & + & \cdots \\ - & + & - & + & - & \cdots \\ + & - & + & - & + & \cdots \\ - & + & - & + & - & \cdots \\ \vdots & \vdots & \vdots & \vdots & \vdots & \end{bmatrix}$$

For example, $C_{11} = M_{11}$, $C_{21} = -M_{21}$, $C_{12} = -M_{12}$, $C_{22} = M_{22}$, and so on.

Consider the general 3×3 matrix

$$A = \begin{bmatrix} a_{11} & a_{12} & a_{13} \\ a_{21} & a_{22} & a_{23} \\ a_{31} & a_{32} & a_{33} \end{bmatrix}$$

In Example 7 we showed

$$\det(A) = a_{11}a_{22}a_{33} + a_{12}a_{23}a_{31} + a_{13}a_{21}a_{32}$$
$$- a_{13}a_{22}a_{31} - a_{12}a_{21}a_{33} - a_{11}a_{23}a_{32} \quad (2.2)$$

which can be rewritten as

$$\det(A) = a_{11}(a_{22}a_{33} - a_{23}a_{32}) + a_{21}(a_{13}a_{32} - a_{12}a_{33}) + a_{31}(a_{12}a_{23} - a_{13}a_{22})$$

Since the expressions in parentheses are just the cofactors C_{11}, C_{21}, and C_{31} (verify), we have

$$\det(A) = a_{11}C_{11} + a_{21}C_{21} + a_{31}C_{31} \quad (2.3)$$

Equation (2.3) shows that the determinant of A can be computed by multiplying the entries in the first column of A by their cofactors and adding the resulting products. This method of evaluating $\det(A)$ is called ***cofactor expansion*** along the first column of A.

Example 27

Let

$$A = \begin{bmatrix} 3 & 1 & 0 \\ -2 & -4 & 3 \\ 5 & 4 & -2 \end{bmatrix}$$

Evaluate $\det(A)$ by cofactor expansion along the first column of A.

Solution. From (2.3)

$$\det(A) = 3 \begin{vmatrix} -4 & 3 \\ 4 & -2 \end{vmatrix} - (-2) \begin{vmatrix} 1 & 0 \\ 4 & -2 \end{vmatrix} + 5 \begin{vmatrix} 1 & 0 \\ -4 & 3 \end{vmatrix}$$
$$= 3(-4) - (-2)(-2) + 5(3) = -1 \quad \blacktriangle$$

By rearranging the terms in (2.2) in various ways, it is possible to obtain other formulas like (2.3). There should be no trouble checking that all of the following are correct (see Exercise 27):

$$\begin{aligned} \det(A) &= a_{11}C_{11} + a_{12}C_{12} + a_{13}C_{13} \\ &= a_{11}C_{11} + a_{21}C_{21} + a_{31}C_{31} \\ &= a_{21}C_{21} + a_{22}C_{22} + a_{23}C_{23} \\ &= a_{12}C_{12} + a_{22}C_{22} + a_{32}C_{32} \\ &= a_{31}C_{31} + a_{32}C_{32} + a_{33}C_{33} \\ &= a_{13}C_{13} + a_{23}C_{23} + a_{33}C_{33} \end{aligned} \quad (2.4)$$

Notice that in each equation the entries and cofactors all come from the same row or column. These equations are called the ***cofactor expansions*** of $\det(A)$.

The results we have just given for 3 × 3 matrices form a special case of the following general theorem, which we state without proof.

Theorem 8. *The determinant of an n × n matrix A can be computed by multiplying the entries in any row (or column) by their cofactors and adding the resulting products; that is, for each $1 \leq i \leq n$ and $1 \leq j \leq n$,*

$$\det(A) = a_{1j}C_{1j} + a_{2j}C_{2j} + \cdots + a_{nj}C_{nj}$$
(cofactor expansion along the jth column)

and

$$\det(A) = a_{i1}C_{i1} + a_{i2}C_{i2} + \cdots + a_{in}C_{in}$$
(cofactor expansion along the ith row)

Example 28

Let A be the matrix in Example 27. Evaluate $\det(A)$ by cofactor expansion along the first row.

Solution.

$$\det(A) = \begin{vmatrix} 3 & 1 & 0 \\ -2 & -4 & 3 \\ 5 & 4 & -2 \end{vmatrix} = 3 \begin{vmatrix} -4 & 3 \\ 4 & -2 \end{vmatrix} - (1) \begin{vmatrix} -2 & 3 \\ 5 & -2 \end{vmatrix} + 0 \begin{vmatrix} -2 & -4 \\ 5 & 4 \end{vmatrix}$$

$$= 3(-4) - (1)(-11) = -1$$

This agrees with the result obtained in Example 27. ▲

REMARK. In this example it was unnecessary to compute the last cofactor, since it was multiplied by zero. In general, the best strategy for evaluating a determinant by cofactor expansion is to expand along a row or column having the largest number of zeros.

Cofactor expansion and row or column operations can sometimes be used in combination to provide an effective method for evaluating determinants. The following example illustrates this idea.

Example 29

Evaluate $\det(A)$ where

$$A = \begin{bmatrix} 3 & 5 & -2 & 6 \\ 1 & 2 & -1 & 1 \\ 2 & 4 & 1 & 5 \\ 3 & 7 & 5 & 3 \end{bmatrix}$$

Solution. By adding suitable multiples of the second row to the remaining rows, we obtain

$$\det(A) = \begin{vmatrix} 0 & -1 & 1 & 3 \\ 1 & 2 & -1 & 1 \\ 0 & 0 & 3 & 3 \\ 0 & 1 & 8 & 0 \end{vmatrix}$$

$$= - \begin{vmatrix} -1 & 1 & 3 \\ 0 & 3 & 3 \\ 1 & 8 & 0 \end{vmatrix}$$

> Cofactor expansion along the first column.

$$= - \begin{vmatrix} -1 & 1 & 3 \\ 0 & 3 & 3 \\ 0 & 9 & 3 \end{vmatrix}$$

> We added the first row to the third row.

$$= -(-1) \begin{vmatrix} 3 & 3 \\ 9 & 3 \end{vmatrix}$$

> Cofactor expansion along the first column.

$$= -18$$

In a cofactor expansion we compute $\det(A)$ by multiplying the entries in a row or column by their cofactors and adding the resulting products. It turns out that if one multiplies the entries in any row by the corresponding cofactors from a *different* row, the sum of these products is always zero. (This result also holds for columns.) Although we omit the general proof, the next example illustrates the idea of the proof in a special case.

Example 30

Let

$$A = \begin{bmatrix} a_{11} & a_{12} & a_{13} \\ a_{21} & a_{22} & a_{23} \\ a_{31} & a_{32} & a_{33} \end{bmatrix}$$

Consider the quantity

$$a_{11}C_{31} + a_{12}C_{32} + a_{13}C_{33}$$

which is formed by multiplying the entries in the first row by the cofactors of the corresponding entries in the third row and adding the resulting products. We now show this quantity is equal to zero by the following trick. Construct a new matrix A' by replacing the third row of A with another copy of the first row. Thus,

$$A' = \begin{bmatrix} a_{11} & a_{12} & a_{13} \\ a_{21} & a_{22} & a_{23} \\ a_{11} & a_{12} & a_{13} \end{bmatrix}$$

Let C'_{31}, C'_{32} and C'_{33} be the cofactors of the entires in the third row of A'. Since the first two rows of A and A' are the same, and since the computation of C_{31}, C_{32},

C_{33}, C'_{31}, C'_{32}, and C'_{33} involve only entries from the first two rows of A and A', it follows that

$$C_{31} = C'_{31}, \quad C_{32} = C'_{32}, \quad C_{33} = C'_{33}$$

Since A' has two identical rows,

$$\det(A') = 0 \tag{2.5}$$

In contrast, evaluating det (A') by cofactor expansion along the third row gives

$$\begin{aligned} \det(A') &= a_{11}C'_{31} + a_{12}C'_{32} + a_{13}C'_{33} \\ &= a_{11}C_{31} + a_{12}C_{32} + a_{13}C_{33} \end{aligned} \tag{2.6}$$

From (2.5) and (2.6) we obtain

$$a_{11}C_{31} + a_{12}C_{32} + a_{13}C_{33} = 0$$

Definition. If A is any $n \times n$ matrix and C_{ij} is the cofactor of a_{ij}, then the matrix

$$\begin{bmatrix} C_{11} & C_{12} & \cdots & C_{1n} \\ C_{21} & C_{22} & \cdots & C_{2n} \\ \vdots & \vdots & & \vdots \\ C_{n1} & C_{n2} & \cdots & C_{nn} \end{bmatrix}$$

is called the *matrix of cofactors from A*. The transpose of this matrix is called the *adjoint of A* and is denoted by adj(A).

Example 31

Let

$$A = \begin{bmatrix} 3 & 2 & -1 \\ 1 & 6 & 3 \\ 2 & -4 & 0 \end{bmatrix}$$

The cofactors of A are

$$\begin{array}{lll} C_{11} = 12 & C_{12} = 6 & C_{13} = -16 \\ C_{21} = 4 & C_{22} = 2 & C_{23} = 16 \\ C_{31} = 12 & C_{32} = -10 & C_{33} = 16 \end{array}$$

so that the matrix of cofactors is

$$\begin{bmatrix} 12 & 6 & -16 \\ 4 & 2 & 16 \\ 12 & -10 & 16 \end{bmatrix}$$

and the adjoint of A is

$$\text{adj}(A) = \begin{bmatrix} 12 & 4 & 12 \\ 6 & 2 & -10 \\ -16 & 16 & 16 \end{bmatrix}$$

We are now in a position to derive a formula for the inverse of an invertible matrix.

Theorem 9. *If A is an invertible matrix, then*

$$A^{-1} = \frac{1}{\det(A)}\, adj(A)$$

Proof. We show first that

$$A\, \text{adj}(A) = \det(A)I$$

Consider the product

$$A\, \text{adj}(A) = \begin{bmatrix} a_{11} & a_{12} & \cdots & a_{1n} \\ a_{21} & a_{22} & \cdots & a_{2n} \\ \vdots & \vdots & & \vdots \\ a_{i1} & a_{i2} & \cdots & a_{in} \\ \vdots & \vdots & & \vdots \\ a_{n1} & a_{n2} & \cdots & a_{nn} \end{bmatrix} \begin{bmatrix} C_{11} & C_{21} & \cdots & C_{j1} & \cdots & C_{n1} \\ C_{12} & C_{22} & \cdots & C_{j2} & & C_{n2} \\ \vdots & \vdots & & \vdots & & \vdots \\ C_{1n} & C_{2n} & \cdots & C_{jn} & \cdots & C_{nn} \end{bmatrix}$$

The entry in the ith row and jth column of $A\, \text{adj}(A)$ is

$$a_{i1}C_{j1} + a_{i2}C_{j2} + \cdots + a_{in}C_{jn} \tag{2.7}$$

(see the shaded lines above).

On the one hand, if $i = j$, then (2.7) is the cofactor expansion of $\det(A)$ along the ith row of A (Theorem 8). On the other hand, if $i \neq j$, then the a's and the cofactors come from different rows of A, so the value of (2.7) is zero. Therefore,

$$A\, \text{adj}(A) = \begin{bmatrix} \det(A) & 0 & \cdots & 0 \\ 0 & \det(A) & \cdots & 0 \\ \vdots & \vdots & & \vdots \\ 0 & 0 & \cdots & \det(A) \end{bmatrix} = \det(A)I \tag{2.8}$$

Since A is invertible, $\det(A) \neq 0$. Therefore, equation (2.8) can be rewritten as

$$\frac{1}{\det(A)}\left[A\, \text{adj}(A)\right] = I$$

or

$$A\left[\frac{1}{\det(A)}\, \text{adj}(A)\right] = I$$

Multiplying both sides on the left by A^{-1} yields

$$A^{-1} = \frac{1}{\det(A)} \text{adj}(A) \tag{2.9}$$

Example 32

Use (2.9) to find the inverse of the matrix A in Example 31.

Solution. The reader can check that $\det(A) = 64$. Thus,

$$A^{-1} = \frac{1}{\det(A)} \text{adj}(A) = \frac{1}{64} \begin{bmatrix} 12 & 4 & 12 \\ 6 & 2 & -10 \\ -16 & 16 & 16 \end{bmatrix}$$

$$= \begin{bmatrix} \frac{12}{64} & \frac{4}{64} & \frac{12}{64} \\ \frac{6}{64} & \frac{2}{64} & -\frac{10}{64} \\ -\frac{16}{64} & \frac{16}{64} & \frac{16}{64} \end{bmatrix}$$

We note that for matrices larger than 3×3 the matrix inversion method in this example is less efficient than the technique given in Section 1.6. However, the inversion method in Section 1.6 is just a computational procedure or algorithm for computing A^{-1} and is not very useful for studying properties of the inverse. Formula 2.9 can often be used to obtain properties of the inverse without actually computing it (Exercise 24).

In a similar vein, it is often useful to have a formula for the solution of a system of equations that can be used to study properties of the solution without solving the system. The next theorem establishes such a formula for systems of n equations in n unknowns. The formula is known as *Cramer's rule*.

Theorem 10 (*Cramer's Rule*). *If $AX = B$ is a system of n linear equations in n unknowns such that $\det(A) \neq 0$, then the system has a unique solution. This solution is*

$$x_1 = \frac{\det(A_1)}{\det(A)}, \quad x_2 = \frac{\det(A_2)}{\det(A)}, \ldots, \quad x_n = \frac{\det(A_n)}{\det(A)}$$

where A_j is the matrix obtained by replacing the entries in the jth column of A by the entries in the matrix

$$B = \begin{bmatrix} b_1 \\ b_2 \\ \vdots \\ b_n \end{bmatrix}$$

Proof. If $\det(A) \neq 0$, then A is invertible and, by Theorem 13 in Section 1.7, $X = A^{-1}B$ is the unique solution of $AX = B$. Therefore, by Theorem 9 we have

$$X = A^{-1}B = \frac{1}{\det(A)} \, \text{adj}(A)B = \frac{1}{\det(A)} \begin{bmatrix} C_{11} & C_{21} & \cdots & C_{n1} \\ C_{12} & C_{22} & \cdots & C_{n2} \\ \vdots & \vdots & & \vdots \\ C_{1n} & C_{2n} & \cdots & C_{nn} \end{bmatrix} \begin{bmatrix} b_1 \\ b_2 \\ \vdots \\ b_n \end{bmatrix}$$

Multiplying the matrices out gives

$$X = \frac{1}{\det(A)} \begin{bmatrix} b_1C_{11} + b_2C_{21} + \cdots + b_nC_{n1} \\ b_1C_{12} + b_2C_{22} + \cdots + b_nC_{n2} \\ \vdots & \vdots & & \vdots \\ b_1C_{1n} + b_2C_{2n} + \cdots + b_nC_{nn} \end{bmatrix}$$

The entry in the jth row of X is therefore

$$x_j = \frac{b_1C_{1j} + b_2C_{2j} + \cdots + b_nC_{nj}}{\det(A)} \tag{2.10}$$

Now let

$$A_j = \begin{bmatrix} a_{11} & a_{12} & \cdots & a_{1j-1} & b_1 & a_{1j+1} & \cdots & a_{1n} \\ a_{21} & a_{22} & \cdots & a_{2j-1} & b_2 & a_{2j+1} & \cdots & a_{2n} \\ \vdots & \vdots & & \vdots & \vdots & \vdots & & \vdots \\ a_{n1} & a_{n2} & \cdots & a_{nj-1} & b_n & a_{nj+1} & \cdots & a_{nn} \end{bmatrix}$$

Since A_j differs from A only in the jth column, the cofactors of entries $b_1, b_2, \ldots, b_n$ in A_j are the same as the cofactors of the corresponding entries in the jth column of A. The cofactor expansion of $\det(A_j)$ along the jth column is therefore

$$\det(A_j) = b_1C_{1j} + b_2C_{2j} + \cdots + b_nC_{nj}$$

Substituting this result in (2.10) gives

$$x_j = \frac{\det(A_j)}{\det(A)}$$

Example 33

Use Cramer's Rule to solve

$$\begin{aligned} x_1 + + 2x_3 &= 6 \\ -3x_1 + 4x_2 + 6x_3 &= 30 \\ -x_1 - 2x_2 + 3x_3 &= 8 \end{aligned}$$

Solution.

$$A = \begin{bmatrix} 1 & 0 & 2 \\ -3 & 4 & 6 \\ -1 & -2 & 3 \end{bmatrix} \qquad A_1 = \begin{bmatrix} 6 & 0 & 2 \\ 30 & 4 & 6 \\ 8 & -2 & 3 \end{bmatrix}$$

$$A_2 = \begin{bmatrix} 1 & 6 & 2 \\ -3 & 30 & 6 \\ -1 & 8 & 3 \end{bmatrix} \qquad A_3 = \begin{bmatrix} 1 & 0 & 6 \\ -3 & 4 & 30 \\ -1 & -2 & 8 \end{bmatrix}$$

Therefore,

$$x_1 = \frac{\det(A_1)}{\det(A)} = \frac{-40}{44} = \frac{-10}{11} \qquad x_2 = \frac{\det(A_2)}{\det(A)} = \frac{72}{44} = \frac{18}{11}$$

$$x_3 = \frac{\det(A_3)}{\det(A)} = \frac{152}{44} = \frac{38}{11}$$

To solve a system of n equations in n unknowns by Cramer's Rule, it is necessary to evaluate $n + 1$ determinants of $n \times n$ matrices. For systems with more than three equations, Gaussian elimination is more efficient, since it is only necessary to reduce one n by $n + 1$ augmented matrix. Cramer's rule, however, gives a formula for the solution.

EXERCISE SET 2.4

1. Let

$$A = \begin{bmatrix} 1 & 6 & -3 \\ -2 & 7 & 1 \\ 3 & -1 & 4 \end{bmatrix}$$

(a) Find all the minors.
(b) Find all the cofactors.

2. Let

$$A = \begin{bmatrix} 4 & 0 & 4 & 4 \\ -1 & 0 & 1 & 1 \\ 1 & -3 & 0 & 3 \\ 6 & 3 & 14 & 2 \end{bmatrix}$$

Find
(a) M_{13} and C_{13} (b) M_{23} and C_{23} (c) M_{22} and C_{22} (d) M_{21} and C_{21}

3. Evaluate the determinant of the matrix in Exercise 1 by a cofactor expansion along
(a) the first row   (b) the first column   (c) the second row
(d) the second column   (e) the third row   (f) the third column

4. For the matrix in Exercise 1, find
(a) adj(A)   (b) A^{-1} using the method of Example 32

In Exercises 5–10 evaluate det(A) by a cofactor expansion along a row or column of your choice.

5. $A = \begin{bmatrix} 0 & 6 & 0 \\ 8 & 6 & 8 \\ 3 & 2 & 2 \end{bmatrix}$

6. $A = \begin{bmatrix} 1 & 3 & 7 \\ 2 & 0 & -8 \\ -1 & -3 & 4 \end{bmatrix}$

7. $A = \begin{bmatrix} 1 & 1 & 1 \\ k & k & k \\ k^2 & k^2 & k^2 \end{bmatrix}$

8. $A = \begin{bmatrix} k-1 & 2 & 3 \\ 2 & k-3 & 4 \\ 3 & 4 & k-4 \end{bmatrix}$

9. $A = \begin{bmatrix} 4 & 4 & 0 & 4 \\ 1 & 1 & 0 & -1 \\ 3 & 0 & -3 & 1 \\ 6 & 14 & 3 & 6 \end{bmatrix}$

10. $A = \begin{bmatrix} 4 & 3 & 1 & 9 & 2 \\ 0 & 3 & 2 & 4 & 2 \\ 0 & 3 & 4 & 6 & 4 \\ 1 & -1 & 2 & 2 & 2 \\ 0 & 0 & 3 & 3 & 3 \end{bmatrix}$

In Exercises 11–14 find A^{-1} using the method of Example 32.

11. $A = \begin{bmatrix} 0 & 1 & 2 \\ 2 & 4 & 3 \\ 3 & 7 & 6 \end{bmatrix}$

12. $A = \begin{bmatrix} 1 & 0 & 1 \\ -1 & 3 & 0 \\ 1 & 0 & 2 \end{bmatrix}$

13. $A = \begin{bmatrix} 2 & -4 & 6 \\ 0 & 1 & -1 \\ 0 & 0 & 2 \end{bmatrix}$

14. $A = \begin{bmatrix} 3 & 0 & 0 \\ 9 & 1 & 0 \\ -4 & 2 & 4 \end{bmatrix}$

15. Let

$$A = \begin{bmatrix} 1 & 3 & 1 & 1 \\ 2 & 5 & 2 & 2 \\ 1 & 3 & 8 & 9 \\ 1 & 3 & 2 & 2 \end{bmatrix}$$

   (a) Evaluate A^{-1} using the method of Example 32.
   (b) Evaluate A^{-1} using the method of Example 31 in Section 1.6.
   (c) Which method involves less computation?

In Exercises 16–21 solve by Cramer's Rule, where it applies.

16. $3x_1 - 4x_2 = -5$
    $2x_1 + x_2 = 4$

17. $4x + 5y = 2$
    $11x + y + 2z = 3$
    $x + 5y + 2z = 1$

18. $x + y - 2z = 1$
    $2x - y + z = 2$
    $x - 2y - 4z = -4$

19. $x_1 - 3x_2 + x_3 = 4$
    $2x_1 - x_2 = -2$
    $4x_1 - 3x_3 = 0$

20. $2x_1 - x_2 + x_3 - 4x_4 = -32$
    $7x_1 + 2x_2 + 9x_3 - x_4 = 14$
    $3x_1 - x_2 + x_3 + x_4 = 11$
    $x_1 + x_2 - 4x_3 - 2x_4 = -4$

21. $2x_1 - x_2 + x_3 = 8$
    $4x_1 + 3x_2 + x_3 = 7$
    $6x_1 + 2x_2 + 2x_3 = 15$

22. Use Cramer's Rule to solve for z without solving for x, y, and w.

$$4x + y + z + w = 6$$
$$3x + 7y - z + w = 1$$
$$7x + 3y - 5z + 8w = -3$$
$$x + y + z + 2w = 3$$

23. Let $AX = B$ be the system in Exercise 22.
(a) Solve by Cramer's Rule.
(b) Solve by Gauss-Jordan elimination.
(c) Which method involves fewer computations?

24. Prove that if $\det(A) = 1$ and all the entries in A are integers, then all the entries in A^{-1} are integers.

25. Let $AX = B$ be a system of n linear equations in n unknowns with integer coefficients and integer constants. Prove that if $\det(A) = 1$, the solution X has integer entries.

26. Prove that if A is an invertible upper triangular matrix, A^{-1} is upper triangular.

27. Derive the first and last cofactor expansions listed in (2.4).

28. Prove: The equation of the line through the distinct points (a_1, b_1) and (a_2, b_2) can be written

$$\begin{vmatrix} x & y & 1 \\ a_1 & b_1 & 1 \\ a_2 & b_2 & 1 \end{vmatrix} = 0$$

29. Prove: (x_1, y_1), (x_2, y_2), and (x_3, y_3) are collinear points if and only if

$$\begin{vmatrix} x_1 & y_1 & 1 \\ x_2 & y_2 & 1 \\ x_3 & y_3 & 1 \end{vmatrix} = 0$$

30. Prove: The equation of the plane through the noncollinear points (a_1, b_1, c_1), (a_2, b_2, c_2), and (a_3, b_3, c_3) can be written:

$$\begin{vmatrix} x & y & z & 1 \\ a_1 & b_1 & c_1 & 1 \\ a_2 & b_2 & c_2 & 1 \\ a_3 & b_3 & c_3 & 1 \end{vmatrix} = 0$$

SUPPLEMENTARY EXERCISES

1. Use Cramer's Rule to solve for x' and y' in terms of x and y.

$$x = \tfrac{3}{5}x' - \tfrac{4}{5}y'$$
$$y = \tfrac{4}{5}x' + \tfrac{3}{5}y'$$

2. Use Cramer's Rule to solve for x' and y' in terms of x and y.

$$x = x' \cos \theta - y' \sin \theta$$
$$y = x' \sin \theta + y' \cos \theta$$

3. By examining the determinant of the coefficient matrix, show that the following system has a nontrivial solution if and only if $\alpha = \beta$.

$$x + y + \alpha z = 0$$
$$x + y + \beta z = 0$$
$$\alpha x + \beta y + z = 0$$

4. Let A be a 3×3 matrix, each of whose entries is 1 or 0. What is the largest possible value for $|A|$?

5. (a) For the triangle below, use trigonometry to show

$$b \cos \gamma + c \cos \beta = a$$
$$c \cos \alpha + a \cos \gamma = b$$
$$a \cos \beta + b \cos \alpha = c$$

and then apply Cramer's Rule to show

$$\cos \alpha = \frac{b^2 + c^2 - a^2}{2bc}$$

(b) Use Cramer's Rule to obtain similar formulas for $\cos \beta$ and $\cos \gamma$.

6. Use determinants to show that for all real values of λ the only solution of

$$x - 2y = \lambda x$$
$$x - y = \lambda y$$

is $x = 0$, $y = 0$.

7. Prove: If A is invertible, then adj(A) is invertible and

$$[\mathrm{adj}(A)]^{-1} = \frac{1}{\det(A)} A = \mathrm{adj}(A^{-1})$$

8. Prove: If A is an $n \times n$ matrix, then

$$\det[\mathrm{adj}(A)] = [\det(A)]^{n-1}$$

9. **(For readers who have studied calculus.)** Show that if $f_1(x)$, $f_2(x)$, $g_1(x)$, and $g_2(x)$ are differentiable functions, and if

$$W = \begin{vmatrix} f_1(x) & f_2(x) \\ g_1(x) & g_2(x) \end{vmatrix}$$

then

$$\frac{dW}{dx} = \begin{vmatrix} f'_1(x) & f'_2(x) \\ g_1(x) & g_2(x) \end{vmatrix} + \begin{vmatrix} f_1(x) & f_2(x) \\ g'_1(x) & g'_2(x) \end{vmatrix}$$

10. (a) In the figure below, the area of triangle ABC is expressible as

$$\text{area } ABC = \text{area } ADEC + \text{area } CEFB - \text{area } ADFB$$

Use this and the fact that the area of a trapezoid equals $\frac{1}{2}$ the altitude times the sum of the parallel sides to show

$$\text{area } ABC = \frac{1}{2}\begin{vmatrix} x_1 & y_1 & 1 \\ x_2 & y_2 & 1 \\ x_3 & y_3 & 1 \end{vmatrix}$$

[*Note.* In the derivation of this formula, the vertices are labeled so the triangle is traced counterclockwise proceeding from (x_1, y_1) to (x_2, y_2) to (x_3, y_3). For a clockwise orientation, the determinant above yields the *negative* of the area.]

(b) Use the result in (a) to find the area of the triangle with vertices $(3, 3), (4, 0), (-2, -1)$.

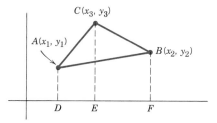

11. Prove: If the entries in each row of an $n \times n$ matrix A add up to zero, then $\det(A) = 0$. [**Hint.** Consider the product AX, where X is the $n \times 1$ matrix, each of whose entries is one.]

12. Let A be an $n \times n$ matrix and B the matrix that results when the rows of A are written in reverse order. How are $\det(A)$ and $\det(B)$ related?

13. How will A^{-1} be affected if
(a) the ith and jth rows of A are interchanged;
(b) the ith row of A is multiplied by a nonzero scalar, c;
(c) c times the ith row of A is added to the jth row?

14. Let A be an $n \times n$ matrix. Suppose that B_1 is obtained by adding the same number t to each entry in the ith row of A and B_2 is obtained by subtracting t from each entry in the ith row of A. Show that

$$\det(A) = \tfrac{1}{2}[\det(B_1) + \det(B_2)]$$

15. Without directly evaluating the determinant, show that

$$\begin{vmatrix} \sin \alpha & \cos \alpha & \sin(\alpha + \delta) \\ \sin \beta & \cos \beta & \sin(\beta + \delta) \\ \sin \gamma & \cos \gamma & \sin(\gamma + \delta) \end{vmatrix} = 0$$

16. Use the fact that 21375, 38798, 34162, 40223, and 79154 are all divisible by 19 to show that

$$\begin{vmatrix} 2 & 1 & 3 & 7 & 5 \\ 3 & 8 & 7 & 9 & 8 \\ 3 & 4 & 1 & 6 & 2 \\ 4 & 0 & 2 & 2 & 3 \\ 7 & 9 & 1 & 5 & 4 \end{vmatrix}$$

is divisible by 19 without directly evaluating the determinant.

CHAPTER THREE

Vectors in 2-Space and 3-Space

Readers familiar with the contents of this chapter can go to Chapter 4 with no loss of continuity.

3.1 INTRODUCTION TO VECTORS (GEOMETRIC)

Many physical quantities, such as area, length, mass, and temperature, are completely described once the magnitude of the quantity is given. Such quantities are called *scalars*. Other physical quantities, called ***vectors***, are not completely determined until both a magnitude and a direction are specified. For example, wind movement is usually described by giving the speed and the direction, say 20 mph northeast. The wind speed and wind direction together form a vector quantity called the wind *velocity*. Other examples of vectors are force and displacement. In this section vectors in 2-space and 3-space will be introduced geometrically, arithmetic operations on vectors will be defined, and some basic properties of these operations will be established.

Vectors can be represented geometrically as directed line segments or arrows in 2-space or 3-space; the direction of the arrow specifies the direction of the vector, and the length of the arrow describes its magnitude. The tail of the arrow is called the ***initial point*** of the vector, and the tip of the arrow the ***terminal point***. We shall denote vectors by lowercase boldface type (for instance, **a**, **k**, **v**, **w**, and **x**). When discussing vectors, we shall refer to numbers as ***scalars***. All our scalars will be real numbers and will be denoted by ordinary lowercase type (for instance, *a*, *k*, *v*, *w*, and *x*).

If, as in Figure 3.1*a*, the initial point of a vector **v** is *A* and the terminal point is *B*, we write

$$\mathbf{v} = \overrightarrow{AB}$$

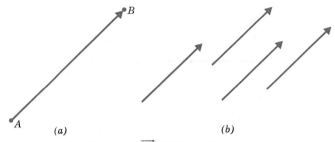

Figure 3.1 (a) The vector $\overrightarrow{AB}$. (b) Equivalent vectors.

Vectors having the same length and same direction, like those in Figure 3.1*b*, are called ***equivalent***. Since we want a vector to be determined solely by its length and direction, equivalent vectors are regarded as ***equal*** even though they may be located in different positions. If **v** and **w** are equivalent, we write

$$\mathbf{v} = \mathbf{w}$$

Definition. If **v** and **w** are any two vectors, then the ***sum* v + w** is the vector determined as follows. Position the vector **w** so that its initial point coincides with the terminal point of **v**. The vector **v** + **w** is represented by the arrow from the initial point of **v** to the terminal point of **w** (Figure 3.2*a*).

In Figure 3.2*b* we have constructed two sums, **v** + **w** (blue arrows) and **w** + **v** (white arrows). It is evident that

$$\mathbf{v} + \mathbf{w} = \mathbf{w} + \mathbf{v}$$

and that the sum coincides with the diagonal of the parallelogram determined by **v** and **w** when these vectors are located so they have the same initial point.

The vector of length zero is called the ***zero vector*** and is denoted by **0**. We define

$$\mathbf{0} + \mathbf{v} = \mathbf{v} + \mathbf{0} = \mathbf{v}$$

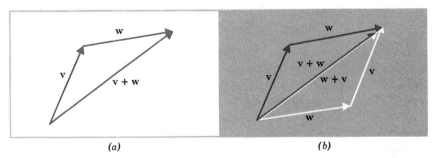

(a)   (b)

Figure 3.2

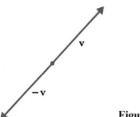

Figure 3.3

for every vector **v**. Since there is no natural direction for the zero vector, we shall agree that it can be assigned any direction that is convenient for the problem being considered. If **v** is any nonzero vector, then −**v** the *negative* of **v**, is defined to be the vector having the same magnitude as **v**, but oppositely directed (Figure 3.3). This vector has the property

$$\mathbf{v} + (-\mathbf{v}) = \mathbf{0}$$

(Why?) In addition, we define −**0** = **0**.

Definition. If **v** and **w** are any two vectors, then *subtraction* of **w** from **v** is defined by

$$\mathbf{v} - \mathbf{w} = \mathbf{v} + (-\mathbf{w})$$

(Figure 3.4*a*).

To obtain the difference **v** − **w** without constructing −**w**, position **v** and **w** so their initial points coincide; the vector from the terminal point of **w** to the terminal point of **v** is then the vector **v** − **w** (Figure 3.4*b*).

Definition. If **v** is a nonzero vector and k is a nonzero real number (scalar), then the *product* $k\mathbf{v}$ is defined to be the vector whose length is $|k|$ times the length of **v** and whose direction is the same as that of **v** if $k > 0$ and opposite to that of **v** if $k < 0$. We define $k\mathbf{v} = \mathbf{0}$ if $k = 0$ *or* **v** = **0**.

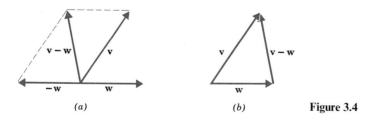

(a)   (b)   **Figure 3.4**

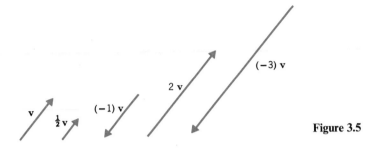

Figure 3.5

Figure 3.5 illustrates the relation between a vector **v** and the vectors $\frac{1}{2}$**v**, (-1)**v**, 2**v**, and (-3)**v**. Note that the vector (-1)**v** has the same length as **v**, but is oppositely directed. Thus, (-1)**v** is just the negative of **v**; that is,

$$(-1)\mathbf{v} = -\mathbf{v}$$

Problems involving vectors can often be simplified by introducing a rectangular coordinate system. For the moment we shall restrict the discussion to vectors in 2-space (the plane). Let **v** be any vector in the plane, and assume, as in Figure 3.6, that **v** has been positioned so its initial point is at the origin of a rectangular coordinate system. The coordinates (v_1, v_2) of the terminal point of **v** are called the ***components of*** **v**, and we write

$$\mathbf{v} = (v_1, v_2)$$

If equivalent vectors, **v** and **w**, are located so their initial points fall at the origin, then it is obvious that their terminal points must coincide (since the vectors have the same length and direction). The vectors thus have the same components. Conversely, vectors with the same components are equivalent since they have the same length and same direction. In summary, two vectors

$$\mathbf{v} = (v_1, v_2) \qquad \text{and} \qquad \mathbf{w} = (w_1, w_2)$$

are equivalent if and only if

$$v_1 = w_1 \qquad \text{and} \qquad v_2 = w_2$$

The operations of vector addition and multiplication by scalars are very easy to carry out in terms of components. As illustrated in Figure 3.7, if

$$\mathbf{v} = (v_1, v_2) \qquad \text{and} \qquad \mathbf{w} = (w_1, w_2)$$

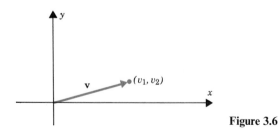

Figure 3.6

then

$$\mathbf{v} + \mathbf{w} = (v_1 + w_1, v_2 + w_2) \tag{3.1a}$$

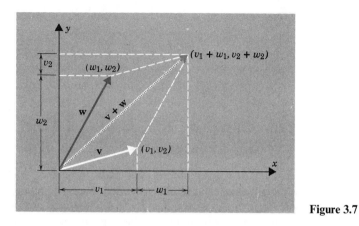

Figure 3.7

If $\mathbf{v} = (v_1, v_2)$ and k is any scalar, then by using a geometric argument involving similar triangles, it can be shown (Exercise 14) that

$$k\mathbf{v} = (kv_1, kv_2) \tag{3.1b}$$

(Figure 3.8).

Thus, for example, if $\mathbf{v} = (1, -2)$ and $\mathbf{w} = (7, 6)$, then

$$\mathbf{v} + \mathbf{w} = (1, -2) + (7, 6) = (1 + 7, -2 + 6) = (8, 4)$$

and

$$4\mathbf{v} = 4(1, -2) = (4(1), 4(-2)) = (4, -8)$$

Since $\mathbf{v} - \mathbf{w} = \mathbf{v} + (-1)\mathbf{w}$, it follows from formulas (3.1a) and (3.1b) that

$$\mathbf{v} - \mathbf{w} = (v_1 - w_1, v_2 - w_2)$$

(Verify.)

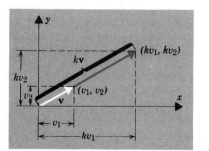

Figure 3.8

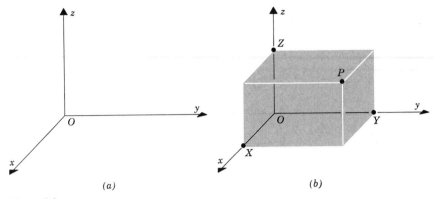

(a)   (b)

Figure 3.9

Just as vectors in the plane can be described by pairs of real numbers, vectors in 3-space can be described by triples of real numbers by introducing a **rectangular coordinate** system. To construct such a coordinate system, select a point O, called the **origin**, and choose three mutually perpendicular lines, called **coordinate axes**, passing through the origin. Label these axes x, y, and z and select a positive direction for each coordinate axis as well as a unit of length for measuring distances (Figure 3.9a). Each pair of coordinate axes determines a plane called a **coordinate plane**. These are referred to as the **xy-plane**, the **xz-plane**, and the **yz-plane**. To each point P in 3-space we assign a triple of numbers (x, y, z) called the **coordinates of P** as follows. Pass three planes through P parallel to the coordinate planes, and denote the points of intersections of these planes with the three coordinate axes by X, Y, and Z (Figure 3.9b). The coordinates of P are defined to be the signed lengths

$$x = OX \qquad y = OY \qquad z = OZ$$

In Figure 3.10 we have constructed the points whose coordinates are $(4, 5, 6)$ and $(-3, 2, -4)$.

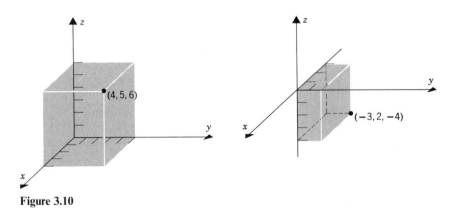

Figure 3.10

Rectangular coordinate systems in 3-space fall into two categories, *left-handed* and *right-handed*. A right-handed system has the property that an ordinary screw pointed in the positive direction on the *z*-axis would be advanced if the positive *x*-axis is rotated 90° toward the positive *y*-axis (Figure 3.11*a*). The system is left-handed if the screw would retract (Figure 3.11*b*).

REMARK. In this book we shall use only right-handed coordinate systems.

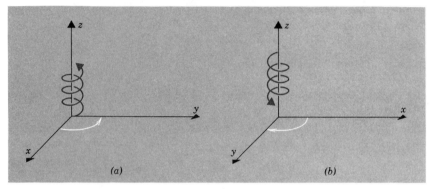

Figure 3.11 (*a*) Right-handed. (*b*) Left-handed.

If, as in Figure 3.12, a vector **v** in 3-space is located so its initial point is at the origin of a rectangular coordinate system, then the coordinates of the terminal point are called the *components* of **v**, and we write

$$\mathbf{v} = (v_1, v_2, v_3)$$

If $\mathbf{v} = (v_1, v_2, v_3)$ and $\mathbf{w} = (w_1, w_2, w_3)$ are two vectors in 3-space, then arguments similar to those used for vectors in a plane can be used to establish the following results.

v and **w** are equivalent if and only if $v_1 = w_1$, $v_2 = w_2$, and $v_3 = w_3$
$\mathbf{v} + \mathbf{w} = (v_1 + w_1, v_2 + w_2, v_3 + w_3)$
$k\mathbf{v} = (kv_1, kv_2, kv_3)$ where k is any scalar

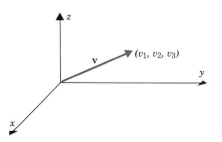

Figure 3.12

Example 1

If $v = (1, -3, 2)$ and $w = (4, 2, 1)$, then

$$v + w = (5, -1, 3), \quad 2v = (2, -6, 4), \quad -w = (-4, -2, -1),$$
$$v - w = v + (-w) = (-3, -5, 1) \quad \blacktriangle$$

Sometimes a vector is positioned so that its initial point is not at the origin. If the vector $\overrightarrow{P_1 P_2}$ has initial point $P_1(x_1, y_1, z_1)$ and terminal point $P_2(x_2, y_2, z_2)$, then

$$\overrightarrow{P_1 P_2} = (x_2 - x_1, y_2 - y_1, z_2 - z_1)$$

That is, the components of $\overrightarrow{P_1 P_2}$ are obtained by subtracting the coordinates of the initial point from the coordinates of the terminal point. This may be seen using Figure 3.13: the vector $\overrightarrow{P_1 P_2}$ is the difference of vectors $\overrightarrow{OP_2}$ and $\overrightarrow{OP_1}$, so

$$\overrightarrow{P_1 P_2} = \overrightarrow{OP_2} - \overrightarrow{OP_1} = (x_2, y_2, z_2) - (x_1, y_1, z_1) = (x_2 - x_1, y_2 - y_1, z_2 - z_1)$$

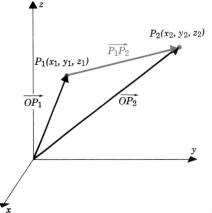

Figure 3.13

Example 2

The components of the vector $v = \overrightarrow{P_1 P_2}$ with initial point $P_1(2, -1, 4)$ and terminal point $P_2(7, 5, -8)$ are

$$v = (7 - 2, 5 - (-1), (-8) - 4) = (5, 6, -12)$$

Analogously, in 2-space the vector with initial point $P_1(x_1, y_1)$ and terminal point $P_2(x_2, y_2)$ is $\overrightarrow{P_1 P_2} = (x_2 - x_1, y_2 - y_1)$. $\quad \blacktriangle$

Example 3

The solutions to many problems can be simplified by translating the coordinate axes to obtain new axes parallel to the original ones.

In Figure 3.14*a* we have translated the xy-coordinate axes to obtain an $x'y'$-coordinate system whose origin O' is at the point $(x, y) = (k, l)$. A point P in 2-space now has both (x, y) coordinates and (x', y') coordinates. To see how the two are related, consider the vector $\overrightarrow{O'P}$ (Figure 3.14*b*). In the xy-system its initial point is at (k, l) and its terminal point is at (x, y), so $\overrightarrow{O'P} = (x - k, y - l)$. In the $x'y'$-system its initial point is at $(0, 0)$ and its terminal point is at (x', y'), so $\overrightarrow{O'P} = (x', y')$. Therefore

$$x' = x - k \qquad y' = y - l$$

These are called the **translation equations**.

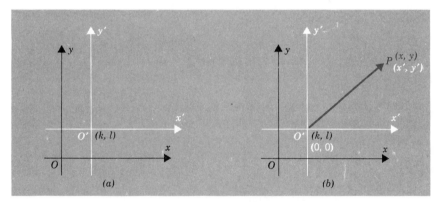

Figure 3.14

As an example, if the new origin is at $(k, l) = (4, 1)$ and the xy-coordinates of a point P are $(2, 0)$, then $x'y'$-coordinates of P are $x' = 2 - 4 = -2$ and $y' = 0 - 1 = -1$.

In 3-space the translation equations are

$$x' = x - k \qquad y' = y - l \qquad z' = z - m$$

where (k, l, m) are the xyz-coordinates of the new origin.

EXERCISE SET 3.1

1. Draw a right-handed coordinate system and locate the points whose coordinates are
   (a) $(2, 3, 4)$ (b) $(-2, 3, 4)$ (c) $(2, -3, 4)$ (d) $(2, 3, -4)$
   (e) $(-2, -3, 4)$ (f) $(-2, 3, -4)$ (g) $(2, -3, -4)$ (h) $(-2, -3, -4)$
   (i) $(0, 2, 0)$ (j) $(0, 0, -2)$ (k) $(2, 0, 2)$ (l) $(-2, 0, 0)$

2. Sketch the following vectors with the initial points located at the origin:
   (a) $\mathbf{v}_1 = (2, 5)$ (b) $\mathbf{v}_2 = (-3, 7)$ (c) $\mathbf{v}_3 = (-5, -4)$
   (d) $\mathbf{v}_4 = (6, -2)$ (e) $\mathbf{v}_5 = (2, 0)$ (f) $\mathbf{v}_6 = (0, -8)$
   (g) $\mathbf{v}_7 = (2, 3, 4)$ (h) $\mathbf{v}_8 = (2, 0, 2)$ (i) $\mathbf{v}_9 = (0, 0, -2)$

3. Find the components of the vector having initial point P_1 and terminal point P_2.
   (a) $P_1(3, 5)$, $P_2(2, 8)$ (b) $P_1(7, -2)$, $P_2(0, 0)$
   (c) $P_1(6, 5, 8)$, $P_2(8, -7, -3)$ (d) $P_1(0, 0, 0)$; $P_2(-8, 7, 4)$

4. Find a vector with initial point $P(2, -1, 4)$ that has the same direction as $\mathbf{v} = (7, 6, -3)$.

5. Find a vector, oppositely directed to $\mathbf{v} = (-2, 4, -1)$, having terminal point $Q(2, 0, -7)$.

6. Let $\mathbf{u} = (1, 2, 3)$, $\mathbf{v} = (2, -3, 1)$, and $\mathbf{w} = (3, 2, -1)$. Find the components of
   (a) $\mathbf{u} - \mathbf{w}$ (b) $7\mathbf{v} + 3\mathbf{w}$ (c) $-\mathbf{w} + \mathbf{v}$
   (d) $3(\mathbf{u} - 7\mathbf{v})$ (e) $-3\mathbf{v} - 8\mathbf{w}$ (f) $2\mathbf{v} - (\mathbf{u} + \mathbf{w})$

7. Let $\mathbf{u}$, $\mathbf{v}$, and $\mathbf{w}$ be the vectors in Exercise 6. Find the components of the vector $\mathbf{x}$ that satisfies $2\mathbf{u} - \mathbf{v} + \mathbf{x} = 7\mathbf{x} + \mathbf{w}$.

8. Let $\mathbf{u}$, $\mathbf{v}$, and $\mathbf{w}$ be the vectors in Exercise 6. Find scalars c_1, c_2, and c_3 such that $c_1\mathbf{u} + c_2\mathbf{v} + c_3\mathbf{w} = (6, 14, -2)$.

9. Show that there do not exist scalars c_1, c_2, and c_3 such that

$$c_1(1, 2, -3) + c_2(5, 7, 1) + c_3(6, 9, -2) = (4, 5, 0).$$

10. Find all scalars c_1, c_2, and c_3 such that

$$c_1(2, 7, 8) + c_2(1, -1, 3) + c_3(3, 6, 11) = (0, 0, 0).$$

11. Let P be the point $(2, 3, -2)$ and Q the point $(7, -4, 1)$.
    (a) Find the midpoint of the line segment connecting P and Q.
    (b) Find the point on the line segment connecting P and Q that is 3/4 of the way from P to Q.

12. Suppose an xy-coordinate system is translated to obtain an $x'y'$-coordinate system whose origin O' has xy-coordinates $(2, -3)$.
    (a) Find the $x'y'$-coordinates of the point P whose xy-coordinates are $(7, 5)$.
    (b) Find the xy-coordinates of the point Q whose $x'y'$-coordinates are $(-3, 6)$.
    (c) Draw the xy and $x'y'$-coordinate axes and locate the points P and Q.

13. Suppose an xyz-coordinate system is translated to obtain an $x'y'z'$-coordinate system. Let $\mathbf{v}$ be a vector whose components are $\mathbf{v} = (v_1, v_2, v_3)$ in the xyz-system. Show that $\mathbf{v}$ has the same components in the $x'y'z'$-system.

14. Prove geometrically that if $\mathbf{v} = (v_1, v_2)$, then $k\mathbf{v} = (kv_1, kv_2)$. (Restrict the proof to the case $k > 0$ illustrated in Figure 3.8. The complete proof would involve many cases depending on the quadrant in which the vector falls and the sign of k.)

3.2 NORM OF A VECTOR; VECTOR ARITHMETIC

In this section we establish the basic rules of vector arithmetic.

Theorem 1. *If* $\mathbf{u}$, $\mathbf{v}$, *and* $\mathbf{w}$ *are vectors in 2- or 3-space and k and l are scalars, then the following relationships hold.*

(*a*) $\mathbf{u} + \mathbf{v} = \mathbf{v} + \mathbf{u}$	(*e*) $k(l\mathbf{u}) = (kl)\mathbf{u}$
(*b*) $(\mathbf{u} + \mathbf{v}) + \mathbf{w} = \mathbf{u} + (\mathbf{v} + \mathbf{w})$	(*f*) $k(\mathbf{u} + \mathbf{v}) = k\mathbf{u} + k\mathbf{v}$
(*c*) $\mathbf{u} + \mathbf{0} = \mathbf{0} + \mathbf{u} = \mathbf{u}$	(*g*) $(k + l)\mathbf{u} = k\mathbf{u} + l\mathbf{u}$
(*d*) $\mathbf{u} + (-\mathbf{u}) = \mathbf{0}$	(*h*) $1\mathbf{u} = \mathbf{u}$

Before discussing the proof, we note that we have developed two approaches to vectors: *geometric*, in which vectors are represented by arrows or directed line segments, and *analytic*, in which vectors are represented by pairs or triples of numbers called components. As a consequence, the results in Theorem 1 can be established either geometrically or analytically. To illustrate, we shall prove part (*b*) both ways. The remaining proofs are left as exercises.

Proof of part (*b*) (*analytic*). We shall give the proof for vectors in 3-space. The proof for 2-space is similar. If $\mathbf{u} = (u_1, u_2, u_3)$, $\mathbf{v} = (v_1, v_2, v_3)$, and $\mathbf{w} = (w_1, w_2, w_3)$ then

$$
\begin{aligned}
(\mathbf{u} + \mathbf{v}) + \mathbf{w} &= \left[(u_1, u_2, u_3) + (v_1, v_2, v_3) \right] + (w_1, w_2, w_3) \\
&= (u_1 + v_1, u_2 + v_2, u_3 + v_3) + (w_1, w_2, w_3) \\
&= \left([u_1 + v_1] + w_1, [u_2 + v_2] + w_2, [u_3 + v_3] + w_3 \right) \\
&= \left(u_1 + [v_1 + w_1], u_2 + [v_2 + w_2], u_3 + [v_3 + w_3] \right) \\
&= (u_1, u_2, u_3) + (v_1 + w_1, v_2 + w_2, v_3 + w_3) \\
&= \mathbf{u} + (\mathbf{v} + \mathbf{w})
\end{aligned}
$$

Proof of part (*b*) (*geometric*). Let $\mathbf{u}$, $\mathbf{v}$, and $\mathbf{w}$ be represented by $\overrightarrow{PQ}$, $\overrightarrow{QR}$, and $\overrightarrow{RS}$ as shown in Figure 3.15. Then

$$
\mathbf{v} + \mathbf{w} = \overrightarrow{QS} \quad \text{and} \quad \mathbf{u} + (\mathbf{v} + \mathbf{w}) = \overrightarrow{PS}
$$

Also,

$$
\mathbf{u} + \mathbf{v} = \overrightarrow{PR} \quad \text{and} \quad (\mathbf{u} + \mathbf{v}) + \mathbf{w} = \overrightarrow{PS}
$$

Therefore,

$$\mathbf{u} + (\mathbf{v} + \mathbf{w}) = (\mathbf{u} + \mathbf{v}) + \mathbf{w}$$

REMARK. In light of part (*b*) of this theorem, the symbol $\mathbf{u} + \mathbf{v} + \mathbf{w}$ is unambiguous since the same result is obtained no matter where parentheses are inserted. Moreover, if the vectors $\mathbf{u}$, $\mathbf{v}$, and $\mathbf{w}$ are placed "tip to tail," then the sum $\mathbf{u} + \mathbf{v} + \mathbf{w}$ is the vector from the initial point of $\mathbf{u}$ to the terminal point of $\mathbf{w}$ (Figure 3.15).

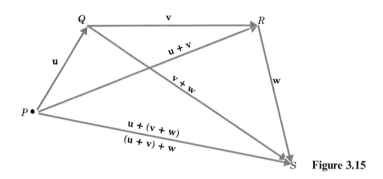

Figure 3.15

The *length* of a vector $\mathbf{v}$ is often called the *norm* of $\mathbf{v}$ and is denoted by $\|\mathbf{v}\|$. It follows from the theorem of Pythagoras that the norm of a vector $\mathbf{v} = (v_1, v_2)$ in 2-space is

$$\|\mathbf{v}\| = \sqrt{v_1^2 + v_2^2}$$

(Figure 3.16*a*). Let $\mathbf{v} = (v_1, v_2, v_3)$ be a vector in 3-space. Using Figure 3.16*b* and two applications of the theorem of Pythagoras, we obtain

$$\|\mathbf{v}\|^2 = (OR)^2 + (RP)^2$$
$$= (OQ)^2 + (OS)^2 + (RP)^2$$
$$= v_1^2 + v_2^2 + v_3^2$$

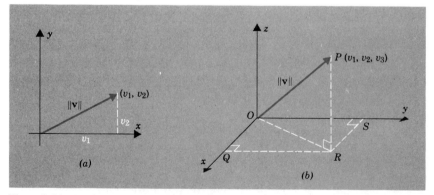

Figure 3.16

Thus,

$$\|\mathbf{v}\| = \sqrt{v_1^2 + v_2^2 + v_3^2} \qquad (3.2)$$

If $P_1(x_1, y_1, z_1)$ and $P_2(x_2, y_2, z_2)$ are two points in 3-space, then the distance d between them is the norm of the vector $\overrightarrow{P_1 P_2}$ (Figure 3.17). Since

$$\overrightarrow{P_1 P_2} = (x_2 - x_1, y_2 - y_1, z_2 - z_1)$$

it follows from (3.2) that

$$d = \sqrt{(x_2 - x_1)^2 + (y_2 - y_1)^2 + (z_2 - z_1)^2}$$

Similarly, if $P_1(x_1, y_1)$ and $P_2(x_2, y_2)$ are points in 2-space, then the distance between them is given by

$$d = \sqrt{(x_2 - x_1)^2 + (y_2 - y_1)^2}$$

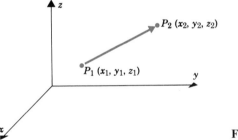

Figure 3.17

Example 4

The norm of the vector $\mathbf{v} = (-3, 2, 1)$ is

$$\|\mathbf{v}\| = \sqrt{(-3)^2 + (2)^2 + (1)^2} = \sqrt{14}$$

The distance d between the points $P_1(2, -1, -5)$ and $P_2(4, -3, 1)$ is

$$d = \sqrt{(4 - 2)^2 + (-3 + 1)^2 + (1 + 5)^2} = \sqrt{44} = 2\sqrt{11}$$

EXERCISE SET 3.2

1. Compute the norm of **v** when
  (a) $\mathbf{v} = (3, 4)$ (b) $\mathbf{v} = (-1, 7)$ (c) $\mathbf{v} = (0, -3)$
  (d) $\mathbf{v} = (1, 1, 1)$ (e) $\mathbf{v} = (-8, 7, 4)$ (f) $\mathbf{v} = (9, 0, 0)$

2. Compute the distance between P_1 and P_2.
   (a) $P_1(2, 3)$, $P_2(4, 6)$ (b) $P_1(-2, 7)$, $P_2(0, -3)$
   (c) $P_1(8, -4, 2)$, $P_2(-6, -1, 0)$ (d) $P_1(1, 1, 1)$, $P_2(6, -7, 3)$

3. Let $\mathbf{u} = (1, -3, 2)$, $\mathbf{v} = (1, 1, 0)$, and $\mathbf{w} = (2, 2, -4)$. Find
   (a) $\|\mathbf{u} + \mathbf{v}\|$ (b) $\|\mathbf{u}\| + \|\mathbf{v}\|$ (c) $\|-2\mathbf{u}\| + 2\|\mathbf{u}\|$

   (d) $\|3\mathbf{u} - 5\mathbf{v} + \mathbf{w}\|$ (e) $\dfrac{1}{\|\mathbf{w}\|}\mathbf{w}$ (f) $\left\|\dfrac{1}{\|\mathbf{w}\|}\mathbf{w}\right\|$

4. Find all scalars k such that $\|k\mathbf{v}\| = 3$, where $\mathbf{v} = (1, 2, 4)$.

5. Verify parts (b), (e), (f), and (g) of Theorem 1 for $\mathbf{u} = (1, -3, 7)$, $\mathbf{v} = (6, 6, 9)$, $\mathbf{w} = (-8, 1, 2)$, $k = -3$, and $l = 6$.

6. Show that if $\mathbf{v}$ is nonzero, then $\dfrac{1}{\|\mathbf{v}\|}\mathbf{v}$ has norm 1.

7. Use Exercise 6 to find a vector of norm 1 having the same direction as $\mathbf{v} = (1, 1, 1)$.

8. Let $\mathbf{p}_0 = (x_0, y_0, z_0)$ and $\mathbf{p} = (x, y, z)$. Describe the set of all points (x, y, z) for which $\|\mathbf{p} - \mathbf{p}_0\| = 1$.

9. Prove geometrically that if $\mathbf{u}$ and $\mathbf{v}$ are vectors in 2- or 3-space, then $\|\mathbf{u} + \mathbf{v}\| \leq \|\mathbf{u}\| + \|\mathbf{v}\|$.

10. Prove parts (a), (c), and (e) of Theorem 1 analytically.

11. Prove parts (d), (g), and (h) of Theorem 1 analytically.

12. Prove part (f) of Theorem 1 geometrically.

3.3 DOT PRODUCT; PROJECTIONS

In this section we introduce a way of multiplying vectors in 2-space and 3-space. Arithmetic properties of this multiplication are established, and some applications are given.

Let $\mathbf{u}$ and $\mathbf{v}$ be two nonzero vectors in 2-space or 3-space, and assume these vectors have been positioned so their initial points coincide. By the *angle between* **u** *and* **v**, we shall mean the angle θ determined by $\mathbf{u}$ and $\mathbf{v}$ that satisfies $0 \leq \theta \leq \pi$ (Figure 3.18).

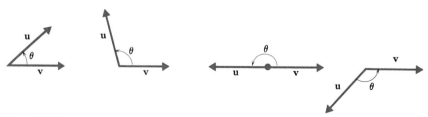

Figure 3.18

Definition. If **u** and **v** are vectors in 2-space or 3-space and θ is the angle between **u** and **v**, then the *dot product* or *Euclidean inner product* **u · v** is defined by

$$\mathbf{u} \cdot \mathbf{v} = \begin{cases} \|\mathbf{u}\| \, \|\mathbf{v}\| \cos \theta & \text{if } \mathbf{u} \neq \mathbf{0} \text{ and } \mathbf{v} \neq \mathbf{0} \\ 0 & \text{if } \mathbf{u} = \mathbf{0} \text{ or } \mathbf{v} = \mathbf{0} \end{cases} \tag{3.3}$$

Example 5

As shown in Figure 3.19, the angle between the vectors **u** = (0, 0, 1) and **v** = (0, 2, 2) is 45°. Thus,

$$\mathbf{u} \cdot \mathbf{v} = \|\mathbf{u}\| \, \|\mathbf{v}\| \cos \theta = (\sqrt{0^2 + 0^2 + 1^2})(\sqrt{0^2 + 2^2 + 2^2}) \left(\frac{1}{\sqrt{2}} \right) = 2$$

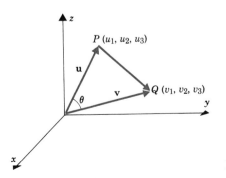

Figure 3.19

For purposes of computation, it is desirable to have a formula that expresses the dot product of two vectors in terms of the components of the vectors. We will derive such a formula for vectors in 3-space; the derivation for vectors in 2-space is similar.

Let **u** = (u_1, u_2, u_3) and **v** = (v_1, v_2, v_3) be two nonzero vectors. If, as in Figure 3.20, θ is the angle between **u** and **v**, then the law of cosines yields

$$\|\overrightarrow{PQ}\|^2 = \|\mathbf{u}\|^2 + \|\mathbf{v}\|^2 - 2\|\mathbf{u}\| \, \|\mathbf{v}\| \cos \theta \tag{3.4}$$

Figure 3.20

Since $\overrightarrow{PQ} = \mathbf{v} - \mathbf{u}$, we can rewrite (3.4) as

$$\|\mathbf{u}\| \, \|\mathbf{v}\| \cos \theta = \tfrac{1}{2}(\|\mathbf{u}\|^2 + \|\mathbf{v}\|^2 - \|\mathbf{v} - \mathbf{u}\|^2)$$

or

$$\mathbf{u} \cdot \mathbf{v} = \tfrac{1}{2}(\|\mathbf{u}\|^2 + \|\mathbf{v}\|^2 - \|\mathbf{v} - \mathbf{u}\|^2)$$

Substituting

$$\|\mathbf{u}\|^2 = u_1^2 + u_2^2 + u_3^2 \qquad \|\mathbf{v}\|^2 = v_1^2 + v_2^2 + v_3^2$$

and

$$\|\mathbf{v} - \mathbf{u}\|^2 = (v_1 - u_1)^2 + (v_2 - u_2)^2 + (v_3 - u_3)^2$$

we obtain after simplifying

$$\mathbf{u} \cdot \mathbf{v} = u_1 v_1 + u_2 v_2 + u_3 v_3 \tag{3.5a}$$

If $\mathbf{u} = (u_1, u_2)$ and $\mathbf{v} = (v_1, v_2)$ are two vectors in 2-space, then the corresponding formula is

$$\mathbf{u} \cdot \mathbf{v} = u_1 v_1 + u_2 v_2 \tag{3.5b}$$

If $\mathbf{u}$ and $\mathbf{v}$ are nonzero vectors, then Formula (3.3) can be written as

$$\cos \theta = \frac{\mathbf{u} \cdot \mathbf{v}}{\|\mathbf{u}\| \, \|\mathbf{v}\|} \tag{3.6}$$

Example 6

Consider the vectors

$$\mathbf{u} = (2, -1, 1) \quad \text{and} \quad \mathbf{v} = (1, 1, 2)$$

Find $\mathbf{u} \cdot \mathbf{v}$ and determine the angle θ between $\mathbf{u}$ and $\mathbf{v}$.

Solution.

$$\mathbf{u} \cdot \mathbf{v} = u_1 v_1 + u_2 v_2 + u_3 v_3 = (2)(1) + (-1)(1) + (1)(2) = 3$$

For the given vectors we have $\|\mathbf{u}\| = \|\mathbf{v}\| = \sqrt{6}$, so that from (3.6)

$$\cos \theta = \frac{\mathbf{u} \cdot \mathbf{v}}{\|\mathbf{u}\| \, \|\mathbf{v}\|} = \frac{3}{\sqrt{6}\sqrt{6}} = \frac{1}{2}$$

Thus, $\theta = 60°$.

Example 7

Find the angle between a diagonal of a cube and one of its edges.

Solution. Let k be the length of an edge and introduce a coordinate system as shown in Figure 3.21.

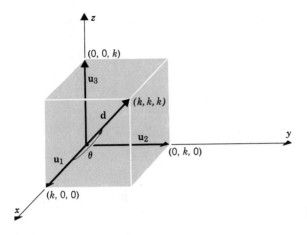

Figure 3.21

If we let $\mathbf{u}_1 = (k, 0, 0)$, $\mathbf{u}_2 = (0, k, 0)$, and $\mathbf{u}_3 = (0, 0, k)$, then the vector

$$\mathbf{d} = (k, k, k) = \mathbf{u}_1 + \mathbf{u}_2 + \mathbf{u}_3$$

is a diagonal of the cube. The angle θ between $\mathbf{d}$ and the edge $\mathbf{u}_1$ satisfies

$$\cos \theta = \frac{\mathbf{u}_1 \cdot \mathbf{d}}{\|\mathbf{u}_1\| \, \|\mathbf{d}\|} = \frac{k^2}{(k)(\sqrt{3k^2})} = \frac{1}{\sqrt{3}}$$

Thus,

$$\theta = \cos^{-1}\left(\frac{1}{\sqrt{3}}\right) \approx 54°44'$$

The next theorem shows how the dot product can be used to obtain information about the angle between two vectors; it also establishes an important relationship between the norm and the dot product.

Theorem 2. *Let* $\mathbf{u}$ *and* $\mathbf{v}$ *be vectors in 2- or 3-space.*

(a) $\mathbf{v} \cdot \mathbf{v} = \|\mathbf{v}\|^2$; *that is,* $\|\mathbf{v}\| = (\mathbf{v} \cdot \mathbf{v})^{1/2}$

(b) If $\mathbf{u}$ *and* $\mathbf{v}$ *are nonzero vectors and* θ *is the angle between them, then*

θ *is acute*	*if and only if*	$\mathbf{u} \cdot \mathbf{v} > 0$
θ *is obtuse*	*if and only if*	$\mathbf{u} \cdot \mathbf{v} < 0$
$\theta = \pi/2$	*if and only if*	$\mathbf{u} \cdot \mathbf{v} = 0$

Proof.

(a) Since the angle θ between $\mathbf{v}$ and $\mathbf{v}$ is 0, we have

$$\mathbf{v} \cdot \mathbf{v} = \|\mathbf{v}\| \, \|\mathbf{v}\| \cos \theta = \|\mathbf{v}\|^2 \cos 0 = \|\mathbf{v}\|^2$$

(b) Since $\|\mathbf{u}\| > 0$, $\|\mathbf{v}\| > 0$, and $\mathbf{u} \cdot \mathbf{v} = \|\mathbf{u}\| \, \|\mathbf{v}\| \cos \theta$, $\mathbf{u} \cdot \mathbf{v}$ has the same sign as $\cos \theta$. Since θ satisfies $0 \leq \theta \leq \pi$, the angle θ is acute if and only if $\cos \theta > 0$; θ is obtuse if and only if $\cos \theta < 0$; and $\theta = \pi/2$ if and only if $\cos \theta = 0$. ∎

Example 8

If $\mathbf{u} = (1, -2, 3)$, $\mathbf{v} = (-3, 4, 2)$, and $\mathbf{w} = (3, 6, 3)$, then

$$\mathbf{u} \cdot \mathbf{v} = (1)(-3) + (-2)(4) + (3)(2) = -5$$
$$\mathbf{v} \cdot \mathbf{w} = (-3)(3) + (4)(6) + (2)(3) = 21$$
$$\mathbf{u} \cdot \mathbf{w} = (1)(3) + (-2)(6) + (3)(3) = 0$$

Therefore, $\mathbf{u}$ and $\mathbf{v}$ make an obtuse angle, $\mathbf{v}$ and $\mathbf{w}$ make an acute angle, and $\mathbf{u}$ and $\mathbf{w}$ are perpendicular. ▲

Perpendicular vectors are also called **orthogonal** vectors. In light of Theorem 2(b), two *nonzero* vectors are orthogonal if and only if their dot product is zero. If we agree to consider $\mathbf{u}$ and $\mathbf{v}$ to be perpendicular when either or both of these vectors is $\mathbf{0}$, then we can state without exception that *two vectors* $\mathbf{u}$ *and* $\mathbf{v}$ *are orthogonal (perpendicular) if and only if* $\mathbf{u} \cdot \mathbf{v} = 0$. To indicate that $\mathbf{u}$ and $\mathbf{v}$ are orthogonal vectors we write $\mathbf{u} \perp \mathbf{v}$.

Example 9

Show that in 2-space the nonzero vector $\mathbf{n} = (a, b)$ is perpendicular to the line $ax + by + c = 0$.

Solution. Let $P_1(x_1, y_1)$ and $P_2(x_2, y_2)$ be distinct points on the line, so that

$$ax_1 + by_1 + c = 0$$
$$ax_2 + by_2 + c = 0 \tag{3.7}$$

Since the vector $\overrightarrow{P_1 P_2} = (x_2 - x_1, y_2 - y_1)$ runs along the line, we need only show that $\mathbf{n}$ and $\overrightarrow{P_1 P_2}$ are perpendicular. But on subtracting the equations in (3.7) we obtain

$$a(x_2 - x_1) + b(y_2 - y_1) = 0$$

which can be expressed in the form

$$(a, b) \cdot (x_2 - x_1, y_2 - y_1) = 0$$

or

$$\mathbf{n} \cdot \overrightarrow{P_1 P_2} = 0$$

so that $\mathbf{n}$ and $\overrightarrow{P_1 P_2}$ are perpendicular. ▲

The following theorem lists the most important properties of the dot product. They are useful in calculations involving vectors.

Theorem 3. *If* **u**, **v**, *and* **w** *are vectors in 2- or 3-space and k is a scalar, then*

(*a*) $\mathbf{u} \cdot \mathbf{v} = \mathbf{v} \cdot \mathbf{u}$
(*b*) $\mathbf{u} \cdot (\mathbf{v} + \mathbf{w}) = \mathbf{u} \cdot \mathbf{v} + \mathbf{u} \cdot \mathbf{w}$
(*c*) $k(\mathbf{u} \cdot \mathbf{v}) = (k\mathbf{u}) \cdot \mathbf{v} = \mathbf{u} \cdot (k\mathbf{v})$
(*d*) $\mathbf{v} \cdot \mathbf{v} > 0$ if $\mathbf{v} \neq \mathbf{0}$ *and* $\mathbf{v} \cdot \mathbf{v} = 0$ *if* $\mathbf{v} = \mathbf{0}$

Proof. We shall prove (*c*) for vectors in 3-space and leave the remaining proofs as exercises. Let $\mathbf{u} = (u_1, u_2, u_3)$ and $\mathbf{v} = (v_1, v_2, v_3)$; then

$$k(\mathbf{u} \cdot \mathbf{v}) = k(u_1 v_1 + u_2 v_2 + u_3 v_3)$$
$$= (ku_1)v_1 + (ku_2)v_2 + (ku_3)v_3$$
$$= (k\mathbf{u}) \cdot \mathbf{v}$$

Similarly,

$$k(\mathbf{u} \cdot \mathbf{v}) = \mathbf{u} \cdot (k\mathbf{v})$$

In many applications it is of interest to "decompose" a vector **u** into a sum of two terms, one parallel to a specified nonzero vector **a** and the other perpendicular to **a**. If **u** and **a** are positioned so their initial points coincide at a point Q, we can decompose the vector **u** as follows (Figure 3.22): Drop a perpendicular from the tip of **u** to the line through **a**, and construct the vector $\mathbf{w}_1$ from Q to the foot of this perpendicular. Next form the difference

$$\mathbf{w}_2 = \mathbf{u} - \mathbf{w}_1$$

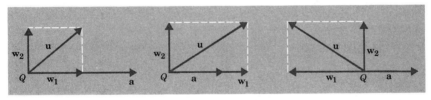

Figure 3.22

As indicated in Figure 3.22, the vector $\mathbf{w}_1$ is parallel to **a**, the vector $\mathbf{w}_2$ is perpendicular to **a**, and

$$\mathbf{w}_1 + \mathbf{w}_2 = \mathbf{w}_1 + (\mathbf{u} - \mathbf{w}_1) = \mathbf{u}$$

The vector $\mathbf{w}_1$ is called the ***orthogonal projection of* u *on* a** or sometimes the ***vector component of* u *along* a**. It is denoted by

$$\text{proj}_\mathbf{a}\, \mathbf{u} \qquad\qquad (3.8)$$

The vector $\mathbf{w}_2$ is called the ***vector component of*** $\mathbf{u}$ ***orthogonal to*** $\mathbf{a}$. Since $\mathbf{w}_2 = \mathbf{u} - \mathbf{w}_1$ this vector can be written in notation (3.8) as

$$\mathbf{w}_2 = \mathbf{u} - \text{proj}_\mathbf{a}\ \mathbf{u}$$

The following theorem gives formulas for calculating the vectors $\text{proj}_\mathbf{a}\ \mathbf{u}$ and $\mathbf{u} - \text{proj}_\mathbf{a}\ \mathbf{u}$.

Theorem 4. *If* $\mathbf{u}$ *and* $\mathbf{a}$ *are vectors in 2-space or 3-space and if* $\mathbf{a} \neq \mathbf{0}$, *then*

$$\text{proj}_\mathbf{a}\ \mathbf{u} = \frac{\mathbf{u} \cdot \mathbf{a}}{\|\mathbf{a}\|^2}\ \mathbf{a} \qquad (\textit{vector component of } \mathbf{u} \textit{ along } \mathbf{a})$$

$$\mathbf{u} - \text{proj}_\mathbf{a}\ \mathbf{u} = \mathbf{u} - \frac{\mathbf{u} \cdot \mathbf{a}}{\|\mathbf{a}\|^2}\ \mathbf{a} \qquad (\textit{vector component of } \mathbf{u} \textit{ orthogonal to } \mathbf{a})$$

Proof. Let $\mathbf{w}_1 = \text{proj}_\mathbf{a}\ \mathbf{u}$ and $\mathbf{w}_2 = \mathbf{u} - \text{proj}_\mathbf{a}\ \mathbf{u}$. Since $\mathbf{w}_1$ is parallel to $\mathbf{a}$, it must be a scalar multiple of $\mathbf{a}$, so it can be written in the form $\mathbf{w}_1 = k\mathbf{a}$. Thus

$$\mathbf{u} = \mathbf{w}_1 + \mathbf{w}_2 = k\mathbf{a} + \mathbf{w}_2 \qquad (3.9)$$

Taking the dot product of both sides of (3.9) with $\mathbf{a}$ and using Theorems 2(*a*) and 3 yields

$$\mathbf{u} \cdot \mathbf{a} = (k\mathbf{a} + \mathbf{w}_2) \cdot \mathbf{a} = k\|\mathbf{a}\|^2 + \mathbf{w}_2 \cdot \mathbf{a} \qquad (3.10)$$

But $\mathbf{w}_2 \cdot \mathbf{a} = 0$ since $\mathbf{w}_2$ is perpendicular to $\mathbf{a}$; so (3.10) yields

$$k = \frac{\mathbf{u} \cdot \mathbf{a}}{\|\mathbf{a}\|^2}$$

Since $\text{proj}_\mathbf{a}\ \mathbf{u} = \mathbf{w}_1 = k\mathbf{a}$, we obtain

$$\text{proj}_\mathbf{a}\ \mathbf{u} = \frac{\mathbf{u} \cdot \mathbf{a}}{\|\mathbf{a}\|^2}\ \mathbf{a} \quad \blacksquare$$

Example 10

Let $\mathbf{u} = (2, -1, 3)$ and $\mathbf{a} = (4, -1, 2)$. Find the vector component of $\mathbf{u}$ along $\mathbf{a}$ and the vector component of $\mathbf{u}$ orthogonal to $\mathbf{a}$.

Solution.

$$\mathbf{u} \cdot \mathbf{a} = (2)(4) + (-1)(-1) + (3)(2) = 15$$
$$\|\mathbf{a}\|^2 = 4^2 + (-1)^2 + 2^2 = 21$$

Thus, the vector component of $\mathbf{u}$ along $\mathbf{a}$ is

$$\text{proj}_\mathbf{a}\ \mathbf{u} = \frac{\mathbf{u} \cdot \mathbf{a}}{\|\mathbf{a}\|^2}\ \mathbf{a} = \frac{15}{21}(4, -1, 2) = \left(\frac{20}{7}, -\frac{5}{7}, \frac{10}{7}\right)$$

and the vector component of **u** orthogonal to **a** is

$$\mathbf{u} - \text{proj}_\mathbf{a}\, \mathbf{u} = (2, -1, 3) - \left(\frac{20}{7}, -\frac{5}{7}, \frac{10}{7}\right)$$

$$= \left(-\frac{6}{7}, -\frac{2}{7}, \frac{11}{7}\right)$$

As a check, the reader may wish to verify that the vectors **u** − proj$_\mathbf{a}$ **u** and **a** are perpendicular by showing that their dot product is zero. ▲

A formula for the length of the vector component of **u** along **a** can be obtained by writing

$$\|\text{proj}_\mathbf{a}\, \mathbf{u}\| = \left\| \frac{\mathbf{u} \cdot \mathbf{a}}{\|\mathbf{a}\|^2}\, \mathbf{a} \right\|$$

$$= \left| \frac{\mathbf{u} \cdot \mathbf{a}}{\|\mathbf{a}\|^2} \right| \|\mathbf{a}\| \qquad \left[\text{Since } \frac{\mathbf{u} \cdot \mathbf{a}}{\|\mathbf{a}\|^2} \text{ is a scalar.} \right]$$

$$= \frac{|\mathbf{u} \cdot \mathbf{a}|}{\|\mathbf{a}\|^2} \|\mathbf{a}\| \qquad [\text{Since } \|\mathbf{a}\|^2 > 0.]$$

which yields

$$\|\text{proj}_\mathbf{a}\, \mathbf{u}\| = \frac{|\mathbf{u} \cdot \mathbf{a}|}{\|\mathbf{a}\|} \tag{3.11}$$

If θ denotes the angle between **u** and **a**, then $\mathbf{u} \cdot \mathbf{a} = \|\mathbf{u}\| \|\mathbf{a}\| \cos\theta$, so that (3.11) can also be written as

$$\|\text{proj}_\mathbf{a}\, \mathbf{u}\| = \|\mathbf{u}\| |\cos\theta| \tag{3.12}$$

(Verify.) A geometric interpretation of this result is given in Figure 3.23.

As an example, we will use vector methods to derive a formula for the distance from a point in the plane to a line.

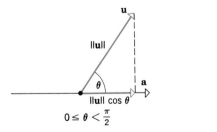

Figure 3.23

Example 11

Find a formula for the distance D between the point $P_0(x_0, y_0)$ and the line $ax + by + c = 0$.

Solution. Let $Q(x_1, y_1)$ be any point on the line and position the vector

$$\mathbf{n} = (a, b)$$

so that its initial point is at Q.

By virtue of Example 9, the vector $\mathbf{n}$ is perpendicular to the line (Figure 3.24). As indicated in the figure, the distance D is equal to the length of the orthogonal projection of $\overrightarrow{QP_0}$ on $\mathbf{n}$; thus, from (3.11),

$$D = \|\text{proj}_{\mathbf{n}} \overrightarrow{QP_0}\| = \frac{|\overrightarrow{QP_0} \cdot \mathbf{n}|}{\|\mathbf{n}\|}$$

But

$$\overrightarrow{QP_0} = (x_0 - x_1, y_0 - y_1)$$
$$\overrightarrow{QP_0} \cdot \mathbf{n} = a(x_0 - x_1) + b(y_0 - y_1)$$
$$\|\mathbf{n}\| = \sqrt{a^2 + b^2}$$

so that

$$D = \frac{|a(x_0 - x_1) + b(y_0 - y_1)|}{\sqrt{a^2 + b^2}} \quad (3.13)$$

Since the point $Q(x_1, y_1)$ lies on the line, its coordinates satisfy the equation of the line, so

$$ax_1 + by_1 + c = 0$$

or

$$c = -ax_1 - by_1$$

Substituting this expression in (3.13) yields the formula

$$D = \frac{|ax_0 + by_0 + c|}{\sqrt{a^2 + b^2}} \quad (3.14)$$

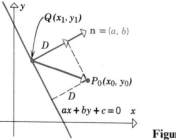

Figure 3.24

As an illustration, the distance D from the point $(1, -2)$ to the line $3x + 4y - 6 = 0$ is

$$D = \frac{|(3)(1) + 4(-2) - 6|}{\sqrt{3^2 + 4^2}} = \frac{|-11|}{\sqrt{25}} = \frac{11}{5}$$

EXERCISE SET 3.3

1. Find $\mathbf{u} \cdot \mathbf{v}$ for
   (a) $\mathbf{u} = (1, 2), \mathbf{v} = (6, -8)$ (b) $\mathbf{u} = (-7, -3), \mathbf{v} = (0, 1)$
   (c) $\mathbf{u} = (1, -3, 7), \mathbf{v} = (8, -2, -2)$ (d) $\mathbf{u} = (-3, 1, 2), \mathbf{v} = (4, 2, -5)$

2. In each part of Exercise 1, find the cosine of the angle θ between $\mathbf{u}$ and $\mathbf{v}$.

3. Determine whether $\mathbf{u}$ and $\mathbf{v}$ make an acute angle, make an obtuse angle, or are orthogonal.
   (a) $\mathbf{u} = (7, 3, 5), \mathbf{v} = (-8, 4, 2)$ (b) $\mathbf{u} = (6, 1, 3), \mathbf{v} = (4, 0, -6)$
   (c) $\mathbf{u} = (1, 1, 1), \mathbf{v} = (-1, 0, 0)$ (d) $\mathbf{u} = (4, 1, 6), \mathbf{v} = (-3, 0, 2)$

4. Find the orthogonal projection of $\mathbf{u}$ on $\mathbf{a}$ if
   (a) $\mathbf{u} = (2, 1), \mathbf{a} = (-3, 2)$ (b) $\mathbf{u} = (2, 6), \mathbf{a} = (-9, 3)$
   (c) $\mathbf{u} = (-7, 1, 3), \mathbf{a} = (5, 0, 1)$ (d) $\mathbf{u} = (0, 0, 1), \mathbf{a} = (8, 3, 4)$

5. In each part of Exercise 4, find the vector component of $\mathbf{u}$ orthogonal to $\mathbf{a}$.

6. In each part find $\|\text{proj}_\mathbf{a} \mathbf{u}\|$.
   (a) $\mathbf{u} = (2, -1), \mathbf{a} = (3, 4)$ (b) $\mathbf{u} = (4, 5), \mathbf{a} = (1, -2)$
   (c) $\mathbf{u} = (2, -1, 3), \mathbf{a} = (1, 2, 2)$ (d) $\mathbf{u} = (4, -1, 7), \mathbf{a} = (2, 3, -6)$

7. Verify Theorem 3 for $\mathbf{u} = (6, -1, 2), \mathbf{v} = (2, 7, 4)$, and $k = -5$.

8. Find two vectors of norm 1 that are orthogonal to $(3, -2)$.

9. Let $\mathbf{u} = (1, 2), \mathbf{v} = (4, -2)$, and $\mathbf{w} = (6, 0)$. Find
   (a) $\mathbf{u} \cdot (7\mathbf{v} + \mathbf{w})$ (b) $\|(\mathbf{u} \cdot \mathbf{w})\mathbf{w}\|$ (c) $\|\mathbf{u}\| (\mathbf{v} \cdot \mathbf{w})$ (d) $(\|\mathbf{u}\| \mathbf{v}) \cdot \mathbf{w}$

10. Explain why each of the following expressions makes no sense.
    (a) $\mathbf{u} \cdot (\mathbf{v} \cdot \mathbf{w})$ (b) $(\mathbf{u} \cdot \mathbf{v}) + \mathbf{w}$ (c) $\|\mathbf{u} \cdot \mathbf{v}\|$ (d) $k \cdot (\mathbf{u} + \mathbf{v})$

11. Use vectors to find the cosines of the interior angles of the triangle with vertices $(-1, 0)$, $(-2, 1)$, and $(1, 4)$.

12. Show that $A(2, -1, 1), B(3, 2, -1)$, and $C(7, 0, -2)$ are vertices of a right triangle. At which vertex is the right angle?

13. Suppose $\mathbf{a} \cdot \mathbf{b} = \mathbf{a} \cdot \mathbf{c}$ and $\mathbf{a} \ne \mathbf{0}$. Does it follow that $\mathbf{b} = \mathbf{c}$? Explain.

14. Let $\mathbf{a} = (k, 1)$ and $\mathbf{b} = (4, 3)$. Find k so that
    (a) $\mathbf{a}$ and $\mathbf{b}$ are orthogonal   (b) the angle between $\mathbf{a}$ and $\mathbf{b}$ is $\pi/4$
    (c) the angle between $\mathbf{a}$ and $\mathbf{b}$ is $\pi/6$ (d) $\mathbf{a}$ and $\mathbf{b}$ are parallel

15. Use Formula (3.14) to calculate the distance between the point and the line.
    (a) $3x + 4y + 7 = 0$; $(1, -2)$ (b) $y = -2x + 1$; $(-3, 5)$ (c) $2x + y = 8$; $(2, 6)$

16. Establish the identity

$$\|\mathbf{u} + \mathbf{v}\|^2 + \|\mathbf{u} - \mathbf{v}\|^2 = 2\|\mathbf{u}\|^2 + 2\|\mathbf{v}\|^2$$

17. Establish the identity

$$\mathbf{u} \cdot \mathbf{v} = \tfrac{1}{4}\|\mathbf{u} + \mathbf{v}\|^2 - \tfrac{1}{4}\|\mathbf{u} - \mathbf{v}\|^2$$

18. Find the angle between a diagonal of a cube and one of its faces.

19. The *direction cosines* of a vector $\mathbf{v}$ in 3-space are the numbers $\cos \alpha$, $\cos \beta$, and $\cos \gamma$, where α, β, and γ are the angles between $\mathbf{v}$ and the positive x, y, and z axes. Show that if $\mathbf{v} = (a, b, c)$, then $\cos \alpha = a/\sqrt{a^2 + b^2 + c^2}$. Find $\cos \beta$ and $\cos \gamma$.

20. Show that if $\mathbf{v}$ is orthogonal to $\mathbf{w}_1$ and $\mathbf{w}_2$, then $\mathbf{v}$ is orthogonal to $k_1\mathbf{w}_1 + k_2\mathbf{w}_2$ for all scalars k_1 and k_2.

21. Let $\mathbf{u}$ and $\mathbf{v}$ be nonzero vectors in 2- or 3-space, and let $k = \|\mathbf{u}\|$ and $l = \|\mathbf{v}\|$. Show that the vector

$$\mathbf{w} = l\mathbf{u} + k\mathbf{v}$$

bisects the angle between $\mathbf{u}$ and $\mathbf{v}$.

3.4 CROSS PRODUCT

In many applications of vectors to problems in geometry, physics, and engineering, it is of interest to construct a vector in 3-space that is perpendicular to two given vectors. In this section we introduce a type of vector multiplication that facilitates this construction.

Definition. If $\mathbf{u} = (u_1, u_2, u_3)$ and $\mathbf{v} = (v_1, v_2, v_3)$ are vectors in 3-space, then the *cross product* $\mathbf{u} \times \mathbf{v}$ is the vector defined by

$$\mathbf{u} \times \mathbf{v} = (u_2 v_3 - u_3 v_2, u_3 v_1 - u_1 v_3, u_1 v_2 - u_2 v_1)$$

or in determinant notation

$$\mathbf{u} \times \mathbf{v} = \left(\begin{vmatrix} u_2 & u_3 \\ v_2 & v_3 \end{vmatrix}, -\begin{vmatrix} u_1 & u_3 \\ v_1 & v_3 \end{vmatrix}, \begin{vmatrix} u_1 & u_2 \\ v_1 & v_2 \end{vmatrix} \right) \tag{3.15}$$

REMARK. There is a pattern in Formula 3.15 that is useful to keep in mind. If we form the 2×3 matrix

$$\begin{bmatrix} u_1 & u_2 & u_3 \\ v_1 & v_2 & v_3 \end{bmatrix}$$

where the entries in the first row are the components of the first factor $\mathbf{u}$ and those in the second row are the components of the second factor $\mathbf{v}$, then the determinant in the first component of $\mathbf{u} \times \mathbf{v}$ is obtained by deleting the first column of the matrix, the determinant in the second component by deleting the second column of the matrix, and the determinant in the third component by deleting the third column of the matrix.

Example 12

Find $\mathbf{u} \times \mathbf{v}$ where $\mathbf{u} = (1, 2, -2)$ and $\mathbf{v} = (3, 0, 1)$.

Solution.

$$\begin{bmatrix} 1 & 2 & -2 \\ 3 & 0 & 1 \end{bmatrix}$$

$$\mathbf{u} \times \mathbf{v} = \left(\begin{vmatrix} 2 & -2 \\ 0 & 1 \end{vmatrix}, -\begin{vmatrix} 1 & -2 \\ 3 & 1 \end{vmatrix}, \begin{vmatrix} 1 & 2 \\ 3 & 0 \end{vmatrix} \right)$$

$$= (2, -7, -6)$$

Whereas the dot product of two vectors is a scalar, the cross product is another vector. The following theorem gives an important relationship between dot product and cross product and also shows that $\mathbf{u} \times \mathbf{v}$ is orthogonal to both $\mathbf{u}$ and $\mathbf{v}$.

Theorem 5.   *If $\mathbf{u}$ and $\mathbf{v}$ are vectors in 3-space, then:*

(a) $\mathbf{u} \cdot (\mathbf{u} \times \mathbf{v}) = 0$ ($\mathbf{u} \times \mathbf{v}$ *is orthogonal to* $\mathbf{u}$)
(b) $\mathbf{v} \cdot (\mathbf{u} \times \mathbf{v}) = 0$ ($\mathbf{u} \times \mathbf{v}$ *is orthogonal to* $\mathbf{v}$)
(c) $\|\mathbf{u} \times \mathbf{v}\|^2 = \|\mathbf{u}\|^2 \|\mathbf{v}\|^2 - (\mathbf{u} \cdot \mathbf{v})^2$ (*Lagrange's identity*)

Proof. Let $\mathbf{u} = (u_1, u_2, u_3)$ and $\mathbf{v} = (v_1, v_2, v_3)$.

(a) $\mathbf{u} \cdot (\mathbf{u} \times \mathbf{v}) = (u_1, u_2, u_3) \cdot (u_2 v_3 - u_3 v_2, u_3 v_1 - u_1 v_3, u_1 v_2 - u_2 v_1)$

$\qquad\qquad\quad = u_1(u_2 v_3 - u_3 v_2) + u_2(u_3 v_1 - u_1 v_3) + u_3(u_1 v_2 - u_2 v_1)$

$\qquad\qquad\quad = 0$

(b) Similar to (a).

(c) Since

$$\|\mathbf{u} \times \mathbf{v}\|^2 = (u_2 v_3 - u_3 v_2)^2 + (u_3 v_1 - u_1 v_3)^2 + (u_1 v_2 - u_2 v_1)^2 \qquad (3.16)$$

and

$$\|\mathbf{u}\|^2 \|\mathbf{v}\|^2 - (\mathbf{u} \cdot \mathbf{v})^2$$

$$= (u_1{}^2 + u_2{}^2 + u_3{}^2)(v_1{}^2 + v_2{}^2 + v_3{}^2) - (u_1 v_1 + u_2 v_2 + u_3 v_3)^2 \qquad (3.17)$$

Lagrange's identity can be established by "multiplying out" the right sides of (3.16) and (3.17) and verifying their equality.

Example 13

Consider the vectors

$$\mathbf{u} = (1, 2, -2) \quad \text{and} \quad \mathbf{v} = (3, 0, 1)$$

In Example 12 we showed that

$$\mathbf{u} \times \mathbf{v} = (2, -7, -6)$$

Since

$$\mathbf{u} \cdot (\mathbf{u} \times \mathbf{v}) = (1)(2) + (2)(-7) + (-2)(-6) = 0$$

and

$$\mathbf{v} \cdot (\mathbf{u} \times \mathbf{v}) = (3)(2) + (0)(-7) + (1)(-6) = 0$$

$\mathbf{u} \times \mathbf{v}$ is orthogonal to both $\mathbf{u}$ and $\mathbf{v}$ as guaranteed by Theorem 5.

The main arithmetic properties of the cross product are listed in the next theorem.

Theorem 6. *If* $\mathbf{u}$, $\mathbf{v}$, *and* $\mathbf{w}$ *are any vectors in 3-space and k is any scalar, then*

(*a*) $\mathbf{u} \times \mathbf{v} = -(\mathbf{v} \times \mathbf{u})$
(*b*) $\mathbf{u} \times (\mathbf{v} + \mathbf{w}) = (\mathbf{u} \times \mathbf{v}) + (\mathbf{u} \times \mathbf{w})$
(*c*) $(\mathbf{u} + \mathbf{v}) \times \mathbf{w} = (\mathbf{u} \times \mathbf{w}) + (\mathbf{v} \times \mathbf{w})$
(*d*) $k(\mathbf{u} \times \mathbf{v}) = (k\mathbf{u}) \times \mathbf{v} = \mathbf{u} \times (k\mathbf{v})$
(*e*) $\mathbf{u} \times \mathbf{0} = \mathbf{0} \times \mathbf{u} = \mathbf{0}$
(*f*) $\mathbf{u} \times \mathbf{u} = \mathbf{0}$

The proofs follow immediately from Formula 3.15 and properties of determinants; for example, (*a*) can be proved as follows:

Proof. (*a*) Interchanging $\mathbf{u}$ and $\mathbf{v}$ in (3.15) interchanges the rows of the three determinants on the right side of (3.15) and thereby changes the sign of each component in the cross product. Thus, $\mathbf{u} \times \mathbf{v} = -(\mathbf{v} \times \mathbf{u})$.

The proofs of the remaining parts are left as exercises.

Example 14

Consider the vectors

$$\mathbf{i} = (1, 0, 0) \quad \mathbf{j} = (0, 1, 0) \quad \mathbf{k} = (0, 0, 1)$$

These vectors each have length 1 and lie along the coordinate axes (Figure 3.25). They are called the **standard unit vectors** in 3-space. Every vector $\mathbf{v} = (v_1, v_2, v_3)$ in 3-space is expressible in terms of $\mathbf{i}$, $\mathbf{j}$, and $\mathbf{k}$ since we can write

$$\mathbf{v} = (v_1, v_2, v_3) = v_1(1, 0, 0) + v_2(0, 1, 0) + v_3(0, 0, 1) = v_1\mathbf{i} + v_2\mathbf{j} + v_3\mathbf{k}$$

For example,

$$(2, -3, 4) = 2\mathbf{i} - 3\mathbf{j} + 4\mathbf{k}$$

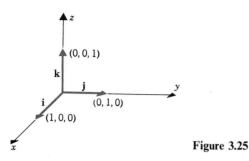

Figure 3.25

From (3.15) we obtain

$$\mathbf{i} \times \mathbf{j} = \left(\begin{vmatrix} 0 & 0 \\ 1 & 0 \end{vmatrix}, -\begin{vmatrix} 1 & 0 \\ 0 & 0 \end{vmatrix}, \begin{vmatrix} 1 & 0 \\ 0 & 1 \end{vmatrix} \right) = (0, 0, 1) = \mathbf{k}$$

The reader should have no trouble obtaining the following results:

$$\mathbf{i} \times \mathbf{i} = \mathbf{j} \times \mathbf{j} = \mathbf{k} \times \mathbf{k} = 0$$
$$\mathbf{i} \times \mathbf{j} = \mathbf{k}, \quad \mathbf{j} \times \mathbf{k} = \mathbf{i}, \quad \mathbf{k} \times \mathbf{i} = \mathbf{j}$$
$$\mathbf{j} \times \mathbf{i} = -\mathbf{k}, \quad \mathbf{k} \times \mathbf{j} = -\mathbf{i}, \quad \mathbf{i} \times \mathbf{k} = -\mathbf{j}$$

The following diagram is helpful for remembering these results.

Referring to this diagram, the cross product of two consecutive vectors going clockwise is the next vector around, and the cross product of two consecutive vectors going counterclockwise is the negative of the next vector around.

It is also worth noting that a cross product can be represented symbolically in the form of a 3×3 determinant:

$$\mathbf{u} \times \mathbf{v} = \begin{vmatrix} \mathbf{i} & \mathbf{j} & \mathbf{k} \\ u_1 & u_2 & u_3 \\ v_1 & v_2 & v_3 \end{vmatrix} = \begin{vmatrix} u_2 & u_3 \\ v_2 & v_3 \end{vmatrix} \mathbf{i} - \begin{vmatrix} u_1 & u_3 \\ v_1 & v_3 \end{vmatrix} \mathbf{j} + \begin{vmatrix} u_1 & u_2 \\ v_1 & v_2 \end{vmatrix} \mathbf{k}$$

For example, if $\mathbf{u} = (1, 2, -2)$ and $\mathbf{v} = (3, 0, 1)$, then

$$\mathbf{u} \times \mathbf{v} = \begin{vmatrix} \mathbf{i} & \mathbf{j} & \mathbf{k} \\ 1 & 2 & -2 \\ 3 & 0 & 1 \end{vmatrix} = 2\mathbf{i} - 7\mathbf{j} - 6\mathbf{k}$$

which agrees with the result obtained in Example 12.

Warning. It is *not* true in general that $\mathbf{u} \times (\mathbf{v} \times \mathbf{w}) = (\mathbf{u} \times \mathbf{v}) \times \mathbf{w}$. For example,

$$\mathbf{i} \times (\mathbf{j} \times \mathbf{j}) = \mathbf{i} \times \mathbf{0} = \mathbf{0}$$

and

$$(\mathbf{i} \times \mathbf{j}) \times \mathbf{j} = \mathbf{k} \times \mathbf{j} = -(\mathbf{j} \times \mathbf{k}) = -\mathbf{i}$$

so that

$$\mathbf{i} \times (\mathbf{j} \times \mathbf{j}) \neq (\mathbf{i} \times \mathbf{j}) \times \mathbf{j}$$

We know from Theorem 5 that $\mathbf{u} \times \mathbf{v}$ is orthogonal to both $\mathbf{u}$ and $\mathbf{v}$. If $\mathbf{u}$ and $\mathbf{v}$ are nonzero vectors, it can be shown that the direction of $\mathbf{u} \times \mathbf{v}$ can be determined using the following "right-hand rule"* (Figure 3.26): Let θ be the angle between $\mathbf{u}$ and $\mathbf{v}$, and suppose $\mathbf{u}$ is rotated through the angle θ until it coincides with $\mathbf{v}$. If the fingers of the right hand are cupped so they point in the direction of rotation, then the thumb indicates (roughly) the direction of $\mathbf{u} \times \mathbf{v}$.

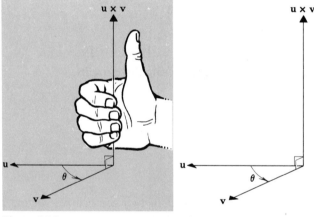

Figure 3.26

The reader may find it instructive to practice this rule with the products

$$\mathbf{i} \times \mathbf{j} = \mathbf{k} \qquad \mathbf{j} \times \mathbf{k} = \mathbf{i} \qquad \mathbf{k} \times \mathbf{i} = \mathbf{j}$$

If $\mathbf{u}$ and $\mathbf{v}$ are nonzero vectors in 3-space, then the norm of $\mathbf{u} \times \mathbf{v}$ has a useful geometric interpretation. Lagrange's identity, given in Theorem 5, states that

$$\|\mathbf{u} \times \mathbf{v}\|^2 = \|\mathbf{u}\|^2 \|\mathbf{v}\|^2 - \mathbf{u} \cdot \mathbf{v} \tag{3.18}$$

If θ denotes the angle between $\mathbf{u}$ and $\mathbf{v}$, then $\mathbf{u} \cdot \mathbf{v} = \|\mathbf{u}\| \|\mathbf{v}\| \cos \theta$, so that (3.18) can

---

* Recall that we agreed to consider only right-handed coordinate systems in this text. Had we used left-handed systems instead, a "left-hand rule" would apply here.

be rewritten as

$$\|\mathbf{u} \times \mathbf{v}\|^2 = \|\mathbf{u}\|^2 \|\mathbf{v}\|^2 - \|\mathbf{u}\|^2 \|\mathbf{v}\|^2 \cos^2 \theta$$
$$= \|\mathbf{u}\|^2 \|\mathbf{v}\|^2 (1 - \cos^2 \theta)$$
$$= \|\mathbf{u}\|^2 \|\mathbf{v}\|^2 \sin^2 \theta$$

Thus,

$$\|\mathbf{u} \times \mathbf{v}\| = \|\mathbf{u}\| \|\mathbf{v}\| \sin \theta \tag{3.19}$$

But $\|\mathbf{v}\| \sin \theta$ is the altitude of the parallelogram determined by $\mathbf{u}$ and $\mathbf{v}$ (Figure 3.27). Thus, from (3.19), the area A of this parallelogram is given by

$$A = (\text{base})(\text{altitude}) = \|\mathbf{u}\| \|\mathbf{v}\| \sin \theta = \|\mathbf{u} \times \mathbf{v}\|$$

In other words, *the norm of* $\mathbf{u} \times \mathbf{v}$ *is equal to the area of the parallelogram determined by* $\mathbf{u}$ *and* $\mathbf{v}$.

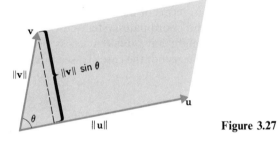

Figure 3.27

Example 15

Find the area of the triangle determined by the points $P_1(2, 2, 0)$, $P_2(-1, 0, 2)$, and $P_3(0, 4, 3)$.

Solution. The area A of the triangle is $1/2$ the area of the parallelogram determined by the vectors $\overrightarrow{P_1 P_2}$ and $\overrightarrow{P_1 P_3}$ (Figure 3.28).

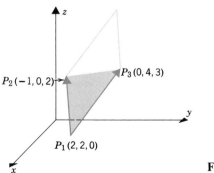

Figure 3.28

Using the method discussed in Example 2 of Section 3.1, $\overrightarrow{P_1P_2} = (-3, -2, 2)$ and $\overrightarrow{P_1P_3} = (-2, 2, 3)$. It follows that

$$\overrightarrow{P_1P_2} \times \overrightarrow{P_1P_3} = (-10, 5, -10)$$

and consequently

$$A = \tfrac{1}{2}\|\overrightarrow{P_1P_2} \times \overrightarrow{P_1P_3}\| = \tfrac{1}{2}(15) = \tfrac{15}{2} \quad \blacktriangle$$

Initially, we defined a vector to be a directed line segment or arrow in 2-space or 3-space; coordinate systems and components were introduced later in order to simplify computations with vectors. Thus, a vector has a "mathematical existence" regardless of whether a coordinate system has been introduced. Further, the components of a vector are not determined by the vector alone; they depend as well on the coordinate system chosen. For example, in Figure 3.29 we have indicated a fixed plane vector **v** and two different coordinate systems. In the xy-coordinate system the components of **v** are $(1, 1)$, and in the $x'y'$-system they are $(\sqrt{2}, 0)$.

This raises an important question about our definition of cross product. Since we defined the cross product $\mathbf{u} \times \mathbf{v}$ in terms of the components of **u** and **v**, and since these components depend on the coordinate system chosen, it seems possible that two *fixed* vectors **u** and **v** might have different cross products in different coordinate systems. Fortunately, this is not the case. To see this, we need only recall that:

(i) $\mathbf{u} \times \mathbf{v}$ is perpendicular to both **u** and **v**.

(ii) The orientation of $\mathbf{u} \times \mathbf{v}$ is determined by the right-hand rule.

(iii) $\|\mathbf{u} \times \mathbf{v}\| = \|\mathbf{u}\| \|\mathbf{v}\| \sin \theta$.

These three properties completely determine the vector $\mathbf{u} \times \mathbf{v}$; properties (i) and (ii) determine the direction, and property (iii) determines the length. Since these properties depend only on the lengths and relative positions of **u** and **v** and not on the particular right-hand coordinate system being used, the vector $\mathbf{u} \times \mathbf{v}$ will remain unchanged if a different right-hand coordinate system is introduced. This fact is described by stating that the definition of $\mathbf{u} \times \mathbf{v}$ is ***coordinate free***. This result is of importance to physicists and engineers who often work with many coordinate systems in the same problem.

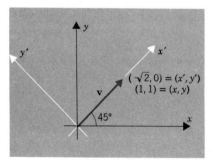

Figure 3.29

Example 16

Consider two perpendicular vectors **u** and **v**, each of length 1 (as shown in Figure 3.30*a*.) If we introduce an *xyz*-coordinate system as shown in Figure 3.30*b*, then

$$\mathbf{u} = (1, 0, 0) = \mathbf{i} \quad \text{and} \quad \mathbf{v} = (0, 1, 0) = \mathbf{j}$$

so that

$$\mathbf{u} \times \mathbf{v} = \mathbf{i} \times \mathbf{j} = \mathbf{k} = (0, 0, 1)$$

However, if we introduce an *x'y'z'*-coordinate system as shown in Figure 3.30*c*, then

$$\mathbf{u} = (0, 0, 1) = \mathbf{k} \quad \text{and} \quad \mathbf{v} = (1, 0, 0) = \mathbf{i}$$

so that

$$\mathbf{u} \times \mathbf{v} = \mathbf{k} \times \mathbf{i} = \mathbf{j} = (0, 1, 0)$$

But it is clear from Figures 3.30*b* and 3.30*c* that the vector (0, 0, 1) in the *xyz*-system is the same as the vector (0, 1, 0) in the *x'y'z'*-system. Thus, we obtain the same vector **u** × **v** whether we compute with coordinates from the *xyz*-system or with coordinates from the *x'y'z'*-system.

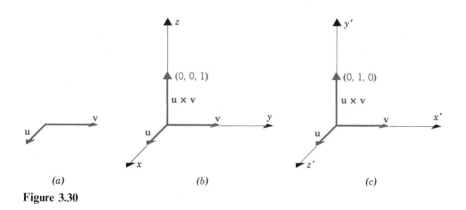

(a)   *(b)*   *(c)*

Figure 3.30

EXERCISE SET 3.4

1. Let $\mathbf{u} = (2, -1, 3)$, $\mathbf{v} = (0, 1, 7)$, and $\mathbf{w} = (1, 4, 5)$. Compute:
(a) $\mathbf{v} \times \mathbf{w}$ (b) $\mathbf{u} \times (\mathbf{v} \times \mathbf{w})$ (c) $(\mathbf{u} \times \mathbf{v}) \times \mathbf{w}$
(d) $(\mathbf{u} \times \mathbf{v}) \times (\mathbf{v} \times \mathbf{w})$ (e) $\mathbf{u} \times (\mathbf{v} - 2\mathbf{w})$ (f) $(\mathbf{u} \times \mathbf{v}) - 2\mathbf{w}$

2. In each part find a vector orthogonal to both **u** and **v**.
(a) $\mathbf{u} = (-7, 3, 1)$ $\mathbf{v} = (2, 0, 4)$
(b) $\mathbf{u} = (-1, -1, -1)$ $\mathbf{v} = (2, 0, 2)$

3. In each part find the area of the triangle having vertices *P*, *Q*, and *R*.
(a) $P(1, 5, -2)$ $Q(0, 0, 0)$ $R(3, 5, 1)$
(b) $P(2, 0, -3)$ $Q(1, 4, 5)$ $R(7, 2, 9)$

4. Verify Theorem 5 for the vectors $\mathbf{u} = (1, -5, 6)$ and $\mathbf{v} = (2, 1, 2)$.

5. Verify Theorem 6 for $\mathbf{u} = (2, 0, -1)$, $\mathbf{v} = (6, 7, 4)$, $\mathbf{w} = (1, 1, 1)$ and $k = -3$.

6. What is wrong with the expression $\mathbf{u} \times \mathbf{v} \times \mathbf{w}$?

7. Let $\mathbf{u} = (-1, 3, 2)$ and $\mathbf{w} = (1, 1, -1)$. Find all vectors $\mathbf{x}$ that satisfy $\mathbf{u} \times \mathbf{x} = \mathbf{w}$.

8. Let $\mathbf{u} = (u_1, u_2, u_3)$, $\mathbf{v} = (v_1, v_2, v_3)$, and $\mathbf{w} = (w_1, w_2, w_3)$. Show that

$$\mathbf{u} \cdot (\mathbf{v} \times \mathbf{w}) = \begin{vmatrix} u_1 & u_2 & u_3 \\ v_1 & v_2 & v_3 \\ w_1 & w_2 & w_3 \end{vmatrix}$$

9. Use the result of Exercise 8 to compute $\mathbf{u} \cdot (\mathbf{v} \times \mathbf{w})$ when $\mathbf{u} = (-1, 4, 7)$, $\mathbf{v} = (6, -7, 3)$, and $\mathbf{w} = (4, 0, 1)$.

10. Let $\mathbf{m}$ and $\mathbf{n}$ be vectors whose components in the xyz-system of Figure 3.30 are $\mathbf{m} = (0, 0, 1)$ and $\mathbf{n} = (0, 1, 0)$.
(a) Find the components of $\mathbf{m}$ and $\mathbf{n}$ in the $x'y'z'$-system of Figure 3.30.
(b) Compute $\mathbf{m} \times \mathbf{n}$ using the components in the xyz-system.
(c) Compute $\mathbf{m} \times \mathbf{n}$ using the components in the $x'y'z'$-system.
(d) Show that the vectors obtained in (b) and (c) are the same.

11. Prove the following identities.
(a) $(\mathbf{u} + k\mathbf{v}) \times \mathbf{v} = \mathbf{u} \times \mathbf{v}$
(b) $(\mathbf{u} \times \mathbf{v}) \cdot \mathbf{z} = \mathbf{u} \cdot (\mathbf{v} \times \mathbf{z})$

12. Let $\mathbf{u}$, $\mathbf{v}$, and $\mathbf{w}$ be nonzero vectors in 3-space with the same initial point, but such that no two of them are collinear. Show:
(a) $\mathbf{u} \times (\mathbf{v} \times \mathbf{w})$ lies in the plane determined by $\mathbf{v}$ and $\mathbf{w}$.
(b) $(\mathbf{u} \times \mathbf{v}) \times \mathbf{w}$ lies in the plane determined by $\mathbf{u}$ and $\mathbf{v}$.

13. Prove that $\mathbf{x} \times (\mathbf{y} \times \mathbf{z}) = (\mathbf{x} \cdot \mathbf{z})\mathbf{y} - (\mathbf{x} \cdot \mathbf{y})\mathbf{z}$. [*Hint.* First prove the result in the case where $\mathbf{z} = \mathbf{i} = (1, 0, 0)$, then when $\mathbf{z} = \mathbf{j} = (0, 1, 0)$, and then when $\mathbf{z} = \mathbf{k} = (0, 0, 1)$. Finally prove it for an arbitrary vector $\mathbf{z} = (z_1, z_2, z_3)$ by writing $\mathbf{z} = z_1\mathbf{i} + z_2\mathbf{j} + z_3\mathbf{k}$.]

14. Prove parts (*a*) and (*b*) of Theorem 6.

15. Prove parts (*c*) and (*d*) of Theorem 6.

16. Prove parts (*e*) and (*f*) of Theorem 6.

3.5 LINES AND PLANES IN 3-SPACE

In this section we shall use vectors to derive equations of lines and planes in 3-space, and we shall use these equations to solve some basic geometric problems.

In plane analytic geometry a line can be specified by giving its slope and one of its points. Similarly, a plane in 3-space can be determined by giving its inclination and specifying one of its points. A convenient method for describing the inclination is to specify a vector (called a *normal*) that is perpendicular to the plane.

Suppose we want the equation of the plane passing through the point $P_0(x_0, y_0, z_0)$ and having the nonzero vector $\mathbf{n} = (a, b, c)$ as a normal. It is evident from Figure 3.31 that the plane consists precisely of those points $P(x, y, z)$ for which the vector $\overrightarrow{P_0P}$ is orthogonal to $\mathbf{n}$; that is, for which

$$\mathbf{n} \cdot \overrightarrow{P_0P} = 0 \tag{3.20}$$

Since $\overrightarrow{P_0P} = (x - x_0, y - y_0, z - z_0)$, equation (3.20) can be written as

$$a(x - x_0) + b(y - y_0) + c(z - z_0) = 0 \tag{3.21}$$

We call this a *point-normal* form of the equation of a plane.

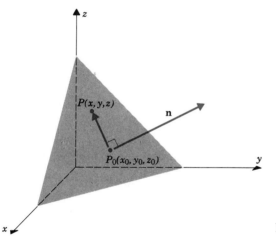

Figure 3.31

Example 17

Find an equation of the plane passing through the point $(3, -1, 7)$ and perpendicular to the vector $\mathbf{n} = (4, 2, -5)$.

Solution. From (3.21) a point-normal form is $4(x - 3) + 2(y + 1) - 5(z - 7) = 0$

By multiplying out and collecting terms, (3.21) can be rewritten in the form

$$ax + by + cz + d = 0 \tag{3.22}$$

where a, b, c and d are constants, and a, b, and c are not all zero. For example, the equation in Example 17 can be rewritten as

$$4x + 2y - 5z + 25 = 0$$

As our next theorem shows, every equation having the form of (3.22) represents a plane in 3-space.

Theorem 7. *If a, b, c, and d are constants and a, b, and c are not all zero, then the graph of the equation*

$$ax + by + cz + d = 0$$

is a plane having the vector $\mathbf{n} = (a, b, c)$ *as a normal.*

Proof. By hypothesis, the coefficients a, b, and c are not all zero. Assume, for the moment, that $a \neq 0$. Then the equation $ax + by + cz + d = 0$ can be rewritten as $a(x + (d/a)) + by + cz = 0$. But this is a point-normal form of the plane passing through the point $(-d/a, 0, 0)$ and having $\mathbf{n} = (a, b, c)$ as a normal.

If $a = 0$, then either $b \neq 0$ or $c \neq 0$. A straightforward modification of the above argument will handle these other cases. ▨

Equation (3.22) is a linear equation in x, y, and z; it is called the **general form** of the equation of a plane.

Just as the solutions to a system of linear equations

$$ax + by = k_1$$
$$cx + dy = k_2$$

correspond to points of intersection of the lines $ax + by = k_1$ and $cx + dy = k_2$ in the xy-plane, so the solutions of a system

$$ax + by + cz = k_1$$
$$dx + ey + fz = k_2 \qquad (3.23)$$
$$gx + hy + iz = k_3$$

correspond to points of intersection of the planes $ax + by + cz = k_1$, $dx + ey + fz = k_2$, and $gx + hy + iz = k_3$.

In Figure 3.32 we have illustrated some of the geometric possibilities that occur when (3.23) has zero, one, or infinitely many solutions.

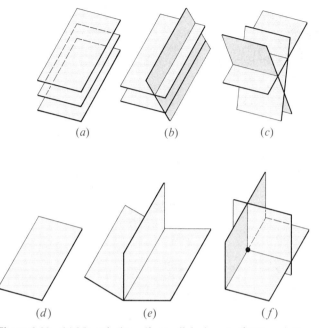

Figure 3.32 (*a*) No solutions (3 parallel planes). (*b*) No solutions (2 planes parallel). (*c*) No solutions (3 planes with no common intersection). (*d*) Infinitely many solutions (3 coincident planes). (*e*) Infinitely many solutions (3 planes intersecting in a line). (*f*) One solution (3 planes intersecting at a point).

Example 18

Find the equation of the plane through the points $P_1(1, 2, -1)$, $P_2(2, 3, 1)$, and $P_3(3, -1, 2)$.

Solution. Since the three points lie in the plane, their coordinates must satisfy the general equation $ax + by + cz + d = 0$ of the plane. Thus,

$$a + 2b - c + d = 0$$
$$2a + 3b + c + d = 0$$
$$3a - b + 2c + d = 0$$

Solving this system gives

$$a = -\tfrac{9}{16}t \qquad b = -\tfrac{1}{16}t \qquad c = \tfrac{5}{16}t \qquad d = t$$

Letting $t = -16$, for example, yields the desired equation

$$9x + y - 5z - 16 = 0$$

We note that any other choice of t gives a multiple of this equation, so that any value of $t \neq 0$ would do equally well.

Alternative solution. Since $P_1(1, 2, -1)$, $P_2(2, 3, 1)$, and $P_3(3, -1, 2)$ lie in the plane, the vectors $\overrightarrow{P_1P_2} = (1, 1, 2)$ and $\overrightarrow{P_1P_3} = (2, -3, 3)$ are parallel to the plane. Therefore, $\overrightarrow{P_1P_2} \times \overrightarrow{P_1P_3} = (9, 1, -5)$ is normal to the plane, since it is perpendicular to both $\overrightarrow{P_1P_2}$ and $\overrightarrow{P_1P_3}$. From this and the fact that P_1 lies in the plane, a point-normal form for the equation of the plane is

$$9(x - 1) + (y - 2) - 5(z + 1) = 0$$

or

$$9x + y - 5z - 16 = 0$$

We shall now show how to obtain equations for lines in 3-space. Suppose l is the line in 3-space through the point $P_0(x_0, y_0, z_0)$ and parallel to the nonzero vector $\mathbf{v} = (a, b, c)$. It is clear (Figure 3.33) that l consists precisely of those points $P(x, y, z)$ for which the vector $\overrightarrow{P_0P}$ is parallel to $\mathbf{v}$, that is, for which there is a scalar t such that

$$\overrightarrow{P_0P} = t\mathbf{v} \tag{3.24}$$

In terms of components (3.24) can be written

$$(x - x_0, y - y_0, z - z_0) = (ta, tb, tc)$$

from which it follows that

$$
\begin{aligned}
x &= x_0 + ta \\
y &= y_0 + tb \qquad \text{where } -\infty < t < +\infty \\
z &= z_0 + tc
\end{aligned}
$$

These equations are called **parametric equations** for l since the line l is traced out by $P(x, y, z)$ as the parameter t varies from $-\infty$ to $+\infty$.

REMARK. For brevity, we shall often omit the phrase "where $-\infty < t < +\infty$".

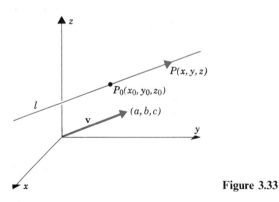

Figure 3.33

Example 19

The line through the point $(1, 2, -3)$ and parallel to the vector $\mathbf{v} = (4, 5, -7)$ has parametric equations

$$x = 1 + 4t$$
$$y = 2 + 5t \qquad \text{where} -\infty < t < +\infty$$
$$z = -3 - 7t \quad \blacktriangle$$

Example 20

(a) Find parametric equations for the line l passing through the points $P_1(2, 4, -1)$ and $P_2(5, 0, 7)$.
(b) Where does the line intersect the xy-plane?

Solution. (a) Since the vector $\overrightarrow{P_1P_2} = (3, -4, 8)$ is parallel to l and $P_1(2, 4, -1)$ lies on l, the line l is given by

$$x = 2 + 3t$$
$$y = 4 - 4t \qquad \text{where} -\infty < t < +\infty$$
$$z = -1 + 8t$$

(b) The line intersects the xy-plane at the point where $z = -1 + 8t = 0$, that is, where $t = 1/8$. Substituting this value of t in the parametric equations for l yields as the point of intersection

$$(x, y, z) = (\tfrac{19}{8}, \tfrac{7}{2}, 0) \quad \blacktriangle$$

Example 21

Find parametric equations for the line of intersection of the planes

$$3x + 2y - 4z - 6 = 0 \qquad \text{and} \qquad x - 3y - 2z - 4 = 0$$

Solution. The line of intersection consists of all points (x, y, z) that satisfy the two equations in the system

$$3x + 2y - 4z = 6$$
$$x - 3y - 2z = 4$$

Solving this system gives

$$x = \tfrac{26}{11} + \tfrac{16}{11}t \qquad y = -\tfrac{6}{11} - \tfrac{2}{11}t \qquad z = t$$

The parametric equations for l are therefore

$$x = \tfrac{26}{11} + \tfrac{16}{11}t$$
$$y = -\tfrac{6}{11} - \tfrac{2}{11}t \qquad \text{where} -\infty < t < +\infty$$
$$z = t \quad \blacktriangle$$

In some problems, a line

$$x = x_0 + at$$
$$y = y_0 + bt \qquad \text{where } -\infty < t < +\infty \tag{3.25}$$
$$z = z_0 + ct$$

is given, and it is of interest to find two planes whose intersection is the given line. Since there are infinitely many planes through the line, there are always infinitely many such pairs of planes. To find two such planes when a, b, and c are all different from zero, we can rewrite each equation in (3.25) in the form

$$\frac{x - x_0}{a} = t \qquad \frac{y - y_0}{b} = t \qquad \frac{z - z_0}{c} = t$$

Eliminating the parameter t shows that the line consists of all points (x, y, z) that satisfy the equations

$$\frac{x - x_0}{a} = \frac{y - y_0}{b} = \frac{z - z_0}{c}$$

called **symmetric equations** for the line. Therefore, the line can be viewed as the intersection of the planes

$$\frac{x - x_0}{a} = \frac{y - y_0}{b} \qquad \text{and} \qquad \frac{y - y_0}{b} = \frac{z - z_0}{c}$$

or as the intersection of

$$\frac{x - x_0}{a} = \frac{z - z_0}{c} \qquad \text{and} \qquad \frac{y - y_0}{b} = \frac{z - z_0}{c}$$

and so forth.

Example 22

Find two planes whose intersection is the line

$$x = 3 + 2t$$
$$y = -4 + 7t \qquad \text{where } -\infty < t < +\infty$$
$$z = 1 + 3t$$

Solution. Since the symmetric equations for this line are

$$\frac{x - 3}{2} = \frac{y + 4}{7} = \frac{z - 1}{3} \tag{3.26}$$

it is the intersection of the planes

$$\frac{x - 3}{2} = \frac{y + 4}{7} \qquad \text{and} \qquad \frac{y + 4}{7} = \frac{z - 1}{3}$$

or equivalently

$$7x - 2y - 29 = 0 \qquad \text{and} \qquad 3y - 7z + 19 = 0$$

Other solutions can be obtained by choosing different pairs of equations from (3.26).

We conclude this section by discussing two basic "distance problems" in 3-space:

(a) Find the distance between a point and a plane.
(b) Find the distance between two parallel planes.

The two problems are related. If we can find the distance between a point and a plane, then we can find the distance between parallel planes by computing the distance between one of the planes and an arbitrary point P_0 in the other (Figure 3.34).

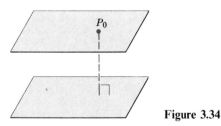

Figure 3.34

Theorem 8. *The distance D between a point $P_0(x_0, y_0, z_0)$ and the plane $ax + by + cz + d = 0$ is*

$$D = \frac{|ax_0 + by_0 + cz_0 + d|}{\sqrt{a^2 + b^2 + c^2}} \qquad (3.27)$$

Proof. Let $Q(x_1, y_1, z_1)$ be any point in the plane. Position the normal $\mathbf{n} = (a, b, c)$ so that its initial point is at Q. As illustrated in Figure 3.35, the distance D is

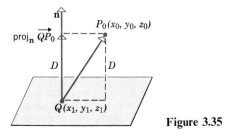

Figure 3.35

equal to the length of the orthogonal projection of QP_0 on $\mathbf{n}$. Thus, from (3.11) of Section 3.3,

$$D = \|\text{proj}_{\mathbf{n}}\, \overrightarrow{QP_0}\| = \frac{|\overrightarrow{QP_0} \cdot \mathbf{n}|}{\|\mathbf{n}\|}$$

But

$$\overrightarrow{QP_0} = (x_0 - x_1, y_0 - y_1, z_0 - z_1)$$
$$\overrightarrow{QP_0} \cdot \mathbf{n} = a(x_0 - x_1) + b(y_0 - y_1) + c(z_0 - z_1)$$
$$\|\mathbf{n}\| = \sqrt{a^2 + b^2 + c^2}$$

Thus,

$$D = \frac{|a(x_0 - x_1) + b(y_0 - y_1) + c(z_0 - z_1)|}{\sqrt{a^2 + b^2 + c^2}} \tag{3.28}$$

Since the point $Q(x_1, y_1, z_1)$ lies in the plane, its coordinates satisfy the equation of the plane, so that

$$ax_1 + by_1 + cz_1 + d = 0$$

or

$$d = -ax_1 - by_1 - cz_1$$

Substituting this expression in (3.28) yields (3.27).

REMARK. Note the similarity between (3.27) and the formula for the distance between a point and a line in 2-space [(3.14) of Section 3.3].

Example 23

Find the distance D between the point $(1, -4, -3)$ and the plane $2x - 3y + 6z = -1$.

Solution. To apply (3.27), we first rewrite the equation of the plane in the form

$$2x - 3y + 6z + 1 = 0$$

Then

$$D = \frac{|(2)(1) + (-3)(-4) + 6(-3) + 1|}{\sqrt{2^2 + (-3)^2 + 6^2}} = \frac{|-3|}{7} = \frac{3}{7}$$

Example 24

The planes

$$x + 2y - 2z = 3 \qquad \text{and} \qquad 2x + 4y - 4z = 7$$

are parallel since their normals, $(1, 2, -2)$ and $(2, 4, -4)$, are parallel vectors. Find the distance between these planes.

Solution. To find the distance D between the planes, we may select an arbitrary point in one of the planes and compute its distance to the other plane. By setting $y = z = 0$ in the equation $x + 2y - 2z = 3$, we obtain the point $P_0(3, 0, 0)$ in this plane. From (3.27), the distance between P_0 and the plane $2x + 4y - 4z = 7$ is

$$D = \frac{|(2)(3) + 4(0) + (-4)(0) - 7|}{\sqrt{2^2 + 4^2 + (-4)^2}} = \frac{1}{6}$$

EXERCISE SET 3.5

1. In each part find a point-normal form of the equation of the plane passing through P and having $\mathbf{n}$ as a normal.
    (a) $P(2, 6, 1)$; $\mathbf{n} = (1, 4, 2)$ (b) $P(-1, -1, 2)$; $\mathbf{n} = (-1, 7, 6)$
    (c) $P(1, 0, 0)$; $\mathbf{n} = (0, 0, 1)$ (d) $P(0, 0, 0)$; $\mathbf{n} = (2, 3, 4)$

2. Write the equations of the planes in Exercise 1 in general form.

3. Find a point-normal form of:
    (a) $2x - 3y + 7z - 10 = 0$ (b) $x + 3z = 0$

4. In each part find an equation for the plane passing through the given points.
    (a) $(-2, 1, 1)$ $(0, 2, 3)$ $(1, 0, -1)$
    (b) $(3, 2, 1)$ $(2, 1, -1)$ $(-1, 3, 2)$

5. Determine whether the planes are parallel.
    (a) $3x - 2y + z = 4$ and $6x - 4y + 3z = 7$
    (b) $2x - 8y - 6z - 2 = 0$ and $-x + 4y + 3z - 5 = 0$
    (c) $y \times 4x - 2z + 3$ and $x = \frac{1}{4}y + \frac{1}{2}z$

6. Determine whether the line and plane are parallel.
    (a) $x = 4 + 2t$, $y = -t$, $z = -1 - 4t$; $3x + 2y + z - 7 = 0$
    (b) $x = t$, $y = 2t$, $z = 3t$; $x - y + 2z = 5$

7. Determine whether the planes are perpendicular.
    (a) $x - y + 3z - 2 = 0$, $2x + z = 1$
    (b) $3x - 2y + z = 1$, $4x + 5y - 2z = 4$

8. Determine whether the line and plane are perpendicular.
    (a) $x = -1 + 2t$, $y = 4 + t$, $z = 1 - t$; $4x + 2y - 2z = 7$
    (b) $x = 3 - t$, $y = 2 + t$, $z = 1 - 3t$; $2x + 2y - 5 = 0$

9. In each part find parametric equations for the line passing through P and parallel to $\mathbf{n}$.
    (a) $P(2, 4, 6)$; $\mathbf{n} = (1, 2, 5)$ (b) $P(-3, 2, -4)$; $\mathbf{n} = (5, -7, -3)$
    (c) $P(1, 1, 5)$; $\mathbf{n} = (0, 0, 1)$ (d) $P(0, 0, 0)$; $\mathbf{n} = (1, 1, 1)$

10. Find symmetric equations for the lines in parts (a) and (b) of Exercise 9.

11. In each part find parametric equations for the line passing through the given points.
    (a) $(6, -1, 5), (7, 2, -4)$ (b) $(0, 0, 0), (-1, -1, -1)$

12. In each part find parametric equations for the line of intersection of the given planes.
(a) $-2x + 3y + 7z + 2 = 0$ and $x + 2y - 3z + 5 = 0$
(b) $3x - 5y + 2z = 0$ and $z = 0$

13. In each part find equations for two planes whose intersection is the given line.

(a) $\begin{aligned} x &= 3 + 4t \\ y &= -7 + 2t \\ z &= 6 - t \end{aligned}$ $-\infty < t < +\infty$ (b) $\begin{aligned} x &= 5t \\ y &= 3t \\ z &= 6t \end{aligned}$ $-\infty < t < +\infty$

14. Show that the line

$$x = 0$$
$$y = t \quad -\infty < t < +\infty$$
$$z = t$$

(a) lies in the plane $6x + 4y - 4z = 0$
(b) is parallel to and below the plane $5x - 3y + 3z = 1$
(c) is parallel to and above the plane $6x + 2y - 2z = 3$

15. Find an equation of the plane through $(-1, 4, -3)$ and perpendicular to the line $x - 2 = t, y + 3 = 2t, z = -t$.

16. Find an equation of
(a) the xy-plane   (b) the xz-plane   (c) the yz-plane

17. Find an equation of the plane that contains the point (x_0, y_0, z_0) and is
(a) parallel to the xy-plane
(b) parallel to the yz-plane
(c) parallel to the xz-plane

18. Find an equation of the plane that passes through the origin and is parallel to the plane $4x - 2y + 7z + 12 = 0$.

19. Find an equation of the plane passing through the point $(2, -7, 6)$ and parallel to the plane $5x - 2y + z - 9 = 0$.

20. Find the point of intersection of the line

$$x - 4 = 5t$$
$$y + 2 = t$$
$$z - 4 = -t$$

and the plane $3x - y + 7z + 8 = 0$.

21. Find an equation of the plane containing the line $x = -2 + 3t, y = 4 + 2t, z = 3 - t$ and perpendicular to the plane $x - 2y + z = 5$.

22. Find an equation of the plane through $(-1, 4, 2)$ and containing the line of intersection of the planes $4x - y + z - 2 = 0$ and $2x + y - 2z - 3 = 0$.

23. Show that the points $(1, 0, -1)$, $(0, 2, 3)$, $(-2, 1, 1)$, and $(4, 2, 3)$ lie in the same plane.

24. Find parametric equations for the line through $(5, 0, -2)$ that is parallel to the planes $x - 4y + 2z = 0$ and $2x + 3y - z + 1 = 0$.

25. Find an equation of the plane through $(-1, 2, -5)$ and perpendicular to the planes $2x - y + z = 1$ and $x + y - 2z = 3$.

26. Find an equation of the plane through $(1, 2, -1)$ and perpendicular to the line of intersection of the planes $2x + y + z = 2$ and $x + 2y + z = 3$.

27. Find an equation of the plane through the points $P_1(-2, 1, 4)$, $P_2(1, 0, 3)$ and perpendicular to the plane $4x - y + 3z = 2$.

28. Show that the lines

$$
\begin{array}{ll}
x = -2 + t & x = 3 - t \\
y = 3 + 2t \quad \text{and} & y = 4 - 2t \\
z = 4 - t & z = t
\end{array}
$$

are parallel, and find an equation of the plane they determine.

29. Find an equation of the plane that contains the point $(2, 0, 3)$ and the line $x = -1 + t$, $y = t$, $z = -4 + 2t$.

30. Find an equation of the plane, each of whose points is equidistant from $(2, -1, 1)$ and $(3, 1, 5)$.

31. Find an equation of the plane containing the line $x = 3t$, $y = 1 + t$, $z = 2t$ and parallel to the intersection of the planes $2x - y + z = 0$ and $y + z + 1 = 0$.

32. Show that the line

$$
\begin{array}{ll}
x - 4 = 2t \\
y = -t & -\infty < t < +\infty \\
z + 1 = -4t
\end{array}
$$

is parallel to the plane $3x + 2y + z - 7 = 0$.

33. Show that the lines

$$
\begin{array}{ll}
x + 1 = 4t & x + 13 = 12t \\
y - 3 = t \quad \text{and} & y - 1 = 6t \\
z - 1 = 0 & z - 2 = 3t
\end{array}
$$

intersect. Find the point of intersection.

34. Find an equation for the plane determined by the lines in Exercise 33.

35. Find parametric equations for the line of intersection of the planes
(a) $-2x + 3y + 7z + 2 = 0$ and $x + 2y - 3z + 5 = 0$
(b) $3x - 5y + 2z = 0$ and $z = 0$

36. Show that the plane whose intercepts with the coordinate axes are $x = a$, $y = b$, and $z = c$ has equation

$$
\frac{x}{a} + \frac{y}{b} + \frac{z}{c} = 1
$$

provided a, b, and c are nonzero.

37. In each part, find the distance between the point and the plane.
  (a) $(1, -2, 3)$; $2x - 2y + z = 4$
  (b) $(0, 1, 5)$; $3x + 6y - 2z - 5 = 0$
  (c) $(7, 2, -1)$; $20x - 4y - 5z = 0$

38. In each part, find the distance between the given parallel planes.
  (a) $2x - 3y + 4z = 7$, $4x - 6y + 8z = 3$
  (b) $-2x + y + z = 0$, $6x - 3y - 3z - 5 = 0$
  (c) $x + y + z = 1$, $x + y + z = -1$

CHAPTER FOUR

Vector Spaces

4.1 EUCLIDEAN n-SPACE

The idea of using pairs of numbers to locate points in the plane and triples of numbers to locate points in 3-space was first clearly spelled out in the mid-seventeenth century. By the latter part of the nineteenth century mathematicians and physicists began to realize that there was no need to stop with triples. It was recognized that quadruples of numbers (a_1, a_2, a_3, a_4) could be regarded as points in "4-dimensional" space, quintuples $(a_1, a_2, \ldots, a_5)$ as points in "5-dimensional" space, and so on. Although our geometric visualization does not extend beyond 3-space, it is nevertheless possible to extend many familiar ideas beyond 3-space by working with analytic or numerical properties of points and vectors rather than the geometric properties. In this section we shall make these ideas more precise.

> **Definition.** If n is a positive integer, then an ***ordered-n-tuple*** is a sequence of n real numbers $(a_1, a_2, \ldots, a_n)$. The set of all ordered n-tuples is called ***n-space*** and is denoted by R^n.

When $n = 2$ or 3, it is usual to use the terms ***ordered pair*** and ***ordered triple*** rather than ordered 2-tuple and 3-tuple. When $n = 1$, each ordered n-tuple consists of one real number, and so R^1 may be viewed as the set of real numbers. It is usual to write R rather than R^1 for this set.

It might have occurred to the reader in the study of 3-space that the symbol (a_1, a_2, a_3) has two different geometric interpretations. It can be interpreted as a point, in which case a_1, a_2, and a_3 are the coordinates (Figure 4.1a), or it can be

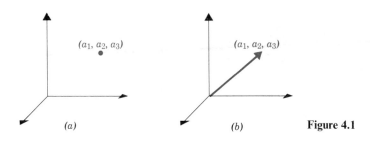

(a_1, a_2, a_3)

(a_1, a_2, a_3)

(a)

(b)

Figure 4.1

interpreted as a vector, in which case a_1, a_2, and a_3 are the components (Figure 4.1b). It follows, therefore, that an ordered n-tuple $(a_1, a_2, \ldots, a_n)$ can be viewed either as a "generalized point" or a "generalized vector"—the distinction is mathematically unimportant. Thus, we are free to describe the 5-tuple $(-2, 4, 0, 1, 6)$ as either a point in R^5 or a vector in R^5. We will use both descriptions.

Definition. Two vectors $\mathbf{u} = (u_1, u_2, \ldots, u_n)$ and $\mathbf{v} = (v_1, v_2, \ldots, v_n)$ in R^n are called **equal** if

$$u_1 = v_1, u_2 = v_2, \ldots, u_n = v_n$$

The **sum** $\mathbf{u} + \mathbf{v}$ is defined by

$$\mathbf{u} + \mathbf{v} = (u_1 + v_1, u_2 + v_2, \ldots, u_n + v_n)$$

and if k is any scalar, the **scalar multiple** $k\mathbf{u}$ is defined by

$$k\mathbf{u} = (ku_1, ku_2, \ldots, ku_n)$$

The operations of addition and scalar multiplication in this definition are called the **standard operations** on R^n.

We define the **zero vector** in R^n to be the vector

$$\mathbf{0} = (0, 0, \ldots, 0)$$

If $\mathbf{u} = (u_1, u_2, \ldots, u_n)$ is any vector in R^n, then the **negative** (or **additive inverse**) of $\mathbf{u}$ is denoted by $-\mathbf{u}$ and is defined by

$$-\mathbf{u} = (-u_1, -u_2, \ldots, -u_n)$$

We define subtraction of vectors in R^n by $\mathbf{v} - \mathbf{u} = \mathbf{v} + (-\mathbf{u})$ or, in terms of components,

$$\mathbf{v} - \mathbf{u} = (v_1 - u_1, v_2 - u_2, \ldots, v_n - u_n)$$

The most important arithmetic properties of addition and scalar multiplication of vectors in R^n are listed in the next theorem. The proofs are all easy and are left as exercises.

Theorem 1. *If* $\mathbf{u} = (u_1, u_2, \ldots, u_n)$, $\mathbf{v} = (v_1, v_2, \ldots, v_n)$, *and* $\mathbf{w} = (w_1, w_2, \ldots, w_n)$ *are vectors in* R^n *and* k *and* l *are scalars, then:*

(a) $\mathbf{u} + \mathbf{v} = \mathbf{v} + \mathbf{u}$
(b) $\mathbf{u} + (\mathbf{v} + \mathbf{w}) = (\mathbf{u} + \mathbf{v}) + \mathbf{w}$
(c) $\mathbf{u} + \mathbf{0} = \mathbf{0} + \mathbf{u} = \mathbf{u}$
(d) $\mathbf{u} + (-\mathbf{u}) = \mathbf{0}$, *that is,* $\mathbf{u} - \mathbf{u} = \mathbf{0}$
(e) $k(l\mathbf{u}) = (kl)\mathbf{u}$
(f) $k(\mathbf{u} + \mathbf{v}) = k\mathbf{u} + k\mathbf{v}$
(g) $(k + l)\mathbf{u} = k\mathbf{u} + l\mathbf{u}$
(h) $1\mathbf{u} = \mathbf{u}$

This theorem enables us to manipulate vectors in R^n without expressing the vectors in terms of components in much the same way that we manipulate real numbers. For example, to solve the vector equation $\mathbf{x} + \mathbf{u} = \mathbf{v}$ for $\mathbf{x}$, we can add $-\mathbf{u}$ to both sides and proceed as follows.

$$(\mathbf{x} + \mathbf{u}) + (-\mathbf{u}) = \mathbf{v} + (-\mathbf{u})$$
$$\mathbf{x} + (\mathbf{u} - \mathbf{u}) = \mathbf{v} - \mathbf{u}$$
$$\mathbf{x} + \mathbf{0} = \mathbf{v} - \mathbf{u}$$
$$\mathbf{x} = \mathbf{v} - \mathbf{u}$$

The reader will find it useful to name the parts of Theorem 1 that justify each of the steps in this computation.

To extend the notions of distance, norm, and angle to R^n, we begin with the following generalization of the dot product on R^2 and R^3 (Section 3.3).

Definition. If $\mathbf{u} = (u_1, u_2, \ldots, u_n)$ and $\mathbf{v} = (v_1, v_2, \ldots, v_n)$ are any vectors in R^n, then the **Euclidean inner product** $\mathbf{u} \cdot \mathbf{v}$ is defined by

$$\mathbf{u} \cdot \mathbf{v} = u_1 v_1 + u_2 v_2 + \cdots + u_n v_n$$

Observe that when $n = 2$ or 3, the Euclidean inner product is the ordinary dot product (Section 3.3).

Example 1
The Euclidean inner product of the vectors

$$\mathbf{u} = (-1, 3, 5, 7) \quad \text{and} \quad \mathbf{v} = (5, -4, 7, 0)$$

in R^4 is

$$\mathbf{u} \cdot \mathbf{v} = (-1)(5) + (3)(-4) + (5)(7) + (7)(0) = 18$$

The four main arithmetic properties of the Euclidean inner product are listed in the next theorem.

Theorem 2. *If* **u**, **v**, *and* **w** *are vectors in* R^n *and* k *is any scalar, then:*

(a) $\mathbf{u} \cdot \mathbf{v} = \mathbf{v} \cdot \mathbf{u}$
(b) $(\mathbf{u} + \mathbf{v}) \cdot \mathbf{w} = \mathbf{u} \cdot \mathbf{w} + \mathbf{v} \cdot \mathbf{w}$
(c) $(k\mathbf{u}) \cdot \mathbf{v} = k(\mathbf{u} \cdot \mathbf{v})$
(d) $\mathbf{v} \cdot \mathbf{v} \geq 0$. *Further,* $\mathbf{v} \cdot \mathbf{v} = 0$ *if and only if* $\mathbf{v} = \mathbf{0}$

We shall prove parts (b) and (d) and leave proofs of the rest as exercises.

Proof (b). Let $\mathbf{u} = (u_1, u_2, \ldots, u_n)$, $\mathbf{v} = (v_1, v_2, \ldots, v_n)$, and $\mathbf{w} = (w_1, w_2, \ldots, w_n)$. Then

$$
\begin{aligned}
(\mathbf{u} + \mathbf{v}) \cdot \mathbf{w} &= (u_1 + v_1, u_2 + v_2, \ldots, u_n + v_n) \cdot (w_1, w_2, \ldots, w_n) \\
&= (u_1 + v_1)w_1 + (u_2 + v_2)w_2 + \cdots + (u_n + v_n)w_n \\
&= (u_1 w_1 + u_2 w_2 + \cdots + u_n w_n) + (v_1 w_1 + v_2 w_2 + \cdots + v_n w_n) \\
&= \mathbf{u} \cdot \mathbf{w} + \mathbf{v} \cdot \mathbf{w}
\end{aligned}
$$

(d). $\mathbf{v} \cdot \mathbf{v} = v_1^2 + v_2^2 + \cdots + v_n^2 \geq 0$. Further, equality holds if and only if $v_1 = v_2 = \cdots = v_n = 0$; that is, if and only if $\mathbf{v} = \mathbf{0}$.

Example 2

Theorem 2 allows us to perform computations with Euclidean inner products in much the same way that we perform them with ordinary arithmetic products. For example,

$$
\begin{aligned}
(3\mathbf{u} + 2\mathbf{v}) \cdot (4\mathbf{u} + \mathbf{v}) &= (3\mathbf{u}) \cdot (4\mathbf{u} + \mathbf{v}) + (2\mathbf{v}) \cdot (4\mathbf{u} + \mathbf{v}) \\
&= (3\mathbf{u}) \cdot (4\mathbf{u}) + (3\mathbf{u}) \cdot \mathbf{v} + (2\mathbf{v}) \cdot (4\mathbf{u}) + (2\mathbf{v}) \cdot \mathbf{v} \\
&= 12(\mathbf{u} \cdot \mathbf{u}) + 3(\mathbf{u} \cdot \mathbf{v}) + 8(\mathbf{v} \cdot \mathbf{u}) + 2(\mathbf{v} \cdot \mathbf{v}) \\
&= 12(\mathbf{u} \cdot \mathbf{u}) + 11(\mathbf{u} \cdot \mathbf{v}) + 2(\mathbf{v} \cdot \mathbf{v})
\end{aligned}
$$

The reader should determine which parts of Theorem 2 were used in each step.

By analogy with the familiar formulas in R^2 and R^3, we define the *Euclidean norm* (or *Euclidean length*) of a vector $\mathbf{u} = (u_1, u_2, \ldots, u_n)$ in R^n by

$$\|\mathbf{u}\| = (\mathbf{u} \cdot \mathbf{u})^{1/2} = \sqrt{u_1^2 + u_2^2 + \cdots + u_n^2}$$

Similarly, the **Euclidean distance** between the points $\mathbf{u} = (u_1, u_2, \ldots, u_n)$ and $\mathbf{v} = (v_1, v_2, \ldots, v_n)$ in R^n is defined by

$$d(\mathbf{u}, \mathbf{v}) = \|\mathbf{u} - \mathbf{v}\| = \sqrt{(u_1 - v_1)^2 + (u_2 - v_2)^2 + \cdots + (u_n - v_n)^2}$$

Example 3

If $\mathbf{u} = (1, 3, -2, 7)$ and $\mathbf{v} = (0, 7, 2, 2)$ then

$$\|\mathbf{u}\| = \sqrt{(1)^2 + (3)^2 + (-2)^2 + (7)^2} = \sqrt{63} = 3\sqrt{7}$$

and

$$d(\mathbf{u}, \mathbf{v}) = \sqrt{(1 - 0)^2 + (3 - 7)^2 + (-2 - 2)^2 + (7 - 2)^2} = \sqrt{58} \;\; ▲$$

Since so many of the familiar ideas from 2-space and 3-space carry over, it is common to refer to R^n with the operations of addition, scalar multiplication, and inner product that we have defined here as **Euclidean n-space**.

We conclude this section by noting that it is possible to use the matrix notation

$$\mathbf{u} = \begin{bmatrix} u_1 \\ u_2 \\ \vdots \\ u_n \end{bmatrix}$$

rather than the horizontal notation $\mathbf{u} = (u_1, u_2, \ldots, u_n)$ to denote vectors in R^n. This is justified because the matrix operations

$$\mathbf{u} + \mathbf{v} = \begin{bmatrix} u_1 \\ u_2 \\ \vdots \\ u_n \end{bmatrix} + \begin{bmatrix} v_1 \\ v_2 \\ \vdots \\ v_n \end{bmatrix} = \begin{bmatrix} u_1 + v_1 \\ u_2 + v_2 \\ \vdots \\ u_n + v_n \end{bmatrix}$$

$$k\mathbf{u} = k\begin{bmatrix} u_1 \\ u_2 \\ \vdots \\ u_n \end{bmatrix} = \begin{bmatrix} ku_1 \\ ku_2 \\ \vdots \\ ku_n \end{bmatrix}$$

produce the same results as the vector operations

$$\mathbf{u} + \mathbf{v} = (u_1, u_2, \ldots, u_n) + (v_1, v_2, \ldots, v_n) = (u_1 + v_1, u_2 + v_2, \ldots, u_n + v_n)$$
$$k\mathbf{u} = k(u_1, u_2, \ldots, u_n) = (ku_1, ku_2, \ldots, ku_n)$$

The only difference is that results are displayed vertically in one case and horizontally in the other. We will use both notations at various times. However, from here on, we shall denote $n \times 1$ matrices by lowercase boldface letters. Thus, a system of linear equations will be written

$$A\mathbf{x} = \mathbf{b}$$

rather than $AX = B$ as before.

If we use matrix notation for the vectors

$$\mathbf{u} = \begin{bmatrix} u_1 \\ u_2 \\ \vdots \\ u_n \end{bmatrix} \quad \text{and} \quad \mathbf{v} = \begin{bmatrix} v_1 \\ v_2 \\ \vdots \\ v_n \end{bmatrix}$$

and omit the brackets on 1×1 matrices, then it follows that

$$\mathbf{v}^t\mathbf{u} = \begin{bmatrix} v_1 v_2 \dots v_n \end{bmatrix} \begin{bmatrix} u_1 \\ u_2 \\ \vdots \\ u_n \end{bmatrix} = [u_1 v_1 + u_2 v_2 + \cdots + u_n v_n] = [\mathbf{u} \cdot \mathbf{v}] = \mathbf{u} \cdot \mathbf{v}$$

Thus, for vectors in vertical notation we have the matrix formula

$$\mathbf{v}^t\mathbf{u} = \mathbf{u} \cdot \mathbf{v}$$

for the Euclidean inner product. For example, if

$$\mathbf{u} = \begin{bmatrix} -1 \\ 3 \\ 5 \\ 7 \end{bmatrix} \quad \text{and} \quad \mathbf{v} = \begin{bmatrix} 5 \\ -4 \\ 7 \\ 0 \end{bmatrix}$$

then

$$\mathbf{u} \cdot \mathbf{v} = \mathbf{v}^t\mathbf{u} = \begin{bmatrix} 5 & -4 & 7 & 0 \end{bmatrix} \begin{bmatrix} -1 \\ 3 \\ 5 \\ 7 \end{bmatrix} = [18] = 18$$

EXERCISE SET 4.1

1. Let $\mathbf{u} = (2, 0, -1, 3)$, $\mathbf{v} = (5, 4, 7, -1)$, and $\mathbf{w} = (6, 2, 0, 9)$. Find
   (a) $\mathbf{u} - \mathbf{v}$ (b) $7\mathbf{v} + 3\mathbf{w}$ (c) $-\mathbf{w} + \mathbf{v}$
   (d) $3(\mathbf{u} - 7\mathbf{v})$ (e) $-3\mathbf{v} - 8\mathbf{w}$ (f) $2\mathbf{v} - (\mathbf{u} + \mathbf{w})$

2. Let $\mathbf{u}$, $\mathbf{v}$, and $\mathbf{w}$ be the vectors in Exercise 1. Find the vector $\mathbf{x}$ that satisfies $2\mathbf{u} - \mathbf{v} + \mathbf{x} = 7\mathbf{x} + \mathbf{w}$.

3. Let $u_1 = (-1, 3, 2, 0)$, $u_2 = (2, 0, 4, -1)$, $u_3 = (7, 1, 1, 4)$, and $u_4 = (6, 3, 1, 2)$. Find scalars c_1, c_2, c_3, and c_4 such that $c_1 u_1 + c_2 u_2 + c_3 u_3 + c_4 u_4 = (0, 5, 6, -3)$.

4. Show that there do not exist scalars c_1, c_2, and c_3 such that $c_1(1, 0, -2, 1) + c_2(2, 0, 1, 2) + c_3(1, -2, 2, 3) = (1, 0, 1, 0)$.

5. Compute the Euclidean norm of v when
   (a) $v = (4, -3)$ (b) $v = (1, -1, 3)$ (c) $v = (2, 0, 3, -1)$ (d) $v = (-1, 1, 1, 3, 6)$

6. Let $u = (3, 0, 1, 2)$, $v = (-1, 2, 7, -3)$, and $w = (2, 0, 1, 1)$. Find
   (a) $\|u + v\|$ (b) $\|u\| + \|v\|$ (c) $\|-2u\| + 2\|u\|$

   (d) $\|3u - 5v + w\|$ (e) $\dfrac{1}{\|w\|} w$ (f) $\left\|\dfrac{1}{\|w\|} w\right\|$

7. Show that if v is a nonzero vector in R^n, then $(1/\|v\|)v$ has norm 1.

8. Find all scalars k such that $\|kv\| = 3$, where $v = (-1, 2, 0, 3)$.

9. Find the Euclidean inner product $u \cdot v$ when
   (a) $u = (-1, 3)$, $v = (7, 2)$ (b) $u = (3, 7, 1)$, $v = (-1, 0, 2)$
   (c) $u = (1, -1, 2, 3)$, $v = (3, 3, -6, 4)$ (d) $u = (1, 3, 2, 6, -1)$, $v = (0, 0, 2, 4, 1)$

10. (a) Find two vectors in R^2 with Euclidean norm 1 whose Euclidean inner products with $(-2, 4)$ are zero.
    (b) Show that there are infinitely many vectors in R^3 with Euclidean norm 1 whose Euclidean inner product with $(-1, 7, 2)$ is zero.

11. Find the Euclidean distance between u and v when:
    (a) $u = (2, -1)$, $v = (3, 2)$ (b) $u = (1, 1, -1)$, $v = (2, 6, 0)$
    (c) $u = (2, 0, 1, 3)$, $v = (-1, 4, 6, 6)$ (d) $u = (6, 0, 1, 3, 0)$, $v = (-1, 4, 2, 8, 3)$

12. Establish the identity

$$\|u + v\|^2 + \|u - v\|^2 = 2\|u\|^2 + 2\|v\|^2$$

for vectors in R^n. Interpret this result geometrically in R^2.

13. Establish the identity

$$u \cdot v = \tfrac{1}{4}\|u + v\|^2 - \tfrac{1}{4}\|u - v\|^2$$

for vectors in R^n.

14. Verify parts (b), (e), (f) and (g) of Theorem 1 when $u = (1, 0, -1, 2)$, $v = (3, -1, 2, 4)$, $w = (2, 7, 3, 0)$, $k = 6$, and $l = -2$.

15. Verify parts (b) and (c) of Theorem 2 for the values of u, v, w, and k in Exercise 14.

16. Prove (a) through (d) of Theorem 1.

17. Prove (e) through (h) of Theorem 1.

18. Prove (a) and (c) of Theorem 2.

19. Prove: If u, v, and w are vectors in R^n and k is any scalar, then
    (a) $u \cdot (kv) = k(u \cdot v)$
    (b) $u \cdot (v + w) = u \cdot v + u \cdot w$

4.2 GENERAL VECTOR SPACES

In this section we generalize the concept of a vector still further. We shall state a set of axioms which, if satisfied by a class of objects, will entitle those objects to be called "vectors." The axioms will be chosen by abstracting the most important properties of vectors in R^n; as a consequence, vectors in R^n will automatically satisfy these axioms. Thus, our new concept of a vector will include our old vectors and many new kinds of vectors as well.

Definition. Let V be an arbitrary nonempty set of objects on which two operations are defined, addition and multiplication by scalars (real numbers). By addition we mean a rule for associating with each pair of objects **u** and **v** in V an element **u** + **v**, called the *sum* of **u** and **v**; by scalar multiplication we mean a rule for associating with each scalar k and each object **u** in V an element $k\mathbf{u}$, called the *scalar multiple* of **u** by k. If the following axioms are satisfied by all objects **u**, **v**, **w** in V and all scalars k and l, then we call V a *vector space* and we called the objects in V *vectors*:

(1) If **u** and **v** are objects in V, then **u** + **v** is in V.
(2) $\mathbf{u} + \mathbf{v} = \mathbf{v} + \mathbf{u}$
(3) $\mathbf{u} + (\mathbf{v} + \mathbf{w}) = (\mathbf{u} + \mathbf{v}) + \mathbf{w}$
(4) There is an object **0** in V such that $\mathbf{0} + \mathbf{u} = \mathbf{u} + \mathbf{0} = \mathbf{u}$ for all **u** in V.
(5) For each **u** in V, there is an object $-\mathbf{u}$ in V called the *negative* of **u** such that $\mathbf{u} + (-\mathbf{u}) = (-\mathbf{u}) + \mathbf{u} = \mathbf{0}$.
(6) If k is any scalar and **u** is any object in V, then $k\mathbf{u}$ is in V.
(7) $k(\mathbf{u} + \mathbf{v}) = k\mathbf{u} + k\mathbf{v}$
(8) $(k + l)\mathbf{u} = k\mathbf{u} + l\mathbf{u}$
(9) $k(l\mathbf{u}) = (kl)(\mathbf{u})$
(10) $1\mathbf{u} = \mathbf{u}$

The vector **0** in Axiom 4 is called the *zero vector* for V.

For some applications it is necessary to consider vector spaces where the scalars are complex numbers rather than real numbers. Vector spaces in which the scalars are complex numbers are called *complex vector spaces*, and those in which the scalars must be real are called *real vector spaces*. In Chapter 9 we shall discuss complex vector spaces; until then, *all our scalars will be real*.

The reader should keep in mind that, in the definition of a vector space, neither the nature of the vectors nor the operations is specified. Any kinds of objects whatsoever can serve as vectors; all that is required is that the vector-space axioms be satisfied. The following examples will give some idea of the diversity of possible vector spaces.

Example 4

The set $V = R^n$ with the standard operations of addition and scalar multiplication defined in the previous section is a vector space. Axioms 1 and 6 follow from the definitions of the standard operations on R^n; the remaining axioms follow from Theorem 1. ▲

Example 5

Let V be any plane through the origin in R^3. We shall show that the points in V form a vector space under the standard addition and scalar multiplication operations for vectors in R^3.

From Example 4, we know that R^3 itself is a vector space under these operations. Thus Axioms 2, 3, 7, 8, 9, and 10 hold for all points in R^3 and consequently for all points in the plane V. We therefore need only show that Axioms 1, 4, 5, and 6 are satisfied.

Since the plane V passes through the origin, it has an equation of the form

$$ax + by + cz = 0 \tag{4.1}$$

(Theorem 7 in Chapter 3). Thus, if $\mathbf{u} = (u_1, u_2, u_3)$ and $\mathbf{v} = (v_1, v_2, v_3)$ are points in V, then $au_1 + bu_2 + cu_3 = 0$ and $av_1 + bv_2 + cv_3 = 0$. Adding these equations gives

$$a(u_1 + v_1) + b(u_2 + v_2) + c(u_3 + v_3) = 0$$

This equality tells us that the coordinates of the point $\mathbf{u} + \mathbf{v} = (u_1 + v_1, u_2 + v_2 + u_3 + v_3)$ satisfy (4.1); thus, $\mathbf{u} + \mathbf{v}$ lies in the plane V. This proves that Axiom 1 is satisfied. Multiplying $au_1 + bu_2 + cu_3 = 0$ through by -1 gives

$$a(-u_1) + b(-u_2) + c(-u_3) = 0$$

Thus, $-\mathbf{u} = (-u_1, -u_2, -u_3)$ lies in V. This establishes Axiom 5. The verifications of Axioms 4 and 6 are left as exercises. ▲

Example 6

The points on a line V passing through the origin in R^3 form a vector space under the standard addition and scalar multiplication operations for vectors in R^3.

The argument is similar to that used in Example 5 and is based on the fact that the points of V satisfy parametric equations of the form

$$\begin{aligned} x &= at \\ y &= bt \qquad -\infty < t < +\infty \\ z &= ct \end{aligned}$$

(Section 3.5). The details are left as an exercise. ▲

Example 7

The set V of all $m \times n$ matrices with real entries, together with the operations of matrix addition and scalar multiplication, is a vector space. The $m \times n$ zero matrix is the zero vector **0**, and if **u** is the $m \times n$ matrix A, then the matrix $-A$ is the vector $-\mathbf{u}$ in Axiom 5. Most of the remaining axioms are satisfied by virtue of Theorem 2 in Section 1.5. We shall denote this vector space by the symbol M_{mn}.

Example 8

Let V be the set of real-valued functions defined on the entire real line. If $\mathbf{f} = f(x)$ and $\mathbf{g} = g(x)$ are two such functions and k is any real number, define the sum function $\mathbf{f} + \mathbf{g}$ and the scalar multiple $k\mathbf{f}$ by

$$(\mathbf{f} + \mathbf{g})(x) = f(x) + g(x)$$
$$(k\mathbf{f})(x) = kf(x)$$

In other words, the value of the function $\mathbf{f} + \mathbf{g}$ at x is obtained by adding together the values of $\mathbf{f}$ and $\mathbf{g}$ at x (Figure 4.2a). Similarly, the value of $k\mathbf{f}$ at x is k times the value of $\mathbf{f}$ at x (Figure 4.2b). The set V is a vector space under these operations.

The zero vector in this space is the zero constant function, that is, the function whose graph is a horizontal line through the origin. The verification of the remaining axioms is an exercise.

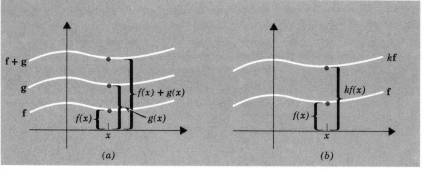

(a)   (b)

Figure 4.2

Example 9

Let V be the set of all points (x, y) in R^2 that lie in the first quadrant, that is, such that $x \geq 0$ and $y \geq 0$. The set V is *not* a vector space under the standard operations on R^2, since Axioms 5 and 6 are not satisfied. To see this, observe that $\mathbf{v} = (1, 1)$ lies in V, but $(-1)\mathbf{v} = -\mathbf{v} = (-1, -1)$ does not (Figure 4.3).

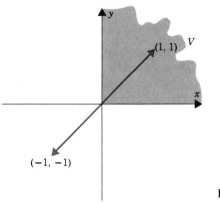

Figure 4.3

Example 10

Let V consist of a single object, which we denote by **0**, and define

$$\mathbf{0} + \mathbf{0} = \mathbf{0}$$
$$k\mathbf{0} = \mathbf{0}$$

for all scalars k. It is easy to check that all the vector-space axioms are satisfied. We call this the *zero vector space*.

As we progress, we shall add more examples of vector spaces to our list. We conclude this section with a theorem that gives a useful list of vector properties.

Theorem 3. *Let V be a vector space, $\mathbf{u}$ a vector in V, and k a scalar; then:*

(*a*) $0\mathbf{u} = \mathbf{0}$
(*b*) $k\mathbf{0} = \mathbf{0}$
(*c*) $(-1)\mathbf{u} = -\mathbf{u}$
(*d*) *If $k\mathbf{u} = \mathbf{0}$, then $k = 0$ or $\mathbf{u} = \mathbf{0}$*

Proof. We shall prove parts (*a*) and (*c*) and leave proofs of the remaining parts as exercises.
(*a*) We can write

$$0\mathbf{u} + 0\mathbf{u} = (0 + 0)\mathbf{u} \qquad \qquad \text{(Axiom 8)}$$
$$= 0\mathbf{u} \qquad \qquad \text{(Property of the number 0)}$$

By Axiom 5 the vector $0\mathbf{u}$ has a negative, $-0\mathbf{u}$. Adding this negative to both sides above yields

$$[0\mathbf{u} + 0\mathbf{u}] + (-0\mathbf{u}) = 0\mathbf{u} + (-0\mathbf{u})$$

or

$$0\mathbf{u} + [0\mathbf{u} + (-0\mathbf{u})] = 0\mathbf{u} + (-0\mathbf{u}) \qquad \text{(Axiom 3)}$$

or

$$0\mathbf{u} + \mathbf{0} = \mathbf{0} \qquad \text{(Axiom 5)}$$

or

$$0\mathbf{u} = \mathbf{0} \qquad \text{(Axiom 4)}$$

(c) To show $(-1)\mathbf{u} = -\mathbf{u}$, we must demonstrate that $\mathbf{u} + (-1)\mathbf{u} = \mathbf{0}$. To see this, observe that

$$\begin{aligned} \mathbf{u} + (-1)\mathbf{u} &= 1\mathbf{u} + (-1)\mathbf{u} & \text{(Axiom 10)} \\ &= (1 + (-1))\mathbf{u} & \text{(Axiom 8)} \\ &= 0\mathbf{u} & \text{(Property of numbers)} \\ &= \mathbf{0} & \text{(Part } (a) \text{ above)} \end{aligned}$$

EXERCISE SET 4.2

In Exercises 1–14 a set of objects is given together with operations of addition and scalar multiplication. Determine which sets are vector spaces under the given operations. For those that are not, list all axioms that fail to hold.

1. The set of all triples of real numbers (x, y, z) with the operations $(x, y, z) + (x', y', z') = (x + x', y + y', z + z')$ and $k(x, y, z) = (kx, y, z)$.

2. The set of all triples of real numbers (x, y, z) with the operations $(x, y, z) + (x', y', z') = (x + x', y + y', z + z')$ and $k(x, y, z) = (0, 0, 0)$.

3. The set of all pairs of real numbers (x, y) with the operations $(x, y) + (x', y') = (x + x', y + y')$ and $k(x, y) = (2kx, 2ky)$.

4. The set of all real numbers x with the standard operations of addition and multiplication.

5. The set of all pairs of real numbers of the form $(x, 0)$ with the standard operations on R^2.

6. The set of all pairs of real numbers of the form (x, y) where $x \geq 0$, with the standard operations on R^2.

7. The set of all n-tuples of real numbers of the form $(x, x, \ldots, x)$ with the standard operations on R^n.

8. The set of all pairs of real numbers (x, y) with the operations $(x, y) + (x', y') = (x + x' + 1, y + y' + 1)$ and $k(x, y) = (kx, ky)$.

9. The set of all positive real numbers x with the operations $x + x' = xx'$ and $kx = x^k$.

10. The set of all 2×2 matrices of the form

$$\begin{bmatrix} a & 1 \\ 1 & b \end{bmatrix}$$

with matrix addition and scalar multiplication.

11. The set of all 2×2 matrices of the form

$$\begin{bmatrix} a & 0 \\ 0 & b \end{bmatrix}$$

 with matrix addition and scalar multiplication.

12. The set of all real-valued functions f defined everywhere on the real line and such that $f(1) = 0$, with the operations defined in Example 8.

13. The set of all 2×2 matrices of the form

$$\begin{bmatrix} a & a+b \\ a+b & b \end{bmatrix}$$

 with matrix addition and scalar multiplication.

14. The set whose only element is the moon. The operations are moon + moon = moon and k(moon) = moon, where k is a real number.

15. Prove that a line passing through the origin in R^3 is a vector space under the standard operations on R^3.

16. Complete the unfinished details in Example 5.

17. Complete the unfinished details in Example 8.

18. Prove part (*b*) of Theorem 3.

19. Prove part (*d*) of Theorem 3.

20. Prove that a vector space cannot have more than one zero vector.

21. Prove that a vector has exactly one negative.

4.3 SUBSPACES

 It is possible for one vector space to be contained within a larger vector space. For example, lines through the origin and planes through the origin are vector spaces that are contained within the larger vector space R^3 (Examples 5 and 6). In this section we shall discuss this idea in detail.

> **Definition.** A subset W of a vector space V is called a *subspace* of V if W is itself a vector space under the addition and scalar multiplication defined on V.

 In general, one must verify the ten vector space axioms to show that a set W with addition and scalar multiplication forms a vector space. However, if W is part of a larger set V that is already known to be a vector space, then certain axioms need not be verified for W because they are "inherited" from V. For example, there is no need to check that $\mathbf{u} + \mathbf{v} = \mathbf{v} + \mathbf{u}$ (Axiom 2) for W because this holds for all vectors in V and consequently for all vectors in W. Other axioms inherited by W from V are 3, 7, 8, 9, and 10. Thus, to show that a set W is a subspace of a vector space V, we need only verify Axioms 1, 4, 5, and 6. The following theorem shows that even Axioms 4 and 5 can be dispensed with.

Theorem 4. *If W is a set of one or more vectors from a vector space V, then W is a subspace of V if and only if the following conditions hold.*

(a) *If $\mathbf{u}$ and $\mathbf{v}$ are vectors in W, then $\mathbf{u} + \mathbf{v}$ is in W.*
(b) *If k is any scalar and $\mathbf{u}$ is any vector in W, then $k\mathbf{u}$ is in W.*

[Conditions (a) and (b) are often described by saying that W is **closed under addition** and **closed under scalar multiplication**.]

Proof. If W is a subspace of V, then all the vector-space axioms are satisfied; in particular Axioms 1 and 6 hold. But these are precisely conditions (a) and (b).

Conversely, assume conditions (a) and (b) hold. Since these conditions are vector-space Axioms 1 and 6, we need only show that W satisfies the remaining eight axioms. Axioms 2, 3, 7, 8, 9, and 10 are automatically satisfied by the vectors in W since they are satisfied by all vectors in V. Therefore, to complete the proof, we need only verify that Axioms 4 and 5 are satisfied by W.

Let $\mathbf{u}$ be any vector in W. By condition (b), $k\mathbf{u}$ is in W for every scalar k. Setting $k = 0$, it follows that $0\mathbf{u} = \mathbf{0}$ is in W, and setting $k = -1$, it follows that $(-1)\mathbf{u} = -\mathbf{u}$ is in W.

Every vector space V has at least two subspaces. V itself is a subspace, and the set $\{\mathbf{0}\}$ consisting of just the zero vector in V is a subspace called the **zero subspace**. The following examples provide less trivial illustrations of subspaces.

Example 11

In Example 5 of Section 4.2 we showed that all vectors in any plane through the origin of R^3 form a vector space; that is, planes through the origin are subspaces of R^3. We can also prove this result geometrically using Theorem 4.

Let W be any plane through the origin and let $\mathbf{u}$ and $\mathbf{v}$ be any vectors in W. Then $\mathbf{u} + \mathbf{v}$ must lie in W because it is the diagonal of the parallelogram determined by $\mathbf{u}$ and $\mathbf{v}$ (Figure 4.4), and $k\mathbf{u}$ must lie in W for any scalar k because $k\mathbf{u}$ lies on a line through $\mathbf{u}$. Thus, W is a subspace of R^3. ▲

REMARK. A geometric argument like the one in this example can be used to show that lines through the origin are subspaces of R^3. It can be shown (Exercise 20,

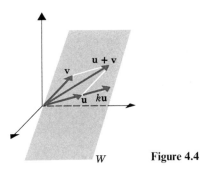

Figure 4.4

Section 4.5) that the only subspaces of R^3 are $\{0\}$, R^3, lines through the origin, and planes through the origin. Also, the only subspaces of R^2 are $\{0\}$, R^2, and lines through the origin.

Example 12

Show that the set W of all 2×2 matrices having zeros on the main diagonal is a subspace of the vector space M_{22} of all 2×2 matrices.

Solution. Let

$$A = \begin{bmatrix} 0 & a_{12} \\ a_{21} & 0 \end{bmatrix} \qquad B = \begin{bmatrix} 0 & b_{12} \\ b_{21} & 0 \end{bmatrix}$$

be any two matrices in W and k any scalar. Then

$$kA = \begin{bmatrix} 0 & ka_{12} \\ ka_{21} & 0 \end{bmatrix} \quad \text{and} \quad A + B = \begin{bmatrix} 0 & a_{12} + b_{12} \\ a_{21} + b_{21} & 0 \end{bmatrix}$$

Since kA and $A + B$ have zeros on the main diagonal, they lie in W. Thus, W is a subspace of M_{22}.

Example 13

Let n be a positive integer, and let W consist of the zero function and all real polynomial functions having degree $\leq n$. Thus, W is the set of all functions expressible in the form

$$p(x) = a_0 + a_1 x + \cdots + a_n x^n \tag{4.2}$$

where $a_0, \ldots, a_n$ are real numbers. The set W is a subspace of the vector space of all real-valued functions discussed in Example 8. To see this, let $\mathbf{p}$ and $\mathbf{q}$ be the polynomials

$$p(x) = a_0 + a_1 x + \cdots + a_n x^n$$

and

$$q(x) = b_0 + b_1 x + \cdots + b_n x^n$$

Then

$$(\mathbf{p} + \mathbf{q})(x) = p(x) + q(x) = (a_0 + b_0) + (a_1 + b_1)x + \cdots + (a_n + b_n)x^n$$

and

$$(k\mathbf{p})(x) = kp(x) = (ka_0) + (ka_1)x + \cdots + (ka_n)x^n$$

have the form given in (4.2). Therefore, $\mathbf{p} + \mathbf{q}$ and $k\mathbf{p}$ lie in W. We shall denote the vector space W in this example by the symbol P_n. ▲

Example 14

(For readers who have studied calculus.)
Recall from calculus that if $\mathbf{f}$ and $\mathbf{g}$ are continuous functions and k is a constant, then $\mathbf{f} + \mathbf{g}$ and $k\mathbf{f}$ are also continuous functions. It follows that the set of all continuous functions is a subspace of the vector space of all real-valued functions. This space is denoted by $C(-\infty, +\infty)$. A closely related example is the vector space of all functions that are continuous on a closed interval $a \leq x \leq b$. This space is denoted by $C[a, b]$. ▲

Example 15

Consider a system of m linear equations in n unknowns

$$
\begin{aligned}
a_{11}x_1 + a_{12}x_2 + \cdots + a_{1n}x_n &= b_1 \\
a_{21}x_1 + a_{22}x_2 + \cdots + a_{2n}x_n &= b_2 \\
&\vdots \\
a_{m1}x_1 + a_{m2}x_2 + \cdots + a_{mn}x_n &= b_m
\end{aligned}
$$

or, in matrix notation, $A\mathbf{x} = \mathbf{b}$. A vector*

$$
\mathbf{s} = \begin{bmatrix} s_1 \\ s_2 \\ \vdots \\ s_n \end{bmatrix}
$$

in R^n is called a **solution vector** of the system if $x_1 = s_1, x_2 = s_2, \ldots, x_n = s_n$ is a solution of the system. In this example we shall show that the set of solution vectors of a *homogeneous* system is a subspace of R^n.

Let $A\mathbf{x} = \mathbf{0}$ be a homogeneous system of m linear equations in n unknowns, let W be the set of solution vectors, and let $\mathbf{s}$ and $\mathbf{s}'$ be vectors in W. To show that W is closed under addition and scalar multiplication, we must demonstrate that $\mathbf{s} + \mathbf{s}'$ and $k\mathbf{s}$ are also solution vectors, where k is any scalar. Since $\mathbf{s}$ and $\mathbf{s}'$ are solution vectors we have

$$A\mathbf{s} = \mathbf{0} \qquad \text{and} \qquad A\mathbf{s}' = \mathbf{0}$$

---

* In this example we are using the matrix notation for vectors in R^n.

Therefore,

$$A(\mathbf{s} + \mathbf{s}') = A\mathbf{s} + A\mathbf{s}' = \mathbf{0} + \mathbf{0} = \mathbf{0}$$

and

$$A(k\mathbf{s}) = k(A\mathbf{s}) = k\mathbf{0} = \mathbf{0}$$

These equations show that $\mathbf{s} + \mathbf{s}'$ and $k\mathbf{s}$ satisfy the equation $A\mathbf{x} = \mathbf{0}$. Thus, $\mathbf{s} + \mathbf{s}'$ and $k\mathbf{s}$ are solution vectors.

The subspace W in this example is called the **solution space** of the system $A\mathbf{x} = \mathbf{0}$.

The following definition is one of the most fundamental concepts in the study of vectors.

Definition. A vector $\mathbf{w}$ is called a **linear combination** of the vectors $\mathbf{v}_1, \mathbf{v}_2, \ldots, \mathbf{v}_r$ if it can be expressed in the form

$$\mathbf{w} = k_1\mathbf{v}_1 + k_2\mathbf{v}_2 + \cdots + k_r\mathbf{v}_r$$

where $k_1, k_2, \ldots, k_r$ are scalars.

Example 16

Consider the vectors $\mathbf{u} = (1, 2, -1)$ and $\mathbf{v} = (6, 4, 2)$ in R^3. Show that $\mathbf{w} = (9, 2, 7)$ is a linear combination of $\mathbf{u}$ and $\mathbf{v}$ and that $\mathbf{w}' = (4, -1, 8)$ is *not* a linear combination of $\mathbf{u}$ and $\mathbf{v}$.

Solution. In order for $\mathbf{w}$ to be a linear combination of $\mathbf{u}$ and $\mathbf{v}$, there must be scalars k_1 and k_2 such that $\mathbf{w} = k_1\mathbf{u} + k_2\mathbf{v}$; that is,

$$(9, 2, 7) = k_1(1, 2, -1) + k_2(6, 4, 2)$$

or

$$(9, 2, 7) = (k_1 + 6k_2, 2k_1 + 4k_2, -k_1 + 2k_2)$$

Equating corresponding components gives

$$k_1 + 6k_2 = 9$$
$$2k_1 + 4k_2 = 2$$
$$-k_1 + 2k_2 = 7$$

Solving this system yields $k_1 = -3$, $k_2 = 2$ so that

$$\mathbf{w} = -3\mathbf{u} + 2\mathbf{v}$$

Similarly, for $\mathbf{w}'$ to be a linear combination of $\mathbf{u}$ and $\mathbf{v}$, there must be scalars k_1 and k_2 such that $\mathbf{w}' = k_1\mathbf{u} + k_2\mathbf{v}$; that is,

$$(4, -1, 8) = k_1(1, 2, -1) + k_2(6, 4, 2)$$

or

$$(4, -1, 8) = (k_1 + 6k_2, 2k_1 + 4k_2, -k_1 + 2k_2)$$

Equating corresponding components gives

$$k_1 + 6k_2 = 4$$
$$2k_1 + 4k_2 = -1$$
$$-k_1 + 2k_2 = 8$$

This system of equations is inconsistent (verify), so that no such scalars exist. Consequently, $\mathbf{w}'$ is not a linear combination of $\mathbf{u}$ and $\mathbf{v}$.

Definition. If $\mathbf{v}_1, \mathbf{v}_2, \ldots, \mathbf{v}_r$ are vectors in a vector space V and if every vector in V is expressible as a linear combination of $\mathbf{v}_1, \mathbf{v}_2, \ldots, \mathbf{v}_r$, then we say that these vectors *span* V.

Example 17

The vectors $\mathbf{i} = (1, 0, 0)$, $\mathbf{j} = (0, 1, 0)$, and $\mathbf{k} = (0, 0, 1)$ span R^3 because every vector (a, b, c) in R^3 can be written as

$$(a, b, c) = a\mathbf{i} + b\mathbf{j} + c\mathbf{k}$$

which is a linear combination of $\mathbf{i}$, $\mathbf{j}$, and $\mathbf{k}$.

Example 18

The polynomials $1, x, x^2, \ldots, x^n$ span the vector space P_n (see Example 13) since each polynomial $\mathbf{p}$ in P_n can be written as

$$\mathbf{p} = a_0 + a_1 x + \cdots + a_n x^n$$

which is a linear combination of $1, x, x^2, \ldots, x^n$.

Example 19

Determine whether $\mathbf{v}_1 = (1, 1, 2)$, $\mathbf{v}_2 = (1, 0, 1)$, and $\mathbf{v}_3 = (2, 1, 3)$ span R^3.

Solution. We must determine whether an arbitrary vector $\mathbf{b} = (b_1, b_2, b_3)$ in R^3 can be expressed as a linear combination

$$\mathbf{b} = k_1 \mathbf{v}_1 + k_2 \mathbf{v}_2 + k_3 \mathbf{v}_3$$

of the vectors $\mathbf{v}_1$, $\mathbf{v}_2$, and $\mathbf{v}_3$. Expressing this equation in terms of components gives

$$(b_1, b_2, b_3) = k_1(1, 1, 2) + k_2(1, 0, 1) + k_3(2, 1, 3)$$

or

$$(b_1, b_2, b_3) = (k_1 + k_2 + 2k_3, \quad k_1 + k_3, \quad 2k_1 + k_2 + 3k_3)$$

or

$$k_1 + k_2 + 2k_3 = b_1$$
$$k_1 \quad\;\; + \; k_3 = b_2$$
$$2k_1 + k_2 + 3k_3 = b_3$$

The problem thus reduces to determining whether this system is consistent for all values of b_1, b_2, and b_3. By parts (*a*) and (*d*) of Theorem 15 in Section 1.7, this system will be consistent for all b_1, b_2, and b_3 if and only if the matrix of coefficients

$$A = \begin{bmatrix} 1 & 1 & 2 \\ 1 & 0 & 1 \\ 2 & 1 & 3 \end{bmatrix}$$

is invertible. But $\det(A) = 0$ (verify), so that A is not invertible; consequently, $\mathbf{v}_1$, $\mathbf{v}_2$, and $\mathbf{v}_3$ do not span R^3.

In general, a given set of vectors $\{\mathbf{v}_1, \mathbf{v}_2, \ldots, \mathbf{v}_r\}$ in a vector space V may or may not span V. If they span, then every vector in V is expressible as a linear combination of $\mathbf{v}_1, \mathbf{v}_2, \ldots, \mathbf{v}_r$, and if they do not span, then some vectors are so expressible while others are not. The following theorem shows that if we group together all vectors in V that are expressible as linear combinations of $\mathbf{v}_1, \mathbf{v}_2, \ldots, \mathbf{v}_r$, then we obtain a subspace of V. It is called the **linear space spanned** by $\{\mathbf{v}_1, \mathbf{v}_2, \ldots, \mathbf{v}_r\}$, or more simply the **space spanned** by $\{\mathbf{v}_1, \mathbf{v}_2, \ldots, \mathbf{v}_r\}$.

Theorem 5. *If* $\mathbf{v}_1, \mathbf{v}_2, \ldots, \mathbf{v}_r$ *are vectors in a vector space* V, *then:*

(a) *The set* W *of all linear combinations of* $\mathbf{v}_1, \mathbf{v}_2, \ldots, \mathbf{v}_r$ *is a subspace of* V.
(b) W *is the smallest subspace of* V *that contains* $\mathbf{v}_1, \mathbf{v}_2, \ldots, \mathbf{v}_r$ *in the sense that every other subspace of* V *that contains* $\mathbf{v}_1, \mathbf{v}_2, \ldots, \mathbf{v}_r$ *must contain* W.

Proof.

(*a*) To show that W is a subspace of V, we must prove it is closed under addition and scalar multiplication. If $\mathbf{u}$ and $\mathbf{v}$ are vectors in W, then

$$\mathbf{u} = c_1\mathbf{v}_1 + c_2\mathbf{v}_2 + \cdots + c_r\mathbf{v}_r$$

and

$$\mathbf{v} = k_1\mathbf{v}_1 + k_2\mathbf{v}_2 + \cdots + k_r\mathbf{v}_r$$

where $c_1, c_2, \ldots, c_r, k_1, k_2, \ldots, k_r$ are scalars. Therefore,

$$\mathbf{u} + \mathbf{v} = (c_1 + k_1)\mathbf{v}_1 + (c_2 + k_2)\mathbf{v}_2 + \cdots + (c_r + k_r)\mathbf{v}_r$$

and, for any scalar k,

$$k\mathbf{u} = (kc_1)\mathbf{v}_1 + (kc_2)\mathbf{v}_2 + \cdots + (kc_r)\mathbf{v}_r$$

Thus, $\mathbf{u} + \mathbf{v}$ and $k\mathbf{u}$ are linear combinations of $\mathbf{v}_1, \mathbf{v}_2, \ldots, \mathbf{v}_r$ and consequently lie in W. Therefore, W is closed under addition and scalar multiplication.

(b) Each vector v_i is a linear combination of $v_1, v_2, \ldots, v_r$ since we can write

$$v_i = 0v_1 + 0v_2 + \cdots + 1v_i + \cdots + 0v_r$$

The subspace W therefore contains each of the vectors $v_1, v_2, \ldots, v_r$. Let W' be any other subspace that contains $v_1, v_2, \ldots, v_r$. Since W' is closed under addition and scalar multiplication, it must contain all linear combinations

$$c_1v_1 + c_2v_2 + \cdots + c_rv_r \qquad \text{of} \qquad v_1, v_2, \ldots, v_r$$

Thus, W' contains each vector of W.

The linear space W spanned by a set of vectors $S = \{v_1, v_2, \ldots, v_r\}$ will be denoted by

$$\text{lin}(S) \qquad \text{or} \qquad \text{lin}\{v_1, v_2, \ldots, v_r\}$$

Example 20

If v_1 and v_2 are noncollinear vectors in R^3 with initial points at the origin, then $\text{lin}\{v_1, v_2\}$, which consists of all linear combinations $k_1v_1 + k_2v_2$, is the plane determined by v_1 and v_2 (Figure 4.5a).

Similarly, if v is a nonzero vector in R^2 or R^3, then $\text{lin}\{v\}$, which is the set of all scalar multiples kv, is the line determined by v (Figure 4.5b).

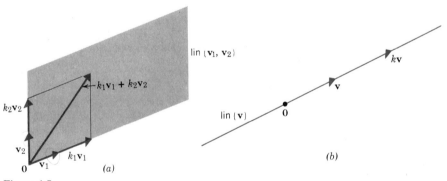

Figure 4.5

EXERCISE SET 4.3

1. Use Theorem 4 to determine which of the following are subspaces of R^3.
   (a) all vectors of the form $(a, 0, 0)$
   (b) all vectors of the form $(a, 1, 1)$
   (c) all vectors of the form (a, b, c), where $b = a + c$
   (d) all vectors of the form (a, b, c), where $b = a + c + 1$

2. Use Theorem 4 to determine which of the following are subspaces of M_{22}.
(a) all matrices of the form

$$\begin{bmatrix} a & b \\ c & d \end{bmatrix}$$

where a, b, c, and d are integers
(b) all matrices of the form

$$\begin{bmatrix} a & b \\ c & d \end{bmatrix}$$

where $a + d = 0$
(c) all 2×2 matrices A such that $A = A^t$
(d) all 2×2 matrices A such that $\det(A) = 0$

3. Use Theorem 4 to determine which of the following are subspaces of P_3.
(a) all polynomials $a_0 + a_1 x + a_2 x^2 + a_3 x^3$ for which $a_0 = 0$
(b) all polynomials $a_0 + a_1 x + a_2 x^2 + a_3 x^3$ for which $a_0 + a_1 + a_2 + a_3 = 0$
(c) all polynomials $a_0 + a_1 x + a_2 x^2 + a_3 x^3$ for which a_0, a_1, a_2, and a_3 are integers
(d) all polynomials of the form $a_0 + a_1 x$, where a_0 and a_1 are real numbers

4. Use Theorem 4 to determine which of the following are subspaces of the space of all real-valued functions f defined on the entire real line.
(a) all f such that $f(x) \le 0$ for all x
(b) all f such that $f(0) = 0$
(c) all f such that $f(0) = 2$
(d) all constant functions
(e) all f of the form $k_1 + k_2 \sin x$, where k_1 and k_2 are real numbers

5. Which of the following are linear combinations of $\mathbf{u} = (1, -1, 3)$ and $\mathbf{v} = (2, 4, 0)$?
(a) $(3, 3, 3)$ (b) $(4, 2, 6)$ (c) $(1, 5, 6)$ (d) $(0, 0, 0)$

6. Express the following as linear combinations of $\mathbf{u} = (2, 1, 4)$, $\mathbf{v} = (1, -1, 3)$, and $\mathbf{w} = (3, 2, 5)$.
(a) $(5, 9, 5)$ (b) $(2, 0, 6)$ (c) $(0, 0, 0)$ (d) $(2, 2, 3)$

7. Express the following as linear combinations of $\mathbf{p}_1 = 2 + x + 4x^2$, $\mathbf{p}_2 = 1 - x + 3x^2$, and $\mathbf{p}_3 = 3 + 2x + 5x^2$.
(a) $5 + 9x + 5x^2$ (b) $2 + 6x^2$ (c) 0 (d) $2 + 2x + 3x^2$

8. Which of the following are linear combinations of

$$A = \begin{bmatrix} 1 & 2 \\ -1 & 3 \end{bmatrix}, \quad B = \begin{bmatrix} 0 & 1 \\ 2 & 4 \end{bmatrix}, \quad \text{and } C = \begin{bmatrix} 4 & -2 \\ 0 & -2 \end{bmatrix}?$$

(a) $\begin{bmatrix} 6 & 3 \\ 0 & 8 \end{bmatrix}$ (b) $\begin{bmatrix} -1 & 7 \\ 5 & 1 \end{bmatrix}$ (c) $\begin{bmatrix} 0 & 0 \\ 0 & 0 \end{bmatrix}$ (d) $\begin{bmatrix} 6 & -1 \\ -8 & -8 \end{bmatrix}$

9. In each part determine whether the given vectors span R^3.
(a) $\mathbf{v}_1 = (1, 1, 1)$, $\mathbf{v}_2 = (2, 2, 0)$, $\mathbf{v}_3 = (3, 0, 0)$
(b) $\mathbf{v}_1 = (2, -1, 3)$, $\mathbf{v}_2 = (4, 1, 2)$, $\mathbf{v}_3 = (8, -1, 8)$
(c) $\mathbf{v}_1 = (3, 1, 4)$, $\mathbf{v}_2 = (2, -3, 5)$, $\mathbf{v}_3 = (5, -2, 9)$, $\mathbf{v}_4 = (1, 4, -1)$
(d) $\mathbf{v}_1 = (1, 3, 3)$, $\mathbf{v}_2 = (1, 3, 4)$, $\mathbf{v}_3 = (1, 4, 3)$, $\mathbf{v}_4 = (6, 2, 1)$

10. Determine which of the following lie in the space spanned by

$$\mathbf{f} = \cos^2 x \qquad \text{and} \qquad \mathbf{g} = \sin^2 x$$

(a) $\cos 2x$ (b) $3 + x^2$ (c) 1 (d) $\sin x$

11. Determine whether the following polynomials span P_2.

$$\mathbf{p}_1 = 1 + 2x - x^2 \qquad \mathbf{p}_2 = 3 + x^2$$
$$\mathbf{p}_3 = 5 + 4x - x^2 \qquad \mathbf{p}_4 = -2 + 2x - 2x^2$$

12. Let $\mathbf{v}_1 = (2, 1, 0, 3)$, $\mathbf{v}_2 = (3, -1, 5, 2)$, and $\mathbf{v}_3 = (-1, 0, 2, 1)$. Which of the following vectors are in $\text{lin}\{\mathbf{v}_1, \mathbf{v}_2, \mathbf{v}_3\}$?
(a) $(2, 3, -7, 3)$ (b) $(0, 0, 0, 0)$ (c) $(1, 1, 1, 1)$ (d) $(-4, 6, -13, 4)$

13. Find an equation for the plane spanned by the vectors $\mathbf{u} = (1, 1, -1)$ and $\mathbf{v} = (2, 3, 5)$.

14. Find parametric equations for the line spanned by the vector $\mathbf{u} = (2, 7, -1)$.

15. Show that the solution vectors of a consistent nonhomogeneous system of m linear equations in n unknowns do not form a subspace of R^n.

16. **(For readers who have studied calculus.)** Show that the following sets of functions are subspaces of the vector space in Example 8.
(a) all everywhere continuous functions
(b) all everywhere differentiable functions
(c) all everywhere differentiable functions that satisfy $\mathbf{f}' + 2\mathbf{f} = \mathbf{0}$

4.4 LINEAR INDEPENDENCE

From Section 4.3, a vector space V is spanned by a set of vectors $S = \{\mathbf{v}_1, \mathbf{v}_2, \dots, \mathbf{v}_r\}$ if each vector in V is a linear combination of $\mathbf{v}_1, \mathbf{v}_2, \dots, \mathbf{v}_r$. Spanning sets are useful in a variety of problems, since it is often possible to study a vector space V by first studying the vectors in a spanning set S and then extending the results to the rest of V. Therefore, it is desirable to keep the spanning set S as small as possible. The problem of finding smallest spanning sets for a vector space depends on the notion of linear independence, which we shall study in this section.

Definition. If $S = \{\mathbf{v}_1, \mathbf{v}_2, \dots, \mathbf{v}_r\}$ is a set of vectors, then the vector equation

$$k_1 \mathbf{v}_1 + k_2 \mathbf{v}_2 + \cdots + k_r \mathbf{v}_r = \mathbf{0}$$

has at least one solution, namely

$$k_1 = 0, \qquad k_2 = 0, \dots, k_r = 0$$

If this is the only solution, then S is called a *linearly independent* set. If there are other solutions, then S is called a *linearly dependent* set.

Example 21

The set of vectors $S = \{v_1, v_2, v_3\}$, where $v_1 = (2, -1, 0, 3)$, $v_2 = (1, 2, 5, -1)$, and $v_3 = (7, -1, 5, 8)$ is linearly dependent, since $3v_1 + v_2 - v_3 = 0$.

$R_1 = 3 \quad R_2 = 1 \quad R_3 = -1$

Example 22

The polynomials $p_1 = 1 - x$, $p_2 = 5 + 3x - 2x^2$, and $p_3 = 1 + 3x - x^2$ form a linearly dependent set in P_2 since $3p_1 - p_2 + 2p_3 = 0$.

$R_1 = 3, \quad R_2 = -1, \quad R_3 = 2$

Example 23

Consider the vectors $i = (1, 0, 0)$, $j = (0, 1, 0)$, and $k = (0, 0, 1)$ in R^3. In terms of components the vector equation

$$k_1 i + k_2 j + k_3 k = 0$$

becomes

$$k_1(1, 0, 0) + k_2(0, 1, 0) + k_3(0, 0, 1) = (0, 0, 0)$$

or equivalently

$$(k_1, k_2, k_3) = (0, 0, 0)$$

Thus, $k_1 = 0, k_2 = 0, k_3 = 0$, so the set $S = \{i, j, k\}$ is linearly independent. A similar argument can be used to show that the vectors $e_1 = (1, 0, 0, \dots, 0)$, $e_2 = (0, 1, 0, \dots, 0), \dots, e_n = (0, 0, 0, \dots, 1)$ form a linearly independent set in R^n.

Example 24

Determine whether the vectors

$$v_1 = (1, -2, 3) \qquad v_2 = (5, 6, -1) \qquad v_3 = (3, 2, 1)$$

form a linearly dependent or linearly independent set.

Solution. In terms of components the vector equation

$$k_1 v_1 + k_2 v_2 + k_3 v_3 = 0$$

becomes

$$k_1(1, -2, 3) + k_2(5, 6, -1) + k_3(3, 2, 1) = (0, 0, 0)$$

or equivalently

$$(k_1 + 5k_2 + 3k_3, -2k_1 + 6k_2 + 2k_3, 3k_1 - k_2 + k_3) = (0, 0, 0)$$

Equating corresponding components gives

$$k_1 + 5k_2 + 3k_3 = 0$$
$$-2k_1 + 6k_2 + 2k_3 = 0$$
$$3k_1 - k_2 + k_3 = 0$$

Thus, v_1, v_2, and v_3 form a linearly dependent set if this system has a nontrivial solution, or a linearly independent set if it has only the trivial solution. Solving this system yields

$$k_1 = -\tfrac{1}{2}t \qquad k_2 = -\tfrac{1}{2}t \qquad k_3 = t$$

Thus, the system has nontrivial solutions and v_1, v_2, and v_3 form a linearly dependent set. Alternatively, we could show the existence of nontrivial solutions without solving the system by showing that the coefficient matrix has determinant zero and consequently is not invertible (verify). ▲

The term "linearly dependent" suggests that the vectors "depend" on each other in some way. The following theorem shows that this is in fact the case.

Theorem 6. *A set S with two or more vectors is*

(a) *linearly dependent if and only if at least one of the vectors in S is expressible as a linear combination of the other vectors in S;*
(b) *linearly independent if and only if no vector in S is expressible as a linear combination of the other vectors in S.*

Proof. We shall prove part (*a*) and leave the proof of part (*b*) as an exercise.
(a) Let $S = \{v_1, v_2, \ldots, v_r\}$ be a set with two or more vectors. If we assume that S is linearly dependent, then there are scalars $k_1, k_2, \ldots, k_r$, not all zero, such that

$$k_1 v_1 + k_2 v_2 + \cdots + k_r v_r = 0 \tag{4.3}$$

To be specific, suppose that $k_1 \neq 0$. Then (4.3) can be rewritten as

$$v_1 = \left(-\frac{k_2}{k_1}\right) v_2 + \cdots + \left(-\frac{k_r}{k_1}\right) v_r$$

which expresses v_1 as a linear combination of the other vectors in S. Similarly, if $k_j \neq 0$ in (4.3) for some $j = 2, 3, \ldots, r$, then v_j is expressible as a linear combination of the other vectors in S.

Conversely, let us assume that at least one of the vectors in S is expressible as a linear combination of the other vectors. To be specific, suppose that

$$v_1 = c_2 v_2 + c_3 v_3 + \cdots + c_r v_r$$

so

$$v_1 - c_2 v_2 - c_3 v_3 - \cdots - c_r v_r = 0$$

It follows that S is linearly dependent since the equation

$$k_1 v_1 + k_2 v_2 + \cdots + k_r v_r = 0$$

is satisfied by

$$k_1 = 1, k_2 = -c_2, \ldots, k_r = -c_r$$

which are not all zero. The proof in the case where some vector other than v_1 is expressible as a linear combination of the other vectors in S is similar. ▮

Example 25

In Example 21 we saw that the vectors $v_1 = (2, -1, 0, 3)$, $v_2 = (1, 2, 5, -1)$, and $v_3 = (7, -1, 5, 8)$ form a linearly dependent set. It follows from Theorem 6 that at least one of these vectors is expressible as a linear combination of the other two. In this example each vector is expressible as a linear combination of the other two since it follows from the equation $3v_1 + v_2 - v_3 = 0$ (see Example 21) that

$$v_1 = -\tfrac{1}{3}v_2 + \tfrac{1}{3}v_3, \; v_2 = -3v_1 + v_3, \text{ and } v_3 = 3v_1 + v_2 \quad \blacktriangle$$

Example 26

In Example 23 we saw that the vectors $i = (1, 0, 0)$, $j = (0, 1, 0)$, and $k = (0, 0, 1)$ form a linearly independent set. Thus, it follows from Theorem 6 that none of these vectors is expressible as a linear combination of the other two. To see that this is so, suppose that k is expressible as

$$k = k_1 i + k_2 j$$

Then, in terms of components,

$$(0, 0, 1) = k_1(1, 0, 0) + k_2(0, 1, 0)$$

or

$$(0, 0, 1) = (k_1, k_2, 0)$$

But this equation is not satisfied by any values of k_1 and k_2, so k cannot be expressed as a linear combination of i and j. Similarly, i is not expressible as a linear combination of j and k, and j is not expressible as a linear combination of i and k. ▲

The following theorem gives two simple facts about linear independence that are useful to know. The proof is left as an exercise.

Theorem 7

(a) *If a set contains the zero vector, it is linearly dependent.*
(b) *A set with exactly two vectors is linearly dependent if and only if at least one of the vectors is a scalar multiple of the other.*

The next two examples give a geometric interpretation of linear dependence in R^2 and R^3.

Example 27

In R^2 or R^3 one vector is a scalar multiple of another if and only if the two vectors lie on the same line through the origin when they are placed with their initial points at the origin. Thus, it follows from part (*b*) of Theorem 7 that in R^2 or R^3 two vectors form a linearly dependent set if and only if they lie on the same line through the origin when they are placed with their initial points at the origin (Figure 4.6).

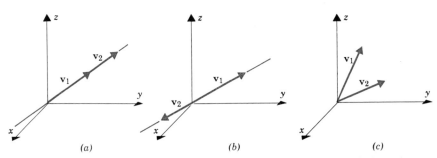

Figure 4.6 (*a*) Linearly dependent. (*b*) Linearly dependent. (*c*) Linearly independent.

Example 28

If v_1, v_2, and v_3 are three vectors in R^3, then the set $S = \{v_1, v_2, v_3\}$ is linearly dependent if and only if the three vectors lie in the same plane through the origin when the vectors are placed with initial points at the origin (Figure 4.7). To see this, recall that v_1, v_2, and v_3 are linearly dependent if and only if at least one of the vectors is a linear combination of the remaining two, or equivalently if and only if at least one of the vectors is in the space spanned by the remaining two. But the space spanned by any two vectors in R^3 is either a line through the origin, a plane through the origin, or the origin itself (Exercise 17). In any case, the space spanned by two vectors in R^3 always lies in a plane through the origin.

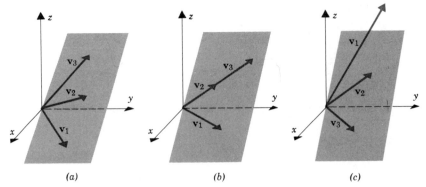

Figure 4.7 (*a*) Linearly dependent. (*b*) Linearly dependent. (*c*) Linearly independent.

We conclude this section with a theorem that shows that a linearly independent set in R^n can contain at most n vectors.

> **Theorem 8.** *Let* $S = \{v_1, v_2, \ldots, v_r\}$ *be a set of vectors in* R^n. *If* $r > n$, *then* S *is linearly dependent.*

Proof. Suppose

$$v_1 = (v_{11}, v_{12}, \ldots, v_{1n})$$
$$v_2 = (v_{21}, v_{22}, \ldots, v_{2n})$$
$$\vdots$$
$$v_r = (v_{r1}, v_{r2}, \ldots, v_{rn})$$

Consider the equation

$$k_1 v_1 + k_2 v_2 + \cdots + k_r v_r = 0$$

If, as illustrated in Example 24, we express both sides of this equation in terms of components and then equate corresponding components, we obtain the system

$$v_{11}k_1 + v_{21}k_2 + \cdots + v_{r1}k_r = 0$$
$$v_{12}k_1 + v_{22}k_2 + \cdots + v_{r2}k_r = 0$$
$$\vdots \qquad \vdots \qquad \qquad \vdots \qquad \vdots$$
$$v_{1n}k_1 + v_{2n}k_2 + \cdots + v_{rn}k_r = 0$$

This is a homogeneous system of n equations in the r unknowns $k_1, \ldots, k_r$. Since $r > n$, it follows from Theorem 1 in Section 1.3 that the system has nontrivial solutions. Therefore, $S = \{v_1, v_2, \ldots, v_r\}$ is a linearly dependent set. ▌

In particular, this theorem tells us that a set in R^2 with more than two vectors is linearly dependent, and a set in R^3 with more than three vectors is linearly dependent.

EXERCISE SET 4.4

1. Explain why the following are linearly dependent sets of vectors. (Solve this problem by inspection.)
  (a) $u_1 = (1, 2)$ and $u_2 = (-3, -6)$ in R^2
  (b) $u_1 = (2, 3)$, $u_2 = (-5, 8)$, $u_3 = (6, 1)$ in R^2
  (c) $p_1 = 2 + 3x - x^2$ and $p_2 = 6 + 9x - 3x^2$ in P_2

  (d) $A = \begin{bmatrix} 1 & 3 \\ 2 & 0 \end{bmatrix}$ and $B = \begin{bmatrix} -1 & -3 \\ -2 & 0 \end{bmatrix}$ in M_{22}

2. Which of the following sets of vectors in R^3 are linearly dependent?
(a) $(2, -1, 4)$, $(3, 6, 2)$, $(2, 10, -4)$
(b) $(3, 1, 1)$, $(2, -1, 5)$, $(4, 0, -3)$
(c) $(6, 0, -1)$, $(1, 1, 4)$
(d) $(1, 3, 3)$, $(0, 1, 4)$, $(5, 6, 3)$, $(7, 2, -1)$

3. Which of the following sets of vectors in R^4 are linearly dependent?
(a) $(1, 2, 1, -2)$, $(0, -2, -2, 0)$, $(0, 2, 3, 1)$, $(3, 0, -3, 6)$
(b) $(4, -4, 8, 0)$, $(2, 2, 4, 0)$, $(6, 0, 0, 2)$, $(6, 3, -3, 0)$
(c) $(4, 4, 0, 0)$, $(0, 0, 6, 6)$, $(-5, 0, 5, 5)$
(d) $(3, 0, 4, 1)$, $(6, 2, -1, 2)$, $(-1, 3, 5, 1)$, $(-3, 7, 8, 3)$

4. Which of the following sets of vectors in P_2 are linearly dependent?
(a) $2 - x + 4x^2$, $3 + 6x + 2x^2$, $2 + 10x - 4x^2$
(b) $3 + x + x^2$, $2 - x + 5x^2$, $4 - 3x^2$
(c) $6 - x^2$, $1 + x + 4x^2$
(d) $1 + 3x + 3x^2$, $x + 4x^2$, $5 + 6x + 3x^2$, $7 + 2x - x^2$

5. Let V be the vector space of all real-valued functions defined on the entire real line. Which of the following sets of vectors in V are linearly dependent?
(a) $2, 4\sin^2 x, \cos^2 x$ (b) $x, \cos x$
(c) $1, \sin x, \sin 2x$ (d) $\cos 2x, \sin^2 x, \cos^2 x$
(e) $(1 + x)^2, x^2 + 2x, 3$ (f) $0, x, x^2$

6. Assume that v_1, v_2, and v_3 are vectors in R^3 that have their initial points at the origin. In each part determine whether the three vectors lie in a plane.
(a) $v_1 = (1, 0, -2)$, $v_2 = (3, 1, 2)$, $v_3 = (1, -1, 0)$
(b) $v_1 = (2, -1, 4)$, $v_2 = (4, 2, 3)$, $v_3 = (2, 7, -6)$

7. Assume that v_1, v_2, and v_3 are vectors in R^3 that have their initial points at the origin. In each part determine whether the three vectors lie on the same line.
(a) $v_1 = (3, -6, 9)$, $v_2 = (2, -4, 6)$, $v_3 = (1, 1, 1)$
(b) $v_1(2, -1, 4)$, $v_2 = (4, 2, 3)$, $v_3 = (2, 7, -6)$
(c) $v_1 = (4, 6, 8)$, $v_2 = (2, 3, 4)$, $v_3 = (-2, -3, -4)$

8. For which real values of λ do the following vectors form a linearly dependent set in R^3?

$$v_1 = (\lambda, -\tfrac{1}{2}, -\tfrac{1}{2}), \quad v_2 = (-\tfrac{1}{2}, \lambda, -\tfrac{1}{2}), \quad v_3 = (-\tfrac{1}{2}, -\tfrac{1}{2}, \lambda)$$

9. (a) Show that the vectors $v_1 = (0, 3, 1, -1)$, $v_2 = (6, 0, 5, 1)$, and $v_3 = (4, -7, 1, 3)$ form a linearly dependent set in R^4.
(b) Express each vector as a linear combination of the other two.

10. If $\{v_1, v_2, v_3\}$ is a linearly independent set of vectors, show that $\{v_1, v_2\}$, $\{v_1, v_3\}$, $\{v_2, v_3\}$, $\{v_1\}$, $\{v_2\}$, and $\{v_3\}$ are also linearly independent sets.

11. If $S = \{v_1, v_2, \ldots, v_n\}$ is a linearly independent set of vectors, show that every subset of S with one or more vectors is also linearly independent.

12. If $\{v_1, v_2, v_3\}$ is a linearly dependent set of vectors in a vector space V, show that $\{v_1, v_2, v_3, v_4\}$ is also linearly dependent, where v_4 is any other vector in V.

13. If $\{\mathbf{v}_1, \mathbf{v}_2, \ldots, \mathbf{v}_r\}$ is a linearly dependent set of vectors in a vector space V, show that $\{\mathbf{v}_1, \mathbf{v}_2, \ldots, \mathbf{v}_{r+1}, \ldots, \mathbf{v}_n\}$ is also linearly dependent, where $\mathbf{v}_{r+1}, \ldots, \mathbf{v}_n$ are any other vectors in V.

14. Show that every set with more than three vectors from P_2 is linearly dependent.

15. Show that if $\{\mathbf{v}_1, \mathbf{v}_2\}$ is linearly independent and $\mathbf{v}_3$ does not lie in $\lin\{\mathbf{v}_1, \mathbf{v}_2\}$, then $\{\mathbf{v}_1, \mathbf{v}_2, \mathbf{v}_3\}$ is linearly independent.

16. Prove: For any vectors $\mathbf{u}$, $\mathbf{v}$, and $\mathbf{w}$, the vectors $\mathbf{u} - \mathbf{v}$, $\mathbf{v} - \mathbf{w}$, and $\mathbf{w} - \mathbf{u}$ form a linearly dependent set.

17. Prove: The space spanned by two vectors in R^3 is either a line through the origin, a plane through the origin, or the origin itself.

18. **(For readers who have studied calculus.)** Let V be the vector space of real-valued functions defined on the entire real line. If $\mathbf{f}$, $\mathbf{g}$, and $\mathbf{h}$ are vectors in V that are twice differentiable, then the function $\mathbf{w} = w(x)$ defined by

$$w(x) = \begin{vmatrix} f(x) & g(x) & h(x) \\ f'(x) & g'(x) & h'(x) \\ f''(x) & g''(x) & h''(x) \end{vmatrix}$$

is called the **Wronskian** of $\mathbf{f}$, $\mathbf{g}$, and $\mathbf{h}$. Prove that $\mathbf{f}$, $\mathbf{g}$, and $\mathbf{h}$ form a linearly independent set if the Wronskian is not the zero vector in V [i.e., $w(x)$ is not identically zero].

19. **(For readers who have studied calculus.)** Use the Wronskian (Exercise 18) to show that the following sets of vectors are linearly independent.
   (a) $1, x, e^x$ (b) $\sin x, \cos x, x \sin x$ (c) e^x, xe^x, x^2e^x (d) $1, x, x^2$

20. Under what conditions is a set with one vector linearly independent?

21. Use part (a) of Theorem 6 to prove part (b).

22. Prove part (a) of Theorem 7.

23. Prove part (b) of Theorem 7.

4.5   BASIS AND DIMENSION

We usually think of a line as being one dimensional, a plane as two dimensional, and space around us as three dimensional. It is the primary purpose of this section to make this intuitive notion of dimension precise.

Definition. If V is any vector space and $S = \{\mathbf{v}_1, \mathbf{v}_2, \ldots, \mathbf{v}_r\}$ is a finite set of vectors in V, then S is called a **basis** for V if

(i) S is linearly independent
(ii) S spans V.

Example 29

Let $e_1 = (1, 0, 0, \ldots, 0)$, $e_2 = (0, 1, 0, \ldots, 0)$, $\ldots$, $e_n = (0, 0, 0, \ldots, 1)$. In Example 23 we pointed out that $S = \{e_1, e_2, \ldots, e_n\}$ is a linearly independent set in R^n. Since any vector $v = (v_1, v_2, \ldots, v_n)$ in R^n can be written as $v = v_1 e_1 + v_2 e_2 + \cdots + v_n e_n$, S spans R^n and is therefore a basis. It is called the ***standard basis for*** R^n. △

Example 30

Let $v_1 = (1, 2, 1)$, $v_2 = (2, 9, 0)$, and $v_3 = (3, 3, 4)$. Show that the set $S = \{v_1, v_2, v_3\}$ is a basis for R^3.

Solution. To show that S spans R^3, we must show that an arbitrary vector $b = (b_1, b_2, b_3)$ can be expressed as a linear combination

$$b = k_1 v_1 + k_2 v_2 + k_3 v_3$$

of the vectors in S. Expressing this equation in terms of components gives

$$(b_1, b_2, b_3) = k_1(1, 2, 1) + k_2(2, 9, 0) + k_3(3, 3, 4)$$

or

$$(b_1, b_2, b_3) = (k_1 + 2k_2 + 3k_3, 2k_1 + 9k_2 + 3k_3, k_1 + 4k_3)$$

or

$$\begin{aligned} k_1 + 2k_2 + 3k_3 &= b_1 \\ 2k_1 + 9k_2 + 3k_3 &= b_2 \\ k_1 \qquad\quad + 4k_3 &= b_3 \end{aligned} \qquad (4.4)$$

Thus, to show S spans V, we must demonstrate that system (4.4) has a solution for all choices of $b = (b_1, b_2, b_3)$. To prove that S is linearly independent, we must show that the only solution of

$$k_1 v_1 + k_2 v_2 + k_3 v_3 = 0 \qquad (4.5)$$

is $k_1 = k_2 = k_3 = 0$.

As above, if (4.5) is expressed in terms of components, the verification of independence reduces to showing the homogeneous system

$$\begin{aligned} k_1 + 2k_2 + 3k_3 &= 0 \\ 2k_1 + 9k_2 + 3k_3 &= 0 \\ k_1 \qquad\quad + 4k_3 &= 0 \end{aligned} \qquad (4.6)$$

has only the trivial solution. Observe that systems (4.4) and (4.6) have the same coefficient matrix. Thus, by parts (*a*), (*b*), and (*d*) of Theorem 15 in Section 1.7, we can simultaneously prove S is linearly independent and spans R^3 by demonstrating that the matrix of coefficients

$$A = \begin{bmatrix} 1 & 2 & 3 \\ 2 & 9 & 3 \\ 1 & 0 & 4 \end{bmatrix}$$

in systems (4.4) and (4.6) is invertible. Since

$$\det(A) = \begin{vmatrix} 1 & 2 & 3 \\ 2 & 9 & 3 \\ 1 & 0 & 4 \end{vmatrix} = -1$$

it follows from Theorem 7 in Section 2.3 that A is invertible. Thus, S is a basis for R^3. ▲

Example 31

The set $S = \{1, x, x^2, \ldots, x^n\}$ is a basis for the vector space P_n introduced in Example 13. From Example 18, the vectors in S span P_n. To see that S is linearly independent, assume that some linear combination of vectors in S is the zero vector; that is,

$$c_0 + c_1 x + \cdots + c_n x^n = 0 \qquad \text{(for all } x) \tag{4.7}$$

We must show that $c_0 = c_1 = \cdots = c_n = 0$. From algebra a nonzero polynomial of degree n has at most n distinct roots. Since (4.7) holds for all x, every value of x is a root of the left-hand side. This implies that $c_1 = c_2 = \cdots = c_n = 0$; otherwise, $c_0 + c_1 x + \cdots + c_n x^n$ could have at most n roots. The set S is therefore linearly independent. ▲

The basis S in the last example is called the ***standard basis for P_n***.

Example 32

Let

$$M_1 = \begin{bmatrix} 1 & 0 \\ 0 & 0 \end{bmatrix} \qquad M_2 = \begin{bmatrix} 0 & 1 \\ 0 & 0 \end{bmatrix} \qquad M_3 = \begin{bmatrix} 0 & 0 \\ 1 & 0 \end{bmatrix} \qquad M_4 = \begin{bmatrix} 0 & 0 \\ 0 & 1 \end{bmatrix}$$

The set $S = \{M_1, M_2, M_3, M_4\}$ is a basis for the vector space M_{22} of 2×2 matrices. To see that S spans M_{22}, note that a typical vector (matrix)

$$\begin{bmatrix} a & b \\ c & d \end{bmatrix}$$

can be written as

$$\begin{bmatrix} a & b \\ c & d \end{bmatrix} = a \begin{bmatrix} 1 & 0 \\ 0 & 0 \end{bmatrix} + b \begin{bmatrix} 0 & 1 \\ 0 & 0 \end{bmatrix} + c \begin{bmatrix} 0 & 0 \\ 1 & 0 \end{bmatrix} + d \begin{bmatrix} 0 & 0 \\ 0 & 1 \end{bmatrix}$$

$$= aM_1 + bM_2 + cM_3 + dM_4$$

To see that S is linearly independent, assume that

$$aM_1 + bM_2 + cM_3 + dM_4 = 0$$

That is,

$$a \begin{bmatrix} 1 & 0 \\ 0 & 0 \end{bmatrix} + b \begin{bmatrix} 0 & 1 \\ 0 & 0 \end{bmatrix} + c \begin{bmatrix} 0 & 0 \\ 1 & 0 \end{bmatrix} + d \begin{bmatrix} 0 & 0 \\ 0 & 1 \end{bmatrix} = \begin{bmatrix} 0 & 0 \\ 0 & 0 \end{bmatrix}$$

Then

$$\begin{bmatrix} a & b \\ c & d \end{bmatrix} = \begin{bmatrix} 0 & 0 \\ 0 & 0 \end{bmatrix}$$

Thus, $a = b = c = d = 0$ so that S is linearly independent. ▲

The basis S in this example is called the *standard basis for* M_{22}. More generally, the **standard basis for** M_{mn} consists of the mn different matrices with a single 1 and zeros for the remaining entries.

Example 33

If $S = \{v_1, v_2, \ldots, v_r\}$ is a *linearly independent* set in a vector space V, then S is a basis for the subspace lin(S) since S is independent, and by definition of lin(S), S spans lin(S). ▲

> **Definition.** A nonzero vector space V is called **finite-dimensional** if it contains a finite set of vectors $\{v_1, v_2, \ldots, v_n\}$ that forms a basis. If no such set exists, V is called **infinite dimensional**. In addition, we shall regard the zero vector space as finite-dimensional even though it has no linearly independent sets and consequently no basis.

Example 34

By Examples 29, 31, and 32, R^n, P_n, and M_{22} are finite-dimensional vector spaces.

▲

The next theorem provides the key to the concept of dimension for vector spaces. From it we shall obtain some of the most important results in linear algebra.

> **Theorem 9.** *If* $S = \{v_1, v_2, \ldots, v_n\}$ *is a basis for a vector space* V, *then every set with more than* n *vectors is linearly dependent.*

Proof. Let $S' = \{w_1, w_2, \ldots, w_m\}$ be any set of m vectors in V where $m > n$. We wish to show that S' is linearly dependent. Since $S = \{v_1, v_2, \ldots, v_n\}$ is a basis, each w_i can be expressed as a linear combination of the vectors in S, say,

$$\begin{aligned}
w_1 &= a_{11}v_1 + a_{21}v_2 + \cdots + a_{n1}v_n \\
w_2 &= a_{12}v_1 + a_{22}v_2 + \cdots + a_{n2}v_n \\
&\;\;\vdots \qquad \vdots \qquad \vdots \qquad\quad \vdots \\
w_m &= a_{1m}v_1 + a_{2m}v_2 + \cdots + a_{nm}v_n
\end{aligned} \tag{4.8}$$

To show that S' is linearly dependent, we must find scalars $k_1, k_2, \ldots, k_m$, not all zero, such that

$$k_1 w_1 + k_2 w_2 + \cdots + k_m w_m = 0 \tag{4.9}$$

Using the equations in (4.8), we can rewrite (4.9) as

$$(k_1 a_{11} + k_2 a_{12} + \cdots + k_m a_{1m})\mathbf{v}_1$$
$$+ (k_1 a_{21} + k_2 a_{22} + \cdots + k_m a_{2m})\mathbf{v}_2$$
$$+ (k_1 a_{n1} + k_2 a_{n2} + \cdots + k_m a_{nm})\mathbf{v}_n = \mathbf{0}$$

The problem of proving that S' is a linearly dependent set thus reduces to showing there are scalars $k_1, k_2, \ldots, k_m$, not all zero, which satisfy

$$\begin{aligned}
a_{11}k_1 + a_{12}k_2 + \cdots + a_{1m}k_m &= 0 \\
a_{21}k_1 + a_{22}k_2 + \cdots + a_{2m}k_m &= 0 \\
\vdots \qquad \vdots \qquad\quad \vdots \qquad\quad \vdots & \\
a_{n1}k_1 + a_{n2}k_2 + \cdots + a_{nm}k_m &= 0
\end{aligned} \qquad (4.10)$$

Since (4.10) has more unknowns than equations, the proof is complete, since Theorem 1 in Section 1.3 guarantees the existence of nontrivial solutions.

As a consequence, we obtain the following result.

Theorem 10. *Any two bases for a finite-dimensional vector space have the same number of vectors.*

Proof. Let $S = \{\mathbf{v}_1, \mathbf{v}_2, \ldots, \mathbf{v}_n\}$ and $S' = \{\mathbf{w}_1, \mathbf{w}_2, \ldots, \mathbf{w}_m\}$ be two bases for a finite-dimensional vector space V. Since S is a basis and S' is a linearly independent set, Theorem 9 implies that $m \le n$. Similarly, since S' is a basis and S is linearly independent, we also have $n \le m$. Therefore, $m = n$.

Example 35

The standard basis for R^n contains n vectors (Example 29). Therefore, every basis for R^n contains n vectors.

Example 36

The standard basis for P_n (Example 31) contains $n + 1$ vectors; thus, every basis for P_n contains $n + 1$ vectors.

The number of vectors in a basis for a finite-dimensional vector space is a particularly important quantity. By Example 35, every basis for R^2 has two vectors and every basis for R^3 has three vectors. Since R^2 (the plane) is intuitively two-dimensional and R^3 is intuitively three-dimensional, the dimension of these spaces is the same as the number of vectors that occur in their bases. This suggests the following definition.

> **Definition.** The **dimension** of a finite-dimensional vector space V is defined to be the number of vectors in a basis for V. In addition, we define the zero vector space to have dimension zero.

From Examples 35 and 36, R^n is an n-dimensional vector space and P_n is an $n + 1$-dimensional vector space.

Example 37

Determine a basis for and the dimension of the solution space of the homogeneous system

$$
\begin{aligned}
2x_1 + 2x_2 - x_3 \qquad\quad + x_5 &= 0 \\
-x_1 - x_2 + 2x_3 - 3x_4 + x_5 &= 0 \\
x_1 + x_2 - 2x_3 \qquad\quad - x_5 &= 0 \\
x_3 + x_4 + x_5 &= 0
\end{aligned}
$$

Solution. In Example 9 of Section 1.3 it was shown that the solutions are given by

$$
x_1 = -s - t \qquad x_2 = s \qquad x_3 = -t \qquad x_4 = 0 \qquad x_5 = t
$$

Therefore, the solution vectors can be written

$$
\begin{bmatrix} x_1 \\ x_2 \\ x_3 \\ x_4 \\ x_5 \end{bmatrix} = \begin{bmatrix} -s-t \\ s \\ -t \\ 0 \\ t \end{bmatrix} = \begin{bmatrix} -s \\ s \\ 0 \\ 0 \\ 0 \end{bmatrix} + \begin{bmatrix} -t \\ 0 \\ -t \\ 0 \\ t \end{bmatrix} = s \begin{bmatrix} -1 \\ 1 \\ 0 \\ 0 \\ 0 \end{bmatrix} + t \begin{bmatrix} -1 \\ 0 \\ -1 \\ 0 \\ 1 \end{bmatrix}
$$

which shows that the vectors

$$
\mathbf{v}_1 = \begin{bmatrix} -1 \\ 1 \\ 0 \\ 0 \\ 0 \end{bmatrix} \qquad \text{and} \qquad \mathbf{v}_2 = \begin{bmatrix} -1 \\ 0 \\ -1 \\ 0 \\ 1 \end{bmatrix}
$$

span the solution space. Since they are also linearly independent (verify), $\{\mathbf{v}_1, \mathbf{v}_2\}$ is a basis, and the solution space is two-dimensional. ▲

In general, to show that a set of vectors $\{\mathbf{v}_1, \mathbf{v}_2, \ldots, \mathbf{v}_n\}$ is a basis for a vector space V, we must show that the vectors are linearly independent and span V. However, if we happen to know that V has dimension n (so that $\{\mathbf{v}_1, \mathbf{v}_2, \ldots, \mathbf{v}_n\}$ contains the right number of vectors for a basis), then it suffices to check *either* linear independence *or* spanning—the remaining condition will hold automatically. This is the content of parts (*a*) and (*b*) of the following theorem. Part (*c*) of this theorem states that every linearly independent set forms part of some basis for V.

Theorem 11.

(a) *If* $S = \{v_1, v_2, \ldots, v_n\}$ *is a set of n linearly independent vectors in an n-dimensional space V, then S is a basis for V.*

(b) *If* $S = \{v_1, v_2, \ldots, v_n\}$ *is a set of n vectors that spans an n-dimensional space V, then S is a basis for V.*

(c) *If* $S = \{v_1, v_2, \ldots, v_r\}$ *is a linearly independent set in an n-dimensional space V and* $r < n$, *then S can be enlarged to a basis for V; that is, there are vectors* $v_{r+1}, \ldots, v_n$ *such that* $\{v_1, v_2, \ldots, v_r, v_{r+1}, \ldots, v_n\}$ *is a basis for V.*

The proofs are left as exercises.

Example 38

Show that $v_1 = (-3, 7)$ and $v_2 = (5, 5)$ form a basis for R^2.

Solution. Since neither vector is a scalar multiple of the other, $S = \{v_1, v_2\}$ is linearly independent. Since R^2 is two-dimensional, S is a basis for R^2 by part (a) of Theorem 11. ▲

EXERCISE SET 4.5

1. Explain why the following sets of vectors are *not* bases for the indicated vector spaces. (Solve this problem by inspection.)
   (a) $u_1 = (1, 2)$, $u_2 = (0, 3)$, $u_3 = (2, 7)$ for R^2
   (b) $u_1 = (-1, 3, 2)$, $u_2 = (6, 1, 1)$ for R^3
   (c) $p_1 = 1 + x + x^2$, $p_2 = x - 1$ for P_2

   (d) $A = \begin{bmatrix} 1 & 1 \\ 2 & 3 \end{bmatrix}$ $B = \begin{bmatrix} 6 & 0 \\ -1 & 4 \end{bmatrix}$ $C = \begin{bmatrix} 3 & 0 \\ 1 & 7 \end{bmatrix}$

   $D = \begin{bmatrix} 5 & 1 \\ 4 & 2 \end{bmatrix}$ $E = \begin{bmatrix} 7 & 1 \\ 2 & 9 \end{bmatrix}$ for M_{22}

2. Which of the following sets of vectors are bases for R^2?
   (a) $(2, 1), (3, 0)$ (b) $(4, 1), (-7, -8)$ (c) $(0, 0), (1, 3)$ (d) $(3, 9), (-4, -12)$

3. Which of the following sets of vectors are bases for R^3?
   (a) $(1, 0, 0), (2, 2, 0), (3, 3, 3)$ (b) $(3, 1, -4), (2, 5, 6), (1, 4, 8)$
   (c) $(2, -3, 1), (4, 1, 1), (0, -7, 1)$ (d) $(1, 6, 4), (2, 4, -1), (-1, 2, 5)$

4. Which of the following sets of vectors are bases for P_2?
   (a) $1 - 3x + 2x^2$, $1 + x + 4x^2$, $1 - 7x$
   (b) $4 + 6x + x^2$, $-1 + 4x + 2x^2$, $5 + 2x - x^2$
   (c) $1 + x + x^2$, $x + x^2$, x^2
   (d) $-4 + x + 3x^2$, $6 + 5x + 2x^2$, $8 + 4x + x^2$

5. Show that the following set of vectors is a basis for M_{22}.

$$\begin{bmatrix} 3 & 6 \\ 3 & -6 \end{bmatrix} \quad \begin{bmatrix} 0 & -1 \\ -1 & 0 \end{bmatrix} \quad \begin{bmatrix} 0 & -8 \\ -12 & -4 \end{bmatrix} \quad \begin{bmatrix} 1 & 0 \\ -1 & 2 \end{bmatrix}$$

6. Let V be the space spanned by $v_1 = \cos^2 x$, $v_2 = \sin^2 x$, $v_3 = \cos 2x$.
   (a) Show that $S = \{v_1, v_2, v_3\}$ is not a basis for V.
   (b) Find a basis for V.

In Exercises 7–12 determine the dimension of and a basis for the solution space of the system.

7. $\quad x_1 + x_2 - x_3 = 0$
$\quad -2x_1 - x_2 + 2x_3 = 0$
$\quad -x_1 \quad\quad + x_3 = 0$

8. $3x_1 + x_2 + x_3 + x_4 = 0$
$\quad 5x_1 - x_2 + x_3 - x_4 = 0$

9. $x_1 - 4x_2 + 3x_3 - x_4 = 0$
$\quad 2x_1 - 8x_2 + 6x_3 - 2x_4 = 0$

10. $x_1 - 3x_2 + x_3 = 0$
$\quad 2x_1 - 6x_2 + 2x_3 = 0$
$\quad 3x_1 - 9x_2 + 3x_3 = 0$

11. $2x_1 + x_2 + 3x_3 = 0$
$\quad x_1 \quad\quad + 5x_3 = 0$
$\quad\quad x_2 + x_3 = 0$

12. $\quad x + y + z = 0$
$\quad 3x + 2y - 2z = 0$
$\quad 4x + 3y - z = 0$
$\quad 6x + 5y + z = 0$

13. Determine bases for the following subspaces of R^3.
   (a) the plane $3x - 2y + 5z = 0$
   (b) the plane $x - y = 0$
   (c) the line $x = 2t, y = -t, z = 4t$
   (d) all vectors of the form (a, b, c) where $b = a + c$

14. Determine the dimensions of the following subspaces of R^4.
   (a) all vectors of the form $(a, b, c, 0)$
   (b) all vectors of the form (a, b, c, d), where $d = a + b$ and $c = a - b$
   (c) all vectors of the form (a, b, c, d), where $a = b = c = d$

15. Determine the dimension of the subspace of P_3 consisting of all polynomials $a_0 + a_1x + a_2x^2 + a_3x^3$ for which $a_0 = 0$.

16. Let $\{v_1, v_2, v_3\}$ be a basis for a vector space V. Show that $\{u_1, u_2, u_3\}$ is also a basis, where $u_1 = v_1$, $u_2 = v_1 + v_2$, and $u_3 = v_1 + v_2 + v_3$.

17. Show that the vector space of all real-valued functions defined on the entire real line is infinite-dimensional. (**Hint.** Assume it is finite-dimensional with dimension n, and obtain a contradiction by producing $n + 1$ linearly independent vectors.)

18. Show that a subspace of a finite-dimensional vector space is finite-dimensional.

19. Let V be a subspace of a finite-dimensional vector space W. Show that $\dim(V) \leq \dim(W)$. (**Hint.** V is finite dimensional by Exercise 18.)

20. Show that the only subspaces of R^3 are lines through the origin, planes through the origin, the zero subspace, and R^3 itself. (**Hint.** By Exercise 19 the subspaces of R^3 must be 0-dimensional, 1-dimensional, 2-dimensional, or 3-dimensional.)

21. Prove part (*a*) of Theorem 11.

22. Prove part (*b*) of Theorem 11.

23. Prove part (*c*) of Theorem 11.

4.6 ROW AND COLUMN SPACE; RANK; FINDING BASES

In this section we shall study certain vector spaces associated with matrices. Our results will provide a simple procedure for finding bases by reducing an appropriate matrix to row-echelon form.

Definition. For an $m \times n$ matrix

$$A = \begin{bmatrix} a_{11} & a_{12} & \cdots & a_{1n} \\ a_{21} & a_{22} & \cdots & a_{2n} \\ \vdots & \vdots & & \vdots \\ a_{m1} & a_{m2} & \cdots & a_{mn} \end{bmatrix}$$

the vectors

$$\mathbf{r}_1 = (a_{11}, a_{12}, \ldots, a_{1n})$$
$$\mathbf{r}_2 = (a_{21}, a_{22}, \ldots, a_{2n})$$
$$\vdots \qquad \qquad \vdots$$
$$\mathbf{r}_m = (a_{m1}, a_{m2}, \ldots, a_{mn})$$

formed from the rows of A are called the *row vectors* of A, and the vectors

$$\mathbf{c}_1 = \begin{bmatrix} a_{11} \\ a_{21} \\ \vdots \\ a_{m1} \end{bmatrix}, \quad \mathbf{c}_2 = \begin{bmatrix} a_{12} \\ a_{22} \\ \vdots \\ a_{m2} \end{bmatrix}, \ldots, \quad \mathbf{c}_n = \begin{bmatrix} a_{1n} \\ a_{2n} \\ \vdots \\ a_{mn} \end{bmatrix}$$

formed from the columns of A are called the *column vectors* of A. The subspace of R^n spanned by the row vectors is called the *row space* of A, and the subspace of R^m spanned by the column vectors is called the *column space* of A.

Example 39

Let

$$A = \begin{bmatrix} 2 & 1 & 0 \\ 3 & -1 & 4 \end{bmatrix}$$

The row vectors of A are

$$\mathbf{r}_1 = (2, 1, 0) \quad \text{and} \quad \mathbf{r}_2 = (3, -1, 4)$$

and the column vectors of A are

$$\mathbf{c}_1 = \begin{bmatrix} 2 \\ 3 \end{bmatrix} \quad \mathbf{c}_2 = \begin{bmatrix} 1 \\ -1 \end{bmatrix} \quad \text{and} \quad \mathbf{c}_3 = \begin{bmatrix} 0 \\ 4 \end{bmatrix}$$

The next theorem will help us to find bases for vector spaces. We shall defer its proof to the end of the section.

Theorem 12. *Elementary row operations do not change the row space of a matrix.*

It follows from this theorem that a matrix and all of its row-echelon forms have the same row space. However, the nonzero row vectors of a matrix in row-echelon form are always linearly independent (Exercise 18) so that these nonzero row vectors form a basis for the row space. Thus, we have the following result.

Theorem 13. *The nonzero row vectors in a row-echelon form of a matrix A form a basis for the row space of A.*

Example 40

Find a basis for the space spanned by the vectors

$$\mathbf{v}_1 = (1, -2, 0, 0, 3) \qquad \mathbf{v}_2 = (2, -5, -3, -2, 6) \qquad \mathbf{v}_3 = (0, 5, 15, 10, 0)$$
$$\mathbf{v}_4 = (2, 6, 18, 8, 6)$$

Solution. The space spanned by these vectors is the row space of the matrix

$$\begin{bmatrix} 1 & -2 & 0 & 0 & 3 \\ 2 & -5 & -3 & -2 & 6 \\ 0 & 5 & 15 & 10 & 0 \\ 2 & 6 & 18 & 8 & 6 \end{bmatrix}$$

Reducing this matrix to row-echelon form we obtain (verify):

$$\begin{bmatrix} 1 & -2 & 0 & 0 & 3 \\ 0 & 1 & 3 & 2 & 0 \\ 0 & 0 & 1 & 1 & 0 \\ 0 & 0 & 0 & 0 & 0 \end{bmatrix}$$

The nonzero row vectors in this matrix are

$$\mathbf{w}_1 = (1, -2, 0, 0, 3) \qquad \mathbf{w}_2 = (0, 1, 3, 2, 0) \qquad \mathbf{w}_3 = (0, 0, 1, 1, 0)$$

These form a basis for the row space and consequently a basis for the space spanned by $\mathbf{v}_1, \mathbf{v}_2, \mathbf{v}_3$, and $\mathbf{v}_4$.

REMARK. We have been writing the row vectors of a matrix in horizontal notation and the column vectors in vertical (matrix) notation because it seems natural to do so. However, there is no reason why the row vectors cannot be written in vertical notation and the column vectors in horizontal notation if it is convenient.

In light of the above remark, it is clear that, except for a change from vertical to horizontal notation, the column space of a matrix is the same as the row space of its transpose. Thus, we can find a basis for the column space of a matrix A by finding a basis for the row space of A^t and then converting back to vertical notation, if desired.

Example 41

Find a basis for the column space of

$$A = \begin{bmatrix} 1 & 0 & 1 & 1 \\ 3 & 2 & 5 & 1 \\ 0 & 4 & 4 & -4 \end{bmatrix}$$

Solution. Transposing, we obtain

$$A^t = \begin{bmatrix} 1 & 3 & 0 \\ 0 & 2 & 4 \\ 1 & 5 & 4 \\ 1 & 1 & -4 \end{bmatrix}$$

and reducing to row-echelon form yields (verify)

$$\begin{bmatrix} 1 & 3 & 0 \\ 0 & 1 & 2 \\ 0 & 0 & 0 \\ 0 & 0 & 0 \end{bmatrix}$$

Thus, the vectors $(1, 3, 0)$ and $(0, 1, 2)$ form a basis for the row space of A^t, or equivalently

$$\mathbf{w}_1 = \begin{bmatrix} 1 \\ 3 \\ 0 \end{bmatrix} \qquad \text{and} \qquad \mathbf{w}_2 = \begin{bmatrix} 0 \\ 1 \\ 2 \end{bmatrix}$$

form a basis for the column space of A.

In Example 40, we used row reduction to find a basis for the space spanned by a set of vectors. However, that method can produce basis vectors that are not among the vectors in the original set. We will now show how to produce a basis that is a *subset* of the given set of vectors. The idea is to delete vectors from the given set that are linear combinations of the rest, thereby leaving a set of linearly independent vectors that span the same space as the original set of vectors. The method, which applies only to vectors in R^n, will be explained by example.

Example 42

Find a subset of the vectors

$$\mathbf{v}_1 = (1, -2, 0, 3), \qquad \mathbf{v}_2 = (2, -5, -3, 6)$$
$$\mathbf{v}_3 = (0, 1, 3, 0), \qquad \mathbf{v}_4 = (2, -1, 4, -7), \qquad \mathbf{v}_5 = (5, -8, 1, 2)$$

that forms a basis for the space spanned by these vectors.

Solution. We begin by solving the vector equation

$$c_1\mathbf{v}_1 + c_2\mathbf{v}_2 + c_3\mathbf{v}_3 + c_4\mathbf{v}_4 + c_5\mathbf{v}_5 = \mathbf{0} \qquad (4.11)$$

Substituting components and then equating the corresponding components on the two sides of (4.11) yields the homogeneous system

$$
\begin{aligned}
c_1 + 2c_2 \qquad\quad + 2c_4 + 5c_5 &= 0 \\
-2c_1 - 5c_2 + c_3 - c_4 - 8c_5 &= 0 \\
- 3c_2 + 3c_3 + 4c_4 + c_5 &= 0 \\
3c_1 + 6c_2 \qquad\quad\; - 7c_4 + 2c_5 &= 0
\end{aligned}
$$

Solving this system by Gauss-Jordan elimination yields (verify):

$$c_1 = -2s - t, \qquad c_2 = s - t, \qquad c_3 = s, \qquad c_4 = -t, \qquad c_5 = t \quad (4.12)$$

where s and t are arbitrary. Substituting (4.12) in (4.11), we obtain

$$(-2s - t)\mathbf{v}_1 + (s - t)\mathbf{v}_2 + s\mathbf{v}_3 - t\mathbf{v}_4 + t\mathbf{v}_5 = \mathbf{0}$$

which can be rewritten as

$$s(-2\mathbf{v}_1 + \mathbf{v}_2 + \mathbf{v}_3) + t(-\mathbf{v}_1 - \mathbf{v}_2 - \mathbf{v}_4 + \mathbf{v}_5) = \mathbf{0} \qquad (4.13)$$

Since s and t are arbitrary, we can set $s = 1$, $t = 0$ and then $s = 0$, $t = 1$. This yields the **dependency equations**

$$
\begin{aligned}
-2\mathbf{v}_1 + \mathbf{v}_2 + \mathbf{v}_3 &= \mathbf{0} \\
-\mathbf{v}_1 - \mathbf{v}_2 - \mathbf{v}_4 + \mathbf{v}_5 &= \mathbf{0}
\end{aligned}
\qquad (4.14)
$$

With these equations we can express the vectors $\mathbf{v}_3$ and $\mathbf{v}_5$ as linear combinations of *preceding* vectors (i.e., vectors with smaller subscripts). We obtain

$$
\begin{aligned}
\mathbf{v}_3 &= 2\mathbf{v}_1 - \mathbf{v}_2 \\
\mathbf{v}_5 &= \mathbf{v}_1 + \mathbf{v}_2 + \mathbf{v}_4
\end{aligned}
\qquad (4.15)
$$

which shows that v_3 and v_5 may be discarded from the original set without affecting the space spanned.

We will now show that the remaining vectors, v_1, v_2, and v_4, are linearly independent and therefore form a basis for the space spanned by v_1, v_2, v_3, v_4, and v_5. If

$$c_1 v_1 + c_2 v_2 + c_4 v_4 = 0$$

then (4.11) holds with $c_3 = 0$ and $c_5 = 0$. Thus, (4.12) holds with $s = c_3 = 0$ and $t = c_5 = 0$. From the remaining equations in (4.12), it follows that

$$c_1 = 0, \qquad c_2 = 0, \qquad \text{and} \qquad c_4 = 0$$

which proves that v_1, v_2, and v_4 are linearly independent. ▲

REMARK. Although we will omit the proof, it can be shown that after the dependency equations are used to eliminate those vectors that depend on preceding vectors, the remaining vectors will always be linearly independent. Thus, it is really not necessary to check the independence as we did in the last example.

Our next theorem is one of the most fundamental results in linear algebra. The proof is deferred to the end of this section.

Theorem 14. *If A is any matrix, then the row space and column space of A have the same dimension.*

Example 43

In Example 41 we saw that the matrix

$$A = \begin{bmatrix} 1 & 0 & 1 & 1 \\ 3 & 2 & 5 & 1 \\ 0 & 4 & 4 & -4 \end{bmatrix}$$

has a two-dimensional column space. Thus, Theorem 14 asserts that the row space is also two-dimensional. To see that this is in fact the case we reduce A to row-echelon form, obtaining (verify)

$$\begin{bmatrix} 1 & 0 & 1 & 1 \\ 0 & 1 & 1 & -1 \\ 0 & 0 & 0 & 0 \end{bmatrix}$$

Since this matrix has two nonzero rows, the row space of A is two-dimensional. ▲

Definition. The dimension of the row and column space of a matrix A is called the **rank** of A and is denoted by rank(A).

For example, the matrix A in Example 43 has rank 2.

The next theorem adds three more results to those in Theorem 15 of Section 1.7 and Theorem 7 of Section 2.3.

Theorem 15. *If A is an n × n matrix, then the following statements are equivalent.*

(*a*) *A is invertible.*
(*b*) $A\mathbf{x} = \mathbf{0}$ *has only the trivial solution.*
(*c*) *A is row equivalent to* I_n.
(*d*) $A\mathbf{x} = \mathbf{b}$ *is consistent for every n × 1 matrix* **b**.
(*e*) $\det(A) \neq 0$
(*f*) *A has rank n.*
(*g*) *The row vectors of A are linearly independent.*
(*h*) *The column vectors of A are linearly independent.*

Proof. We shall show that (*c*), (*f*), (*g*), and (*h*) are equivalent by proving the sequence of implications (*c*)$\Rightarrow$(*f*)$\Rightarrow$(*g*)$\Rightarrow$(*h*)$\Rightarrow$(*c*). This will complete the proof, since we already know that (*c*) is equivalent to (*a*), (*b*), (*d*), and (*e*).

(*c*) $\Rightarrow$ (*f*) Since A is row equivalent to I_n, and I_n has n nonzero rows, the row space of A is n-dimensional by Theorem 13. Thus, A has rank n.

(*f*)$\Rightarrow$(*g*) Since A has rank n, the row space of A is n-dimensional. Since the n row vectors of A span the row space of A, it follows from Theorem 11 in Section 4.5 that the row vectors of A are linearly independent.

(*g*)$\Rightarrow$(*h*) Assume the row vectors of A are linearly independent. Thus, the row space of A is n-dimensional. By Theorem 14 the column space of A is also n-dimensional. Since the column vectors of A span the column space, the column vectors of A are linearly independent by Theorem 11 in Section 4.5.

(*h*) $\Rightarrow$ (*c*) Assume the column vectors of A are linearly independent. Thus, the column space of A is n-dimensional, and consequently the row space of A is n-dimensional by Theorem 14. This means that the reduced row-echelon form of A has n nonzero rows; that is, all rows are nonzero. As noted in Example 24 of Section 2.3 this implies that the reduced row-echelon form of A is I_n. Thus, A is row equivalent to I_n.

It is interesting to note that Theorem 15 ties together all the major topics we have studied so far—matrices, systems of equations, determinants, and vector spaces.

We shall now consider some additional results about systems of linear equations. Consider a system of linear equations

$$A\mathbf{x} = \mathbf{b}$$

or equivalently

$$\begin{bmatrix} a_{11} & a_{12} & \cdots & a_{1n} \\ a_{21} & a_{22} & \cdots & a_{2n} \\ \vdots & \vdots & & \vdots \\ a_{m1} & a_{m2} & \cdots & a_{mn} \end{bmatrix} \begin{bmatrix} x_1 \\ x_2 \\ \vdots \\ x_n \end{bmatrix} = \begin{bmatrix} b_1 \\ b_2 \\ \vdots \\ b_m \end{bmatrix}$$

By multiplying out the matrices on the left side, this system can be rewritten

$$\begin{bmatrix} a_{11}x_1 + a_{12}x_2 + \cdots + a_{1n}x_n \\ a_{21}x_1 + a_{22}x_2 + \cdots + a_{2n}x_n \\ \vdots & \vdots & & \vdots \\ a_{m1}x_1 + a_{m2}x_2 + \cdots + a_{mn}x_n \end{bmatrix} = \begin{bmatrix} b_1 \\ b_2 \\ \vdots \\ b_m \end{bmatrix}$$

or

$$x_1 \begin{bmatrix} a_{11} \\ a_{21} \\ \vdots \\ a_{m1} \end{bmatrix} + x_2 \begin{bmatrix} a_{12} \\ a_{22} \\ \vdots \\ a_{m2} \end{bmatrix} + \cdots + x_n \begin{bmatrix} a_{1n} \\ a_{2n} \\ \vdots \\ a_{mn} \end{bmatrix} = \begin{bmatrix} b_1 \\ b_2 \\ \vdots \\ b_m \end{bmatrix} \tag{4.16}$$

Since the left side of this equation is a linear combination of the column vectors of A, it follows that the system $A\mathbf{x} = \mathbf{b}$ is consistent if and only if $\mathbf{b}$ is a linear combination of the column vectors of A. Thus, we have the following useful theorem.

Theorem 16. *A system of linear equations* $A\mathbf{x} = \mathbf{b}$ *is consistent if and only if* $\mathbf{b}$ *is in the column space of* A.

Example 44

Let $A\mathbf{x} = \mathbf{b}$ be the linear system

$$\begin{bmatrix} -1 & 3 & 2 \\ 1 & 2 & -3 \\ 2 & 1 & -2 \end{bmatrix} \begin{bmatrix} x_1 \\ x_2 \\ x_3 \end{bmatrix} = \begin{bmatrix} 1 \\ -9 \\ -3 \end{bmatrix}$$

Solve the system and use the result to express $\mathbf{b}$ as a linear combination of the column vectors of A.

Solution. Solving the system by Gaussian elimination yields (verify):

$$x_1 = 2, \quad x_2 = -1, \quad x_3 = 3$$

Thus, from equation (4.16),

$$2\begin{bmatrix} -1 \\ 1 \\ 2 \end{bmatrix} - \begin{bmatrix} 3 \\ 2 \\ 1 \end{bmatrix} + 3\begin{bmatrix} 2 \\ -3 \\ -2 \end{bmatrix} = \begin{bmatrix} 1 \\ -9 \\ -3 \end{bmatrix}$$

We shall conclude this section by exploring some important relationships between the notion of rank and systems of linear equations. The following theorem is a direct consequence of Theorem 16. We leave the proof as an exercise.

> **Theorem 17.** *A system of linear equations* $A\mathbf{x} = \mathbf{b}$ *is consistent if and only if the rank of the coefficient matrix A is the same as the rank of the augmented matrix* $[A \,|\, \mathbf{b}]$.

This result is fairly obvious if one remembers that the rank of a matrix is equal to the number of nonzero rows in its reduced row-echelon form. For example, the augmented matrix for the system

$$
\begin{aligned}
x_1 - 2x_2 - 3x_3 + 2x_4 &= -4 \\
-3x_1 + 7x_2 - x_3 + x_4 &= -3 \\
2x_1 - 5x_2 + 4x_3 - 3x_4 &= 7 \\
-3x_1 + 6x_2 + 9x_3 - 6x_4 &= -1
\end{aligned}
$$

is

$$
\left[\begin{array}{cccc|c}
1 & -2 & -3 & 2 & -4 \\
-3 & 7 & -1 & 1 & -3 \\
2 & -5 & 4 & -3 & 7 \\
-3 & 6 & 9 & -6 & -1
\end{array}\right]
$$

which has the following reduced row-echelon form (verify):

$$
\left[\begin{array}{cccc|c}
1 & 0 & -23 & 16 & 0 \\
0 & 1 & -10 & 7 & 0 \\
0 & 0 & 0 & 0 & 1 \\
0 & 0 & 0 & 0 & 0
\end{array}\right]
$$

Because of the row

$$
\begin{array}{ccccc}
0 & 0 & 0 & 0 & 1
\end{array}
$$

we see that the system is inconsistent. However, it is also because of this row that the reduced row-echelon form of the augmented matrix has fewer zero rows than the reduced row-echelon form of the coefficient matrix. Thus, the coefficient matrix and the augmented matrix for the system have different ranks.

The rank of the coefficient matrix of a consistent linear system can be used to determine the number of parameters that appear in the solution. To see how, suppose that A is an $m \times n$ matrix with rank r, and $A\mathbf{x} = \mathbf{b}$ is a consistent linear system of m equations in n unknowns. It follows from Theorem 17 that the augmented matrix $[A\,|\,b]$ has r nonzero rows. Since each of these rows contains a leading 1, it follows that r of the n unknowns are leading variables that can be expressed in terms of the remaining $n - r$ unknowns, which can be assigned arbitrary values. Thus, we have the following result.

Theorem 18. *If $Ax = b$ is a consistent linear system of m equations in n unknowns, and if A has rank r, then the solution of the system contains $n - r$ parameters.*

Example 45

If A is a 5×7 matrix with rank 4, and if $Ax = b$ is a consistent linear system, then the solution of the system contains $7 - 4 = 3$ parameters. ▲

OPTIONAL

Proof of Theorem 12. Suppose that the row vectors of a matrix A are $\mathbf{r}_1, \mathbf{r}_2, \ldots, \mathbf{r}_m$ and let B be obtained from A by performing an elementary row operation. We shall show that every vector in the row space of B is also in the row space of A, and conversely that every vector in the row space of A is in the row space of B. We can then conclude that A and B have the same row space.

Consider the possibilities: if the row operation is a row interchange, then B and A have the same row vectors and consequently the same row space. If the row operation is multiplication of a row by a scalar or addition of a multiple of one row to another, then the row vectors $\mathbf{r}'_1, \mathbf{r}'_2, \ldots, \mathbf{r}'_m$ of B are linear combinations of $\mathbf{r}_1, \mathbf{r}_2, \ldots, \mathbf{r}_m$; thus, they lie in the row space of A. Since a vector space is closed under addition and scalar multiplication, all linear combinations of $\mathbf{r}'_1, \mathbf{r}'_2, \ldots, \mathbf{r}'_m$ will also lie in the row space of A. Therefore, each vector in the row space of B is in the row space of A.

Since B is obtained from A by performing a row operation, A can be obtained from B by performing the inverse operation (Section 1.6). Thus, the argument above shows that the row space of A is contained in the row space of B.

OPTIONAL

Proof of Theorem 14. Denote the row vectors of

$$
A = \begin{bmatrix}
a_{11} & a_{12} & \cdots & a_{1n} \\
a_{21} & a_{22} & \cdots & a_{2n} \\
\vdots & \vdots & & \vdots \\
a_{m1} & a_{m2} & \cdots & a_{mn}
\end{bmatrix}
$$

by

$$\mathbf{r}_1, \mathbf{r}_2, \ldots, \mathbf{r}_m$$

Suppose the row space of A has dimension k and that $S = \{\mathbf{b}_1, \mathbf{b}_2, \ldots, \mathbf{b}_k\}$ is a basis for the row space, where $\mathbf{b}_i = (b_{i1}, b_{i2}, \ldots, b_{in})$. Since S is a basis for the row space, each row vector is expressible as a linear combination of $\mathbf{b}_1, \mathbf{b}_2, \ldots, \mathbf{b}_k$; thus,

$$\begin{aligned}
\mathbf{r}_1 &= c_{11}\mathbf{b}_1 + c_{12}\mathbf{b}_2 + \cdots + c_{1k}\mathbf{b}_k \\
\mathbf{r}_2 &= c_{21}\mathbf{b}_1 + c_{22}\mathbf{b}_2 + \cdots + c_{2k}\mathbf{b}_k \\
&\;\;\vdots \qquad \vdots \qquad\quad \vdots \\
\mathbf{r}_m &= c_{m1}\mathbf{b}_1 + c_{m2}\mathbf{b}_2 + \cdots + c_{mk}\mathbf{b}_k
\end{aligned}$$

Since two vectors in R^n are equal if and only if corresponding components are equal, we can equate the jth component on each side of these equations to obtain

$$\begin{aligned}
a_{1j} &= c_{11}b_{1j} + c_{12}b_{2j} + \cdots + c_{1k}b_{kj} \\
a_{2j} &= c_{21}b_{1j} + c_{22}b_{2j} + \cdots + c_{2k}b_{kj} \\
&\;\;\vdots \qquad \vdots \qquad\;\; \vdots \qquad\qquad \vdots \\
a_{mj} &= c_{m1}b_{1j} + c_{m2}b_{2j} + \cdots + c_{mk}b_{kj}
\end{aligned}$$

or equivalently

$$\begin{bmatrix} a_{1j} \\ a_{2j} \\ \vdots \\ a_{mj} \end{bmatrix} = b_{1j} \begin{bmatrix} c_{11} \\ c_{21} \\ \vdots \\ c_{m1} \end{bmatrix} + b_{2j} \begin{bmatrix} c_{12} \\ c_{22} \\ \vdots \\ c_{m2} \end{bmatrix} + \cdots + b_{kj} \begin{bmatrix} c_{1k} \\ c_{2k} \\ \vdots \\ c_{mk} \end{bmatrix} \tag{4.17}$$

The left side of this equation is the jth column vector of A, and $j = 1, 2, \ldots, n$ is arbitrary; therefore, each column vector of A lies in the space spanned by the k vectors on the right side of (4.17). Thus, the column space of A has dimension $\leq k$. Since

$$k = \dim(\text{row space of } A)$$

we have

$$\dim(\text{column space of } A) \leq \dim(\text{row space of } A) \tag{4.18}$$

Since the matrix A is completely arbitrary, this same conclusion applies to A^t; that is,

$$\dim(\text{column space of } A^t) \leq \dim(\text{row space of } A^t) \tag{4.19}$$

But transposing a matrix converts columns to rows and rows to columns, so that

$$\text{column space of } A^t = \text{row space of } A$$

and

$$\text{row space of } A^t = \text{column space of } A$$

Thus, (4.19) can be rewritten as

$$\dim(\text{row space of } A) \leq \dim(\text{column space of } A)$$

From this result and (4.18) we conclude that

$$\dim(\text{row space of } A) = \dim(\text{column space of } A)$$

EXERCISE SET 4.6

1. List the row vectors and column vectors of the matrix

$$\begin{bmatrix} 2 & -1 & 0 & 1 \\ 3 & 5 & 7 & -1 \\ 1 & 4 & 2 & 7 \end{bmatrix}$$

In Exercises 2–5 find: (a) a basis for the row space; (b) a basis for the column space; (c) the rank of the matrix.

2. $\begin{bmatrix} 1 & -3 \\ 2 & -6 \end{bmatrix}$ **3.** $\begin{bmatrix} 1 & 2 & -1 \\ 2 & 4 & 6 \\ 0 & 0 & -8 \end{bmatrix}$ **4.** $\begin{bmatrix} 1 & 1 & 2 & 1 \\ 1 & 0 & 1 & 2 \\ 2 & 1 & 3 & 4 \end{bmatrix}$

5. $\begin{bmatrix} 1 & -3 & 2 & 2 & 1 \\ 0 & 3 & 6 & 0 & -2 \\ 2 & -3 & -2 & 4 & 4 \\ 3 & -3 & 6 & 6 & 3 \\ 5 & -3 & 10 & 10 & 5 \end{bmatrix}$

6. Find a basis for the subspace of R^4 spanned by the given vectors.
(a) $(1, 1, -4, -3)$, $(2, 0, 2, -2)$, $(2, -1, 3, 2)$
(b) $(-1, 1, -2, 0)$, $(3, 3, 6, 0)$, $(9, 0, 0, 3)$
(c) $(1, 1, 0, 0)$, $(0, 0, 1, 1)$, $(-2, 0, 2, 2)$, $(0, -3, 0, 3)$

7. Verify that the row space and column space have the same dimension (as guaranteed by Theorem 14).

(a) $\begin{bmatrix} 2 & 0 & 2 & 2 \\ 3 & -4 & -1 & -9 \\ 1 & 2 & 3 & 7 \\ -3 & 1 & -2 & 0 \end{bmatrix}$ (b) $\begin{bmatrix} 2 & 3 & 5 & 7 & 4 \\ -1 & 2 & 1 & 0 & -2 \\ 4 & 1 & 5 & 9 & 8 \end{bmatrix}$

In Exercises 8 and 9, use the method of Example 42 to: (a) find a subset of the vectors that forms a basis for the space spanned by the vectors; (b) express each v_i that is not in the basis as a linear combination of the basis vectors.

8. $v_1 = (1, -1, 5, 2)$, $v_2 = (-2, 3, 1, 0)$, $v_3 = (4, -5, 9, 4)$, $v_4 = (0, 4, 2, -3)$, $v_5 = (-7, 18, 2, -8)$

9. $v_1 = (1, 0, 1, 1)$, $v_2 = (-3, 3, 7, 1)$, $v_3 = (-1, 3, 9, 3)$, $v_4 = (-5, 3, 5, -1)$

10. In each part, find a basis for the row space consisting entirely of row vectors, and find a basis for the column space consisting entirely of column vectors.

(a) $\begin{bmatrix} 1 & -2 & -4 & 3 \\ -3 & 6 & 12 & -9 \\ 1 & 0 & 1 & 3 \end{bmatrix}$ (b) $\begin{bmatrix} 2 & -4 & 6 & 8 \\ 2 & -1 & 3 & -5 \\ 4 & -5 & 9 & 3 \\ 0 & 1 & 1 & -2 \end{bmatrix}$

11. (a) If A is a 3×5 matrix, what is the largest possible value for the rank of A?

(b) If A is an $m \times n$ matrix, what is the largest possible value for the rank of A?

12. In each part determine if **b** lies in the column space of A. If it does, express **b** as a linear combination of the column vectors.

(a) $A = \begin{bmatrix} 1 & 3 \\ 4 & -6 \end{bmatrix} \qquad \mathbf{b} = \begin{bmatrix} -2 \\ 10 \end{bmatrix}$

(b) $A = \begin{bmatrix} 1 & -4 \\ 2 & -8 \end{bmatrix} \qquad \mathbf{b} = \begin{bmatrix} 0 \\ 1 \end{bmatrix}$

(c) $A = \begin{bmatrix} 1 & -1 & 1 \\ 1 & 1 & -1 \\ -1 & -1 & 1 \end{bmatrix} \qquad \mathbf{b} = \begin{bmatrix} 2 \\ 0 \\ 0 \end{bmatrix}$

13. In each part, use the information in the table to determine whether the linear system $A\mathbf{x} = \mathbf{b}$ is consistent. If it is, state the number of parameters in the solution.

	Size of A	Rank(A)	Rank$[A \vert \mathbf{b}]$
(a)	3×3	3	3
(b)	3×3	2	3
(c)	3×3	1	1
(d)	5×9	2	2
(e)	5×9	2	3
(f)	4×4	0	0
(g)	6×2	2	2

14. (a) Prove: If A is a 3×5 matrix, then the column vectors of A are linearly dependent.

(b) Prove: If A is a 5×3 matrix, then the row vectors of A are linearly dependent.

15. Prove: If A is a matrix that is not square, then either the row vectors of A or the column vectors of A are linearly dependent.

16. Let

$$A = \begin{bmatrix} a_{11} & a_{12} & a_{13} \\ a_{21} & a_{22} & a_{23} \end{bmatrix}$$

Prove: A has rank 2 if and only if one or more of the determinants

$$\begin{vmatrix} a_{11} & a_{12} \\ a_{21} & a_{22} \end{vmatrix}, \quad \begin{vmatrix} a_{11} & a_{13} \\ a_{21} & a_{23} \end{vmatrix}, \quad \begin{vmatrix} a_{12} & a_{13} \\ a_{22} & a_{23} \end{vmatrix}$$

is nonzero.

17. Prove Theorem 17.

18. Prove Theorem 13.

19. Prove that the row vectors in an $n \times n$ invertible matrix A form a basis for R^n.

4.7 INNER PRODUCT SPACES

In Section 4.1 we studied the Euclidean inner product on the vector space R^n. In this section we introduce the notion of an inner product on any real vector space. As a consequence of our work, we will be able to define useful notions of angle, length, and distance in more general vector spaces.

In Theorem 2 of Section 4.1, we collected the most important properties of the Euclidean inner product. In a general real vector space, an inner product is defined axiomatically using these properties as axioms.

Definition. An ***inner product*** on a real vector space V is a function that associates a real number $\langle \mathbf{u}, \mathbf{v} \rangle$ with each pair of vectors $\mathbf{u}$ and $\mathbf{v}$ in V in such a way that the following axioms are satisfied for all vectors $\mathbf{u}$, $\mathbf{v}$, and $\mathbf{w}$ in V and all scalars k.

(1) $\langle \mathbf{u}, \mathbf{v} \rangle = \langle \mathbf{v}, \mathbf{u} \rangle$	(symmetry axiom)
(2) $\langle \mathbf{u} + \mathbf{v}, \mathbf{w} \rangle = \langle \mathbf{u}, \mathbf{w} \rangle + \langle \mathbf{v}, \mathbf{w} \rangle$	(additivity axiom)
(3) $\langle k\mathbf{u}, \mathbf{v} \rangle = k\langle \mathbf{u}, \mathbf{v} \rangle$	(homogeneity axiom)
(4) $\langle \mathbf{v}, \mathbf{v} \rangle \geq 0$; and $\langle \mathbf{v}, \mathbf{v} \rangle = 0$ if and only if $\mathbf{v} = \mathbf{0}$	(positivity axiom)

A real vector space with an inner product is called a ***real inner product space.***

REMARK. In Chapter 9 we shall study complex inner products, that is, inner products whose values are complex numbers. Until that time we shall use the term "inner product space" to mean "real inner product space."

Example 46

If $\mathbf{u} = (u_1, u_2, \ldots, u_n)$ and $\mathbf{v} = (v_1, v_2, \ldots, v_n)$ are vectors in R^n, then the formula

$$\langle \mathbf{u}, \mathbf{v} \rangle = \mathbf{u} \cdot \mathbf{v} = u_1 v_1 + u_2 v_2 + \cdots + u_n v_n$$

defines $\langle \mathbf{u}, \mathbf{v} \rangle$ to be the Euclidean inner product on R^n. The four inner product axioms are satisfied by Theorem 2 of Section 4.1. ▲

The Euclidean inner product is the most important inner product on R^n. However, there are various applications in which it is desirable to modify the Euclidean inner product by *weighting* its terms differently. More precisely, if

$$w_1, w_2, \ldots, w_n$$

are *positive* real numbers, which we shall call ***weights***, and if $\mathbf{u} = (u_1, u_2, \ldots, u_n)$ and $\mathbf{v} = (v_1, v_2, \ldots, v_n)$ are vectors in R^n, then it can be shown (Exercise 14) that the formula

$$\langle \mathbf{u}, \mathbf{v} \rangle = w_1 u_1 v_1 + w_2 u_2 v_2 + \cdots + w_n u_n v_n$$

defines as inner product on R^n; it is called the **weighted Euclidean inner product with weights** $w_1, w_2, \ldots, w_n$.

Example 47

Let $\mathbf{u} = (u_1, u_2)$ and $\mathbf{v} = (v_1, v_2)$ be vectors in R^2. Verify that the weighted Euclidean inner product

$$\langle \mathbf{u}, \mathbf{v} \rangle = 3u_1v_1 + 2u_2v_2$$

satisfies the four inner product axioms.

Solution. Note first that if $\mathbf{u}$ and $\mathbf{v}$ are interchanged in this equation, the right side remains the same. Therefore,

$$\langle \mathbf{u}, \mathbf{v} \rangle = \langle \mathbf{v}, \mathbf{u} \rangle$$

If $\mathbf{w} = (w_1, w_2)$, then

$$
\begin{aligned}
\langle \mathbf{u} + \mathbf{v}, \mathbf{w} \rangle &= 3(u_1 + v_1)w_1 + 2(u_2 + v_2)w_2 \\
&= (3u_1w_1 + 2u_2w_2) + (3v_1w_1 + 2v_2w_2) \\
&= \langle \mathbf{u}, \mathbf{w} \rangle + \langle \mathbf{v}, \mathbf{w} \rangle
\end{aligned}
$$

which establishes the second axiom.

Next,

$$
\begin{aligned}
\langle k\mathbf{u}, \mathbf{v} \rangle &= 3(ku_1)v_1 + 2(ku_2)v_2 \\
&= k(3u_1v_1 + 2u_2v_2) \\
&= k\langle \mathbf{u}, \mathbf{v} \rangle
\end{aligned}
$$

which establishes the third axiom.

Finally,

$$\langle \mathbf{v}, \mathbf{v} \rangle = 3v_1v_1 + 2v_2v_2 = 3v_1^2 + 2v_2^2$$

Obviously, $\langle \mathbf{v}, \mathbf{v} \rangle = 3v_1^2 + 2v_2^2 \geq 0$. Further, $\langle \mathbf{v}, \mathbf{v} \rangle = 3v_1^2 + 2v_2^2 = 0$ if and only if $v_1 = v_2 = 0$; that is, if and only if $\mathbf{v} = (v_1, v_2) = \mathbf{0}$. Thus, the fourth axiom is satisfied.

The Euclidean inner product and the weighted Euclidean inner products are special cases of a general class of inner products on R^n, which we shall now describe. Let

$$
\mathbf{u} = \begin{bmatrix} u_1 \\ u_2 \\ \vdots \\ u_n \end{bmatrix}
\quad \text{and} \quad
\mathbf{v} = \begin{bmatrix} v_1 \\ v_2 \\ \vdots \\ v_n \end{bmatrix}
$$

be vectors in R^n (expressed as $n \times 1$ matrices), and let A be an *invertible* $n \times n$ matrix. It can be shown (Exercise 18) that if $\mathbf{u} \cdot \mathbf{v}$ is the Euclidean inner product on

R^n, then the formula

$$\langle \mathbf{u}, \mathbf{v} \rangle = A\mathbf{u} \cdot A\mathbf{v} \qquad (4.20)$$

defines an inner product; it is called the ***inner product on R^n generated by A***.

Recalling that the Euclidean inner product $\mathbf{u} \cdot \mathbf{v}$ can be written as the matrix product $\mathbf{v}^t\mathbf{u}$ (see the last paragraph in Section 4.1), it follows that (4.20) can be written in the alternative form

$$\langle \mathbf{u}, \mathbf{v} \rangle = (A\mathbf{v})^t A\mathbf{u}$$

or equivalently

$$\langle \mathbf{u}, \mathbf{v} \rangle = \mathbf{v}^t A^t A\, \mathbf{u} \qquad (4.21)$$

Example 48

The inner product on R^n generated by the $n \times n$ identity matrix is the Euclidean inner product, since substituting $A = I$ in (4.20) yields

$$\langle \mathbf{u}, \mathbf{v} \rangle = I\mathbf{u} \cdot I\mathbf{v} = \mathbf{u} \cdot \mathbf{v}$$

The weighted Euclidean inner product $\langle \mathbf{u}, \mathbf{v} \rangle = 3u_1v_1 + 2u_2v_2$ discussed in Example 47 is the inner product on R^2 generated by

$$A = \begin{bmatrix} \sqrt{3} & 0 \\ 0 & \sqrt{2} \end{bmatrix}$$

because substituting this matrix in (4.21) yields

$$\langle \mathbf{u}, \mathbf{v} \rangle = \begin{bmatrix} v_1 & v_2 \end{bmatrix} \begin{bmatrix} \sqrt{3} & 0 \\ 0 & \sqrt{2} \end{bmatrix} \begin{bmatrix} \sqrt{3} & 0 \\ 0 & \sqrt{2} \end{bmatrix} \begin{bmatrix} u_1 \\ u_2 \end{bmatrix}$$

$$= \begin{bmatrix} v_1 & v_2 \end{bmatrix} \begin{bmatrix} 3 & 0 \\ 0 & 2 \end{bmatrix} \begin{bmatrix} u_1 \\ u_2 \end{bmatrix}$$

$$= 3u_1v_1 + 2u_2v_2$$

In general, the weighted Euclidean inner product $\langle \mathbf{u}, \mathbf{v} \rangle = w_1u_1v_1 + w_2u_2v_2 + \cdots + w_nu_nv_n$ is the inner product on R^n generated by

$$A = \begin{bmatrix} \sqrt{w_1} & 0 & 0 & \cdots & 0 \\ 0 & \sqrt{w_2} & 0 & \cdots & 0 \\ \vdots & \vdots & \vdots & & \vdots \\ 0 & 0 & 0 & \cdots & \sqrt{w_n} \end{bmatrix} \qquad (4.22)$$

(verify.)

In the following examples we shall describe some inner products on vector spaces other than R^n.

Example 49

If

$$U = \begin{bmatrix} u_1 & u_2 \\ u_3 & u_4 \end{bmatrix} \quad \text{and} \quad V = \begin{bmatrix} v_1 & v_2 \\ v_3 & v_4 \end{bmatrix}$$

are any two 2×2 matrices, then the following formula defines an inner product on M_{22} (verify):

$$\langle U, V \rangle = u_1 v_1 + u_2 v_2 + u_3 v_3 + u_4 v_4$$

For example, if

$$U = \begin{bmatrix} 1 & 2 \\ 3 & 4 \end{bmatrix} \quad \text{and} \quad V = \begin{bmatrix} -1 & 0 \\ 3 & 2 \end{bmatrix}$$

then

$$\langle U, V \rangle = 1(-1) + 2(0) + 3(3) + 4(2) = 16 \quad \blacktriangle$$

Example 50

If

$$\mathbf{p} = a_0 + a_1 x + a_2 x^2 \quad \text{and} \quad \mathbf{q} = b_0 + b_1 x + b_2 x^2$$

are any two vectors in P_2, then the following formula defines an inner product on P_2 (verify):

$$\langle \mathbf{p}, \mathbf{q} \rangle = a_0 b_0 + a_1 b_1 + a_2 b_2 \quad \blacktriangle$$

Example 51 (For readers who have studied calculus.)

Let $\mathbf{p} = p(x)$ and $\mathbf{q} = q(x)$ be two polynomials in P_n, and define

$$\langle \mathbf{p}, \mathbf{q} \rangle = \int_a^b p(x) q(x) \, dx \tag{4.23}$$

where a and b are any fixed real numbers such that $a < b$. We shall now show that this formula defines an inner product on P_n.

(1) $\langle \mathbf{p}, \mathbf{q} \rangle = \int_a^b p(x) q(x) \, dx = \int_a^b q(x) p(x) \, dx = \langle \mathbf{q}, \mathbf{p} \rangle$

which proves that Axiom 1 holds.

(2) $\langle \mathbf{p} + \mathbf{q}, \mathbf{s} \rangle = \int_a^b (p(x) + q(x)) s(x) \, dx$

$\qquad = \int_a^b p(x) s(x) \, dx + \int_a^b q(x) s(x) \, dx$

$\qquad = \langle \mathbf{p}, \mathbf{s} \rangle + \langle \mathbf{q}, \mathbf{s} \rangle$

which proves that Axiom 2 holds.

(3) $\langle k\mathbf{p}, \mathbf{q} \rangle = \int_a^b k p(x) q(x) \, dx = k \int_a^b p(x) q(x) \, dx = k \langle \mathbf{p}, \mathbf{q} \rangle$

which proves that Axiom 3 holds.

(4) If $\mathbf{p} = p(x)$ is any polynomial in P_n, then $p^2(x) \geq 0$ for all x; therefore,

$$\langle \mathbf{p}, \mathbf{p} \rangle = \int_a^b p^2(x)\, dx \geq 0$$

Further, since $p^2(x) \geq 0$ and polynomials are continuous functions, $\int_a^b p^2(x)\, dx = 0$ if and only if $p(x) = 0$ for all x satisfying $a \leq x \leq b$. Therefore, $\langle \mathbf{p}, \mathbf{p} \rangle = \int_a^b p^2(x)\, dx = 0$ if and only if $\mathbf{p} = \mathbf{0}$. This establishes Axiom 4. ▲

Example 52 (For readers who have studied calculus.)

Arguments similar to those in Example 51 can be used to show that the formula

$$\langle \mathbf{f}, \mathbf{g} \rangle = \int_a^b f(x)g(x)\, dx$$

defines an inner product on the vector space $C[a, b]$ discussed in Example 14. ▲

The following theorem gives three properties of inner products that follow immediately from the four inner product axioms.

Theorem 19. *If $\mathbf{u}$, $\mathbf{v}$, and $\mathbf{w}$ are vectors in a real inner product space, and k is any scalar, then*

(a) $\langle \mathbf{0}, \mathbf{v} \rangle = \langle \mathbf{v}, \mathbf{0} \rangle = 0$
(b) $\langle \mathbf{u}, \mathbf{v} + \mathbf{w} \rangle = \langle \mathbf{u}, \mathbf{v} \rangle + \langle \mathbf{u}, \mathbf{w} \rangle$
(c) $\langle \mathbf{u}, k\mathbf{v} \rangle = k\langle \mathbf{u}, \mathbf{v} \rangle$

Proof. We shall prove part (b) and leave the proofs of (a) and (c) as exercises.

$$
\begin{aligned}
(b)\quad \langle \mathbf{u}, \mathbf{v} + \mathbf{w} \rangle &= \langle \mathbf{v} + \mathbf{w}, \mathbf{u} \rangle && \text{(by symmetry)}\\
&= \langle \mathbf{v}, \mathbf{u} \rangle + \langle \mathbf{w}, \mathbf{u} \rangle && \text{(by additivity)}\\
&= \langle \mathbf{u}, \mathbf{v} \rangle + \langle \mathbf{u}, \mathbf{w} \rangle && \text{(by symmetry)} \ \blacksquare
\end{aligned}
$$

Example 53

In the exercises we ask the reader to verify that the inner products in Examples 46 through 52 satisfy the properties in Theorem 19. As an illustration, we shall verify that the inner product (4.21) satisfies property (b).

$$
\begin{aligned}
\langle \mathbf{u}, \mathbf{v} + \mathbf{w} \rangle &= (\mathbf{v} + \mathbf{w})^t A^t A \mathbf{u}\\
&= (\mathbf{v}^t + \mathbf{w}^t) A^t A \mathbf{u} && \text{(property of transpose)}\\
&= (\mathbf{v}^t A^t A \mathbf{u}) + (\mathbf{w}^t A^t A \mathbf{u}) && \text{(property of matrix multiplication)}\\
&= \langle \mathbf{u}, \mathbf{v} \rangle + \langle \mathbf{u}, \mathbf{w} \rangle \quad ▲
\end{aligned}
$$

EXERCISE SET 4.7

1. Let $\langle \mathbf{u}, \mathbf{v} \rangle$ be the Euclidean inner product on R^2, and let $\mathbf{u} = (2, -1)$, $\mathbf{v} = (-1, 3)$, $\mathbf{w} = (0, -5)$, and $k = -3$. Verify that
   (a) $\langle \mathbf{u}, \mathbf{v} \rangle = \langle \mathbf{v}, \mathbf{u} \rangle$
   (b) $\langle \mathbf{u} + \mathbf{v}, \mathbf{w} \rangle = \langle \mathbf{u}, \mathbf{w} \rangle + \langle \mathbf{v}, \mathbf{w} \rangle$
   (c) $\langle k\mathbf{u}, \mathbf{v} \rangle = k\langle \mathbf{u}, \mathbf{v} \rangle$

2. Let $\langle \mathbf{u}, \mathbf{v} \rangle$ be the Euclidean inner product on R^2, and let $\mathbf{u}$, $\mathbf{v}$, $\mathbf{w}$, and k be the vectors and scalar in Exercise 1. Verify that
   (a) $\langle \mathbf{0}, \mathbf{v} \rangle = \langle \mathbf{v}, \mathbf{0} \rangle = 0$
   (b) $\langle \mathbf{u}, \mathbf{v} + \mathbf{w} \rangle = \langle \mathbf{u}, \mathbf{v} \rangle + \langle \mathbf{u}, \mathbf{w} \rangle$
   (c) $\langle \mathbf{u}, k\mathbf{v} \rangle = k\langle \mathbf{u}, \mathbf{v} \rangle$

3. Compute $\langle \mathbf{u}, \mathbf{v} \rangle$ using the inner product in Example 49.

   (a) $\mathbf{u} = \begin{bmatrix} 2 & -1 \\ 3 & 7 \end{bmatrix}$ $\mathbf{v} = \begin{bmatrix} 0 & 4 \\ 2 & 2 \end{bmatrix}$ (b) $\mathbf{u} = \begin{bmatrix} 1 & 2 \\ -3 & 5 \end{bmatrix}$ $\mathbf{v} = \begin{bmatrix} 4 & 6 \\ 0 & 8 \end{bmatrix}$

4. Compute $\langle \mathbf{p}, \mathbf{q} \rangle$ using the inner product in Example 50.
   (a) $\mathbf{p} = -1 + 2x + x^2$ $\mathbf{q} = 2 - 4x^2$
   (b) $\mathbf{p} = -3 + 2x + x^2$ $\mathbf{q} = 2 + 4x - 2x^2$

5. (a) Use formula (4.21) to show that $\langle \mathbf{u}, \mathbf{v} \rangle = 9u_1v_1 + 4u_2v_2$ is the inner product on R^2 generated by
   $$A = \begin{bmatrix} 3 & 0 \\ 0 & 2 \end{bmatrix}$$
   (b) Use the inner product in (a) to compute $\langle \mathbf{u}, \mathbf{v} \rangle$ if $\mathbf{u} = (-2, 1)$ and $\mathbf{v} = (2, 9)$.

6. (a) Use formula (4.21) to show that $\langle \mathbf{u}, \mathbf{v} \rangle = 5u_1v_1 - u_1v_2 - u_2v_1 + 10u_2v_2$ is the inner product on R^2 generated by
   $$A = \begin{bmatrix} 2 & 1 \\ -1 & 3 \end{bmatrix}$$
   (b) Use the inner product in (a) to compute $\langle \mathbf{u}, \mathbf{v} \rangle$ if $\mathbf{u} = (0, -1)$ and $\mathbf{v} = (1, 4)$.

7. Let $\mathbf{u} = (u_1, u_2)$ and $\mathbf{v} = (v_1, v_2)$. In each part, the given expression is an inner product on R^2. Find a matrix that generates it.
   (a) $\langle \mathbf{u}, \mathbf{v} \rangle = 4u_1v_1 + 16u_2v_2$ (b) $\langle \mathbf{u}, \mathbf{v} \rangle = 2u_1v_1 + 7u_2v_2$

8. Let $\mathbf{u} = (u_1, u_2)$ and $\mathbf{v} = (v_1, v_2)$. Show that the following are inner products on R^2 by verifying that the inner product axioms hold.
   (a) $\langle \mathbf{u}, \mathbf{v} \rangle = 6u_1v_1 + 2u_2v_2$
   (b) $\langle \mathbf{u}, \mathbf{v} \rangle = 2u_1v_1 + u_2v_1 + u_1v_2 + 2u_2v_2$

9. Let $\mathbf{u} = (u_1, u_2, u_3)$ and $\mathbf{v} = (v_1, v_2, v_3)$. Determine which of the following are inner products on R^3. For those that are not, list the axioms which do not hold.
   (a) $\langle \mathbf{u}, \mathbf{v} \rangle = u_1v_1 + u_3v_3$
   (b) $\langle \mathbf{u}, \mathbf{v} \rangle = u_1^2v_1^2 + u_2^2v_2^2 + u_3^2v_3^2$
   (c) $\langle \mathbf{u}, \mathbf{v} \rangle = 2u_1v_1 + u_2v_2 + 4u_3v_3$
   (d) $\langle \mathbf{u}, \mathbf{v} \rangle = u_1v_1 - u_2v_2 + u_3v_3$

10. Let $U = \begin{bmatrix} u_1 & u_2 \\ u_3 & u_4 \end{bmatrix}$ and $V = \begin{bmatrix} v_1 & v_2 \\ v_3 & v_4 \end{bmatrix}$

Determine whether $\langle U, V \rangle = u_1 v_1 + u_2 v_3 + u_3 v_2 + u_4 v_4$ is an inner product on M_{22}.

11. Let $\mathbf{p} = p(x)$ and $\mathbf{q} = q(x)$ be polynomials in P_2. Show that

$$\langle \mathbf{p}, \mathbf{q} \rangle = p(0)q(0) + p(1/2)q(1/2) + p(1)q(1)$$

is an inner product on P_2.

12. Prove: If $\langle \mathbf{u}, \mathbf{v} \rangle$ is the Euclidean inner product on R^n, and if A is an $n \times n$ matrix, then

$$\langle \mathbf{u}, A\mathbf{v} \rangle = \langle A^t\mathbf{u}, \mathbf{v} \rangle$$

[**Hint.** Use the fact that $\langle \mathbf{u}, \mathbf{v} \rangle = \mathbf{u} \cdot \mathbf{v} = \mathbf{v}^t\mathbf{u}$.]

13. Verify the result in Exercise 12 for the Euclidean inner product on R^3 and

$$\mathbf{u} = \begin{bmatrix} -1 \\ 2 \\ 4 \end{bmatrix} \quad \mathbf{v} = \begin{bmatrix} 3 \\ 0 \\ -2 \end{bmatrix} \quad A = \begin{bmatrix} 1 & -2 & 1 \\ 3 & 4 & 0 \\ 5 & -1 & 2 \end{bmatrix}$$

14. Let $w_1, w_2, \ldots, w_n$ be positive real numbers and let $\mathbf{u} = (u_1, u_2, \ldots, u_n)$ and $\mathbf{v} = (v_1, v_2, \ldots, v_n)$. Show that $\langle \mathbf{u}, \mathbf{v} \rangle = w_1 u_1 v_1 + w_2 u_2 v_2 + \cdots + w_n u_n v_n$ is an inner product on R^n.

15. (**For readers who have studied calculus.**) Use the inner product

$$\langle \mathbf{p}, \mathbf{q} \rangle = \int_{-1}^{1} p(x)q(x) \, dx$$

to compute $\langle \mathbf{p}, \mathbf{q} \rangle$ for the vectors $\mathbf{p} = p(x)$ and $\mathbf{q} = q(x)$ in P_3.
(a) $\mathbf{p} = 1 - x + x^2 + 5x^3$ $\quad$ $\mathbf{q} = x - 3x^2$
(b) $\mathbf{p} = x - 5x^3$ $\quad$ $\mathbf{q} = 2 + 8x^2$

16. (**For readers who have studied calculus.**) Use the inner product

$$\langle \mathbf{f}, \mathbf{g} \rangle = \int_{0}^{1} f(x)g(x) \, dx$$

to compute $\langle \mathbf{f}, \mathbf{g} \rangle$ for the vectors $\mathbf{f} = f(x)$ and $\mathbf{g} = g(x)$ in $C[0, 1]$.
(a) $\mathbf{f} = \cos 2\pi x$ $\quad$ $\mathbf{g} = \sin 2\pi x$
(b) $\mathbf{f} = x$ $\quad$ $\mathbf{g} = e^x$

(c) $\mathbf{f} = \tan \dfrac{\pi}{4} x$ $\quad$ $\mathbf{g} = 1$

17. The *trace* of a square matrix A is denoted by $\mathrm{tr}(A)$ and is defined to be the sum of the entries on the main diagonal of A. Show that the inner product in Example 49 can be written as

$$\langle U, V \rangle = \mathrm{tr}(U^t V)$$

18. Prove that formula (4.20) defines an inner product on R^n. [**Hint.** Use the alternate version of formula (4.20) given by (4.21).]

19. Show that matrix (4.22) generates the weighted Euclidean inner product $\langle \mathbf{u}, \mathbf{v} \rangle = w_1 u_1 v_1 + w_2 u_2 v_2 + \cdots + w_n u_n v_n$ on R^n.

20. Prove part (*a*) of Theorem 19.

21. Prove part (*c*) of Theorem 19.

4.8 LENGTH AND ANGLE IN INNER PRODUCT SPACES

In this section we shall develop notions of length, distance, and angle in general inner product spaces.

In R^2 the length of a vector $\mathbf{u} = (u_1, u_2)$ is given by

$$\|\mathbf{u}\| = \sqrt{u_1^2 + u_2^2}$$

which can be written in terms of the dot product as

$$\|\mathbf{u}\| = \sqrt{\mathbf{u} \cdot \mathbf{u}} = (\mathbf{u} \cdot \mathbf{u})^{1/2}$$

Similarly, if $\mathbf{u} = (u_1, u_2, u_3)$ is a vector in R^3, then

$$\|\mathbf{u}\| = \sqrt{u_1^2 + u_2^2 + u_3^2} = \sqrt{\mathbf{u} \cdot \mathbf{u}} = (\mathbf{u} \cdot \mathbf{u})^{1/2}$$

Motivated by these results, we make the following definition.

Definition. If V is an inner product space, then the *norm* (or *length*) of a vector $\mathbf{u}$ is denoted by $\|\mathbf{u}\|$ and defined by

$$\|\mathbf{u}\| = \langle \mathbf{u}, \mathbf{u} \rangle^{1/2}$$

In R^2, the distance between two points $\mathbf{u} = (u_1, u_2)$ and $\mathbf{v} = (v_1, v_2)$ is given by

$$d(\mathbf{u}, \mathbf{v}) = \sqrt{(u_1 - v_1)^2 + (u_2 - v_2)^2} = \|\mathbf{u} - \mathbf{v}\|$$

Similarly, in R^3 the distance between two points $\mathbf{u} = (u_1, u_2, u_3)$ and $\mathbf{v} = (v_1, v_2, v_3)$ is given by

$$d(\mathbf{u}, \mathbf{v}) = \sqrt{(u_1 - v_1)^2 + (u_2 - v_2)^2 + (u_3 - v_3)^2} = \|\mathbf{u} - \mathbf{v}\|$$

Motivated by these results, we make the following definition.

Definition. If V is an inner product space, then the ***distance*** between two
points (vectors) $\mathbf{u}$ and $\mathbf{v}$ is denoted by $d(\mathbf{u}, \mathbf{v})$ and is defined by

$$d(\mathbf{u}, \mathbf{v}) = \|\mathbf{u} - \mathbf{v}\|$$

Example 54

If $\mathbf{u} = (u_1, u_2, \ldots, u_n)$ and $\mathbf{v} = (v_1, v_2, \ldots, v_n)$ are vectors in R^n with the Euclidean
inner product, then

$$\|\mathbf{u}\| = \langle \mathbf{u}, \mathbf{u} \rangle^{1/2} = \sqrt{u_1^2 + u_2^2 + \cdots + u_n^2}$$

and

$$d(\mathbf{u}, \mathbf{v}) = \|\mathbf{u} - \mathbf{v}\| = \langle \mathbf{u} - \mathbf{v}, \mathbf{u} - \mathbf{v} \rangle^{1/2}$$
$$= \sqrt{(u_1 - v_1)^2 + (u_2 - v_2)^2 + \cdots + (u_n - v_n)^2}$$

Observe that these are simply the standard formulas for the Euclidean norm and
distance discussed in Section 4.1. ▲

Example 55

It is important to keep in mind that norm and distance depend on the inner product being used. If the inner product is changed, then the norms and distances
between vectors also change. For example, for the vectors $\mathbf{u} = (1, 0)$ and $\mathbf{v} = (0, 1)$
in R^2 with the Euclidean inner product, we have

$$\|\mathbf{u}\| = \sqrt{1^2 + 0^2} = 1$$

and

$$d(\mathbf{u}, \mathbf{v}) = \|\mathbf{u} - \mathbf{v}\| = \|(1, -1)\| = \sqrt{1^2 + (-1)^2} = \sqrt{2}$$

However, if we change to the weighted Euclidean inner product $\langle \mathbf{u}, \mathbf{v} \rangle = 3u_1 v_1 + 2u_2 v_2$, then we obtain

$$\|\mathbf{u}\| = \langle \mathbf{u}, \mathbf{u} \rangle^{1/2} = [3(1)(1) + 2(0)(0)]^{1/2} = \sqrt{3}$$

and

$$d(\mathbf{u}, \mathbf{v}) = \|\mathbf{u} - \mathbf{v}\| = \langle (1, -1), (1, -1) \rangle^{1/2}$$
$$= [3(1)(1) + 2(-1)(-1)]^{1/2} = \sqrt{5} \ ▲$$

It would not be unreasonable for the reader to feel uncomfortable with the
results in the last example. For even though our definitions of length and distance
reduce to the standard definitions when applied to R^2 or R^3 with the Euclidean

inner product, it does require a stretch of the imagination to state that the length of the vector $\mathbf{u} = (1, 0)$ is $\sqrt{3}$. However, even though new inner products lead us to unfamiliar values for lengths and distances, there are some theorems that show that these new quantities enjoy many of the familiar geometric properties of Euclidean lengths and distances. For example, it is a basic fact in Euclidean geometry that the sum of the lengths of two sides of a triangle is at least as large as the length of the third side [Figure 4.8(*a*)]. We shall see below that this familiar result holds in all inner product spaces, regardless of how unusual the inner product might be. As another example, recall the theorem from Euclidean geometry that states that the sum of the squares of the diagonals of a parallelogram is equal to the sum of the squares of the four sides [Figure 4.8(*b*)]. This result also holds in all inner product spaces, regardless of the inner product (Exercise 19).

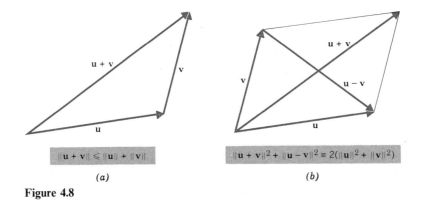

(*a*)   (*b*)

Figure 4.8

As we progress we shall encounter more and more properties of Euclidean geometry that carry over to all inner product spaces, so that these new spaces with their strange values for lengths and distances will eventually seem quite natural.

If $\mathbf{u}$ and $\mathbf{v}$ are nonzero vectors in R^3, then $\mathbf{u} \cdot \mathbf{v} = \|\mathbf{u}\| \, \|\mathbf{v}\| \cos \theta$, where θ is the angle between $\mathbf{u}$ and $\mathbf{v}$ (Section 3.3). If we square both sides of this inequality and use the relationships $\|\mathbf{u}\|^2 = \mathbf{u} \cdot \mathbf{u}$, $\|\mathbf{v}\|^2 = \mathbf{v} \cdot \mathbf{v}$, and $\cos^2 \theta \leq 1$, we obtain the inequality

$$(\mathbf{u} \cdot \mathbf{v})^2 \leq (\mathbf{u} \cdot \mathbf{u})(\mathbf{v} \cdot \mathbf{v})$$

The following theorem shows that this inequality can be generalized to any inner product space. The resulting inequality, called the *Cauchy-Schwarz inequality*, will enable us to define angles in general inner product spaces.

Theorem 20 (*Cauchy-Schwarz* Inequality*). *If* **u** *and* **v** *are vectors in an inner product space, then*

$$\langle \mathbf{u}, \mathbf{v} \rangle^2 \leq \langle \mathbf{u}, \mathbf{u} \rangle \langle \mathbf{v}, \mathbf{v} \rangle \qquad (4.24)$$

Proof. We warn the reader in advance that the proof presented here depends on a clever trick that is not easy to motivate. If $\mathbf{u} = \mathbf{0}$, then $\langle \mathbf{u}, \mathbf{v} \rangle = \langle \mathbf{u}, \mathbf{u} \rangle = 0$, so that the equality clearly holds. Assume now that $\mathbf{u} \neq \mathbf{0}$. Let $a = \langle \mathbf{u}, \mathbf{u} \rangle$, $b = 2\langle \mathbf{u}, \mathbf{v} \rangle$, $c = \langle \mathbf{v}, \mathbf{v} \rangle$, and let t be any real number. By the positivity axiom, the inner product of any vector with itself is always nonnegative. Therefore,

$$0 \leq \langle (t\mathbf{u} + \mathbf{v}), (t\mathbf{u} + \mathbf{v}) \rangle = \langle \mathbf{u}, \mathbf{u} \rangle t^2 + 2\langle \mathbf{u}, \mathbf{v} \rangle t + \langle \mathbf{v}, \mathbf{v} \rangle$$
$$= at^2 + bt + c$$

This inequality implies that the quadratic polynomial $at^2 + bt + c$ has either no real roots or a repeated real root. Therefore, its discriminant must satisfy $b^2 - 4ac \leq 0$. Expressing the coefficients a, b, and c in terms of **u** and **v** gives $4\langle \mathbf{u}, \mathbf{v} \rangle^2 - 4\langle \mathbf{u}, \mathbf{u} \rangle \langle \mathbf{v}, \mathbf{v} \rangle \leq 0$, or equivalently, $\langle \mathbf{u}, \mathbf{v} \rangle^2 \leq \langle \mathbf{u}, \mathbf{u} \rangle \langle \mathbf{v}, \mathbf{v} \rangle$. ▪

We note that the Cauchy-Schwarz inequality (4.24) can be written in two useful alternative forms. Since $\|\mathbf{u}\|^2 = \langle \mathbf{u}, \mathbf{u} \rangle$ and $\|\mathbf{v}\|^2 = \langle \mathbf{v}, \mathbf{v} \rangle$, it follows from (4.24) that

$$\langle \mathbf{u}, \mathbf{v} \rangle^2 \leq \|\mathbf{u}\|^2 \|\mathbf{v}\|^2 \qquad (4.24a)$$

or, upon taking square roots, that

$$|\langle \mathbf{u}, \mathbf{v} \rangle| \leq \|\mathbf{u}\| \, \|\mathbf{v}\| \qquad (4.24b)$$

Example 56

If $\mathbf{u} = (u_1, u_2, \ldots, u_n)$ and $\mathbf{v} = (v_1, v_2, \ldots, v_n)$ are any two vectors in R^n with the Euclidean inner product, then the Cauchy-Schwarz inequality (4.24b) applied to **u** and **v** yields

$$|u_1 v_1 + u_2 v_2 + \cdots + u_n v_n| \leq (u_1^2 + u_2^2 + \cdots + u_n^2)^{1/2}(v_1^2 + v_2^2 + \cdots + v_n^2)^{1/2}$$

This is called *Cauchy's inequality*. ▲

Through the years mathematicians have isolated what are considered to be the most important properties of Euclidean length and distance in R^2 and R^3. They are listed in the following table.

---

* *Augustin Louis (Baron de) Cauchy* (1789–1857), sometimes called the father of modern analysis, helped put calculus on a firm mathematical footing. He was a partisan of the Bourbons and spent several years in exile for his political involvements.

Hermann Amandus Schwarz (1843–1921) was a German mathematician.

Basic *Properties of Length*	Basic *Properties of Distance*
*L*1. $\|\mathbf{u}\| \geq 0$	*D*1. $d(\mathbf{u}, \mathbf{v}) \geq 0$
*L*2. $\|\mathbf{u}\| = 0$ if and only if $\mathbf{u} = \mathbf{0}$	*D*2. $d(\mathbf{u}, \mathbf{v}) = 0$ if and only if $\mathbf{u} = \mathbf{v}$
*L*3. $\|k\mathbf{u}\| = \|k\| \|\mathbf{u}\|$	*D*3. $d(\mathbf{u}, \mathbf{v}) = d(\mathbf{v}, \mathbf{u})$
*L*4. $\|\mathbf{u} + \mathbf{v}\| \leq \|\mathbf{u}\| + \|\mathbf{v}\|$ (triangle inequality)	*D*4. $d(\mathbf{u}, \mathbf{v}) \leq d(\mathbf{u}, \mathbf{w}) + d(\mathbf{w}, \mathbf{v})$ (triangle inequality)

The next theorem is a strong justification for our definitions of norm and distance in inner product spaces.

Theorem 21. *If V is an inner product space, then the norm $\|\mathbf{u}\| = \langle \mathbf{u}, \mathbf{u} \rangle^{1/2}$ and the distance $d(\mathbf{u}, \mathbf{v}) = \|\mathbf{u} - \mathbf{v}\|$ satisfy all the properties listed in the table above.*

We shall prove property *L*4 and leave the proofs of the remaining parts as exercises.

Proof of Property L4. By definition

$$\begin{aligned}
\|\mathbf{u} + \mathbf{v}\|^2 &= \langle \mathbf{u} + \mathbf{v}, \mathbf{u} + \mathbf{v} \rangle \\
&= \langle \mathbf{u}, \mathbf{u} \rangle + 2\langle \mathbf{u}, \mathbf{v} \rangle + \langle \mathbf{v}, \mathbf{v} \rangle \\
&\leq \langle \mathbf{u}, \mathbf{u} \rangle + 2|\langle \mathbf{u}, \mathbf{v} \rangle| + \langle \mathbf{v}, \mathbf{v} \rangle \\
&\leq \langle \mathbf{u}, \mathbf{u} \rangle + 2\|\mathbf{u}\| \|\mathbf{v}\| + \langle \mathbf{v}, \mathbf{v} \rangle \qquad \text{(by 4.24b)} \\
&= \|\mathbf{u}\|^2 + 2\|\mathbf{u}\| \|\mathbf{v}\| + \|\mathbf{v}\|^2 \\
&= (\|\mathbf{u}\| + \|\mathbf{v}\|)^2.
\end{aligned}$$

Taking square roots gives

$$\|\mathbf{u} + \mathbf{v}\| \leq \|\mathbf{u}\| + \|\mathbf{v}\|. \quad \blacksquare$$

We shall now show how the Cauchy-Schwarz inequality can be used to define angles in general inner product spaces. Suppose that $\mathbf{u}$ and $\mathbf{v}$ are nonzero vectors in an inner product space V. If we divide both sides of formula (4.24a) by $\|\mathbf{u}\|^2\|\mathbf{v}\|^2$, we obtain

$$\left(\frac{\langle \mathbf{u}, \mathbf{v} \rangle}{\|\mathbf{u}\| \|\mathbf{v}\|} \right)^2 \leq 1$$

or equivalently

$$-1 \leq \frac{\langle \mathbf{u}, \mathbf{v} \rangle}{\|\mathbf{u}\| \|\mathbf{v}\|} \leq 1 \tag{4.25}$$

Now if θ is an angle whose radian measure varies from 0 to π, then $\cos \theta$ assumes every value between -1 and 1 inclusive exactly once (Figure 4.9).

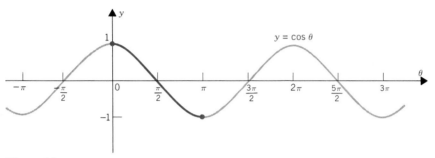

Figure 4.9

Thus, from (4.25) there is a unique angle θ such that

$$\cos \theta = \frac{\langle \mathbf{u}, \mathbf{v} \rangle}{\|\mathbf{u}\| \|\mathbf{v}\|} \qquad \text{and} \qquad 0 \leq \theta \leq \pi \tag{4.26}$$

We define θ to be the ***angle between* u *and* v**. Observe that in R^2 or R^3 with the Euclidean inner product, (4.26) agrees with the usual formula for the cosine of the angle between two nonzero vectors (Section 3.3).

Example 57

Find the cosine of the angle θ between the vectors

$$\mathbf{u} = (4, 3, 1, -2) \qquad \text{and} \qquad \mathbf{v} = (-2, 1, 2, 3)$$

where the vector space is R^4 with the Euclidean inner product.

Solution.

$$\|\mathbf{u}\| = \sqrt{30} \qquad \|\mathbf{v}\| = \sqrt{18} \qquad \text{and} \qquad \langle \mathbf{u}, \mathbf{v} \rangle = -9$$

so that

$$\cos \theta = \frac{\langle \mathbf{u}, \mathbf{v} \rangle}{\|\mathbf{u}\| \|\mathbf{v}\|} = -\frac{9}{\sqrt{30}\sqrt{18}} = -\frac{3}{2\sqrt{15}} \quad \triangle$$

Example 57 is primarily a mathematical exercise, for there is relatively little need to find angles between vectors except in R^2 and R^3 with the Euclidean inner product. However, a problem of major importance in all inner product spaces is to determine whether two vectors are *orthogonal*—that is, whether they have an angle of $\theta = \pi/2$ between them.

Example 58

Show that if M_{22} has the inner product of Example 49, the angle between the matrices

$$U = \begin{bmatrix} 1 & 0 \\ 1 & 1 \end{bmatrix} \quad \text{and} \quad V = \begin{bmatrix} 0 & 2 \\ 0 & 0 \end{bmatrix}$$

is $\pi/2$.

Solution.

$$\cos \theta = \frac{U \cdot V}{\|U\| \|V\|} = \frac{1(0) + 0(2) + 1(0) + 1(0)}{\|U\| \|V\|} = 0$$

Thus, $\theta = \pi/2$.

It follows from (4.26) that if **u** and **v** are *nonzero* vectors in an inner product space and θ is the angle between them, then $\cos \theta = 0$ if and only if $\langle \mathbf{u}, \mathbf{v} \rangle = 0$. Equivalently, $\theta = \pi/2$ if and only if $\langle \mathbf{u}, \mathbf{v} \rangle = 0$. If we agree to consider the angle between **u** and **v** to be $\pi/2$ when either or both of these vectors is **0**, then we can state without exception that the angle between **u** and **v** is $\pi/2$ if and only if $\langle \mathbf{u}, \mathbf{v} \rangle = 0$. This suggests the following definition.

Definition. In an inner product space, two vectors **u** and **v** are called *orthogonal* if $\langle \mathbf{u}, \mathbf{v} \rangle = 0$. Further, if **u** is orthogonal to each vector in a set W, we say that **u** is *orthogonal to W*.

We emphasize that orthogonality depends on the selection of the inner product. Two vectors can be orthogonal with respect to one inner product but not another.

Example 59 (For readers who have studied calculus.)

Let P_2 have the inner product

$$\langle \mathbf{p}, \mathbf{q} \rangle = \int_{-1}^{1} p(x)q(x)\,dx$$

(see Example 51), and let

$$\mathbf{p} = x, \qquad \mathbf{q} = x^2$$

Then

$$\|\mathbf{p}\| = \langle \mathbf{p}, \mathbf{p} \rangle^{1/2} = \left[\int_{-1}^{1} xx\, dx \right]^{1/2} = \left[\int_{-1}^{1} x^2\, dx \right]^{1/2} = \sqrt{\frac{2}{3}}$$

$$\|\mathbf{q}\| = \langle \mathbf{q}, \mathbf{q} \rangle^{1/2} = \left[\int_{-1}^{1} x^2 x^2\, dx \right]^{1/2} = \left[\int_{-1}^{1} x^4\, dx \right]^{1/2} = \sqrt{\frac{2}{5}}$$

$$\langle \mathbf{p}, \mathbf{q} \rangle = \int_{-1}^{1} xx^2\, dx = \int_{-1}^{1} x^3\, dx = 0$$

Because $\langle \mathbf{p}, \mathbf{q} \rangle = 0$ the vectors $\mathbf{p} = x$ and $\mathbf{q} = x^2$ are orthogonal relative to the given inner product. ▲

We conclude this section with an interesting and useful generalization of a familiar result.

Theorem 22 (*Generalized Theorem of Pythagoras*). *If* **u** *and* **v** *are orthogonal vectors in an inner product space, then*

$$\|\mathbf{u} + \mathbf{v}\|^2 = \|\mathbf{u}\|^2 + \|\mathbf{v}\|^2.$$

Proof.
$$\|\mathbf{u} + \mathbf{v}\|^2 = \langle (\mathbf{u} + \mathbf{v}), (\mathbf{u} + \mathbf{v}) \rangle = \|\mathbf{u}\|^2 + 2\langle \mathbf{u}, \mathbf{v} \rangle + \|\mathbf{v}\|^2$$
$$= \|\mathbf{u}\|^2 + \|\mathbf{v}\|^2.$$

Note that in R^2 or R^3 with the Euclidean inner product this theorem reduces to the ordinary Pythagorean theorem (Figure 4.10).

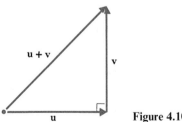

Figure 4.10

EXERCISE SET 4.8

1. In each part use the given inner product on R^2 to find $\|\mathbf{w}\|$, where $\mathbf{w} = (-1, 3)$.
   (a) the Euclidean inner product
   (b) the weighted Euclidean inner product $\langle \mathbf{u}, \mathbf{v} \rangle = 3u_1 v_1 + 2u_2 v_2$, where $\mathbf{u} = (u_1, u_2)$ and $\mathbf{v} = (v_1, v_2)$
   (c) the inner product generated by the matrix

$$A = \begin{bmatrix} 1 & 2 \\ -1 & 3 \end{bmatrix}$$

2. Use the inner products in Exercise 1 to find $d(\mathbf{u}, \mathbf{v})$, where $\mathbf{u} = (-1, 2)$ and $\mathbf{v} = (2, 5)$.

3. Let P_2 have the inner product in Example 50. Find $\|\mathbf{p}\|$ when
   (a) $\mathbf{p} = -1 + 2x + x^2$ (b) $\mathbf{p} = 3 - 4x^2$

4. Let M_{22} have the inner product in Example 49. Find $\|A\|$ when

   (a) $A = \begin{bmatrix} -1 & 7 \\ 6 & 2 \end{bmatrix}$ (b) $A = \begin{bmatrix} 0 & 0 \\ 0 & 0 \end{bmatrix}$

5. Let P_2 have the inner product in Example 50. Find $d(\mathbf{p}, \mathbf{q})$ when

$$\mathbf{p} = 2 - x + x^2, \qquad \mathbf{q} = 1 + 5x^2$$

6. Let M_{22} have the inner product in Example 49. Find $d(A, B)$ when

   (a) $A = \begin{bmatrix} 1 & 5 \\ 8 & 3 \end{bmatrix}$ $B = \begin{bmatrix} -5 & 0 \\ 7 & -3 \end{bmatrix}$

   (b) $A = \begin{bmatrix} 6 & 3 \\ 2 & 1 \end{bmatrix}$ $B = \begin{bmatrix} 6 & 3 \\ 2 & 1 \end{bmatrix}$

7. In each part determine whether the given vectors are orthogonal with respect to the Euclidean inner product.
   (a) $\mathbf{u} = (-1, 2, 4)$, $\mathbf{v} = (2, 3, -1)$
   (b) $\mathbf{u} = (1, 1, 1)$, $\mathbf{v} = (-1, -1, -1)$
   (c) $\mathbf{u} = (a, b, c)$, $\mathbf{v} = (0, 0, 0)$
   (d) $\mathbf{u} = (-2, 3, -5, 1)$, $\mathbf{v} = (2, 1, -2, -9)$
   (e) $\mathbf{u} = (0, -1, 2, 5)$, $\mathbf{v} = (1, -2, 3, 0)$
   (f) $\mathbf{u} = (a, b)$, $\mathbf{v} = (-b, a)$

8. Let R^4 have the Euclidean inner product, and let $\mathbf{u} = (-1, 1, 0, 2)$. Determine whether $\mathbf{u}$ is orthogonal to the set of vectors $W = \{\mathbf{w}_1, \mathbf{w}_2, \mathbf{w}_3\}$, where $\mathbf{w}_1 = (0, 0, 0, 0)$, $\mathbf{w}_2 = (1, -1, 3, 0)$, and $\mathbf{w}_3 = (4, 0, 9, 2)$

9. Let R^2, R^3, and R^4 have the Euclidean inner product. In each part find the cosine of the angle between $\mathbf{u}$ and $\mathbf{v}$.
   (a) $\mathbf{u} = (1, -3)$, $\mathbf{v} = (2, 4)$ (b) $\mathbf{u} = (-1, 0)$, $\mathbf{v} = (3, 8)$
   (c) $\mathbf{u} = (-1, 5, 2)$, $\mathbf{v} = (2, 4, -9)$ (d) $\mathbf{u} = (4, 1, 8)$, $\mathbf{v} = (1, 0, -3)$
   (e) $\mathbf{u} = (1, 0, 1, 0)$, $\mathbf{v} = (-3, -3, -3, -3)$ (f) $\mathbf{u} = (2, 1, 7, -1)$, $\mathbf{v} = (4, 0, 0, 0)$

10. Let P_2 have the inner product in Example 50. Find the cosine of the angle between $\mathbf{p}$ and $\mathbf{q}$.
   (a) $\mathbf{p} = -1 + 5x + 2x^2$ $\mathbf{q} = 2 + 4x - 9x^2$
   (b) $\mathbf{p} = x - x^2$ $\mathbf{q} = 7 + 3x + 3x^2$

11. Let M_{22} have the inner product in Example 49. Find the cosine of the angle between A and B.

   (a) $A = \begin{bmatrix} 2 & 6 \\ 1 & -3 \end{bmatrix}$ $B = \begin{bmatrix} 3 & 2 \\ 1 & 0 \end{bmatrix}$

   (b) $A = \begin{bmatrix} 2 & 4 \\ -1 & 3 \end{bmatrix}$ $B = \begin{bmatrix} -3 & 1 \\ 4 & 2 \end{bmatrix}$

12. Let R^3 have the Euclidean inner product. For which values of k are **u** and **v** orthogonal?
    (a) $\mathbf{u} = (2, 1, 3)$ $\mathbf{v} = (1, 7, k)$
    (b) $\mathbf{u} = (k, k, 1)$ $\mathbf{v} = (k, 5, 6)$

13. Let P_2 have the inner product in Example 50. Show that $\mathbf{p} = 1 - x + 2x^2$ and $\mathbf{q} = 2x + x^2$ are orthogonal.

14. Let M_{22} have the inner product in Example 49. Determine which of the following are orthogonal to

$$A = \begin{bmatrix} 2 & 1 \\ -1 & 3 \end{bmatrix}$$

(a) $\begin{bmatrix} -3 & 0 \\ 0 & 2 \end{bmatrix}$ (b) $\begin{bmatrix} 1 & 1 \\ 0 & -1 \end{bmatrix}$ (c) $\begin{bmatrix} 0 & 0 \\ 0 & 0 \end{bmatrix}$ (d) $\begin{bmatrix} 2 & 1 \\ 5 & 2 \end{bmatrix}$

15. Let R^4 have the Euclidean inner product. Find two vectors of norm 1 orthogonal to all of the vectors $\mathbf{u} = (2, 1, -4, 0)$, $\mathbf{v} = (-1, -1, 2, 2)$, and $\mathbf{w} = (3, 2, 5, 4)$.

16. In each part verify that the Cauchy-Schwarz inequality holds for the given vectors using the Euclidean inner product.
    (a) $\mathbf{u} = (2, 1)$, $\mathbf{v} = (1, -3)$
    (b) $\mathbf{u} = (3, -1, 2)$, $\mathbf{v} = (0, 1, -3)$
    (c) $\mathbf{u} = (1, 2, -4)$, $\mathbf{v} = (-2, -4, 8)$
    (d) $\mathbf{u} = (1, 1, -1, -1)$, $\mathbf{v} = (1, 2, -2, 0)$

17. In each part verify that the Cauchy-Schwarz inequality holds for the given vectors.
    (a) $\mathbf{u} = (-2, 1)$, $\mathbf{v} = (1, 0)$ using the inner product of Example 47
    (b) $U = \begin{bmatrix} -1 & 2 \\ 6 & 1 \end{bmatrix}$ and $V = \begin{bmatrix} 1 & 0 \\ 3 & 3 \end{bmatrix}$

    using the inner product in Example 49
    (c) $\mathbf{p} = -1 + 2x + x^2$ and $\mathbf{q} = 2 - 4x^2$ using the inner product in Example 50.

18. Let V be an inner product space. Show that if **u** and **v** are orthogonal vectors in V such that $\|\mathbf{u}\| = \|\mathbf{v}\| = 1$, then $\|\mathbf{u} - \mathbf{v}\| = \sqrt{2}$.

19. Let V be an inner product space. Establish the identity

$$\|\mathbf{u} + \mathbf{v}\|^2 + \|\mathbf{u} - \mathbf{v}\|^2 = 2\|\mathbf{u}\|^2 + 2\|\mathbf{v}\|^2$$

for vectors in V.

20. Let V be an inner product space. Establish the identity

$$\langle \mathbf{u}, \mathbf{v} \rangle = \tfrac{1}{4}\|\mathbf{u} + \mathbf{v}\|^2 - \tfrac{1}{4}\|\mathbf{u} - \mathbf{v}\|^2$$

for vectors in V.

21. Let $\{\mathbf{v}_1, \mathbf{v}_2, \ldots, \mathbf{v}_r\}$ be a basis for an inner product space V. Show that the zero vector is the only vector in V that is orthogonal to all of the basis vectors.

22. Let **v** be a vector in an inner product space V.
    (a) Show that the set of all vectors in V orthogonal to **v** forms a subspace of V.
    (b) Describe the subspace geometrically in R^2 and R^3 with the Euclidean inner product.

23. Prove the following generalization of Theorem 22. If $v_1, v_2, \ldots, v_r$ are pairwise orthogonal vectors in an inner product space V, then

$$\|v_1 + v_2 + \cdots + v_r\|^2 = \|v_1\|^2 + \|v_2\|^2 + \cdots + \|v_r\|^2$$

24. Prove the following parts of Theorem 21.
(a) part $L1$ (b) part $L2$ (c) part $L3$ (d) part $D1$
(e) part $D2$ (f) part $D3$ (g) part $D4$

25. Use vector methods to prove that a triangle inscribed in a circle and having a diameter for a side must be a right triangle. (***Hint.*** Express the vectors $\overrightarrow{AB}$ and $\overrightarrow{BC}$ in the following figure in terms of **u** and **v**.)

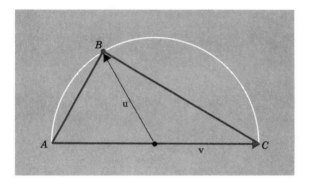

26. Let V be an inner product space. Show that if **w** is orthogonal to both u_1 and u_2, it is orthogonal to $k_1 u_1 + k_2 u_2$ for all scalars k_1 and k_2. Interpret this result geometrically in R^3 with the Euclidean inner product.

27. Let V be an inner product space. Show that if **w** is orthogonal to each of the vectors $u_1, u_2, \ldots, u_r$, then it is orthogonal to every vector in $\lin\{u_1, u_2, \ldots, u_r\}$.

28. Prove: If **u** and **v** are $n \times 1$ matrices and A is an invertible $n \times n$ matrix, then

$$[v^t A^t A u]^2 \leq (u^t A^t A u)(v^t A^t A v)$$

29. Use the Cauchy-Schwarz inequality to prove that for all real values of a, b, and θ,

$$[a \cos \theta + b \sin \theta]^2 \leq a^2 + b^2$$

30. Prove: If $w_1, w_2, \ldots, w_n$ are positive real numbers and $u = (u_1, u_2, \ldots, u_n)$ and $v = (v_1, v_2, \ldots, v_n)$ are any two vectors in R^n, then

$$|w_1 u_1 v_1 + w_2 u_2 v_2 + \cdots + w_n u_n v_n| \leq (w_1 u_1^2 + w_2 u_2^2 + \cdots + w_n u_n^2)^{1/2} \cdot$$
$$(w_1 v_1^2 + w_2 v_2^2 + \cdots + w_n v_n^2)^{1/2}$$

31. Show that equality holds in the Cauchy-Schwarz inequality if and only if **u** and **v** are linearly dependent.

32. **(For readers who have studied calculus.)** Let the vector space P_2 have the inner product

$$\langle \mathbf{p}, \mathbf{q} \rangle = \int_{-1}^{1} p(x)q(x)\, dx$$

   (a) Find $\|\mathbf{p}\|$ for $\mathbf{p} = 1$, $\mathbf{p} = x$, and $\mathbf{p} = x^2$.
   (b) Find $d(\mathbf{p}, \mathbf{q})$ if $\mathbf{p} = 1$ and $\mathbf{q} = x$.

33. **(For readers who have studied calculus.)** Let $C[0, \pi]$ have the inner product

$$\langle \mathbf{f}, \mathbf{g} \rangle = \int_{0}^{\pi} f(x)g(x)\, dx$$

   and let $\mathbf{f}_n = \cos nx$ $(n = 0, 1, 2, \ldots)$. Show that if $k \neq l$, then $\mathbf{f}_k$ and $\mathbf{f}_l$ are orthogonal with respect to the given inner product.

34. **(For readers who have studied calculus.)** Let $f(x)$ and $g(x)$ be continuous functions on $[0, 1]$. Prove

   (a) $\left[\int_0^1 f(x)g(x)\, dx \right]^2 \leq \left[\int_0^1 f^2(x)\, dx \right]\left[\int_0^1 g^2(x)\, dx \right]$

   (b) $\left[\int_0^1 [f(x) + g(x)]^2\, dx \right]^{1/2} \leq \left[\int_0^1 f^2(x)\, dx \right]^{1/2} + \left[\int_0^1 g^2(x)\, dx \right]^{1/2}$

   (**Hint.** Use the Cauchy-Schwarz inequality.)

4.9 ORTHONORMAL BASES; GRAM-SCHMIDT PROCESS

In many problems concerned with vector spaces, the selection of a basis for the space is at the discretion of the problem solver. Naturally, the best strategy is to choose the basis to simplify the solution of the problem at hand. In inner product spaces, it is often the case that the best choice is a basis in which all the vectors are orthogonal to one another. In this section we show how such bases can be constructed.

> **Definition.** A set of vectors in an inner product space is called an **orthogonal set** if all pairs of distinct vectors in the set are orthogonal. An orthogonal set in which each vector has norm 1 is called **orthonormal**.

Example 60
Let

$$\mathbf{v}_1 = (0, 1, 0) \qquad \mathbf{v}_2 = \left(\frac{1}{\sqrt{2}}, 0, \frac{1}{\sqrt{2}} \right) \qquad \mathbf{v}_3 = \left(\frac{1}{\sqrt{2}}, 0, -\frac{1}{\sqrt{2}} \right)$$

The set $S = \{\mathbf{v}_1, \mathbf{v}_2, \mathbf{v}_3\}$ is orthonormal if R^3 has the Euclidean inner product, since

$$\langle \mathbf{v}_1, \mathbf{v}_2 \rangle = \langle \mathbf{v}_1, \mathbf{v}_3 \rangle = \langle \mathbf{v}_2, \mathbf{v}_3 \rangle = 0$$

and

$$\|\mathbf{v}_1\| = \|\mathbf{v}_2\| = \|\mathbf{v}_3\| = 1 \quad \blacktriangle$$

If $\mathbf{v}$ is a nonzero vector in an inner product space, then by property *L3* of Theorem 21 the vector

$$\frac{1}{\|\mathbf{v}\|} \mathbf{v}$$

has norm 1, since

$$\left\| \frac{1}{\|\mathbf{v}\|} \mathbf{v} \right\| = \frac{1}{\|\mathbf{v}\|} \|\mathbf{v}\| = 1$$

This process of multiplying a nonzero vector $\mathbf{v}$ by the reciprocal of its length to obtain a vector of norm 1 is called ***normalizing*** $\mathbf{v}$. An orthogonal set of *nonzero* vectors can always be converted to an orthonormal set by normalizing each of its vectors.

Example 61

The set of vectors $S = \{\mathbf{u}_1, \mathbf{u}_2, \mathbf{u}_3\}$, where

$$\mathbf{u}_1 = (0, 1, 0), \quad \mathbf{u}_2 = (1, 0, 1), \quad \mathbf{u}_3 = (1, 0, -1)$$

is orthogonal, since $\langle \mathbf{u}_1, \mathbf{u}_2 \rangle = \langle \mathbf{u}_1, \mathbf{u}_3 \rangle = \langle \mathbf{u}_2, \mathbf{u}_3 \rangle = 0$. Since $\|\mathbf{u}_1\| = 1$, $\|\mathbf{u}_2\| = \sqrt{2}$, and $\|\mathbf{u}_3\| = \sqrt{2}$, normalizing each vector yields the orthonormal set in Example 60. $\blacktriangle$

The interest in finding orthonormal bases for inner product spaces is in part motivated by the next theorem, which shows that it is exceptionally simple to express a vector in terms of an orthonormal basis.

Theorem 23. *If* $S = \{\mathbf{v}_1, \mathbf{v}_2, \dots, \mathbf{v}_n\}$ *is an orthonormal basis for an inner product space* V, *and* $\mathbf{u}$ *is any vector in* V, *then*

$$\mathbf{u} = \langle \mathbf{u}, \mathbf{v}_1 \rangle \mathbf{v}_1 + \langle \mathbf{u}, \mathbf{v}_2 \rangle \mathbf{v}_2 + \cdots + \langle \mathbf{u}, \mathbf{v}_n \rangle \mathbf{v}_n$$

Proof. Since $S = \{\mathbf{v}_1, \mathbf{v}_2, \dots, \mathbf{v}_n\}$ is a basis, a vector $\mathbf{u}$ can be expressed in the form

$$\mathbf{u} = k_1 \mathbf{v}_1 + k_2 \mathbf{v}_2 + \cdots + k_n \mathbf{v}_n$$

We shall complete the proof by showing $k_i = \langle \mathbf{u}, \mathbf{v}_i \rangle$ for $i = 1, 2, \ldots, n$. For each vector $\mathbf{v}_i$ in S we have

$$\langle \mathbf{u}, \mathbf{v}_i \rangle = \langle k_1 \mathbf{v}_1 + k_2 \mathbf{v}_2 + \cdots + k_n \mathbf{v}_n, \mathbf{v}_i \rangle$$
$$= k_1 \langle \mathbf{v}_1, \mathbf{v}_i \rangle + k_2 \langle \mathbf{v}_2, \mathbf{v}_i \rangle + \cdots + k_n \langle \mathbf{v}_n, \mathbf{v}_i \rangle$$

Since $S = \{\mathbf{v}_1, \mathbf{v}_2, \ldots, \mathbf{v}_n\}$ is an orthonormal set, we have

$$\langle \mathbf{v}_i, \mathbf{v}_i \rangle = \|\mathbf{v}_i\|^2 = 1 \quad \text{and} \quad \langle \mathbf{v}_i, \mathbf{v}_j \rangle = 0 \quad \text{if } j \neq i$$

Therefore, the above equation simplifies to

$$\langle \mathbf{u}, \mathbf{v}_i \rangle = k_i$$

Example 62

Let

$$\mathbf{v}_1 = (0, 1, 0), \quad \mathbf{v}_2 = (-\tfrac{4}{5}, 0, \tfrac{3}{5}), \quad \mathbf{v}_3 = (\tfrac{3}{5}, 0, \tfrac{4}{5})$$

It is easy to check that $S = \{\mathbf{v}_1, \mathbf{v}_2, \mathbf{v}_3\}$ is an orthonormal basis for R^3 with the Euclidean inner product. Express the vector $\mathbf{u} = (1, 1, 1)$ as a linear combination of the vectors in S.

Solution

$$\langle \mathbf{u}, \mathbf{v}_1 \rangle = 1 \quad \langle \mathbf{u}, \mathbf{v}_2 \rangle = -\tfrac{1}{5} \quad \text{and} \quad \langle \mathbf{u}, \mathbf{v}_3 \rangle = \tfrac{7}{5}$$

Therefore, by Theorem 23

$$\mathbf{u} = \mathbf{v}_1 - \tfrac{1}{5}\mathbf{v}_2 + \tfrac{7}{5}\mathbf{v}_3$$

that is,

$$(1, 1, 1) = (0, 1, 0) - \tfrac{1}{5}(-\tfrac{4}{5}, 0, \tfrac{3}{5}) + \tfrac{7}{5}(\tfrac{3}{5}, 0, \tfrac{4}{5})$$

The usefulness of Theorem 23 should be evident from this example if it is kept in mind that for nonorthonormal bases, it is usually necessary to solve a system of equations in order to express a vector in terms of a basis.

Theorem 24. *If* $S = \{\mathbf{v}_1, \mathbf{v}_2, \ldots, \mathbf{v}_n\}$ *is an orthogonal set of nonzero vectors in an inner product space, then S is linearly independent.*

Proof. Assume

$$k_1 \mathbf{v}_1 + k_2 \mathbf{v}_2 + \cdots + k_n \mathbf{v}_n = \mathbf{0} \tag{4.27}$$

To demonstrate that $S = \{\mathbf{v}_1, \mathbf{v}_2, \ldots, \mathbf{v}_n\}$ is linearly independent, we must prove that $k_1 = k_2 = \cdots = k_n = 0$.

For each $\mathbf{v}_i$ in S, it follows from (4.27) that

$$\langle k_1 \mathbf{v}_1 + k_2 \mathbf{v}_2 + \cdots + k_n \mathbf{v}_n, \mathbf{v}_i \rangle = \langle \mathbf{0}, \mathbf{v}_i \rangle = 0$$

or equivalently

$$k_1 \langle \mathbf{v}_1, \mathbf{v}_i \rangle + k_2 \langle \mathbf{v}_2, \mathbf{v}_i \rangle + \cdots + k_n \langle \mathbf{v}_n, \mathbf{v}_i \rangle = 0$$

From the orthogonality of S, $\langle \mathbf{v}_j, \mathbf{v}_i \rangle = 0$ when $j \neq i$, so that this equation reduces to

$$k_i \langle \mathbf{v}_i, \mathbf{v}_i \rangle = 0$$

Since the vectors in S are assumed to be nonzero, $\langle \mathbf{v}_i, \mathbf{v}_i \rangle \neq 0$ by the positivity axiom for inner products. Therefore, $k_i = 0$. Since the subscript i is arbitrary, we have $k_1 = k_2 = \cdots = k_n = 0$; thus, S is linearly independent.

Example 63

In Example 60 we showed that

$$\mathbf{v}_1 = (0, 1, 0), \quad \mathbf{v}_2 = \left(\frac{1}{\sqrt{2}}, 0, \frac{1}{\sqrt{2}} \right), \quad \text{and} \quad \mathbf{v}_3 = \left(\frac{1}{\sqrt{2}}, 0, -\frac{1}{\sqrt{2}} \right)$$

form an orthonormal set with respect to the Euclidean inner product on R^3. By Theorem 24, these vectors form a linearly independent set. Therefore, since R^3 is three-dimensional, $S = \{ \mathbf{v}_1, \mathbf{v}_2, \mathbf{v}_3 \}$ is an orthonormal basis for R^3.

We now turn to the problem of constructing orthonormal bases for inner product spaces. The proof of the following preliminary result is discussed in the exercises at the end of this section.

Theorem 25. *Let V be an inner product space and $\{ \mathbf{v}_1, \mathbf{v}_2, \ldots, \mathbf{v}_r \}$ an orthonormal set of vectors in V. If W denotes the space spanned by $\mathbf{v}_1, \mathbf{v}_2, \ldots, \mathbf{v}_r$, then every vector $\mathbf{u}$ in V can be expressed in the form*

$$\mathbf{u} = \mathbf{w}_1 + \mathbf{w}_2$$

where $\mathbf{w}_1$ is in W and $\mathbf{w}_2$ is orthogonal to W by letting

$$\mathbf{w}_1 = \langle \mathbf{u}, \mathbf{v}_1 \rangle \mathbf{v}_1 + \langle \mathbf{u}, \mathbf{v}_2 \rangle \mathbf{v}_2 + \cdots + \langle \mathbf{u}, \mathbf{v}_r \rangle \mathbf{v}_r \tag{4.28}$$

and

$$\mathbf{w}_2 = \mathbf{u} - \langle \mathbf{u}, \mathbf{v}_1 \rangle \mathbf{v}_1 - \langle \mathbf{u}, \mathbf{v}_2 \rangle \mathbf{v}_2 - \cdots - \langle \mathbf{u}, \mathbf{v}_r \rangle \mathbf{v}_r \tag{4.29}$$

(See Figure 4.11 for an illustration in R^3.)

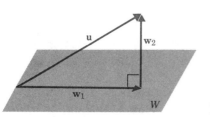

Figure 4.11

Motivated by Figure 4.11, we call $\mathbf{w}_1$ the ***orthogonal projection of*** $\mathbf{u}$ ***on*** W and denote it by $\text{proj}_W \mathbf{u}$. The vector $\mathbf{w}_2 = \mathbf{u} - \text{proj}_W \mathbf{u}$ is called the ***component of*** $\mathbf{u}$ ***orthogonal to*** W. With this notation formulas (4.28) and (4.29) can be written

$$\text{proj}_W \mathbf{u} = \langle \mathbf{u}, \mathbf{v}_1 \rangle \mathbf{v}_1 + \langle \mathbf{u}, \mathbf{v}_2 \rangle \mathbf{v}_2 + \cdots + \langle \mathbf{u}, \mathbf{v}_r \rangle \mathbf{v}_r \qquad (4.30)$$

(orthogonal projection of $\mathbf{u}$ on W)

$$\mathbf{u} - \text{proj}_W \mathbf{u} = \mathbf{u} - \langle \mathbf{u}, \mathbf{v}_1 \rangle \mathbf{v}_1 - \langle \mathbf{u}, \mathbf{v}_2 \rangle \mathbf{v}_2 - \cdots - \langle \mathbf{u}, \mathbf{v}_r \rangle \mathbf{v}_r \qquad (4.31)$$

(component of $\mathbf{u}$ orthogonal to W)

Example 64

Let R^3 have the Euclidean inner product, and let W be the subspace spanned by the orthonormal vectors $\mathbf{v}_1 = (0, 1, 0)$ and $\mathbf{v}_2 = (-\frac{4}{5}, 0, \frac{3}{5})$. The orthogonal projection of $\mathbf{u} = (1, 1, 1)$ on W is

$$-\frac{84}{125} + 0 + \frac{84}{125} = 0$$

$$\begin{aligned}
\text{proj}_W \mathbf{u} &= \langle \mathbf{u}, \mathbf{v}_1 \rangle \mathbf{v}_1 + \langle \mathbf{u}, \mathbf{v}_2 \rangle \mathbf{v}_2 \\
&= (1)(0, 1, 0) + (-\tfrac{1}{5})(-\tfrac{4}{5}, 0, \tfrac{3}{5}) \\
&= (\tfrac{4}{25}, 1, -\tfrac{3}{25})
\end{aligned}$$

The component of $\mathbf{u}$ orthogonal to W is

$$\mathbf{u} - \text{proj}_W \mathbf{u} = (1, 1, 1) - (\tfrac{4}{25}, 1, -\tfrac{3}{25}) = (\tfrac{21}{25}, 0, \tfrac{28}{25})$$

Observe that $\mathbf{u} - \text{proj}_W \mathbf{u}$ is orthogonal to both $\mathbf{v}_1$ and $\mathbf{v}_2$ so that this vector is orthogonal to each vector in the space W spanned by $\mathbf{v}_1$ and $\mathbf{v}_2$ as it should be.

We are now in a position to prove the main result of this section.

> **Theorem 26.** *Every nonzero finite-dimensional inner product space has an orthonormal basis.*

Proof. Let V be any nonzero, n-dimensional inner product space, and let $S = \{\mathbf{u}_1, \mathbf{u}_2, \ldots, \mathbf{u}_n\}$ be any basis for V. The following sequence of steps will produce an orthonormal basis $\{\mathbf{v}_1, \mathbf{v}_2, \ldots, \mathbf{v}_n\}$ for V.

Step 1. Let $\mathbf{v}_1 = \mathbf{u}_1 / \|\mathbf{u}_1\|$. The vector $\mathbf{v}_1$ has norm 1.

Step 2. To construct a vector $\mathbf{v}_2$ of norm 1 that is orthogonal to $\mathbf{v}_1$, we compute the component of $\mathbf{u}_2$ orthogonal to the space W_1 spanned by $\mathbf{v}_1$ and then normalize it; that is,

$$\mathbf{v}_2 = \frac{\mathbf{u}_2 - \text{proj}_{W_1} \mathbf{u}_2}{\|\mathbf{u}_2 - \text{proj}_{W_1} \mathbf{u}_2\|} = \frac{\mathbf{u}_2 - \langle \mathbf{u}_2, \mathbf{v}_1 \rangle \mathbf{v}_1}{\|\mathbf{u}_2 - \langle \mathbf{u}_2, \mathbf{v}_1 \rangle \mathbf{v}_1\|}$$

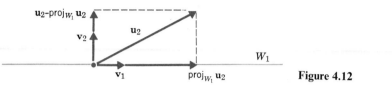

Figure 4.12

(Figure 4.12). Of course, if $\mathbf{u}_2 - \langle \mathbf{u}_2, \mathbf{v}_1 \rangle \mathbf{v}_1 = \mathbf{0}$, then we cannot carry out the normalization. But this cannot happen, since we would then have

$$\mathbf{u}_2 = \langle \mathbf{u}_2, \mathbf{v}_1 \rangle \mathbf{v}_1 = \frac{\langle \mathbf{u}_2, \mathbf{v}_1 \rangle}{\|\mathbf{u}_1\|} \mathbf{u}_1$$

which says that $\mathbf{u}_2$ is a multiple of $\mathbf{u}_1$, contradicting the linear independence of the basis $S = \{\mathbf{u}_1, \mathbf{u}_2, \ldots, \mathbf{u}_n\}$.

Step 3. To construct a vector $\mathbf{v}_3$ of norm 1 that is orthogonal to both $\mathbf{v}_1$ and $\mathbf{v}_2$, we compute the component of $\mathbf{u}_3$ orthogonal to the space W_2 spanned by $\mathbf{v}_1$ and $\mathbf{v}_2$ and normalize it (Figure 4.13); that is,

$$\mathbf{v}_3 = \frac{\mathbf{u}_3 - \text{proj}_{W_2} \mathbf{u}_3}{\|\mathbf{u}_3 - \text{proj}_{W_2} \mathbf{u}_3\|} = \frac{\mathbf{u}_3 - \langle \mathbf{u}_3, \mathbf{v}_1 \rangle \mathbf{v}_1 - \langle \mathbf{u}_3, \mathbf{v}_2 \rangle \mathbf{v}_2}{\|\mathbf{u}_3 - \langle \mathbf{u}_3, \mathbf{v}_1 \rangle \mathbf{v}_1 - \langle \mathbf{u}_3, \mathbf{v}_2 \rangle \mathbf{v}_2\|}$$

As in Step 2, the linear independence of $\{\mathbf{u}_1, \mathbf{u}_2, \ldots, \mathbf{u}_n\}$ assures that $\mathbf{u}_3 - \langle \mathbf{u}_3, \mathbf{v}_1 \rangle \mathbf{v}_1 - \langle \mathbf{u}_3, \mathbf{v}_2 \rangle \mathbf{v}_2 \neq \mathbf{0}$ so that the normalization can always be carried out. We leave the details as an exercise.

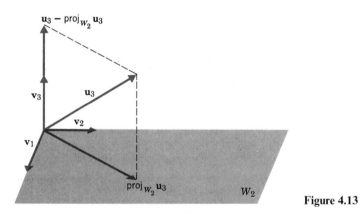

Figure 4.13

Step 4. To determine a vector $\mathbf{v}_4$ of norm 1 that is orthogonal to $\mathbf{v}_1$, $\mathbf{v}_2$, and $\mathbf{v}_3$, we compute the component of $\mathbf{u}_4$ orthogonal to the space W_3 spanned by $\mathbf{v}_1$, $\mathbf{v}_2$, and $\mathbf{v}_3$ and normalize it. Thus,

$$\mathbf{v}_4 = \frac{\mathbf{u}_4 - \text{proj}_{W_3} \mathbf{u}_4}{\|\mathbf{u}_4 - \text{proj}_{W_3} \mathbf{u}_4\|} = \frac{\mathbf{u}_4 - \langle \mathbf{u}_4, \mathbf{v}_1 \rangle \mathbf{v}_1 - \langle \mathbf{u}_4, \mathbf{v}_2 \rangle \mathbf{v}_2 - \langle \mathbf{u}_4, \mathbf{v}_3 \rangle \mathbf{v}_3}{\|\mathbf{u}_4 - \langle \mathbf{u}_4, \mathbf{v}_1 \rangle \mathbf{v}_1 - \langle \mathbf{u}_4, \mathbf{v}_2 \rangle \mathbf{v}_2 - \langle \mathbf{u}_4, \mathbf{v}_3 \rangle \mathbf{v}_3\|}$$

Continuing in this way, we will obtain an orthonormal set of vectors, $\{v_1, v_2, \ldots, v_n\}$. Since V is n-dimensional and every orthonormal set is linearly independent, the set $\{v_1, v_2, \ldots, v_n\}$ will be an orthonormal basis for V. ■

The above step-by-step construction for converting an arbitrary basis into an orthonormal basis is called the ***Gram-Schmidt* process***. It can be shown that at each stage in this process, the vectors $v_1, v_2, \ldots, v_k$ form an orthonormal basis for the subspace spanned by $u_1, u_2, \ldots, u_k$.

Example 65

Consider the vector space R^3 with the Euclidean inner product. Apply the Gram-Schmidt process to transform the basis $u_1 = (1, 1, 1)$, $u_2 = (0, 1, 1)$, $u_3 = (0, 0, 1)$ into an orthonormal basis.

Solution.

Step 1. $v_1 = \dfrac{u_1}{\|u_1\|} = \dfrac{(1, 1, 1)}{\sqrt{3}} = \left(\dfrac{1}{\sqrt{3}}, \dfrac{1}{\sqrt{3}}, \dfrac{1}{\sqrt{3}} \right)$

Step 2. $u_2 - \mathrm{proj}_{W_1} u_2 = u_2 - \langle u_2, v_1 \rangle v_1$

$$= (0, 1, 1) - \frac{2}{\sqrt{3}} \left(\frac{1}{\sqrt{3}}, \frac{1}{\sqrt{3}}, \frac{1}{\sqrt{3}} \right)$$

$$= \left(-\frac{2}{3}, \frac{1}{3}, \frac{1}{3} \right)$$

Therefore,

$$v_2 = \frac{u_2 - \mathrm{proj}_{W_1} u_2}{\|u_2 - \mathrm{proj}_{W_1} u_2\|} = \frac{3}{\sqrt{6}} \left(-\frac{2}{3}, \frac{1}{3}, \frac{1}{3} \right) = \left(-\frac{2}{\sqrt{6}}, \frac{1}{\sqrt{6}}, \frac{1}{\sqrt{6}} \right)$$

Step 3. $u_3 - \mathrm{proj}_{W_2} u_3 = u_3 - \langle u_3, v_1 \rangle v_1 - \langle u_3, v_2 \rangle v_2$

$$= (0, 0, 1) - \frac{1}{\sqrt{3}} \left(\frac{1}{\sqrt{3}}, \frac{1}{\sqrt{3}}, \frac{1}{\sqrt{3}} \right) - \frac{1}{\sqrt{6}} \left(-\frac{2}{\sqrt{6}}, \frac{1}{\sqrt{6}}, \frac{1}{\sqrt{6}} \right)$$

$$= \left(0, -\frac{1}{2}, \frac{1}{2} \right)$$

Therefore,

$$v_3 = \frac{u_3 - \mathrm{proj}_{W_2} u_3}{\|u_3 - \mathrm{proj}_{W_2} u_3\|} = \sqrt{2} \left(0, -\frac{1}{2}, \frac{1}{2} \right) = \left(0, -\frac{1}{\sqrt{2}}, \frac{1}{\sqrt{2}} \right)$$

---

* *Jörgen Pederson Gram* (1850–1916) was a Danish actuary,
Erhardt Schmidt (1876–1959) was a German mathematician.

Thus,

$$\mathbf{v}_1 = \left(\frac{1}{\sqrt{3}}, \frac{1}{\sqrt{3}}, \frac{1}{\sqrt{3}} \right), \; \mathbf{v}_2 = \left(-\frac{2}{\sqrt{6}}, \frac{1}{\sqrt{6}}, \frac{1}{\sqrt{6}} \right), \; \mathbf{v}_3 = \left(0, -\frac{1}{\sqrt{2}}, \frac{1}{\sqrt{2}} \right)$$

form an orthonormal basis for R^3. ▲

OPTIONAL

The following consequences of the Gram-Schmidt process have numerous applications, some of which are discussed later in the text. The reader will have the background to read these applications after completing this optional section.

> **Theorem 27 (*Projection Theorem*).** *If W is a finite-dimensional subspace of an inner product space V, then every vector $\mathbf{u}$ in V can be expressed in exactly one way as*
>
> $$\mathbf{u} = \mathbf{w}_1 + \mathbf{w}_2$$
>
> *where $\mathbf{w}_1$ is in W and $\mathbf{w}_2$ is orthogonal to W.*

Proof. There are two parts to the proof. First we must find vectors $\mathbf{w}_1$ and $\mathbf{w}_2$ with the stated properties, and then we must show that these are the only such vectors.

By the Gram-Schmidt process there is an orthonormal basis $\{\mathbf{v}_1, \mathbf{v}_2, \ldots, \mathbf{v}_r\}$ for W, so that $W = \mathrm{lin}\{\mathbf{v}_1, \mathbf{v}_2, \ldots, \mathbf{v}_r\}$. Therefore, by Theorem 25, the vectors

$$\mathbf{w}_1 = \mathrm{proj}_W \mathbf{u} \quad \text{and} \quad \mathbf{w}_2 = \mathbf{u} - \mathrm{proj}_W \mathbf{u}$$

will have the properties stated in this theorem. To see that these are the only vectors with these properties, suppose that we can also write

$$\mathbf{u} = \mathbf{w}_1' + \mathbf{w}_2' \tag{4.32}$$

where $\mathbf{w}_1'$ is in W and $\mathbf{w}_2'$ is orthogonal to W. If we subtract from (4.32) the equation

$$\mathbf{u} = \mathbf{w}_1 + \mathbf{w}_2$$

we obtain

$$\mathbf{0} = (\mathbf{w}_1' - \mathbf{w}_1) + (\mathbf{w}_2' - \mathbf{w}_2)$$

or

$$\mathbf{w}_1 - \mathbf{w}_1' = \mathbf{w}_2' - \mathbf{w}_2 \tag{4.33}$$

Since $\mathbf{w}_2$ and $\mathbf{w}_2'$ are orthogonal to W, their difference will also be orthogonal to W, since for any vector $\mathbf{w}$ in W we can write

$$\langle \mathbf{w}, \mathbf{w}_2' - \mathbf{w}_2 \rangle = \langle \mathbf{w}, \mathbf{w}_2' \rangle - \langle \mathbf{w}, \mathbf{w}_2 \rangle = 0 - 0 = 0$$

But $\mathbf{w}_2' - \mathbf{w}_2$ is itself a vector in W, since from (4.33) it is a difference of two vectors in the subspace W. Thus, $\mathbf{w}_2' - \mathbf{w}_2$ must be orthogonal to itself; that is,

$$\langle \mathbf{w}_2' - \mathbf{w}_2, \mathbf{w}_2' - \mathbf{w}_2 \rangle = 0$$

But this implies that $\mathbf{w}_2' - \mathbf{w}_2 = 0$ by Axiom 4 for inner products. Thus, $\mathbf{w}_2' = \mathbf{w}_2$, and by (4.33), $\mathbf{w}_1' = \mathbf{w}_1$. ∎

If P is a point in ordinary 3-space and W is a plane through the origin, then the point Q in W, closest to P, is obtained by dropping a perpendicular from P to W (Figure 4.14a). Therefore, if we let $\mathbf{u} = \overrightarrow{OP}$, the distance between P and W is given by

$$\|\mathbf{u} - \text{proj}_W \, \mathbf{u}\|$$

In other words, among all vectors $\mathbf{w}$ in W, the vector $\mathbf{w} = \text{proj}_W \, \mathbf{u}$ minimizes the distance $\|\mathbf{u} - \mathbf{w}\|$ (Figure 4.14b).

There is another way of thinking about this idea. View $\mathbf{u}$ as a fixed vector that we would like to approximate by a vector in W. Any such approximation $\mathbf{w}$ will result in an "error vector,"

$$\mathbf{u} - \mathbf{w}$$

which, unless $\mathbf{u}$ is in W, cannot be made equal to $\mathbf{0}$. However, by choosing

$$\mathbf{w} = \text{proj}_W \, \mathbf{u}$$

we can make the length of the error vector

$$\|\mathbf{u} - \mathbf{w}\| = \|\mathbf{u} - \text{proj}_W \, \mathbf{u}\|$$

as small as possible. Thus, we can describe $\text{proj}_W \, \mathbf{u}$ as the "best approximation" to $\mathbf{u}$ by vectors in W. The following theorem will make these intuitive ideas precise.

Theorem 28 (*Best Approximation Theorem*). *If W is a finite-dimensional subspace of an inner product space V, and if $\mathbf{u}$ is a vector in V, then $\text{proj}_W \, \mathbf{u}$ is the **best approximation** to $\mathbf{u}$ from W in the sense that*

$$\|\mathbf{u} - \text{proj}_W \, \mathbf{u}\| < \|\mathbf{u} - \mathbf{w}\|$$

for every vector $\mathbf{w}$ in W different from $\text{proj}_W \, \mathbf{u}$.

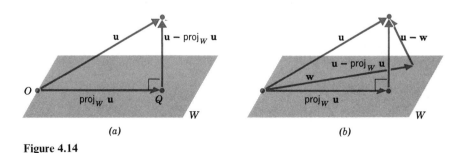

(a)   (b)

Figure 4.14

Proof. For any vector **w** in W we can write

$$\mathbf{u} - \mathbf{w} = (\mathbf{u} - \text{proj}_W \, \mathbf{u}) + (\text{proj}_W \, \mathbf{u} - \mathbf{w}) \qquad (4.34)$$

But $\text{proj}_W \, \mathbf{u} - \mathbf{w}$, being a difference of vectors in W, is in W; and $\mathbf{u} - \text{proj}_W \, \mathbf{u}$ is orthogonal to W, so that the two terms on the right side of (4.34) are orthogonal. Thus, by the theorem of Pythagoras (Theorem 22 of Section 4.8),

$$\|\mathbf{u} - \mathbf{w}\|^2 = \|\mathbf{u} - \text{proj}_W \, \mathbf{u}\|^2 + \|\text{proj}_W \, \mathbf{u} - \mathbf{w}\|^2$$

If $\mathbf{w} \neq \text{proj}_W \, \mathbf{u}$, then the second term in this sum will be positive, so that

$$\|\mathbf{u} - \mathbf{w}\|^2 > \|\mathbf{u} - \text{proj}_W \, \mathbf{u}\|^2$$

or equivalently

$$\|\mathbf{u} - \mathbf{w}\| > \|\mathbf{u} - \text{proj}_W \, \mathbf{u}\| \quad \blacksquare$$

Applications of the last two theorems will be given later in the text.

EXERCISE SET 4.9

1. Let R^2 have the Euclidean inner product. Which of the following form orthonormal sets?

(a) $(1, 0), (0, 2)$

(b) $\left(\frac{1}{\sqrt{2}}, -\frac{1}{\sqrt{2}}\right), \left(\frac{1}{\sqrt{2}}, \frac{1}{\sqrt{2}}\right)$

(c) $\left(\frac{1}{\sqrt{2}}, \frac{1}{\sqrt{2}}\right), \left(-\frac{1}{\sqrt{2}}, -\frac{1}{\sqrt{2}}\right)$

(d) $(1, 0), (0, 0)$

2. Let R^3 have the Euclidean inner product. Which of the following form orthonormal sets?

(a) $\left(\frac{1}{\sqrt{2}}, 0, \frac{1}{\sqrt{2}}\right), \left(\frac{1}{\sqrt{3}}, \frac{1}{\sqrt{3}}, -\frac{1}{\sqrt{3}}\right), \left(-\frac{1}{\sqrt{2}}, 0, \frac{1}{\sqrt{2}}\right)$

(b) $(\frac{2}{3}, -\frac{2}{3}, \frac{1}{3}), \ (\frac{2}{3}, \frac{1}{3}, -\frac{2}{3}), \ (\frac{1}{3}, \frac{2}{3}, \frac{2}{3})$

(c) $(1, 0, 0), \left(0, \frac{1}{\sqrt{2}}, \frac{1}{\sqrt{2}}\right), (0, 0, 1)$

(d) $\left(\frac{1}{\sqrt{6}}, \frac{1}{\sqrt{6}}, -\frac{2}{\sqrt{6}}\right), \left(\frac{1}{\sqrt{2}}, -\frac{1}{\sqrt{2}}, 0\right)$

3. Let P_2 have the inner product in Example 50. Which of the following form orthonormal sets?

(a) $\frac{2}{3} - \frac{2}{3}x + \frac{1}{3}x^2, \ \frac{2}{3} + \frac{1}{3}x - \frac{2}{3}x^2, \ \frac{1}{3} + \frac{2}{3}x + \frac{2}{3}x^2$

(b) $1, \ \frac{1}{\sqrt{2}}x + \frac{1}{\sqrt{2}}x^2, \ x^2$

4. Let M_{22} have the inner product in Example 49. Which of the following form orthonormal sets?

(a) $\begin{bmatrix} 1 & 0 \\ 0 & 0 \end{bmatrix}$ $\begin{bmatrix} 0 & \frac{2}{3} \\ \frac{1}{3} & -\frac{2}{3} \end{bmatrix}$ $\begin{bmatrix} 0 & \frac{2}{3} \\ -\frac{2}{3} & \frac{1}{3} \end{bmatrix}$ $\begin{bmatrix} 0 & \frac{1}{3} \\ \frac{2}{3} & \frac{2}{3} \end{bmatrix}$

(b) $\begin{bmatrix} 1 & 0 \\ 0 & 0 \end{bmatrix}$ $\begin{bmatrix} 0 & 1 \\ 0 & 0 \end{bmatrix}$ $\begin{bmatrix} 0 & 0 \\ 1 & 1 \end{bmatrix}$ $\begin{bmatrix} 0 & 0 \\ 1 & -1 \end{bmatrix}$

5. Let $\mathbf{x} = \left(\dfrac{1}{\sqrt{5}}, -\dfrac{1}{\sqrt{5}} \right)$ and $\mathbf{y} = \left(\dfrac{2}{\sqrt{30}}, \dfrac{3}{\sqrt{30}} \right)$.

Show that $\{\mathbf{x}, \mathbf{y}\}$ is orthonormal if R^2 has the inner product $\langle \mathbf{u}, \mathbf{v} \rangle = 3u_1 v_1 + 2u_2 v_2$, but is not orthonormal if R^2 has the Euclidean inner product.

6. (a) Show that

$$\mathbf{u}_1 = (1, 0, 0, 1), \quad \mathbf{u}_2 = (-1, 0, 2, 1), \quad \mathbf{u}_3 = (2, 3, 2, -2), \quad \mathbf{u}_4 = (-1, 2, -1, 1)$$

is an orthogonal set in R^4 with the Euclidean inner product. By normalizing each of these vectors, obtain an orthonormal set.

(b) Use Theorem 23 to express the vector $(1, 1, 1, 1)$ as a linear combination of the vectors in the orthonormal set obtained in part (a).

7. Let R^2 have the Euclidean inner product. Use the Gram-Schmidt process to transform the basis $\{\mathbf{u}_1, \mathbf{u}_2\}$ into an orthornormal basis.

(a) $\mathbf{u}_1 = (1, -3), \mathbf{u}_2 = (2, 2)$ (b) $\mathbf{u}_1 = (1, 0), \mathbf{u}_2 = (3, -5)$

8. Let R^3 have the Euclidean inner product. Use the Gram-Schmidt process to transform the basis $\{\mathbf{u}_1, \mathbf{u}_2, \mathbf{u}_3\}$ into an orthonormal basis.

(a) $\mathbf{u}_1 = (1, 1, 1), \mathbf{u}_2 = (-1, 1, 0), \mathbf{u}_3 = (1, 2, 1)$
(b) $\mathbf{u}_1 = (1, 0, 0), \mathbf{u}_2 = (3, 7, -2), \mathbf{u}_3 = (0, 4, 1)$

9. Let R^4 have the Euclidean inner product. Use the Gram-Schmidt process to transform the basis $\{\mathbf{u}_1, \mathbf{u}_2, \mathbf{u}_3, \mathbf{u}_4\}$ into an orthonormal basis.

$$\mathbf{u}_1 = (0, 2, 1, 0), \quad \mathbf{u}_2 = (1, -1, 0, 0), \quad \mathbf{u}_3 = (1, 2, 0, -1), \quad \mathbf{u}_4 = (1, 0, 0, 1)$$

10. Let R^3 have the Euclidean inner product. Find an orthonormal basis for the subspace spanned by $(0, 1, 2), (-1, 0, 1)$.

11. Let R^3 have the inner product $\langle \mathbf{u}, \mathbf{v} \rangle = u_1 v_1 + 2u_2 v_2 + 3u_3 v_3$. Use the Gram-Schmidt process to transform

$$\mathbf{u}_1 = (1, 1, 1) \qquad \mathbf{u}_2 = (1, 1, 0) \qquad \mathbf{u}_3 = (1, 0, 0)$$

into an orthonormal basis.

12. The subspace of R^3 spanned by the vectors $\mathbf{u}_1 = (\frac{4}{5}, 0, -\frac{3}{5})$ and $\mathbf{u}_2 = (0, 1, 0)$ is a plane passing through the origin. Express $\mathbf{w} = (1, 2, 3)$ in the form $\mathbf{w} = \mathbf{w}_1 + \mathbf{w}_2$, where $\mathbf{w}_1$ lies in the plane and $\mathbf{w}_2$ is perpendicular to the plane.

13. Repeat Exercise 12 with $\mathbf{u}_1 = (1, 1, 1)$ and $\mathbf{u}_2 = (2, 0, -1)$.

14. Let R^4 have the Euclidean inner product. Express $\mathbf{w} = (-1, 2, 6, 0)$ in the form $\mathbf{w} = \mathbf{w}_1 + \mathbf{w}_2$, where $\mathbf{w}_1$ is in the space W spanned by $\mathbf{u}_1 = (-1, 0, 1, 2)$ and $\mathbf{u}_2 = (0, 1, 0, 1)$, and $\mathbf{w}_2$ is orthogonal to W.

15. Let $\{\mathbf{v}_1, \mathbf{v}_2, \mathbf{v}_3\}$ be an orthonormal basis for an inner product space V. Show that if $\mathbf{w}$ is a vector in V, then $\|\mathbf{w}\|^2 = \langle \mathbf{w}, \mathbf{v}_1 \rangle^2 + \langle \mathbf{w}, \mathbf{v}_2 \rangle^2 + \langle \mathbf{w}, \mathbf{v}_3 \rangle^2$.

16. Let $\{\mathbf{v}_1, \mathbf{v}_2, \ldots, \mathbf{v}_n\}$ be an orthonormal basis for an inner product space V. Show that if $\mathbf{w}$ is a vector in V, then

$$\|\mathbf{w}\|^2 = \langle \mathbf{w}, \mathbf{v}_1 \rangle^2 + \langle \mathbf{w}, \mathbf{v}_2 \rangle^2 + \cdots + \langle \mathbf{w}, \mathbf{v}_n \rangle^2$$

17. In Step 3 of the proof of Theorem 26, it was stated that "the linear independence of $\{\mathbf{u}_1, \mathbf{u}_2, \ldots, \mathbf{u}_n\}$ assures that

$$\mathbf{u}_3 - \langle \mathbf{u}_3, \mathbf{v}_1 \rangle \mathbf{v}_1 - \langle \mathbf{u}_3, \mathbf{v}_2 \rangle \mathbf{v}_2 \neq \mathbf{0}."$$

Prove this statement.

18. Prove Theorem 25.
    (**Hint.** Show that the vector $\mathbf{w}_1$ in (4.28) lies in W, the vector $\mathbf{w}_2$ in (4.29) is orthogonal to W, and $\mathbf{u} = \mathbf{w}_1 + \mathbf{w}_2$.)

19. (**For readers who have studied calculus.**) Let the vector space P_2 have the inner product

$$\langle \mathbf{p}, \mathbf{q} \rangle = \int_{-1}^{1} p(x)q(x)\,dx$$

Apply the Gram-Schmidt process to transform the standard basis $S = \{1, x, x^2\}$ into an orthonormal basis. (The polynomials in the resulting basis are called the first three *normalized Legendre polynomials.*)

20. (**For readers who have studied calculus.**) Use Theorem 23 to express the following as linear combinations of the first three normalized Legendre polynomials (Exercise 19).
    (a) $1 + x + 4x^2$ (b) $2 - 7x^2$ (c) $4 + 3x$

21. (**For readers who have studied calculus.**) Let P_2 have the inner product

$$\langle \mathbf{p}, \mathbf{q} \rangle = \int_{0}^{1} p(x)q(x)\,dx$$

Apply the Gram-Schmidt process to transform the standard basis $S = \{1, x, x^2\}$ into an orthonormal basis.

22. (**For readers who have studied the optional material in this section.**) Find the point Q in the plane $5x - 3y + z = 0$ closest to $P(1, -2, 4)$, and determine the distance between the point P and the plane. (**Hint.** View the plane as a subspace W of R^3 with the Euclidean inner product and apply Theorem 28.)

23. (**For readers who have studied the optional material in this section.**) Find the point Q on the line

$$x = 2t$$
$$y = -t \qquad -\infty < t < +\infty$$
$$z = 4t$$

closest to $P(-4, 8, 1)$. (**Hint.** See the hint in the previous exercise.)

4.10   COORDINATES; CHANGE OF BASIS

There is a close relationship between the notion of a basis and the notion of a coordinate system. In this section we develop this idea and also discuss results about changing bases for vector spaces.

In plane analytic geometry we associate a pair of coordinates (a, b) with a point P in the plane by using two perpendicular coordinate axes. However, coordinates can also be introduced without reference to coordinate axes by using vectors. For example, instead of introducing coordinate axes as in Figure 4.15*a*, consider two perpendicular vectors $\mathbf{v}_1$ and $\mathbf{v}_2$, each of length 1 and having the same initial point O. (These vectors form a basis for R^2.) By dropping perpendiculars from a point P onto the lines determined by $\mathbf{v}_1$ and $\mathbf{v}_2$, we obtain vectors $a\mathbf{v}_1$ and $b\mathbf{v}_2$ such that

$$\overrightarrow{OP} = a\mathbf{v}_1 + b\mathbf{v}_2$$

(Figure 4.15*b*). Clearly, the numbers a and b just obtained are the same as the coordinates of P relative to the coordinate system in Figure 4.15*a*. Thus, we can view the coordinates of P as the numbers needed to express the vector $\overrightarrow{OP}$ in terms of the basis vectors $\mathbf{v}_1$ and $\mathbf{v}_2$.

For the purposes of attaching coordinates to points in the plane, it is not essential that the basis vectors $\mathbf{v}_1$ and $\mathbf{v}_2$ be perpendicular or have length 1; any basis for R^2 will do. For example, using the basis vectors $\mathbf{v}_1$ and $\mathbf{v}_2$ in Figure 4.16, we can attach a unique pair of coordinates to a point P by projecting P parallel to the basis vectors in order to make $\overrightarrow{OP}$ the diagonal of a parallelogram determined by vectors $a\mathbf{v}_1$ and $b\mathbf{v}_2$; thus,

$$\overrightarrow{OP} = a\mathbf{v}_1 + b\mathbf{v}_2$$

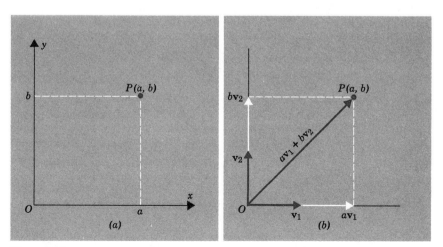

Figure 4.15

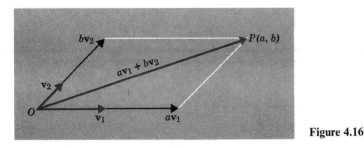

Figure 4.16

We can regard (a, b) as the coordinates of P relative to the basis $\{\mathbf{v}_1, \mathbf{v}_2\}$. This generalized notion of coordinates is important because it can be extended to more general vector spaces. But first we will need some preliminary results.

Suppose $S = \{\mathbf{v}_1, \mathbf{v}_2, \ldots, \mathbf{v}_n\}$ is a basis for a finite dimensional vector space V. Since S spans V, every vector in V is expressible as a linear combination of vectors in S. Moreover, the linear independence of S assures that there is only *one* way to express a vector as a linear combination of vectors in S. To see why, suppose a vector $\mathbf{v}$ can be written as

$$\mathbf{v} = c_1\mathbf{v}_1 + c_2\mathbf{v}_2 + \cdots + c_n\mathbf{v}_n$$

and also

$$\mathbf{v} = k_1\mathbf{v}_1 + k_2\mathbf{v}_2 + \cdots + k_n\mathbf{v}_n$$

Subtracting the second equation from the first gives

$$\mathbf{0} = (c_1 - k_1)\mathbf{v}_1 + (c_2 - k_2)\mathbf{v}_2 + \cdots + (c_n - k_n)\mathbf{v}_n$$

Since the right side of this equation is a linear combination of vectors in S, the linear independence of S implies that

$$c_1 - k_1 = 0, \qquad c_2 - k_2 = 0, \ldots, c_n - k_n = 0$$

That is,

$$c_1 = k_1, \qquad c_2 = k_2, \ldots, c_n = k_n$$

In summary, we have the following result.

Theorem 29. *If $S = \{\mathbf{v}_1, \mathbf{v}_2, \ldots, \mathbf{v}_n\}$ is a basis for a vector space V, then every vector $\mathbf{v}$ in V can be expressed in the form $\mathbf{v} = c_1\mathbf{v}_1 + c_2\mathbf{v}_2 + \cdots + c_n\mathbf{v}_n$ in exactly one way.*

If $S = \{\mathbf{v}_1, \mathbf{v}_2, \ldots, \mathbf{v}_n\}$ is a basis for a finite-dimensional vector space V, and

$$\mathbf{v} = c_1\mathbf{v}_1 + c_2\mathbf{v}_2 + \cdots + c_n\mathbf{v}_n$$

is the expression for $\mathbf{v}$ in terms of the basis S, then the scalars $c_1, c_2, \ldots, c_n$ are called the ***coordinates*** of $\mathbf{v}$ relative to the basis S. The ***coordinate vector*** of $\mathbf{v}$ relative to S is denoted by $(v)_S$ and is the vector in R^n defined by

$$(v)_S = (c_1, c_2, \ldots, c_n)$$

The **coordinate matrix** of v relative to S is denoted by $[v]_S$ and is the $n \times 1$ matrix defined by

$$\begin{bmatrix} c_1 \\ c_2 \\ \vdots \\ c_n \end{bmatrix}$$

Example 66

In Example 30 of Section 4.5 we showed that $S = \{v_1, v_2, v_3\}$ is a basis for R^3, where $v_1 = (1, 2, 1)$, $v_2 = (2, 9, 0)$, and $v_3 = (3, 3, 4)$.

(a) Find the coordinate vector and coordinate matrix of $v = (5, -1, 9)$ with respect to S.
(b) Find the vector v in R^3 whose coordinate vector with respect to S is $(v)_S = (-1, 3, 2)$.

Solution (a). We must find scalars c_1, c_2, c_3 such that

$$v = c_1 v_1 + c_2 v_2 + c_3 v_3$$

or, in terms of components,

$$(5, -1, 9) = c_1(1, 2, 1) + c_2(2, 9, 0) + c_3(3, 3, 4)$$

Equating corresponding components gives

$$\begin{array}{rcr} c_1 + 2c_2 + 3c_3 &=& 5 \\ 2c_1 + 9c_2 + 3c_3 &=& -1 \\ c_1 \qquad\; + 4c_3 &=& 9 \end{array}$$

Solving this system, we obtain $c_1 = 1$, $c_2 = -1$, $c_3 = 2$. Therefore,

$$(v)_S = (1, -1, 2) \quad \text{and} \quad [v]_S = \begin{bmatrix} 1 \\ -1 \\ 2 \end{bmatrix}$$

Solution (b). Using the definition of the coordinate vector $(v)_S$, we obtain

$$v = (-1)v_1 + 3v_2 + 2v_3 = (11, 31, 7) \quad \triangle$$

Coordinate vectors and matrices depend on the order in which the basis vectors are written; a change in the order of the basis vectors results in a corresponding change of order for the entries in the coordinate matrices and coordinate vectors.

Example 67

Consider the basis $S = \{1, x, x^2\}$ for P_2. By inspection, the coordinate vector and coordinate matrix with respect to S for a polynomial $\mathbf{p} = a_0 + a_1 x + a_2 x^2$ are

$$(\mathbf{p})_S = (a_0, a_1, a_2) \quad \text{and} \quad [\mathbf{p}]_S = \begin{bmatrix} a_0 \\ a_1 \\ a_2 \end{bmatrix} \quad \blacktriangle$$

Example 68

Let a rectangular xyz-coordinate system be introduced into 3-space and consider the standard basis $S = \{\mathbf{i}, \mathbf{j}, \mathbf{k}\}$ where

$$\mathbf{i} = (1, 0, 0), \quad \mathbf{j} = (0, 1, 0), \quad \text{and } \mathbf{k} = (0, 0, 1)$$

If, as in Figure 4.17, $\mathbf{v} = (a, b, c)$ is any vector in R^3, then

$$\mathbf{v} = (a, b, c) = a(1, 0, 0) + b(0, 1, 0) + c(0, 0, 1) = a\mathbf{i} + b\mathbf{j} + c\mathbf{k}$$

which means

$$\mathbf{v} = (a, b, c) = (\mathbf{v})_S$$

In other words, the components of a vector $\mathbf{v}$ relative to a rectangular xyz-coordinate system are the same as the coordinates of $\mathbf{v}$ relative to the standard basis $\{\mathbf{i}, \mathbf{j}, \mathbf{k}\}$. $\blacktriangle$

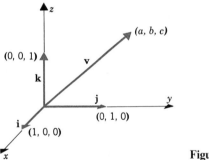

Figure 4.17

Example 69

If $S = \{\mathbf{v}_1, \mathbf{v}_2, \ldots, \mathbf{v}_n\}$ is an orthonormal basis for an inner product space V, then by Theorem 23 of Section 4.9 the expression for a vector $\mathbf{u}$ in terms of the basis S is

$$\mathbf{u} = \langle \mathbf{u}, \mathbf{v}_1 \rangle \mathbf{v}_1 + \langle \mathbf{u}, \mathbf{v}_2 \rangle \mathbf{v}_2 + \cdots + \langle \mathbf{u}, \mathbf{v}_n \rangle \mathbf{v}_n$$

which means that

$$(\mathbf{u})_S = (\langle \mathbf{u}, \mathbf{v}_1 \rangle, \langle \mathbf{u}, \mathbf{v}_2 \rangle, \ldots, \langle \mathbf{u}, \mathbf{v}_n \rangle)$$

and

$$[\mathbf{u}]_S = \begin{bmatrix} \langle \mathbf{u}, \mathbf{v}_1 \rangle \\ \langle \mathbf{u}, \mathbf{v}_2 \rangle \\ \vdots \\ \langle \mathbf{u}, \mathbf{v}_n \rangle \end{bmatrix}$$

For example, if

$$\mathbf{v}_1 = (0, 1, 0), \quad \mathbf{v}_2 = (-\tfrac{4}{5}, 0, \tfrac{3}{5}), \quad \mathbf{v}_3 = (\tfrac{3}{5}, 0, \tfrac{4}{5})$$

then, as observed in Example 62 of Section 4.9, $S = \{\mathbf{v}_1, \mathbf{v}_2, \mathbf{v}_3\}$ is an orthonormal basis for R^3 with the Euclidean inner product. If $\mathbf{u} = (2, -1, 4)$, then

$$\langle \mathbf{u}, \mathbf{v}_1 \rangle = -1, \quad \langle \mathbf{u}, \mathbf{v}_2 \rangle = \tfrac{4}{5}, \quad \langle \mathbf{u}, \mathbf{v}_3 \rangle = \tfrac{22}{5}$$

so that

$$(\mathbf{u})_S = (-1, \tfrac{4}{5}, \tfrac{22}{5}) \quad \text{and} \quad [\mathbf{u}]_S = \begin{bmatrix} -1 \\ \tfrac{4}{5} \\ \tfrac{22}{5} \end{bmatrix}$$

Orthonormal bases for inner product spaces are convenient because, as the following theorem shows, many familiar formulas hold in such spaces.

Theorem 30. *If S is an orthonormal basis for an n-dimensional inner product space and if*

$$(\mathbf{u})_S = (u_1, u_2, \ldots, u_n) \quad \text{and} \quad (\mathbf{v})_S = (v_1, v_2, \ldots, v_n)$$

then

(a) $\|\mathbf{u}\| = \sqrt{u_1^2 + u_2^2 + \cdots + u_n^2}$
(b) $d(\mathbf{u}, \mathbf{v}) = \sqrt{(u_1 - v_1)^2 + (u_2 - v_2)^2 + \cdots + (u_n - v_n)^2}$
(c) $\langle \mathbf{u}, \mathbf{v} \rangle = u_1 v_1 + u_2 v_2 + \cdots + u_n v_n$

The proofs and some numerical examples are discussed in the exercises.
We now turn to the main problem in this section.

Change of Basis Problem. If we change the basis for a vector space from some old basis B to some new basis B', how is the old coordinate matrix $[\mathbf{v}]_B$ of a vector $\mathbf{v}$ related to the new coordinate matrix $[\mathbf{v}]_{B'}$?

For simplicity, we will solve this problem for two-dimensional spaces. The solution for n-dimensional spaces is similar and will be left as an exercise. Let

$$B = \{\mathbf{u}_1, \mathbf{u}_2\} \quad \text{and} \quad B' = \{\mathbf{u}_1', \mathbf{u}_2'\}$$

be the old and new bases, respectively. We will need the coordinate matrices for the new basis vectors relative to the old basis. Suppose they are

$$[\mathbf{u}_1']_B = \begin{bmatrix} a \\ b \end{bmatrix} \quad \text{and} \quad [\mathbf{u}_2']_B = \begin{bmatrix} c \\ d \end{bmatrix} \tag{4.35}$$

That is,

$$\begin{aligned} \mathbf{u}_1' &= a\mathbf{u}_1 + b\mathbf{u}_2 \\ \mathbf{u}_2' &= c\mathbf{u}_1 + d\mathbf{u}_2 \end{aligned} \tag{4.36}$$

Now let **v** be any vector in V and let

$$[\mathbf{v}]_{B'} = \begin{bmatrix} k_1 \\ k_2 \end{bmatrix} \tag{4.37}$$

be the new coordinate matrix, so that

$$\mathbf{v} = k_1\mathbf{u}_1' + k_2\mathbf{u}_2' \tag{4.38}$$

In order to find the old coordinates of **v** we must express **v** in terms of the old basis B. To do this, we substitute (4.36) into (4.38). This yields

$$\mathbf{v} = k_1(a\mathbf{u}_1 + b\mathbf{u}_2) + k_2(c\mathbf{u}_1 + d\mathbf{u}_2)$$

or

$$\mathbf{v} = (k_1 a + k_2 c)\mathbf{u}_1 + (k_1 b + k_2 d)\mathbf{u}_2$$

Thus, the old coordinate matrix for **v** is

$$[\mathbf{v}]_B = \begin{bmatrix} k_1 a + k_2 c \\ k_1 b + k_2 d \end{bmatrix}$$

which can be rewritten

$$[\mathbf{v}]_B = \begin{bmatrix} a & c \\ b & d \end{bmatrix} \begin{bmatrix} k_1 \\ k_2 \end{bmatrix}$$

or, from (4.37),

$$[\mathbf{v}]_B = \begin{bmatrix} a & c \\ b & d \end{bmatrix} [\mathbf{v}]_{B'}$$

This equation states that the old coordinate matrix $[\mathbf{v}]_B$ results when we multiply the new coordinate matrix $[\mathbf{v}]_{B'}$ on the left by the matrix

$$P = \begin{bmatrix} a & c \\ b & d \end{bmatrix}$$

whose columns are the coordinates of the new basis vectors relative to the old basis [see (4.35)]. Thus, we have the following solution of the change-of-basis problem.

Solution of the Change of Basis Problem. If we change the basis for a vector space V from some old basis $B = \{\mathbf{u}_1, \mathbf{u}_2, \ldots, \mathbf{u}_n\}$ to some new basis $B' = \{\mathbf{u}_1', \mathbf{u}_2', \ldots, \mathbf{u}_n'\}$, then the old coordinate matrix $[\mathbf{v}]_B$ of a vector **v** is related to the new coordinate matrix $[\mathbf{v}]_{B'}$ of the same vector by the equation

$$[\mathbf{v}]_B = P[\mathbf{v}]_{B'} \tag{4.39}$$

where the columns of P are the coordinate matrices of the new basis vectors relative to the old basis; that is, the column vectors of P are

$$[\mathbf{u}_1']_B, [\mathbf{u}_2']_B, \dots, [\mathbf{u}_n']_B$$

Symbolically, the matrix P can be written

$$P = \left[[\mathbf{u}_1']_B \mid [\mathbf{u}_2']_B \mid \cdots \mid [\mathbf{u}_n']_B \right]$$

It is called the **transition matrix** from B' to B.

Example 70

Consider the bases $B = \{\mathbf{u}_1, \mathbf{u}_2\}$ and $B' = \{\mathbf{u}_1', \mathbf{u}_2'\}$ for R^2, where

$$\mathbf{u}_1 = (1, 0), \quad \mathbf{u}_2 = (0, 1); \qquad \mathbf{u}_1' = (1, 1), \quad \mathbf{u}_2' = (2, 1)$$

(a) Find the transition matrix from B' to B.
(b) Use (4.39) to find $[\mathbf{v}]_B$ if

$$[\mathbf{v}]_{B'} = \begin{bmatrix} -3 \\ 5 \end{bmatrix}$$

Solution (a). First we must find the coordinate matrices for the new basis vectors $\mathbf{u}_1'$ and $\mathbf{u}_2'$ relative to the old basis B. By inspection

$$\mathbf{u}_1' = \mathbf{u}_1 + \mathbf{u}_2$$
$$\mathbf{u}_2' = 2\mathbf{u}_1 + \mathbf{u}_2$$

so that $[\mathbf{u}_1']_B = \begin{bmatrix} 1 \\ 1 \end{bmatrix}$ and $[\mathbf{u}_2']_B = \begin{bmatrix} 2 \\ 1 \end{bmatrix}$. Thus, the transition matrix from B' to B is

$$P = \begin{bmatrix} 1 & 2 \\ 1 & 1 \end{bmatrix}$$

Solution (b). Using (4.39) and the transition matrix in part (a),

$$[\mathbf{v}]_B = \begin{bmatrix} 1 & 2 \\ 1 & 1 \end{bmatrix} \begin{bmatrix} -3 \\ 5 \end{bmatrix} = \begin{bmatrix} 7 \\ 2 \end{bmatrix}$$

As a check, we should be able to recover the vector $\mathbf{v}$ either from $[\mathbf{v}]_B$ or $[\mathbf{v}]_{B'}$. We leave it for the reader to show that $-3\mathbf{u}_1' + 5\mathbf{u}_2' = 7\mathbf{u}_1 + 2\mathbf{u}_2 = \mathbf{v} = (7, 2)$.

Example 71

Consider the vectors $\mathbf{u}_1 = (1, 0)$, $\mathbf{u}_2 = (0, 1)$, $\mathbf{u}_1' = (1, 1)$, $\mathbf{u}_2' = (2, 1)$. In Example 70 we found the transition matrix from the basis $B' = \{\mathbf{u}_1', \mathbf{u}_2'\}$ for R^2 to the basis $B = \{\mathbf{u}_1, \mathbf{u}_2\}$. However, we can just as well ask for the transition matrix from B to B'. To obtain this matrix, we simply change our point of view and regard B' as the old basis and B as the new basis. As usual, the columns of the transition matrix will be the coordinates of the new basis vectors relative to the old basis.

Following the procedure of Example 66(a), the reader should be able to show that

$$\mathbf{u}_1 = -\mathbf{u}'_1 + \mathbf{u}'_2$$
$$\mathbf{u}_2 = 2\mathbf{u}'_1 - \mathbf{u}'_2$$

so that $[\mathbf{u}_1]_{B'} = \begin{bmatrix} -1 \\ 1 \end{bmatrix}$ and $[\mathbf{u}_2]_{B'} = \begin{bmatrix} 2 \\ -1 \end{bmatrix}$. Thus, the transition matrix from B to B' is

$$Q = \begin{bmatrix} -1 & 2 \\ 1 & -1 \end{bmatrix}$$

If we multiply the transition matrix from B' to B obtained in Example 70 and the transition matrix from B to B' obtained in the last example, we find

$$PQ = \begin{bmatrix} 1 & 2 \\ 1 & 1 \end{bmatrix} \begin{bmatrix} -1 & 2 \\ 1 & -1 \end{bmatrix} = \begin{bmatrix} 1 & 0 \\ 0 & 1 \end{bmatrix} = I$$

which shows that $Q = P^{-1}$. The following theorem shows that this is not accidental.

Theorem 31. *If P is the transition matrix from a basis B' to a basis B, then*

(a) P is invertible

(b) P^{-1} is the transition matrix from B to B'.

(The proof is deferred to the end of the section.) To summarize, if P is the transition matrix from a basis B' to a basis B, then for every vector $\mathbf{v}$

$$[\mathbf{v}]_B = P[\mathbf{v}]_{B'} \tag{4.40a}$$
$$[\mathbf{v}]_{B'} = P^{-1}[\mathbf{v}]_B \tag{4.40b}$$

To illustrate this result, consider the transition matrix

$$P = \begin{bmatrix} 1 & 2 \\ 1 & 1 \end{bmatrix}$$

obtained in Example 70. In order to change from the old coordinates of a vector $\mathbf{v}$ to the new coordinates of $\mathbf{v}$, we first calculate

$$P^{-1} = \begin{bmatrix} -1 & 2 \\ 1 & -1 \end{bmatrix}$$

If $\mathbf{v} = (7, 2)$, then

$$[\mathbf{v}]_B = \begin{bmatrix} 7 \\ 2 \end{bmatrix}$$

by inspection, and

$$[\mathbf{v}]_{B'} = \begin{bmatrix} -1 & 2 \\ 1 & -1 \end{bmatrix} \begin{bmatrix} 7 \\ 2 \end{bmatrix} = \begin{bmatrix} -3 \\ 5 \end{bmatrix}$$

Example 72 (Application to Rotation of Coordinate Axes).

In many problems a rectangular xy-coordinate system is given and a new $x'y'$-coordinate system is obtained by rotating the xy-system counterclockwise about the origin through an angle θ. When this is done, each point Q in the plane has two sets of coordinates: coordinates (x, y) relative to the xy-system and coordinates (x', y') relative to the $x'y'$-system (Figure 4.18*a*).

By introducing unit vectors $\mathbf{u}_1$ and $\mathbf{u}_2$ along the positive x and y axes and unit vectors $\mathbf{u}'_1$ and $\mathbf{u}'_2$ along the positive x' and y' axes, we can regard this rotation as a change from an old basis $B = \{\mathbf{u}_1, \mathbf{u}_2\}$ to a new basis $B' = \{\mathbf{u}'_1, \mathbf{u}'_2\}$ (Figure 4.18*b*). Thus, the new coordinates (x', y') and the old coordinates (x, y) of a point Q will be related by

$$\begin{bmatrix} x' \\ y' \end{bmatrix} = P^{-1} \begin{bmatrix} x \\ y \end{bmatrix} \tag{4.41}$$

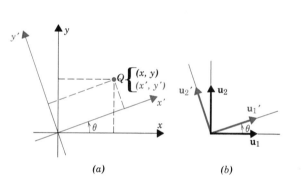

Figure 4.18   (*a*)   (*b*)

where P is the transition from B' to B. To find P we must determine the coordinate matrices of the new basis vectors $\mathbf{u}'_1$ and $\mathbf{u}'_2$ relative to the old basis. As indicated in Figure 4.18*c*, the components of $\mathbf{u}'_1$ in the old basis are $\cos\theta$ and $\sin\theta$ so that

$$[\mathbf{u}'_1]_B = \begin{bmatrix} \cos\theta \\ \sin\theta \end{bmatrix}$$

whereas, as indicated in Figure 4.18*d*, the components of $\mathbf{u}'_2$ in the old basis are $\cos(\theta + \pi/2) = -\sin\theta$ and $\sin(\theta + \pi/2) = \cos\theta$, so that

$$[\mathbf{u}'_2]_B = \begin{bmatrix} -\sin\theta \\ \cos\theta \end{bmatrix}$$

Thus, the transition matrix from B' to B is

$$P = \begin{bmatrix} \cos\theta & -\sin\theta \\ \sin\theta & \cos\theta \end{bmatrix}$$

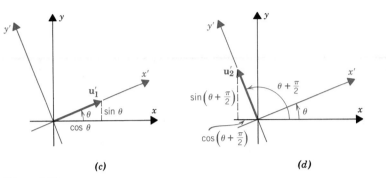

(c)   (d)

Figure 4.18

and it is easy to verify that

$$P^{-1} = \begin{bmatrix} \cos\theta & \sin\theta \\ -\sin\theta & \cos\theta \end{bmatrix}$$

(Use the method of Example 26 in Section 1.5.) Thus, (4.40) becomes

$$\begin{bmatrix} x' \\ y' \end{bmatrix} = \begin{bmatrix} \cos\theta & \sin\theta \\ -\sin\theta & \cos\theta \end{bmatrix}\begin{bmatrix} x \\ y \end{bmatrix}$$ (4.42)

or equivalently

$$x' = x\cos\theta + y\sin\theta$$
$$y' = -x\sin\theta + y\cos\theta$$

For example, if the axes are rotated $\theta = 45°$, then since

$$\sin 45° = \cos 45° = \frac{1}{\sqrt{2}}$$

(4.42) becomes

$$\begin{bmatrix} x' \\ y' \end{bmatrix} = \begin{bmatrix} \dfrac{1}{\sqrt{2}} & \dfrac{1}{\sqrt{2}} \\ -\dfrac{1}{\sqrt{2}} & \dfrac{1}{\sqrt{2}} \end{bmatrix}\begin{bmatrix} x \\ y \end{bmatrix}$$

Thus, if the old coordinates of a point Q are $(x, y) = (2, -1)$, then

$$\begin{bmatrix} x' \\ y' \end{bmatrix} = \begin{bmatrix} \dfrac{1}{\sqrt{2}} & \dfrac{1}{\sqrt{2}} \\ -\dfrac{1}{\sqrt{2}} & \dfrac{1}{\sqrt{2}} \end{bmatrix}\begin{bmatrix} 2 \\ -1 \end{bmatrix} = \begin{bmatrix} \dfrac{1}{\sqrt{2}} \\ -\dfrac{3}{\sqrt{2}} \end{bmatrix}$$

so the new coordinates of Q are $(x', y') = (1/\sqrt{2}, -3/\sqrt{2})$. ▲

Example 73 (Application to Rotation of Axes in 3-space).

Suppose a rectangular xyz-coordinate system is rotated around its z-axis counterclockwise (looking down the positive z-axis) through an angle θ (Figure 4.19). If we introduce unit vectors $\mathbf{u}_1$, $\mathbf{u}_2$, and $\mathbf{u}_3$ along the positive x, y, and z axes and unit vectors $\mathbf{u}_1'$, $\mathbf{u}_2'$, and $\mathbf{u}_3'$ along the positive x', y', and z' axes, we can regard the rotation as a change from the old basis $B = \{\mathbf{u}_1, \mathbf{u}_2, \mathbf{u}_3\}$ to the new basis $B' = \{\mathbf{u}_1', \mathbf{u}_2', \mathbf{u}_3'\}$. In light of Example 72 it should be evident that

$$[\mathbf{u}_1']_B = \begin{bmatrix} \cos\theta \\ \sin\theta \\ 0 \end{bmatrix} \quad \text{and} \quad [\mathbf{u}_2']_B = \begin{bmatrix} -\sin\theta \\ \cos\theta \\ 0 \end{bmatrix}$$

Moreover, since $\mathbf{u}_3'$ extends 1 unit up the positive z' axis

$$[\mathbf{u}_3']_B = \begin{bmatrix} 0 \\ 0 \\ 1 \end{bmatrix}$$

Thus, the transition matrix from B' to B is

$$P = \begin{bmatrix} \cos\theta & -\sin\theta & 0 \\ \sin\theta & \cos\theta & 0 \\ 0 & 0 & 1 \end{bmatrix}$$

and the transition matrix from B to B' is

$$P^{-1} = \begin{bmatrix} \cos\theta & \sin\theta & 0 \\ -\sin\theta & \cos\theta & 0 \\ 0 & 0 & 1 \end{bmatrix}$$

(verify.) Thus the old coordinates (x, y, z) of a point Q are related to its new coordinates (x', y', z') by

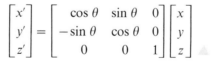

$$\begin{bmatrix} x' \\ y' \\ z' \end{bmatrix} = \begin{bmatrix} \cos\theta & \sin\theta & 0 \\ -\sin\theta & \cos\theta & 0 \\ 0 & 0 & 1 \end{bmatrix} \begin{bmatrix} x \\ y \\ z \end{bmatrix}$$

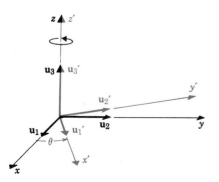

Figure 4.19

The next theorem shows that if P is the transition matrix from one *orthonormal* basis to another, then the inverse of P is especially easy to find.

Theorem 32. *If P is the transition matrix from one orthonormal basis to another orthonormal basis for an inner product space, then*

$$P^{-1} = P^t$$

(We omit the proof.)

Example 74

For the transition matrix

$$P = \begin{bmatrix} \cos\theta & -\sin\theta \\ \sin\theta & \cos\theta \end{bmatrix}$$

in Example 72, we found

$$P^{-1} = \begin{bmatrix} \cos\theta & \sin\theta \\ -\sin\theta & \cos\theta \end{bmatrix}$$

by direct computation. However, P is the transition matrix from one orthonormal basis to another, so we can obtain P^{-1} more simply from Theorem 32:

$$P^{-1} = P^t = \begin{bmatrix} \cos\theta & \sin\theta \\ -\sin\theta & \cos\theta \end{bmatrix}$$

Similarly, the inverse of the matrix P in Example 73 can be found by transposing P. ▲

Matrices whose inverses can be obtained by transposition are sufficiently important that there is some terminology associated with them.

Definition. A square matrix A with the property

$$A^{-1} = A^t$$

is said to be an ***orthogonal matrix***.

To paraphrase this definition, a square matrix A is orthogonal if and only if

$$AA^t = A^tA = I$$

Moreover, using this new terminology Theorem 32 states that *a transition matrix from one orthonormal basis to another is orthogonal.*

The following result, whose proof is discussed in the exercises, makes it easy to determine when an $n \times n$ matrix A is orthogonal.

Theorem 33. *The following are equivalent*:

(a) *A is orthogonal.*
(b) *The row vectors of A form an orthonormal set in R^n with the Euclidean inner product.*
(c) *The column vectors of A form an orthonormal set in R^n with the Euclidean inner product.*

Example 75

Consider the matrix

$$A = \begin{bmatrix} \dfrac{1}{\sqrt{2}} & \dfrac{1}{\sqrt{2}} & 0 \\ 0 & 0 & 1 \\ \dfrac{1}{\sqrt{2}} & -\dfrac{1}{\sqrt{2}} & 0 \end{bmatrix}$$

The row vectors of A are

$$\mathbf{r}_1 = \left(\frac{1}{\sqrt{2}}, \frac{1}{\sqrt{2}}, 0\right), \quad \mathbf{r}_2 = (0, 0, 1), \quad \mathbf{r}_3 = \left(\frac{1}{\sqrt{2}}, -\frac{1}{\sqrt{2}}, 0\right)$$

Relative to the Euclidean inner product we have

$$\|\mathbf{r}_1\| = \|\mathbf{r}_2\| = \|\mathbf{r}_3\| = 1$$

and

$$\mathbf{r}_1 \cdot \mathbf{r}_2 = \mathbf{r}_2 \cdot \mathbf{r}_3 = \mathbf{r}_1 \cdot \mathbf{r}_3 = 0$$

so that the row vectors of A form an orthonormal set in R^3. Thus, A is orthogonal and

$$A^{-1} = A^t = \begin{bmatrix} \dfrac{1}{\sqrt{2}} & 0 & \dfrac{1}{\sqrt{2}} \\ \dfrac{1}{\sqrt{2}} & 0 & -\dfrac{1}{\sqrt{2}} \\ 0 & 1 & 0 \end{bmatrix}$$

(The reader will find it instructive to check that the column vectors of A also form an orthonormal set.) ▲

In Examples 72 and 73 we considered the problem of relating old and new coordinates when a geometric change (rotation) was made in the coordinate axes.

Sometimes the following converse problem arises. A relationship

$$\begin{bmatrix} x \\ y \end{bmatrix} = \begin{bmatrix} a_1 & a_2 \\ b_1 & b_2 \end{bmatrix} \begin{bmatrix} x' \\ y' \end{bmatrix} \tag{4.43}$$

between old and new coordinates is known, where the 2×2 matrix is orthogonal; and it is of interest to determine how the xy-coordinate system and $x'y'$-coordinate system are related geometrically. Equation (4.43) is called an ***orthogonal coordinate transformation*** on R^2. To study the effect of an orthogonal coordinate transformation, consider the vectors

$$\mathbf{u}_1 = \begin{bmatrix} 1 \\ 0 \end{bmatrix}, \quad \mathbf{u}_2 = \begin{bmatrix} 0 \\ 1 \end{bmatrix}, \quad \mathbf{u}_1' = \begin{bmatrix} a_1 \\ b_1 \end{bmatrix}, \quad \mathbf{u}_2' = \begin{bmatrix} a_2 \\ b_2 \end{bmatrix}$$

in the xy-coordinate system and introduce an $x'y'$-coordinate system with positive x'-axis along $\mathbf{u}_1'$ and positive y'-axis along $\mathbf{u}_2'$ (Figure 4.20). Because the 2×2 matrix in (4.43) is assumed orthogonal, the vectors $\mathbf{u}_1'$ and $\mathbf{u}_2'$ are orthogonal, which assures that the x' and y' axes are perpendicular. Since

$$\mathbf{u}_1' = a_1\mathbf{u}_1 + b_1\mathbf{u}_2$$

and

$$\mathbf{u}_2' = a_2\mathbf{u}_1 + b_2\mathbf{u}_2$$

the matrix

$$P = \begin{bmatrix} a_1 & a_2 \\ b_1 & b_2 \end{bmatrix}$$

in (4.43) is the transition matrix from the basis $\{\mathbf{u}_1', \mathbf{u}_2'\}$ to the basis $\{\mathbf{u}_1, \mathbf{u}_2\}$. Clearly, there are two possibilities: either the $x'y'$-coordinate system can be obtained by rotating the xy-coordinate system (Figure 4.21*a*), or the $x'y'$-coordinate system can be obtained by first reflecting the xy-coordinate system about the x-axis and then rotating the reflected coordinate system (Figure 4.21*b*). It is shown in the exercises that the determinant of an orthogonal matrix is always $+1$ or -1.

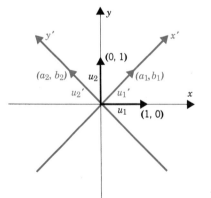

Figure 4.20

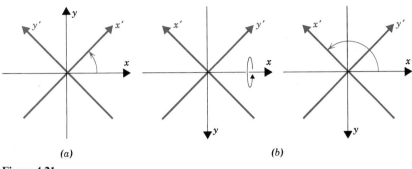

(a)   (b)

Figure 4.21

Moreover, it can be proved that the orthogonal coordinate transformation (4.43) is a rotation if

$$\begin{vmatrix} a_1 & a_2 \\ b_1 & b_2 \end{vmatrix} = 1$$

and is a reflection followed by a rotation if this determinant is -1.

Similarly, an orthogonal coordinate transformation

$$\begin{bmatrix} x \\ y \\ z \end{bmatrix} = \begin{bmatrix} a_1 & a_2 & a_3 \\ b_1 & b_2 & b_3 \\ c_1 & c_2 & c_3 \end{bmatrix} \begin{bmatrix} x' \\ y' \\ z' \end{bmatrix} \qquad \text{on } R^3 \text{ is a rotation if} \qquad \begin{vmatrix} a_1 & a_2 & a_3 \\ b_1 & b_2 & b_3 \\ c_1 & c_2 & c_3 \end{vmatrix} = 1$$

and is a rotation combined with a reflection in one of the coordinate planes if this determinant is -1.

Example 76

The orthogonal coordinate transformation

$$\begin{bmatrix} x \\ y \end{bmatrix} = \begin{bmatrix} \dfrac{1}{\sqrt{2}} & -\dfrac{1}{\sqrt{2}} \\ \dfrac{1}{\sqrt{2}} & \dfrac{1}{\sqrt{2}} \end{bmatrix} \begin{bmatrix} x' \\ y' \end{bmatrix} \qquad \text{is a rotation since} \qquad \begin{vmatrix} \dfrac{1}{\sqrt{2}} & -\dfrac{1}{\sqrt{2}} \\ \dfrac{1}{\sqrt{2}} & \dfrac{1}{\sqrt{2}} \end{vmatrix} = 1$$

The positive x' and y' axes are along the column vectors

$$\mathbf{u}_1' = \begin{bmatrix} \dfrac{1}{\sqrt{2}} \\ \dfrac{1}{\sqrt{2}} \end{bmatrix} \qquad \text{and} \qquad \mathbf{u}_2' = \begin{bmatrix} -\dfrac{1}{\sqrt{2}} \\ \dfrac{1}{\sqrt{2}} \end{bmatrix}$$

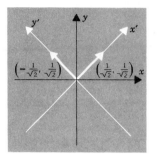

Figure 4.22

OPTIONAL

Proof of Theorem 31. Let Q be the transition matrix from B to B'. We shall show that $PQ = I$ and thus conclude that $Q = P^{-1}$ to complete the proof.

Assume that $B = \{\mathbf{u}_1, \mathbf{u}_2, \ldots, \mathbf{u}_n\}$ and suppose

$$PQ = \begin{bmatrix} c_{11} & c_{12} & \cdots & c_{1n} \\ c_{21} & c_{22} & \cdots & c_{2n} \\ \vdots & \vdots & & \vdots \\ c_{n1} & c_{n2} & \cdots & c_{nn} \end{bmatrix}$$

From (4.39)

$$[\mathbf{x}]_B = P[\mathbf{x}]_{B'}$$

and

$$[\mathbf{x}]_{B'} = Q[\mathbf{x}]_B$$

for all $\mathbf{x}$ in V. Multiplying the bottom equation through on the left by P and substituting the top equation gives

$$[\mathbf{x}]_B = PQ[\mathbf{x}]_B \qquad (4.44)$$

for all $\mathbf{x}$ in V. Letting $\mathbf{x} = \mathbf{u}_1$ in (4.44) gives

$$\begin{bmatrix} 1 \\ 0 \\ 0 \\ \vdots \\ 0 \end{bmatrix} = \begin{bmatrix} c_{11} & c_{12} & \cdots & c_{1n} \\ c_{21} & c_{22} & \cdots & c_{2n} \\ c_{31} & c_{32} & \cdots & c_{3n} \\ \vdots & \vdots & & \vdots \\ c_{n1} & c_{n2} & \cdots & c_{nn} \end{bmatrix} \begin{bmatrix} 1 \\ 0 \\ 0 \\ \vdots \\ 0 \end{bmatrix}$$

or

$$\begin{bmatrix} 1 \\ 0 \\ 0 \\ \vdots \\ 0 \end{bmatrix} = \begin{bmatrix} c_{11} \\ c_{21} \\ c_{31} \\ \vdots \\ c_{n1} \end{bmatrix}$$

Similarly, successively substituting $\mathbf{x} = \mathbf{u}_2, \ldots, \mathbf{u}_n$ in (4.44) yields

$$\begin{bmatrix} c_{12} \\ c_{22} \\ c_{32} \\ \vdots \\ c_{n2} \end{bmatrix} = \begin{bmatrix} 0 \\ 1 \\ 0 \\ \vdots \\ 0 \end{bmatrix}, \ldots, \begin{bmatrix} c_{1n} \\ c_{2n} \\ c_{3n} \\ \vdots \\ c_{nn} \end{bmatrix} = \begin{bmatrix} 0 \\ 0 \\ 0 \\ \vdots \\ 1 \end{bmatrix}$$

Therefore, $PQ = I$.

EXERCISE SET 4.10

1. Find the coordinate matrix and coordinate vector for $\mathbf{w}$ relative to the basis $S = \{\mathbf{u}_1, \mathbf{u}_2\}$.
   (a) $\mathbf{u}_1 = (1, 0)$, $\mathbf{u}_2 = (0, 1)$; $\mathbf{w} = (3, -7)$
   (b) $\mathbf{u}_1 = (2, -4)$, $\mathbf{u}_2 = (3, 8)$; $\mathbf{w} = (1, 1)$
   (c) $\mathbf{u}_1 = (1, 1)$, $\mathbf{u}_2 = (0, 2)$; $\mathbf{w} = (a, b)$

2. Find the coordinate vector and coordinate matrix for $\mathbf{v}$ relative to $S = \{\mathbf{v}_1, \mathbf{v}_2, \mathbf{v}_3\}$.
   (a) $\mathbf{v} = (2, -1, 3)$; $\mathbf{v}_1 = (1, 0, 0)$, $\mathbf{v}_2 = (2, 2, 0)$, $\mathbf{v}_3 = (3, 3, 3)$
   (b) $\mathbf{v} = (5, -12, 3)$; $\mathbf{v}_1 = (1, 2, 3)$, $\mathbf{v}_2 = (-4, 5, 6)$, $\mathbf{v}_3 = (7, -8, 9)$

3. Find the coordinate vector and coordinate matrix for $\mathbf{p}$ relative to $S = \{\mathbf{p}_1, \mathbf{p}_2, \mathbf{p}_3\}$.
   (a) $\mathbf{p} = 4 - 3x + x^2$; $\mathbf{p}_1 = 1$, $\mathbf{p}_2 = x$, $\mathbf{p}_3 = x^2$
   (b) $\mathbf{p} = 2 - x + x^2$; $\mathbf{p}_1 = 1 + x$, $\mathbf{p}_2 = 1 + x^2$, $\mathbf{p}_3 = x + x^2$

4. Find the coordinate vector and coordinate matrix for A relative to $S = \{A_1, A_2, A_3, A_4\}$.

   $$A = \begin{bmatrix} 2 & 0 \\ -1 & 3 \end{bmatrix} \quad A_1 = \begin{bmatrix} -1 & 1 \\ 0 & 0 \end{bmatrix} \quad A_2 = \begin{bmatrix} 1 & 1 \\ 0 & 0 \end{bmatrix} \quad A_3 = \begin{bmatrix} 0 & 0 \\ 1 & 0 \end{bmatrix} \quad A_4 = \begin{bmatrix} 0 & 0 \\ 0 & 1 \end{bmatrix}$$

5. In each part an orthonormal basis relative to the Euclidean inner product is given. Use the method of Example 69 to find the coordinate vector and coordinate matrix of $\mathbf{w}$.

   (a) $\mathbf{w} = (3, 7)$; $\mathbf{u}_1 = \left(\dfrac{1}{\sqrt{2}}, -\dfrac{1}{\sqrt{2}} \right)$, $\mathbf{u}_2 = \left(\dfrac{1}{\sqrt{2}}, \dfrac{1}{\sqrt{2}} \right)$

   (b) $\mathbf{w} = (-1, 0, 2)$; $\mathbf{u}_1 = (\frac{2}{3}, -\frac{2}{3}, \frac{1}{3})$, $\mathbf{u}_2 = (\frac{2}{3}, \frac{1}{3}, -\frac{2}{3})$, $\mathbf{u}_3 = (\frac{1}{3}, \frac{2}{3}, \frac{2}{3})$

6. (a) Find $\mathbf{w}$ if $(\mathbf{w})_S = (6, -1, 4)$ and S is the basis in Exercise 2a.
   (b) Find $\mathbf{q}$ if $(\mathbf{q})_S = (3, 0, 4)$ and S is the basis in Exercise 3a.
   (c) Find B if $(B)_S = (-8, 7, 6, 3)$ and S is the basis in Exercise 4.

7. Let R^2 have the Euclidean inner product and let $S = \{\mathbf{w}_1, \mathbf{w}_2\}$ be the orthonormal basis with $\mathbf{w}_1 = (\frac{3}{5}, -\frac{4}{5})$, $\mathbf{w}_2 = (\frac{4}{5}, \frac{3}{5})$. Let $\mathbf{u}$, $\mathbf{v}$ be the vectors in R^2 for which $(\mathbf{u})_S = (1, 1)$ and $(\mathbf{v})_S = (-1, 4)$.
   (a) Compute $\|\mathbf{u}\|$, $d(\mathbf{u}, \mathbf{v})$, and $\langle \mathbf{u}, \mathbf{v} \rangle$ using Theorem 30.
   (b) Find $\mathbf{u}$ and $\mathbf{v}$ and check the results of part (a) by computing $\|\mathbf{u}\|$, $d(\mathbf{u}, \mathbf{v})$, and $\langle \mathbf{u}, \mathbf{v} \rangle$ directly.

8. Consider the bases $B = \{\mathbf{u}_1, \mathbf{u}_2\}$ and $B' = \{\mathbf{v}_1, \mathbf{v}_2\}$ for R^2, where

$$\mathbf{u}_1 = \begin{bmatrix} 1 \\ 0 \end{bmatrix} \quad \mathbf{u}_2 = \begin{bmatrix} 0 \\ 1 \end{bmatrix} \quad \mathbf{v}_1 = \begin{bmatrix} 2 \\ 1 \end{bmatrix} \quad \text{and} \quad \mathbf{v}_2 = \begin{bmatrix} -3 \\ 4 \end{bmatrix}$$

(a) Find the transition matrix from B' to B.
(b) Find the transition matrix from B to B'.
(c) Compute the coordinate matrix $[\mathbf{w}]_B$, where

$$\mathbf{w} = \begin{bmatrix} 3 \\ -5 \end{bmatrix}$$

and use (4.40b) to compute $[\mathbf{w}]_{B'}$.
(d) Check your work by computing $[\mathbf{w}]_{B'}$ directly.

9. Repeat the directions of Exercise 8 with

$$\mathbf{u}_1 = \begin{bmatrix} 2 \\ 2 \end{bmatrix} \quad \mathbf{u}_2 = \begin{bmatrix} 4 \\ -1 \end{bmatrix} \quad \mathbf{v}_1 = \begin{bmatrix} 1 \\ 3 \end{bmatrix} \quad \mathbf{v}_2 = \begin{bmatrix} -1 \\ -1 \end{bmatrix}$$

10. Consider the bases $B = \{\mathbf{u}_1, \mathbf{u}_2, \mathbf{u}_3\}$ and $B' = \{\mathbf{v}_1, \mathbf{v}_2, \mathbf{v}_3\}$ for R^3, where

$$\mathbf{u}_1 = \begin{bmatrix} -3 \\ 0 \\ -3 \end{bmatrix} \quad \mathbf{u}_2 = \begin{bmatrix} -3 \\ 2 \\ -1 \end{bmatrix} \quad \mathbf{u}_3 = \begin{bmatrix} 1 \\ 6 \\ -1 \end{bmatrix} \quad \mathbf{v}_1 = \begin{bmatrix} -6 \\ -6 \\ 0 \end{bmatrix}$$

$$\mathbf{v}_2 = \begin{bmatrix} -2 \\ -6 \\ 4 \end{bmatrix} \quad \mathbf{v}_3 = \begin{bmatrix} -2 \\ -3 \\ 7 \end{bmatrix}$$

(a) Find the transition matrix from B to B'.
(b) Compute the coordinate matrix $[\mathbf{w}]_B$, where

$$\mathbf{w} = \begin{bmatrix} -5 \\ 8 \\ -5 \end{bmatrix}$$

and use (4.40b) to compute $[\mathbf{w}]_{B'}$.
(c) Check your work by computing $[\mathbf{w}]_{B'}$ directly.

11. Repeat the directions of Exercise 10 with the same vector $\mathbf{w}$, but with

$$\mathbf{u}_1 = \begin{bmatrix} 2 \\ 1 \\ 1 \end{bmatrix} \quad \mathbf{u}_2 = \begin{bmatrix} 2 \\ -1 \\ 1 \end{bmatrix} \quad \mathbf{u}_3 = \begin{bmatrix} 1 \\ 2 \\ 1 \end{bmatrix} \quad \mathbf{v}_1 = \begin{bmatrix} 3 \\ 1 \\ -5 \end{bmatrix} \quad \mathbf{v}_2 = \begin{bmatrix} 1 \\ 1 \\ -3 \end{bmatrix} \quad \mathbf{v}_3 = \begin{bmatrix} -1 \\ 0 \\ 2 \end{bmatrix}$$

12. Consider the bases $B = \{\mathbf{p}_1, \mathbf{p}_2\}$ and $B' = \{\mathbf{q}_1, \mathbf{q}_2\}$ for P_1, where $\mathbf{p}_1 = 6 + 3x$, $\mathbf{p}_2 = 10 + 2x$, $\mathbf{q}_1 = 2$, $\mathbf{q}_2 = 3 + 2x$.
(a) Find the transition matrix from B' to B.
(b) Find the transition matrix from B to B'.

(c) Compute the coordinate matrix $[\mathbf{p}]_B$, where $\mathbf{p} = -4 + x$, and use (4.40b) to compute $[\mathbf{p}]_{B'}$.

(d) Check your work by computing $[\mathbf{p}]_{B'}$ directly.

13. Let V be the space spanned by $\mathbf{f}_1 = \sin x$ and $\mathbf{f}_2 = \cos x$.

(a) Show that $\mathbf{g}_1 = 2 \sin x + \cos x$ and $\mathbf{g}_2 = 3 \cos x$ form a basis for V.

(b) Find the transition matrix from $B' = \{\mathbf{g}_1, \mathbf{g}_2\}$ to $B = \{\mathbf{f}_1, \mathbf{f}_2\}$.

(c) Find the transition matrix from B to B'.

(d) Compute the coordinate matrix $[\mathbf{h}]_B$, where $\mathbf{h} = 2 \sin x - 5 \cos x$, and use (4.40b) to obtain $[\mathbf{h}]_{B'}$.

(e) Check your work by computing $[\mathbf{h}]_{B'}$ directly.

14. Let a rectangular $x'y'$-coordinate system be obtained by rotating a rectangular xy-coordinate system through an angle $\theta = 3\pi/4$.

(a) Find the $x'y'$-coordinates of the point whose xy-coordinates are $(-2, 6)$.

(b) Find the xy-coordinates of the point whose $x'y'$-coordinates are $(5, 2)$.

15. Repeat Exercise 14 with $\theta = \pi/3$.

16. Let a rectangular $x'y'z'$-coordinate system be obtained by rotating a rectangular xyz-coordinate system counterclockwise about the z-axis (looking down the z-axis) through an angle $\theta = \pi/4$.

(a) Find the $x'y'z'$-coordinates of the point whose xyz-coordinates are $(-1, 2, 5)$.

(b) Find the xyz-coordinates of the point whose $x'y'z'$-coordinates are $(1, 6, -3)$.

17. Repeat Exercise 16 for a rotation of $\theta = \pi/3$ counterclockwise about the y-axis (looking along the positive y-axis toward the origin).

18. Repeat Exercise 16 for a rotation of $\theta = 3\pi/4$ counterclockwise about the x-axis (looking along the positive x-axis toward the origin).

19. Use Theorem 33 to determine which of the following are orthogonal.

(a) $\begin{bmatrix} 1 & 0 \\ 0 & 1 \end{bmatrix}$

(b) $\begin{bmatrix} 1/\sqrt{2} & -1/\sqrt{2} \\ 1/\sqrt{2} & 1/\sqrt{2} \end{bmatrix}$

(c) $\begin{bmatrix} 0 & 1 & 1/\sqrt{2} \\ 1 & 0 & 0 \\ 0 & 0 & 1/\sqrt{2} \end{bmatrix}$

(d) $\begin{bmatrix} -1/\sqrt{2} & 1/\sqrt{6} & 1/\sqrt{3} \\ 0 & -2/\sqrt{6} & 1/\sqrt{3} \\ 1/\sqrt{2} & 1/\sqrt{6} & 1/\sqrt{3} \end{bmatrix}$

(e) $\begin{bmatrix} \frac{1}{2} & \frac{1}{2} & \frac{1}{2} & \frac{1}{2} \\ \frac{1}{2} & -\frac{5}{6} & \frac{1}{6} & \frac{1}{6} \\ \frac{1}{2} & \frac{1}{6} & \frac{1}{6} & -\frac{5}{6} \\ \frac{1}{2} & \frac{1}{6} & -\frac{5}{6} & \frac{1}{6} \end{bmatrix}$

(f) $\begin{bmatrix} 1 & 0 & 0 & 0 \\ 0 & 1/\sqrt{3} & -1/2 & 0 \\ 0 & 1/\sqrt{3} & 0 & 1 \\ 0 & 1/\sqrt{3} & 1/2 & 0 \end{bmatrix}$

20. Find the inverse of those matrices in Exercise 19 which are orthogonal.

21. Show that the following are orthogonal matrices for every value of θ.

(a) $\begin{bmatrix} \cos\theta & -\sin\theta \\ \sin\theta & \cos\theta \end{bmatrix}$
(b) $\begin{bmatrix} \cos\theta & -\sin\theta & 0 \\ \sin\theta & \cos\theta & 0 \\ 0 & 0 & 1 \end{bmatrix}$

22. Find the inverses of the matrices in Exercise 21.

23. Consider the orthogonal coordinate transformation

$$\begin{bmatrix} x \\ y \end{bmatrix} = \begin{bmatrix} -\frac{3}{5} & -\frac{4}{5} \\ \frac{4}{5} & -\frac{3}{5} \end{bmatrix}\begin{bmatrix} x' \\ y' \end{bmatrix}$$

Find (x', y') for the points with the following (x, y) coordinates.
(a) $(2, -1)$ (b) $(4, 2)$ (c) $(-7, -8)$ (d) $(0, 0)$

24. Sketch the xy-axes and the $x'y'$-axes for the coordinate transformation in Exercise 23.

25. For which of the following is $\mathbf{x} = P\mathbf{x}'$ a rotation?

(a) $P = \begin{bmatrix} 1/\sqrt{2} & 1/\sqrt{2} \\ -1/\sqrt{2} & 1/\sqrt{2} \end{bmatrix}$
(b) $P = \begin{bmatrix} -1/\sqrt{2} & 1/\sqrt{2} \\ 1/\sqrt{2} & 1/\sqrt{2} \end{bmatrix}$

(c) $P = \begin{bmatrix} \frac{3}{5} & \frac{4}{5} \\ -\frac{4}{5} & \frac{3}{5} \end{bmatrix}$
(d) $P = \begin{bmatrix} -\frac{3}{5} & -\frac{4}{5} \\ -\frac{4}{5} & \frac{3}{5} \end{bmatrix}$

26. Sketch the xy-axes and the $x'y'$-axes for the coordinate transformations in Exercise 25.

27. Consider the orthogonal coordinate transformation

$$\begin{bmatrix} x \\ y \\ z \end{bmatrix} = \begin{bmatrix} \frac{4}{5} & -\frac{3}{5} & 0 \\ \frac{3}{5} & \frac{4}{5} & 0 \\ 0 & 0 & 1 \end{bmatrix}\begin{bmatrix} x' \\ y' \\ z' \end{bmatrix}$$

Find (x', y', z') for the points with the following (x, y, z) coordinates.
(a) $(3, 0, -7)$ (b) $(1, 2, 6)$ (c) $(-9, -2, -3)$ (d) $(0, 0, 0)$

28. Sketch the xyz-axes and the $x'y'z'$-axes for the coordinate transformation in Exercise 27.

29. For which of the following is $\mathbf{x} = P\mathbf{x}'$ a rotation?

(a) $P = \begin{bmatrix} \frac{4}{5} & 0 & -\frac{3}{5} \\ \frac{3}{5} & 0 & \frac{4}{5} \\ 0 & 1 & 0 \end{bmatrix}$
(b) $P = \begin{bmatrix} \frac{6}{7} & \frac{2}{7} & \frac{3}{7} \\ \frac{2}{7} & \frac{3}{7} & -\frac{6}{7} \\ -\frac{3}{7} & \frac{6}{7} & \frac{2}{7} \end{bmatrix}$

30. Sketch the xyz-axes and the $x'y'z'$-axes for the coordinate transformations in Exercise 29.

31. (a) A rectangular $x'y'z'$-coordinate system is obtained by rotating an xyz-coordinate system counterclockwise about the y-axis through an angle θ (looking along the positive y-axis toward the origin). Find a matrix A such that

$$\begin{bmatrix} x' \\ y' \\ z' \end{bmatrix} = A \begin{bmatrix} x \\ y \\ z \end{bmatrix}$$

where (x, y, z) and (x', y', z') are the coordinates of a point in the xyz and $x'y'z'$-systems, respectively.

(b) Repeat part (a) for a rotation about the x-axis.

32. A rectangular $x''y''z''$-coordinate system is obtained by first rotating a rectangular xyz-coordinate system $60°$ counterclockwise about the z-axis (looking down the positive z-axis) to obtain an $x'y'z'$-coordinate system, and then rotating the $x'y'z'$-coordinate system $45°$ counterclockwise about the y'-axis (looking along the positive y'-axis toward the origin). Find a matrix A such that

$$\begin{bmatrix} x'' \\ y'' \\ z'' \end{bmatrix} = A \begin{bmatrix} x \\ y \\ z \end{bmatrix}$$

where (x, y, z) and (x'', y'', z'') are the xyz and $x''y''z''$-coordinates of a point.

33. Prove that if A is an orthogonal matrix, then A^t is also orthogonal.

34. Prove that an $n \times n$ matrix is orthogonal if and only if its rows form an orthonormal set in R^n.

35. Use Exercises 33 and 34 to prove that an $n \times n$ matrix is orthogonal if and only if its columns form an orthonormal set in R^n.

36. Prove that if P is an orthogonal matrix, then $\det(P) = 1$ or -1.

37. Prove Theorem 30a.

38. Prove Theorem 30b.

39. Prove Theorem 30c.

40. Let S be a basis for an n-dimensional vector space V. Show that if $\mathbf{v}_1, \mathbf{v}_2, \ldots, \mathbf{v}_r$ form a linearly independent set of vectors in V, then the coordinate vectors $(\mathbf{v}_1)_S, (\mathbf{v}_2)_S, \ldots, (\mathbf{v}_r)_S$ form a linearly independent set in R^n, and conversely.

41. Using the notation from Exercise 40, show that if $\mathbf{v}_1, \mathbf{v}_2, \ldots, \mathbf{v}_r$ span V, then the coordinate vectors $(\mathbf{v}_1)_S, (\mathbf{v}_2)_S, \ldots, (\mathbf{v}_r)_S$ span R^n, and conversely.

42. Find a basis for the subspace of P_2 spanned by the given vectors.
(a) $-1 + x - 2x^2$, $\quad 3 + 3x + 6x^2$, $\quad 9$
(b) $1 + x$, $\quad x^2$, $\quad -2 + 2x^2$, $\quad -3x$
(c) $1 + x - 3x^2$, $\quad 2 + 2x - 6x^2$, $\quad 3 + 3x - 9x^2$
(**Hint.** Let S be the standard basis for P_2 and work with the coordinate vectors relative to S; note Exercises 40, 41.)

SUPPLEMENTARY EXERCISES

1. In each part, the solution space is a subspace of R^3 and so must be a line through the origin, a plane through the origin, all of R^3, or the origin only. (See the remark following Example 11.) For each system, determine which is the case. If the subspace is a plane, find an equation for it, and if a line, find parametric equations.

(a) $0x + 0y + 0z = 0$

(b) $\begin{aligned} 2x - 3y + z &= 0 \\ 6x - 9y + 3z &= 0 \\ -4x + 6y - 2z &= 0 \end{aligned}$

(c) $\begin{aligned} x - 2y + 7z &= 0 \\ -4x + 8y + 5z &= 0 \\ 2x - 4y + 3z &= 0 \end{aligned}$

(d) $\begin{aligned} x + 4y + 8z &= 0 \\ 2x + 5y + 6z &= 0 \\ 3x + y - 4z &= 0 \end{aligned}$

2. Let R^4 have the Euclidean inner product.
   (a) Find a vector in R^4 that is orthogonal to $\mathbf{u}_1 = (1, 0, 0, 0)$ and $\mathbf{u}_4 = (0, 0, 0, 1)$ and makes equal angles with $\mathbf{u}_2 = (0, 1, 0, 0)$ and $\mathbf{u}_3 = (0, 0, 1, 0)$.
   (b) Find a vector $\mathbf{x} = (x_1, x_2, x_3, x_4)$ of length 1 that is orthogonal to $\mathbf{u}_1$ and $\mathbf{u}_4$ above and such that the cosine of the angle between $\mathbf{x}$ and $\mathbf{u}_2$ is twice the cosine of the angle between $\mathbf{x}$ and $\mathbf{u}_3$.

3. (a) Let A be an $m \times n$ matrix with row vectors $\mathbf{r}_1, \mathbf{r}_2, \ldots, \mathbf{r}_m$ and B an $n \times p$ matrix with column vectors $\mathbf{c}_1, \mathbf{c}_2, \ldots, \mathbf{c}_p$. Show that the entry in row i and column j of AB is the Euclidean inner product of $\mathbf{r}_i$ and $\mathbf{c}_j$.
   (b) Use the result in (a) to show that if A is an $n \times n$ matrix such that $AA^t = I$, then the row vectors of A form an orthonormal set in R^n with the Euclidean inner product.
   (c) Show that if A is an $n \times n$ matrix and $AA^t = I$, then the column vectors of A form an orthonormal set in R^n with the Euclidean inner product. [**Hint.** Show that $A^t A = I$ and apply part (b) to A^t.]

4. Let $A\mathbf{x} = \mathbf{0}$ be a system of m equations in n unknowns. Show that

$$\mathbf{x} = \begin{bmatrix} x_1 \\ x_2 \\ \vdots \\ x_n \end{bmatrix}$$

is a solution of the system if and only if the vector $\mathbf{x} = (x_1, x_2, \ldots, x_n)$ is orthogonal to every row vector of A in the Euclidean inner product on R^n. [**Hint.** Use Supplementary Exercise 3a.]

5. Use the Cauchy-Schwarz inequality to show that if $a_1, a_2, \ldots, a_n$ are positive real numbers, then

$$(a_1 + a_2 + \cdots + a_n)\left(\frac{1}{a_1} + \frac{1}{a_2} + \cdots + \frac{1}{a_n}\right) \geq n^2$$

6. Let W be the space spanned by $\mathbf{f} = \sin x$ and $\mathbf{g} = \cos x$.
   (a) Show that for any value of θ, $\mathbf{f}_1 = \sin(x + \theta)$ and $\mathbf{g}_1 = \cos(x + \theta)$ are vectors in W.
   (b) Show that $\mathbf{f}_1$ and $\mathbf{g}_1$ form a basis for W.

7. (a) Express $(4a, a - b, a + 2b)$ as a linear combination of $(4, 1, 1)$ and $(0, -1, 2)$.

(b) Express $(3a + b + 3c, -a + 4b - c, 2a + b + 2c)$ as a linear combination of $(3, -1, 2)$ and $(1, 4, 1)$.

(c) Express $(2a - b + 4c, 3a - c, 4b + c)$ as a linear combination of three nonzero vectors.

8. (a) Express $\mathbf{v} = (1, 1)$ as a linear combination of $\mathbf{v}_1 = (1, -1)$, $\mathbf{v}_2 = (3, 0)$, $\mathbf{v}_3 = (2, 1)$ in two different ways.

(b) Show that this does not violate Theorem 29.

9. Show that if $\mathbf{x}$ and $\mathbf{y}$ are vectors in an inner product space and c is any scalar, then

$$\|c\mathbf{x} + \mathbf{y}\|^2 = c^2\|\mathbf{x}\|^2 + 2c\langle \mathbf{x}, \mathbf{y}\rangle + \|\mathbf{y}\|^2$$

10. Let R^3 have the Euclidean inner product. Find two vectors of length 1 that are orthogonal to all three of the vectors $\mathbf{u}_1 = (1, 1, -1)$, $\mathbf{u}_2 = (-2, -1, 2)$, and $\mathbf{u}_3 = (-1, 0, 1)$.

11. Find a weighted Euclidean inner product on R^n such that the vectors

$$\mathbf{v}_1 = (1, 0, 0, \ldots, 0)$$
$$\mathbf{v}_2 = (0, \sqrt{2}, 0, \ldots, 0)$$
$$\mathbf{v}_3 = (0, 0, \sqrt{3}, \ldots, 0)$$
$$\vdots$$
$$\mathbf{v}_n = (0, 0, 0, \ldots, \sqrt{n})$$

form an orthonormal set.

12. Is there a weighted Euclidean inner product on R^2 for which the vectors $(1, 2)$ and $(3, -1)$ form an orthonormal set? Justify your answer.

13. (a) Prove: If $\mathbf{v}$ is orthogonal to $\mathbf{u}_1$ and $\mathbf{u}_2$, then $\mathbf{v}$ is orthogonal to every vector in lin $(\mathbf{u}_1, \mathbf{u}_2)$.

(b) Give a geometric interpretation of this result in R^3.

14. If $\mathbf{u}$ and $\mathbf{v}$ are vectors in an inner product space V, then $\mathbf{u}$, $\mathbf{v}$, and $\mathbf{u} - \mathbf{v}$ can be regarded as sides of a "triangle" in V (Figure 4.23). Prove that the law of cosines holds for any such triangle; that is,

$$\|\mathbf{u} - \mathbf{v}\|^2 = \|\mathbf{u}\|^2 + \|\mathbf{v}\|^2 - 2\|\mathbf{u}\|\,\|\mathbf{v}\| \cos \theta$$

where θ is the angle between $\mathbf{u}$ and $\mathbf{v}$.

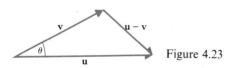

Figure 4.23

15. (a) In R^3 the vectors $(k, 0, 0)$, $(0, k, 0)$, and $(0, 0, k)$ form the edges of a cube with diagonal (k, k, k) (Figure 3.21 of Chapter 3). Similarly, in R^n the vectors

$$(k, 0, 0, \ldots, 0), \quad (0, k, 0, \ldots, 0), \ldots, (0, 0, 0, \ldots, k)$$

can be regarded as edges of a "cube" with diagonal $(k, k, k, \ldots, k)$. Show that each of the above edges makes an angle of θ with the diagonal, where

$$\cos \theta = \frac{1}{\sqrt{n}}$$

(b) **(For readers who have studied calculus.)** What happens to the angle θ in part (a) as the dimension of R^n approaches $+\infty$?

16. Let **u** and **v** be vectors in an inner product space.
    (a) Prove that $\|\mathbf{u}\| = \|\mathbf{v}\|$ if and only if $\mathbf{u} + \mathbf{v}$ and $\mathbf{u} - \mathbf{v}$ are orthogonal.
    (b) Give a geometric interpretation of this result in R^2 with the Euclidean inner product.

17. Let **u** be a vector in an inner product space V, and let $\{\mathbf{v}_1, \mathbf{v}_2, \ldots, \mathbf{v}_n\}$ be an orthonormal basis for V. Show that if α_i is the angle between **u** and $\mathbf{v}_i$, then

$$\cos^2 \alpha_1 + \cos^2 \alpha_2 + \cdots + \cos^2 \alpha_n = 1$$

18. Prove: If $\langle \mathbf{u}, \mathbf{v} \rangle_1$ and $\langle \mathbf{u}, \mathbf{v} \rangle_2$ are two inner products on a vector space V, then $\langle \mathbf{u}, \mathbf{v} \rangle = \langle \mathbf{u}, \mathbf{v} \rangle_1 + \langle \mathbf{u}, \mathbf{v} \rangle_2$ is also an inner product.

19. Show that the inner product on R^n generated by any orthogonal matrix is the Euclidean inner product.

20. Let A be an $n \times n$ matrix and let $\mathbf{v}_1, \mathbf{v}_2, \ldots, \mathbf{v}_n$ be linearly independent vectors in R^n expressed as $n \times 1$ matrices. What must be true about A for $A\mathbf{v}_1, A\mathbf{v}_2, \ldots, A\mathbf{v}_n$ to be linearly independent?

21. Must a basis for P_n contain a polynomial of degree k for each $k = 0, 1, 2, \ldots, n$? Justify your answer.

22. In advanced linear algebra one proves the following determinant criterion for rank: *The rank of a matrix A is r if and only if A has some $r \times r$ submatrix with a nonzero determinant, and all square submatrices of larger size have determinant zero.* (A submatrix of A is any matrix obtained by deleting rows or columns of A. The matrix A itself is also considered to be a submatrix of A.) In each part use this criterion to find the rank of the matrix.

(a) $\begin{bmatrix} 1 & 2 & 0 \\ 2 & 4 & -1 \end{bmatrix}$ (b) $\begin{bmatrix} 1 & 2 & 3 \\ 2 & 4 & 6 \end{bmatrix}$

(c) $\begin{bmatrix} 1 & 0 & 1 \\ 2 & -1 & 3 \\ 3 & -1 & 4 \end{bmatrix}$ (d) $\begin{bmatrix} 1 & -1 & 2 & 0 \\ 3 & 1 & 0 & 0 \\ -1 & 2 & 4 & 0 \end{bmatrix}$

23. Use the result in Exercise 22 to find the possible ranks for matrices of the form

$$\begin{bmatrix} 0 & 0 & 0 & 0 & 0 & a_{16} \\ 0 & 0 & 0 & 0 & 0 & a_{26} \\ 0 & 0 & 0 & 0 & 0 & a_{36} \\ 0 & 0 & 0 & 0 & 0 & a_{46} \\ a_{51} & a_{52} & a_{53} & a_{54} & a_{55} & a_{56} \end{bmatrix}$$

24. Prove: If S is a basis for a vector space V, then for any vectors **u** and **v** in V and any scalar k the following relationships hold:
    (a) $[\mathbf{u} + \mathbf{v}]_S = [\mathbf{u}]_S + [\mathbf{v}]_S$
    (b) $[k\mathbf{u}]_S = k[\mathbf{u}]_S$

CHAPTER FIVE

Linear Transformations

5.1 INTRODUCTION TO LINEAR TRANSFORMATIONS

In this section we begin the study of vector-valued functions of a vector variable—that is, functions having the form $\mathbf{w} = F(\mathbf{v})$, where the independent variable $\mathbf{v}$ and the dependent variable $\mathbf{w}$ are both vectors. We shall concentrate on a special class of vector functions called *linear transformations*. These have many important applications in physics, engineering, social sciences, and various branches of mathematics.

If V and W are vector spaces and F is a function that associates a unique vector in W with each vector in V, we say F *maps* V into W, and write $F: V \rightarrow W$. Further, if F associates the vector $\mathbf{w}$ with the vector $\mathbf{v}$, we write $\mathbf{w} = F(\mathbf{v})$ and say that $\mathbf{w}$ is the *image* of $\mathbf{v}$ under F. The vector space V is called the *domain* of F.

For example, if $\mathbf{v} = (x, y)$ is a vector in R^2, then the formula

$$F(\mathbf{v}) = (x, x + y, x - y) \tag{5.1}$$

defines a function that maps R^2 into R^3. In particular, if $\mathbf{v} = (1, 1)$, then $x = 1$ and $y = 1$, so the image of $\mathbf{v}$ under F is $F(\mathbf{v}) = (1, 2, 0)$. The domain of F is R^2.

Definition. If $F: V \rightarrow W$ is a function from the vector space V into the vector space W, then F is called a *linear transformation* if

(i) $F(\mathbf{u} + \mathbf{v}) = F(\mathbf{u}) + F(\mathbf{v})$ for all vectors $\mathbf{u}$ and $\mathbf{v}$ in V
(ii) $F(k\mathbf{u}) = kF(\mathbf{u})$ for all vectors $\mathbf{u}$ in V and all scalars k.

For example, let $F:R^2 \to R^3$ be the function defined by (5.1). If $\mathbf{u} = (x_1, y_1)$ and $\mathbf{v} = (x_2, y_2)$, then $\mathbf{u} + \mathbf{v} = (x_1 + x_2, y_1 + y_2)$, so that

$$F(\mathbf{u} + \mathbf{v}) = (x_1 + x_2, [x_1 + x_2] + [y_1 + y_2], [x_1 + x_2] - [y_1 + y_2]) \quad \in R^3$$
$$= (x_1, x_1 + y_1, x_1 - y_1) + (x_2, x_2 + y_2, x_2 - y_2) \quad - \text{a vector } R^3$$
$$= F(\mathbf{u}) + F(\mathbf{v})$$

Also, if k is a scalar, $k\mathbf{u} = (kx_1, ky_1)$, so that

$$F(k\mathbf{u}) = (kx_1, kx_1 + ky_1, kx_1 - ky_1) \quad \text{a vector } R^3$$
$$= k(x_1, x_1 + y_1, x_1 - y_1)$$
$$= kF(\mathbf{u})$$

Thus, F is a linear transformation.

If $F:V \to W$ is a linear transformation, then for any $\mathbf{v}_1$ and $\mathbf{v}_2$ in V and any scalars k_1 and k_2, we have

$$F(k_1\mathbf{v}_1 + k_2\mathbf{v}_2) = F(k_1\mathbf{v}_1) + F(k_2\mathbf{v}_2) = k_1 F(\mathbf{v}_1) + k_2 F(\mathbf{v}_2)$$

Similarly, if $\mathbf{v}_1, \mathbf{v}_2, \ldots, \mathbf{v}_n$ are vectors in V and $k_1, k_2, \ldots, k_n$ are scalars, then

$$F(k_1\mathbf{v}_1 + k_2\mathbf{v}_2 + \cdots + k_n\mathbf{v}_n) = k_1 F(\mathbf{v}_1) + k_2 F(\mathbf{v}_2) + \cdots + k_n F(\mathbf{v}_n) \qquad (5.2)$$

We now give some further examples of linear transformations.

Example 1

Let A be a fixed $m \times n$ matrix. If we use matrix notation for vectors in R^m and R^n, then we can define a function $T:R^n \to R^m$ by

$$T(\mathbf{x}) = A\mathbf{x}$$

Observe that if $\mathbf{x}$ is an $n \times 1$ matrix, then the product $A\mathbf{x}$ is an $m \times 1$ matrix; thus T maps R^n into R^m. Moreover, T is linear; to see this, let $\mathbf{u}$ and $\mathbf{v}$ be $n \times 1$ matrices and let k be a scalar. Using properties of matrix multiplication, we obtain

$$A(\mathbf{u} + \mathbf{v}) = A\mathbf{u} + A\mathbf{v} \qquad \text{and} \qquad A(k\mathbf{u}) = k(A\mathbf{u})$$

or equivalently

$$T(\mathbf{u} + \mathbf{v}) = T(\mathbf{u}) + T(\mathbf{v}) \qquad \text{and} \qquad T(k\mathbf{u}) = kT(\mathbf{u})$$

We shall call the linear transformation in this example *multiplication by A*. Linear transformations of this kind are called *matrix transformations*.

Example 2

As a special case of the previous example, let θ be a fixed angle, and let $T:R^2 \to R^2$ be multiplication by the matrix

$$A = \begin{bmatrix} \cos\theta & -\sin\theta \\ \sin\theta & \cos\theta \end{bmatrix}$$

If **v** is the vector

$$\mathbf{v} = \begin{bmatrix} x \\ y \end{bmatrix}$$

then

$$T(\mathbf{v}) = A\mathbf{v} = \begin{bmatrix} \cos\theta & -\sin\theta \\ \sin\theta & \cos\theta \end{bmatrix}\begin{bmatrix} x \\ y \end{bmatrix} = \begin{bmatrix} x\cos\theta - y\sin\theta \\ x\sin\theta + y\cos\theta \end{bmatrix}$$

Geometrically, $T(\mathbf{v})$ is the vector that results if **v** is rotated through an angle θ. To see this, let ϕ be the angle between **v** and the positive x axis, and let

$$\mathbf{v}' = \begin{bmatrix} x' \\ y' \end{bmatrix}$$

be the vector that results when **v** is rotated through an angle θ (Figure 5.1). We shall show $\mathbf{v}' = T(\mathbf{v})$. If r denotes the length of **v**, then

$$x = r\cos\phi \qquad y = r\sin\phi$$

Similarly, since $\mathbf{v}'$ has the same length as **v**, we have

$$x' = r\cos(\theta + \phi) \qquad y' = r\sin(\theta + \phi)$$

Therefore,

$$\mathbf{v}' = \begin{bmatrix} x' \\ y' \end{bmatrix} = \begin{bmatrix} r\cos(\theta + \phi) \\ r\sin(\theta + \phi) \end{bmatrix}$$

$$= \begin{bmatrix} r\cos\theta\cos\phi - r\sin\theta\sin\phi \\ r\sin\theta\cos\phi + r\cos\theta\sin\phi \end{bmatrix}$$

$$= \begin{bmatrix} x\cos\theta - y\sin\theta \\ x\sin\theta + y\cos\theta \end{bmatrix}$$

$$= \begin{bmatrix} \cos\theta & -\sin\theta \\ \sin\theta & \cos\theta \end{bmatrix}\begin{bmatrix} x \\ y \end{bmatrix}$$

$$= A\mathbf{v} = T(\mathbf{v})$$

The linear transformation in this example is called the ***rotation of R^2 through the angle θ***.

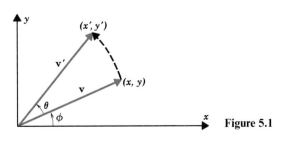

Figure 5.1

Example 3

Let V and W be any two vector spaces. The mapping $T:V \to W$ such that $T(\mathbf{v}) = \mathbf{0}$ for every $\mathbf{v}$ in V is a linear transformation called the *zero transformation*. To see that T is linear, observe that

$$T(\mathbf{u} + \mathbf{v}) = \mathbf{0}, \ T(\mathbf{u}) = \mathbf{0}, \ T(\mathbf{v}) = \mathbf{0}, \quad \text{and} \quad T(k\mathbf{u}) = \mathbf{0}$$

Therefore,

$$T(\mathbf{u} + \mathbf{v}) = T(\mathbf{u}) + T(\mathbf{v}) \quad \text{and} \quad T(k\mathbf{u}) = kT(\mathbf{u})$$

Example 4

Let V be any vector space. The mapping $T:V \to V$ defined by $T(\mathbf{v}) = \mathbf{v}$ is called the *identity transformation* on V. The verification that T is linear is left as an exercise.

If, as in Examples 2 and 4, $T:V \to V$ is a linear transformation from a vector space V into itself, then T is called a *linear operator* on V.

Example 5

Let V be any vector space and k any fixed scalar. We leave it as an exercise to check that the function $T:V \to V$ defined by

$$T(\mathbf{v}) = k\mathbf{v}$$

is a linear operator on V. If $k > 1$, T is called a *dilation* of V and if $0 < k < 1$, then T is called a *contraction* of V. Geometrically, a dilation "stretches" each vector in V by a factor of k, and a contraction of V "compresses" each vector by a factor of k (Figure 5.2).

Example 6

Let V be an inner product space, and suppose W is a finite dimensional subspace of V having

$$S = \{\mathbf{w}_1, \mathbf{w}_2, \ldots, \mathbf{w}_r\}$$

as an orthonormal basis. Let $T:V \to W$ be the function that maps a vector $\mathbf{v}$ in V into its orthogonal projection on W (Section 4.9); that is,

$$T(\mathbf{v}) = \langle \mathbf{v}, \mathbf{w}_1 \rangle \mathbf{w}_1 + \langle \mathbf{v}, \mathbf{w}_2 \rangle \mathbf{w}_2 + \cdots + \langle \mathbf{v}, \mathbf{w}_r \rangle \mathbf{w}_r$$

(See Figure 5.3.) The mapping T is called the *orthogonal projection of V onto W*; its linearity follows from the basic properties of the inner product. For example,

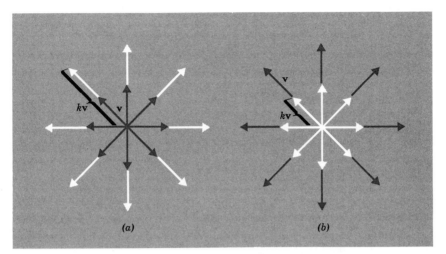

Figure 5.2 (*a*) Dilation of *V*. (*b*) Contraction of *V*.

$$T(\mathbf{u} + \mathbf{v}) = \langle \mathbf{u} + \mathbf{v}, \mathbf{w}_1 \rangle \mathbf{w}_1 + \langle \mathbf{u} + \mathbf{v}, \mathbf{w}_2 \rangle \mathbf{w}_2 + \cdots + \langle \mathbf{u} + \mathbf{v}, \mathbf{w}_r \rangle \mathbf{w}_r$$
$$= \langle \mathbf{u}, \mathbf{w}_1 \rangle \mathbf{w}_1 + \langle \mathbf{u}, \mathbf{w}_2 \rangle \mathbf{w}_2 + \cdots + \langle \mathbf{u}, \mathbf{w}_r \rangle \mathbf{w}_r$$
$$+ \langle \mathbf{v}, \mathbf{w}_1 \rangle \mathbf{w}_1 + \langle \mathbf{v}, \mathbf{w}_2 \rangle \mathbf{w}_2 + \cdots + \langle \mathbf{v}, \mathbf{w}_r \rangle \mathbf{w}_r$$
$$= T(\mathbf{u}) + T(\mathbf{v})$$

Similarly, $T(k\mathbf{u}) = kT(\mathbf{u})$.

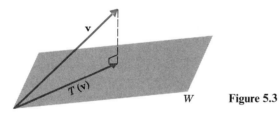

Figure 5.3

Example 7

As a special case of the previous example, let $V = R^3$ have the Euclidean inner product. The vectors $\mathbf{w}_1 = (1, 0, 0)$ and $\mathbf{w}_2 = (0, 1, 0)$ form an orthonormal basis for the xy-plane. Thus, if $\mathbf{v} = (x, y, z)$ is any vector in R^3, the orthogonal projection of R^3 onto the xy-plane is given by

$$T(\mathbf{v}) = \langle \mathbf{v}, \mathbf{w}_1 \rangle \mathbf{w}_1 + \langle \mathbf{v}, \mathbf{w}_2 \rangle \mathbf{w}_2$$
$$= x(1, 0, 0) + y(0, 1, 0)$$
$$= (x, y, 0)$$

(See Figure 5.4.)

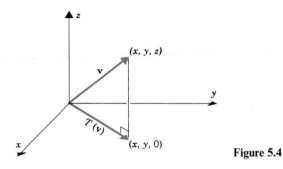

Figure 5.4

Example 8

Let V be an n-dimensional vector space and $S = \{\mathbf{w}_1, \mathbf{w}_2, \ldots, \mathbf{w}_n\}$ a fixed basis for V. By Theorem 29 of Section 4.10 any two vectors $\mathbf{u}$ and $\mathbf{v}$ in V can be written uniquely in the form

$$\mathbf{u} = c_1\mathbf{w}_1 + c_2\mathbf{w}_2 + \cdots + c_n\mathbf{w}_n \quad \text{and} \quad \mathbf{v} = d_1\mathbf{w}_1 + d_2\mathbf{w}_2 + \cdots + d_n\mathbf{w}_n$$

Thus,

$$(\mathbf{u})_S = (c_1, c_2, \ldots, c_n)$$
$$(\mathbf{v})_S = (d_1, d_2, \ldots, d_n)$$

But

$$\mathbf{u} + \mathbf{v} = (c_1 + d_1)\mathbf{w}_1 + (c_2 + d_2)\mathbf{w}_2 + \cdots + (c_n + d_n)\mathbf{w}_n$$
$$k\mathbf{u} = (kc_1)\mathbf{w}_1 + (kc_2)\mathbf{w}_2 + \cdots + (kc_n)\mathbf{w}_n$$

so that

$$(\mathbf{u} + \mathbf{v})_S = (c_1 + d_1, c_2 + d_2, \ldots, c_n + d_n)$$
$$(k\mathbf{u})_S = (kc_1, kc_2, \ldots, kc_n)$$

Therefore,

$$(\mathbf{u} + \mathbf{v})_S = (\mathbf{u})_S + (\mathbf{v})_S \quad \text{and} \quad (k\mathbf{u})_S = k(\mathbf{u})_S \tag{5.3a}$$

Similarly, for coordinate matrices we have

$$[\mathbf{u} + \mathbf{v}]_S = [\mathbf{u}]_S + [\mathbf{v}]_S \quad \text{and} \quad [k\mathbf{u}]_S = k[\mathbf{u}]_S \tag{5.3b}$$

Suppose we let $T: V \to R^n$ be the function that maps a vector $\mathbf{v}$ in V into its coordinate vector with respect to S; that is,

$$T(\mathbf{v}) = (\mathbf{v})_S$$

Then in terms of T, (5.3a) states that

$$T(\mathbf{u} + \mathbf{v}) = T(\mathbf{u}) + T(\mathbf{v})$$

and

$$T(k\mathbf{u}) = kT(\mathbf{u})$$

Thus, T is a linear transformation from V into R^n.

Example 9

Let V be an inner product space and let $\mathbf{v_0}$ be any fixed vector in V. Let $T:V \rightarrow R$ be the transformation that maps a vector $\mathbf{v}$ into its inner product with $\mathbf{v_0}$; that is,

$$T(\mathbf{v}) = \langle \mathbf{v}, \mathbf{v_0} \rangle$$

From the properties of an inner product

$$T(\mathbf{u} + \mathbf{v}) = \langle \mathbf{u} + \mathbf{v}, \mathbf{v_0} \rangle = \langle \mathbf{u}, \mathbf{v_0} \rangle + \langle \mathbf{v}, \mathbf{v_0} \rangle = T(\mathbf{u}) + T(\mathbf{v})$$

and

$$T(k\mathbf{u}) = \langle k\mathbf{u}, \mathbf{v_0} \rangle = k\langle \mathbf{u}, \mathbf{v_0} \rangle = kT(\mathbf{u})$$

so that T is a linear transformation. ▲

Example 10   (For readers who have studied calculus.)

Let $V = C[0, 1]$ be the vector space of all real-valued functions continuous on the interval $0 \leq x \leq 1$, and let W be the subspace of $C[0, 1]$ consisting of all functions with continuous first derivatives on the interval $0 \leq x \leq 1$.

Let $D:W \rightarrow V$ be the transformation that maps $\mathbf{f}$ into its derivative; that is,

$$D(\mathbf{f}) = \mathbf{f}'$$

From the properties of differentiation, we have

$$D(\mathbf{f} + \mathbf{g}) = D(\mathbf{f}) + D(\mathbf{g})$$

and

$$D(k\mathbf{f}) = kD(\mathbf{f})$$

Thus, D is a linear transformation. ▲

Example 11   (For readers who have studied calculus.)

Let $V = C[0, 1]$ be as in the previous example, and let $J:V \rightarrow R$ be defined by

$$J(\mathbf{f}) = \int_0^1 f(x)\, dx$$

For example, if $f(x) = x^2$, then

$$J(\mathbf{f}) = \int_0^1 x^2\, dx = \frac{1}{3}$$

Since

$$\int_0^1 (f(x) + g(x))\, dx = \int_0^1 f(x)\, dx + \int_0^1 g(x)\, dx$$

and

$$\int_0^1 kf(x)\, dx = k \int_0^1 f(x)\, dx$$

for any constant k, it follows that

$$J(\mathbf{f} + \mathbf{g}) = J(\mathbf{f}) + J(\mathbf{g})$$
$$J(k\mathbf{f}) = kJ(\mathbf{f})$$

Thus, J is a linear transformation. ▲

EXERCISE SET 5.1

In Exercises 1–8 a formula is given for a function $F:R^2 \rightarrow R^2$. In each exercise determine whether F is linear.

1. $F(x, y) = (2x, y)$ **2.** $F(x, y) = (x^2, y)$

3. $F(x, y) = (y, x)$ **4.** $F(x, y) = (0, y)$

5. $F(x, y) = (x, y + 1)$ **6.** $F(x, y) = (2x + y, x - y)$

7. $F(x, y) = (y, y)$ **8.** $F(x, y) = (\sqrt[3]{x}, \sqrt[3]{y})$

In Exercises 9–12 a formula is given for a function $F:R^3 \rightarrow R^2$. In each exercise determine whether F is linear.

9. $F(x, y, z) = (x, x + y + z)$ **10.** $F(x, y, z) = (0, 0)$

11. $F(x, y, z) = (1, 1)$ **12.** $F(x, y, z) = (2x + y, 3y - 4z)$

In Exercises 13–16 a formula is given for a function $F:M_{22} \rightarrow R$. In each exercise determine whether F is linear.

13. $F\left(\begin{bmatrix} a & b \\ c & d \end{bmatrix}\right) = a + d$ **14.** $F\left(\begin{bmatrix} a & b \\ c & d \end{bmatrix}\right) = \det\begin{bmatrix} a & b \\ c & d \end{bmatrix}$

15. $F\left(\begin{bmatrix} a & b \\ c & d \end{bmatrix}\right) = 2a + 3b + c - d$ **16.** $F\left(\begin{bmatrix} a & b \\ c & d \end{bmatrix}\right) = a^2 + b^2$

In Exercises 17–20 a formula is given for a function $F:P_2 \rightarrow P_2$. In each exercise determine whether F is linear.

17. $F(a_0 + a_1x + a_2x^2) = a_0 + (a_1 + a_2)x + (2a_0 - 3a_1)x^2$

18. $F(a_0 + a_1x + a_2x^2) = a_0 + a_1(x + 1) + a_2(x + 1)^2$

19. $F(a_0 + a_1x + a_2x^2) = 0$

20. $F(a_0 + a_1x + a_2x^2) = (a_0 + 1) + a_1x + a_2x^2$

21. Let $F:R^2 \rightarrow R^2$ be the function that maps each point in the plane into its reflection about the y-axis. Find a formula for F and show that F is a linear operator on R^2.

22. Let B be a fixed 2×3 matrix. Show that the function $T:M_{22} \rightarrow M_{23}$ defined by $T(A) = AB$ is a linear transformation.

23. Let $T: R^3 \rightarrow R^2$ be a matrix transformation, and suppose

$$T\left(\begin{bmatrix} 1 \\ 0 \\ 0 \end{bmatrix}\right) = \begin{bmatrix} 1 \\ 1 \end{bmatrix}, \quad T\left(\begin{bmatrix} 0 \\ 1 \\ 0 \end{bmatrix}\right) = \begin{bmatrix} 3 \\ 0 \end{bmatrix}, \quad T\left(\begin{bmatrix} 0 \\ 0 \\ 1 \end{bmatrix}\right) = \begin{bmatrix} 4 \\ -7 \end{bmatrix}$$

(a) Find the matrix.

(b) Find $T\left(\begin{bmatrix} 1 \\ 3 \\ 8 \end{bmatrix}\right)$

(c) Find $T\left(\begin{bmatrix} x \\ y \\ z \end{bmatrix}\right)$

24. Let $T: R^3 \rightarrow W$ be the orthogonal projection of R^3 onto the xz-plane W.
   (a) Find a formula for T.
   (b) Find $T(2, 7, -1)$.

25. Let $T: R^3 \rightarrow W$ be the orthogonal projection of R^3 onto the plane W having the equation $x + y + z = 0$.
   (a) Find a formula for T.
   (b) Find $T(3, 8, 4)$.

26. In each part let $T: R^2 \rightarrow R^2$ be the linear operator that rotates each vector in the plane through the angle θ. Find $T(-1, 2)$ and $T(x, y)$ when
   (a) $\theta = \dfrac{\pi}{4}$ (b) $\theta = \pi$ (c) $\theta = \dfrac{\pi}{6}$ (d) $\theta = -\dfrac{\pi}{3}$

27. Prove that if $T: V \rightarrow W$ is a linear transformation, then $T(\mathbf{u} - \mathbf{v}) = T(\mathbf{u}) - T(\mathbf{v})$ for all vectors $\mathbf{u}$ and $\mathbf{v}$ in V.

28. Let $\{\mathbf{v}_1, \mathbf{v}_2, \ldots, \mathbf{v}_n\}$ be a basis for a vector space V and let $T: V \rightarrow W$ be a linear transformation. Show that if $T(\mathbf{v}_1) = T(\mathbf{v}_2) = \cdots = T(\mathbf{v}_n) = \mathbf{0}$, then T is the zero transformation.

29. Let $\{\mathbf{v}_1, \mathbf{v}_2, \ldots, \mathbf{v}_n\}$ be a basis for a vector space V and let $T: V \rightarrow V$ be a linear operator. Show that if $T(\mathbf{v}_1) = \mathbf{v}_1$, $T(\mathbf{v}_2) = \mathbf{v}_2, \ldots, T(\mathbf{v}_n) = \mathbf{v}_n$, then T is the identity transformation on V.

5.2 PROPERTIES OF LINEAR TRANSFORMATIONS: KERNEL AND RANGE

In this section we develop some basic properties of linear transformations. In particular, we show that once the images of the basis vectors under a linear transformation are known, it is possible to find the images of the remaining vectors in the space.

Theorem 1. *If $T:V \to W$ is a linear transformation, then:*

(a) $T(\mathbf{0}) = \mathbf{0}$
(b) $T(-\mathbf{v}) = -T(\mathbf{v})$ *for all* $\mathbf{v}$ *in* V
(c) $T(\mathbf{v} - \mathbf{w}) = T(\mathbf{v}) - T(\mathbf{w})$ *for all* $\mathbf{v}$ *and* $\mathbf{w}$ *in* V.

Proof. Let $\mathbf{v}$ be any vector in V. Since $0\mathbf{v} = \mathbf{0}$ we have

$$T(\mathbf{0}) = T(0\mathbf{v}) = 0T(\mathbf{v}) = \mathbf{0}$$

which proves (a).

Also, $T(-\mathbf{v}) = T((-1)\mathbf{v}) = (-1)T(\mathbf{v}) = -T(\mathbf{v})$, which proves (b).

Finally, $\mathbf{v} - \mathbf{w} = \mathbf{v} + (-1)\mathbf{w}$; thus,

$$\begin{aligned} T(\mathbf{v} - \mathbf{w}) &= T(\mathbf{v} + (-1)\mathbf{w}) \\ &= T(\mathbf{v}) + (-1)T(\mathbf{w}) \\ &= T(\mathbf{v}) - T(\mathbf{w}) \end{aligned}$$

Definition. If $T:V \to W$ is a linear transformation, then the set of vectors in V that T maps into $\mathbf{0}$ is called the **kernel** (or **nullspace**) of T; it is denoted by $\ker(T)$. The set of all vectors in W that are images under T of at least one vector in V is called the **range** of T; it is denoted by $R(T)$.

Example 12

Let $T:V \to W$ be the zero transformation. Since T maps every vector into $\mathbf{0}$, $\ker(T) = V$. Since $\mathbf{0}$ is the only possible image under T, $R(T)$ consists of the zero vector.

Example 13

Let $T:R^n \to R^m$ be multiplication by

$$A = \begin{bmatrix} a_{11} & a_{12} & \cdots & a_{1n} \\ a_{21} & a_{22} & \cdots & a_{2n} \\ \vdots & \vdots & & \vdots \\ a_{m1} & a_{m2} & \cdots & a_{mn} \end{bmatrix}$$

The kernel of T consists of all

$$\mathbf{x} = \begin{bmatrix} x_1 \\ x_2 \\ \vdots \\ x_n \end{bmatrix}$$

that are solution vectors of the homogeneous system

$$A \begin{bmatrix} x_1 \\ x_2 \\ \vdots \\ x_n \end{bmatrix} = \begin{bmatrix} 0 \\ 0 \\ \vdots \\ 0 \end{bmatrix}$$

The range of T consists of vectors

$$\mathbf{b} = \begin{bmatrix} b_1 \\ b_2 \\ \vdots \\ b_m \end{bmatrix}$$

such that the system

$$A \begin{bmatrix} x_1 \\ x_2 \\ \vdots \\ x_n \end{bmatrix} = \begin{bmatrix} b_1 \\ b_2 \\ \vdots \\ b_m \end{bmatrix}$$

is consistent.   ▲

Theorem 2. *If $T : V \rightarrow W$ is a linear transformation, then:*

(a) the kernel of T is a subspace of V
(b) the range of T is a subspace of W.

Proof.
    (*a*) To show that ker(T) is a subspace, we must show it is closed under addition and scalar multiplication. Let $\mathbf{v}_1$ and $\mathbf{v}_2$ be vectors in ker(T), and let k be any

scalar. Then

$$T(\mathbf{v}_1 + \mathbf{v}_2) = T(\mathbf{v}_1) + T(\mathbf{v}_2)$$
$$= 0 + 0 = 0$$

so that $\mathbf{v}_1 + \mathbf{v}_2$ is in $\ker(T)$. Also,

$$T(k\mathbf{v}_1) = kT(\mathbf{v}_1) = k0 = 0$$

so that $k\mathbf{v}_1$ is in $\ker(T)$.

(b) Let $\mathbf{w}_1$ and $\mathbf{w}_2$ be vectors in the range of T. To prove this part we must show that $\mathbf{w}_1 + \mathbf{w}_2$ and $k\mathbf{w}_1$ are in the range of T for any scalar k; that is, we must find vectors $\mathbf{a}$ and $\mathbf{b}$ in V such that $T(\mathbf{a}) = \mathbf{w}_1 + \mathbf{w}_2$ and $T(\mathbf{b}) = k\mathbf{w}_1$.

Since $\mathbf{w}_1$ and $\mathbf{w}_2$ are in the range of T, there are vectors $\mathbf{a}_1$ and $\mathbf{a}_2$ in V such that $T(\mathbf{a}_1) = \mathbf{w}_1$ and $T(\mathbf{a}_2) = \mathbf{w}_2$. Let $\mathbf{a} = \mathbf{a}_1 + \mathbf{a}_2$ and $\mathbf{b} = k\mathbf{a}_1$. Then

$$T(\mathbf{a}) = T(\mathbf{a}_1 + \mathbf{a}_2) = T(\mathbf{a}_1) + T(\mathbf{a}_2) = \mathbf{w}_1 + \mathbf{w}_2$$

and

$$T(\mathbf{b}) = T(k\mathbf{a}_1) = kT(\mathbf{a}_1) = k\mathbf{w}_1$$

which completes the proof.

Example 14

Let $T : R^n \to R^m$ be multiplication by an $m \times n$ matrix A. From Example 13 the kernel of T consists of all solutions of $A\mathbf{x} = 0$; thus, the kernel is the *solution space* of this system. Also from Example 13, the range of T consists of all vectors $\mathbf{b}$ such that $A\mathbf{x} = \mathbf{b}$ is consistent. Thus, by Theorem 16 of Section 4.6, the range of T is the *column space* of the matrix A.

Suppose $\{\mathbf{v}_1, \mathbf{v}_2, \dots, \mathbf{v}_n\}$ is a basis for a vector space V and $T : V \to W$ is a linear transformation. If we happen to know the images of the basis vectors, that is,

$$T(\mathbf{v}_1), T(\mathbf{v}_2), \dots, T(\mathbf{v}_n)$$

then we can obtain the image $T(\mathbf{v})$ of any vector $\mathbf{v}$ by first expressing $\mathbf{v}$ in terms of the basis, say

$$\mathbf{v} = k_1\mathbf{v}_1 + k_2\mathbf{v}_2 + \cdots + k_n\mathbf{v}_n$$

and then using relation (5.2) of Section 5.1 to write

$$T(\mathbf{v}) = k_1 T(\mathbf{v}_1) + k_2 T(\mathbf{v}_2) + \cdots + k_n T(\mathbf{v}_n)$$

In short, *a linear transformation is completely determined by its "values" at a basis.*

Example 15

Consider the basis $S = \{v_1, v_2, v_3\}$ for R^3, where $v_1 = (1, 1, 1)$, $v_2 = (1, 1, 0)$, $v_3 = (1, 0, 0)$; let $T:R^3 \to R^2$ be the linear transformation such that

$$T(v_1) = (1, 0) \qquad T(v_2) = (2, -1) \qquad T(v_3) = (4, 3)$$

Find a formula for $T(x_1, x_2, x_3)$; then use this formula to compute $T(2, -3, 5)$.

Solution. We first express $\mathbf{x} = (x_1, x_2, x_3)$ as a linear combination of $v_1 = (1, 1, 1)$, $v_2 = (1, 1, 0)$, and $v_3 = (1, 0, 0)$. If we write

$$(x_1, x_2, x_3) = k_1(1, 1, 1) + k_2(1, 1, 0) + k_3(1, 0, 0)$$

then on equating corresponding components we obtain

$$
\begin{aligned}
k_1 + k_2 + k_3 &= x_1 \\
k_1 + k_2 \phantom{{}+ k_3} &= x_2 \\
k_1 \phantom{{}+ k_2 + k_3} &= x_3
\end{aligned}
$$

which yields $k_1 = x_3$, $k_2 = x_2 - x_3$, $k_3 = x_1 - x_2$ so that

$$
\begin{aligned}
(x_1, x_2, x_3) &= x_3(1, 1, 1) + (x_2 - x_3)(1, 1, 0) + (x_1 - x_2)(1, 0, 0) \\
&= x_3 v_1 + (x_2 - x_3)v_2 + (x_1 - x_2)v_3
\end{aligned}
$$

Thus,

$$
\begin{aligned}
T(x_1, x_2, x_3) &= x_3 T(v_1) + (x_2 - x_3)T(v_2) + (x_1 - x_2)T(v_3) \\
&= x_3(1, 0) + (x_2 - x_3)(2, -1) + (x_1 - x_2)(4, 3) \\
&= (4x_1 - 2x_2 - x_3, \, 3x_1 - 4x_2 + x_3)
\end{aligned}
$$

From this formula we obtain

$$T(2, -3, 5) = (9, 23)$$

Definition. If $T:V \to W$ is a linear transformation, then the dimension of the range of T is called the **rank of T**, and the dimension of the kernel is called the **nullity of T**.

Example 16

Let $T:R^2 \to R^2$ be the rotation of R^2 through the angle $\pi/4$. It is geometrically obvious that the range of T is all of R^2 and the kernel of T is $\{0\}$. Therefore, T has rank $= 2$ and nullity $= 0$.

Example 17

Let $T:R^n \to R^m$ be multiplication by an $m \times n$ matrix A. In Example 14 we observed that

the range of T is the column space of A

the kernel of T is the solution space of $A\mathbf{x} = \mathbf{0}$

Thus, it follows that

rank of $T = \dim(\text{column space of } A) = \text{rank}(A)$

nullity of $T = \dim(\text{solution space of } A\mathbf{x} = \mathbf{0})$ ▲

Our next theorem establishes a relationship between the rank and nullity of a linear transformation defined on a finite-dimensional vector space. We shall defer the proof to the end of the section.

Theorem 3 (*Dimension Theorem*). *If $T:V \to W$ is a linear transformation from an n-dimensional vector space V to a vector space W, then*

$$(rank \text{ of } T) + (nullity \text{ of } T) = n \qquad\qquad (5.4)$$

In words, this theorem states that *the rank plus the nullity of a linear transformation is equal to the dimension of its domain.*

In the special case where $V = R^n$, $W = R^m$, and $T:V \to W$ is multiplication by an $m \times n$ matrix A, it follows from (5.4) and Example 17 that

rank$(A) + \dim(\text{solution space of } A\mathbf{x} = \mathbf{0}) = n$

Thus, we have the following theorem.

Theorem 4. *If A is an $m \times n$ matrix, then the dimension of the solution space of $A\mathbf{x} = \mathbf{0}$ is*

$$n - rank(A)$$

In words, this theorem states that *the dimension of the solution space of $A\mathbf{x} = \mathbf{0}$ is equal to the number of columns of A minus the rank of A.*

Example 18

In Example 37 of Section 4.5 we showed that the homogeneous system

$$
\begin{aligned}
2x_1 + 2x_2 - x_3 \qquad + x_5 &= 0 \\
-x_1 - x_2 + 2x_3 - 3x_4 + x_5 &= 0 \\
x_1 + x_2 - 2x_3 \qquad - x_5 &= 0 \\
x_3 + x_4 + x_5 &= 0
\end{aligned}
$$

has a two-dimensional solution space by solving the system and finding a basis. Since the coefficient matrix

$$
A = \begin{bmatrix}
2 & 2 & -1 & 0 & 1 \\
-1 & -1 & 2 & -3 & 1 \\
1 & 1 & -2 & 0 & -1 \\
0 & 0 & 1 & 1 & 1
\end{bmatrix}
$$

has five columns, it follows from Theorem 4 that the rank of A must satisfy

$$
2 = 5 - \text{rank}(A)
$$

so that $\text{rank}(A) = 3$. The reader can check this result by reducing A to row-echelon form and showing that the resulting matrix has three nonzero rows.

OPTIONAL

Proof of Theorem 3. We must show that

$$
\dim(R(T)) + \dim(\ker(T)) = n
$$

We shall give the proof for the case where $1 \le \dim(\ker(T)) < n$. The cases $\dim(\ker(T)) = 0$ and $\dim(\ker(T)) = n$ are left as exercises. Assume $\dim(\ker(T)) = r$, and let $v_1, \ldots, v_r$ be a basis for the kernel. Since $\{v_1, \ldots, v_r\}$ is linearly independent, part (c) of Theorem 11 in Section 4.5 states that there are $n - r$ vectors, $v_{r+1}, \ldots, v_n$, such that $\{v_1, \ldots, v_r, v_{r+1}, \ldots, v_n\}$ is a basis for V. To complete the proof, we shall show that the $n - r$ vectors in the set $S = \{T(v_{r+1}), \ldots, T(v_n)\}$ form a basis for the range of T. It will then follow that

$$
\dim(R(T)) + \dim(\ker(T)) = (n - r) + r = n
$$

First we show that S spans the range of T. If b is any vector in the range of T, then $b = T(v)$ for some vector v in V. Since $\{v_1, \ldots, v_r, v_{r+1}, \ldots, v_n\}$ is a basis for V, v can be written in the form

$$
v = c_1 v_1 + \cdots + c_r v_r + c_{r+1} v_{r+1} + \cdots + c_n v_n
$$

Since $v_1, \ldots, v_r$ lie in the kernel of T, $T(v_1) = \cdots = T(v_r) = 0$, so that

$$b = T(v) = c_{r+1}T(v_{r+1}) + \cdots + c_n T(v_n)$$

Thus, S spans the range of T.

Finally, we show that S is a linearly independent set and consequently forms a basis for the range of T. Suppose some linear combination of the vectors in S is zero; that is,

$$k_{r+1}T(v_{r+1}) + \cdots + k_n T(v_n) = 0 \tag{5.5}$$

We must show $k_{r+1} = \cdots = k_n = 0$. Since T is linear, (5.5) can be rewritten as

$$T(k_{r+1}v_{r+1} + \cdots + k_n v_n) = 0$$

which says that $k_{r+1}v_{r+1} + \cdots + k_n v_n$ is in the kernel of T. This vector can therefore be written as a linear combination of the basis vectors $\{v_1, \ldots, v_r\}$, say

$$k_{r+1}v_{r+1} + \cdots + k_n v_n = k_1 v_1 + \cdots + k_r v_r$$

Thus,

$$k_1 v_1 + \cdots + k_r v_r - k_{r+1}v_{r+1} - \cdots - k_n v_n = 0$$

Since $\{v_1, \ldots, v_n\}$ is linearly independent, all the k's are zero; in particular, $k_{r+1} = \cdots = k_n = 0$, which completes the proof.

EXERCISE SET 5.2

1. Let $T:R^2 \to R^2$ be multiplication by

$$\begin{bmatrix} 2 & -1 \\ -8 & 4 \end{bmatrix}$$

Which of the following are in $R(T)$?

(a) $\begin{bmatrix} 1 \\ -4 \end{bmatrix}$ (b) $\begin{bmatrix} 5 \\ 0 \end{bmatrix}$ (c) $\begin{bmatrix} -3 \\ 12 \end{bmatrix}$

2. Let $T:R^2 \to R^2$ be the linear transformation in Exercise 1. Which of the following are in ker(T)?

(a) $\begin{bmatrix} 5 \\ 10 \end{bmatrix}$ (b) $\begin{bmatrix} 3 \\ 2 \end{bmatrix}$ (c) $\begin{bmatrix} 1 \\ 1 \end{bmatrix}$

3. Let $T:R^4 \to R^3$ be multiplication by

$$\begin{bmatrix} 4 & 1 & -2 & -3 \\ 2 & 1 & 1 & -4 \\ 6 & 0 & -9 & 9 \end{bmatrix}$$

Which of the following are in $R(T)$?

(a) $\begin{bmatrix} 0 \\ 0 \\ 6 \end{bmatrix}$ (b) $\begin{bmatrix} 1 \\ 3 \\ 0 \end{bmatrix}$ (c) $\begin{bmatrix} 2 \\ 4 \\ 1 \end{bmatrix}$

4. Let $T:R^4 \rightarrow R^3$ be the linear transformation in Exercise 3. Which of the following are in $\ker(T)$?

(a) $\begin{bmatrix} 3 \\ -8 \\ 2 \\ 0 \end{bmatrix}$ (b) $\begin{bmatrix} 0 \\ 0 \\ 0 \\ 1 \end{bmatrix}$ (c) $\begin{bmatrix} 0 \\ -4 \\ 1 \\ 0 \end{bmatrix}$

5. Let $T:P_2 \rightarrow P_3$ be the linear transformation defined by $T(p(x)) = xp(x)$. Which of the following are in $\ker(T)$?
(a) x^2 (b) 0 (c) $1 + x$

6. Let $T:P_2 \rightarrow P_3$ be the linear transformation in Exercise 5. Which of the following are in $R(T)$?
(a) $x + x^2$ (b) $1 + x$ (c) $3 - x^2$

7. Let V be any vector space, and let $T:V \rightarrow V$ be defined by $T(\mathbf{v}) = 3\mathbf{v}$.
(a) What is the kernel of T?
(b) What is the range of T?

8. Find the rank and nullity of the linear transformation in Exercise 1.

9. Find the rank and nullity of the linear transformation in Exercise 5.

10. Let V be an n-dimensional vector space. Find the rank and nullity of the linear transformation $T:V \rightarrow V$ defined by
(a) $T(\mathbf{x}) = \mathbf{x}$ (b) $T(\mathbf{x}) = \mathbf{0}$ (c) $T(\mathbf{x}) = 3\mathbf{x}$

11. Consider the basis $S = \{\mathbf{v}_1, \mathbf{v}_2, \mathbf{v}_3\}$ for R^3, where $\mathbf{v}_1 = (1, 2, 3)$, $\mathbf{v}_2 = (2, 5, 3)$, and $\mathbf{v}_3 = (1, 0, 10)$, and let $T:R^3 \rightarrow R^2$ be the linear transformation for which $T(\mathbf{v}_1) = (1, 0)$, $T(\mathbf{v}_2) = (1, 0)$, and $T(\mathbf{v}_3) = (0, 1)$. Find a formula for $T(x_1, x_2, x_3)$ and use it to compute $T(1, 1, 1)$.

12. Consider the linear transformation $T:P_2 \rightarrow P_2$ for which $T(1) = 1 + x$, $T(x) = 3 - x^2$, and $T(x^2) = 4 + 2x - 3x^2$. Find a formula for $T(a_0 + a_1x + a_2x^2)$ and use it to compute $T(2 - 2x + 3x^2)$.

13. In each part use the given information to find the nullity of T.
(a) $T:R^5 \rightarrow R^7$ has rank 3.   (b) $T:P_4 \rightarrow P_3$ has rank 1.
(c) The range of $T:R^6 \rightarrow R^3$ is R^3.   (d) $T:M_{22} \rightarrow M_{22}$ has rank 3.

14. Let A be a 7×6 matrix such that $A\mathbf{x} = \mathbf{0}$ has only the trivial solution, and let $T:R^6 \rightarrow R^7$ be multiplication by A. Find the rank and nullity of T.

15. Let A be a 5×7 matrix with rank 4.
(a) What is the dimension of the solution space of $A\mathbf{x} = \mathbf{0}$?
(b) Is $A\mathbf{x} = \mathbf{b}$ consistent for all vectors $\mathbf{b}$ in R^5? Explain.

In Exercises 16–19 let T be multiplication by the given matrix. Find:
  (a) a basis for the range of T
  (b) a basis for the kernel of T
  (c) the rank and nullity of T

16. $\begin{bmatrix} 1 & -1 & 3 \\ 5 & 6 & -4 \\ 7 & 4 & 2 \end{bmatrix}$ **17.** $\begin{bmatrix} 2 & 0 & -1 \\ 4 & 0 & -2 \\ 0 & 0 & 0 \end{bmatrix}$

18. $\begin{bmatrix} 4 & 1 & 5 & 2 \\ 1 & 2 & 3 & 0 \end{bmatrix}$ **19.** $\begin{bmatrix} 1 & 4 & 5 & 0 & 9 \\ 3 & -2 & 1 & 0 & -1 \\ -1 & 0 & -1 & 0 & -1 \\ 2 & 3 & 5 & 1 & 8 \end{bmatrix}$

20. Let $T:R^3 \to V$ be a linear transformation from R^3 to any vector space. Show that the kernel of T is a line through the origin, a plane through the origin, the origin only, or all of R^3.

21. Let $T:V \to R^3$ be a linear transformation from any vector space to R^3. Show that the range of T is a line through the origin, a plane through the origin, the origin only, or all of R^3.

22. Let $T:R^3 \to R^3$ be multiplication by

$$\begin{bmatrix} 1 & 3 & 4 \\ 3 & 4 & 7 \\ -2 & 2 & 0 \end{bmatrix}$$

  (a) Show that the kernel of T is a line through the origin and find parametric equations for it.
  (b) Show that the range of T is a plane through the origin and find an equation for it.

23. Prove: If $\{v_1, v_2, \ldots, v_n\}$ is a basis for V and $w_1, w_2, \ldots, w_n$ are vectors in W, not necessarily distinct, then there exists a linear transformation $T:V \to W$ such that $T(v_1) = w_1$, $T(v_2) = w_2, \ldots, T(v_n) = w_n$.

24. Prove the dimension theorem in the cases:
  (a) $\dim(\ker(T)) = 0$
  (b) $\dim(\ker(T)) = n$

25. Let $T:V \to V$ be a linear operator on a finite dimensional vector space V. Prove that $R(T) = V$ if and only if $\ker(T) = \{0\}$.

26. **(For readers who have studied calculus.)** Let $D:P_3 \to P_2$ be the differentiation transformation $D(p) = p'$. Describe the kernel of D.

27. **(For readers who have studied calculus.)** Let $J:P_1 \to R$ be the integration transformation $J(p) = \int_{-1}^{1} p(x)\,dx$. Describe the kernel of J.

5.3   LINEAR TRANSFORMATIONS FROM R^n TO R^m; GEOMETRY OF LINEAR TRANSFORMATIONS FROM R^2 TO R^2

In this section we study linear transformations of R^n to R^m and obtain geometric properties of linear transformations from R^2 to R^2.

We shall show first that every linear transformation from R^n to R^m is a matrix transformation. More precisely, we shall show that if $T:R^n \rightarrow R^m$ is any linear transformation, we can find an $m \times n$ matrix A such that T is multiplication by A. To see this, let

$$\mathbf{e}_1, \mathbf{e}_2, \ldots, \mathbf{e}_n$$

be the standard basis for R^n, and let A be the $m \times n$ matrix having

$$T(\mathbf{e}_1), T(\mathbf{e}_2), \ldots, T(\mathbf{e}_n)$$

as its column vectors. (We shall assume in this section that all vectors are expressed in matrix notation.) For example, if $T:R^2 \rightarrow R^2$ is given by

$$T\left(\begin{bmatrix} x_1 \\ x_2 \end{bmatrix} \right) = \begin{bmatrix} x_1 + 2x_2 \\ x_1 - x_2 \end{bmatrix}$$

then

$$T(\mathbf{e}_1) = T\left(\begin{bmatrix} 1 \\ 0 \end{bmatrix} \right) = \begin{bmatrix} 1 \\ 1 \end{bmatrix} \quad \text{and} \quad T(\mathbf{e}_2) = T\left(\begin{bmatrix} 0 \\ 1 \end{bmatrix} \right) = \begin{bmatrix} 2 \\ -1 \end{bmatrix}$$

$$A = \begin{bmatrix} 1 & 2 \\ 1 & -1 \end{bmatrix}$$
$$\uparrow \quad \uparrow$$
$$T(\mathbf{e}_1) \quad T(\mathbf{e}_2)$$

More generally, if

$$T(\mathbf{e}_1) = \begin{bmatrix} a_{11} \\ a_{21} \\ \vdots \\ a_{m1} \end{bmatrix}, \quad T(\mathbf{e}_2) = \begin{bmatrix} a_{12} \\ a_{22} \\ \vdots \\ a_{m2} \end{bmatrix}, \ldots, \quad T(\mathbf{e}_n) = \begin{bmatrix} a_{1n} \\ a_{2n} \\ \vdots \\ a_{mn} \end{bmatrix}$$

then

$$A = \begin{bmatrix} a_{11} & a_{12} & \cdots & a_{1n} \\ a_{21} & a_{22} & \cdots & a_{2n} \\ \vdots & \vdots & & \vdots \\ a_{m1} & a_{m2} & \cdots & a_{mn} \end{bmatrix} \tag{5.6}$$
$$\uparrow \quad \uparrow \qquad \uparrow$$
$$T(\mathbf{e}_1) \quad T(\mathbf{e}_2) \quad \cdots \quad T(\mathbf{e}_n)$$

This matrix is called the **standard matrix for T**. We shall show that the linear transformation $T: R^n \to R^m$ is multiplication by A. To see this, observe first that

$$\mathbf{x} = \begin{bmatrix} x_1 \\ x_2 \\ \vdots \\ x_n \end{bmatrix} = x_1\mathbf{e}_1 + x_2\mathbf{e}_2 + \cdots + x_n\mathbf{e}_n$$

Therefore, by the linearity of T,

$$T(\mathbf{x}) = x_1 T(\mathbf{e}_1) + x_2 T(\mathbf{e}_2) + \cdots + x_n T(\mathbf{e}_n) \tag{5.7}$$

In contrast,

$$A\mathbf{x} = \begin{bmatrix} a_{11} & a_{12} & \cdots & a_{1n} \\ a_{21} & a_{22} & \cdots & a_{2n} \\ \vdots & \vdots & & \vdots \\ a_{m1} & a_{m2} & \cdots & a_{mn} \end{bmatrix} \begin{bmatrix} x_1 \\ x_2 \\ \vdots \\ x_n \end{bmatrix} = \begin{bmatrix} a_{11}x_1 + a_{12}x_2 + \cdots + a_{1n}x_n \\ a_{21}x_1 + a_{22}x_2 + \cdots + a_{2n}x_n \\ \vdots & \vdots & & \vdots \\ a_{m1}x_1 + a_{m2}x_2 + \cdots + a_{mn}x_n \end{bmatrix}$$

$$= x_1 \begin{bmatrix} a_{11} \\ a_{21} \\ \vdots \\ a_{m1} \end{bmatrix} + x_2 \begin{bmatrix} a_{12} \\ a_{22} \\ \vdots \\ a_{m2} \end{bmatrix} + \cdots + x_n \begin{bmatrix} a_{1n} \\ a_{2n} \\ \vdots \\ a_{mn} \end{bmatrix}$$

$$= x_1 T(\mathbf{e}_1) + x_2 T(\mathbf{e}_2) + \cdots + x_n T(\mathbf{e}_n) \tag{5.8}$$

Comparing (5.7) and (5.8) yields $T(\mathbf{x}) = A\mathbf{x}$; that is, T is multiplication by A. The foregoing observations are summarized in the following theorem.

Theorem 5. *If $T: R^n \to R^m$ is a linear transformation, and if $\mathbf{e}_1, \mathbf{e}_2, \ldots, \mathbf{e}_n$ is the standard basis for R^n, then T is multiplication by A, where A is the matrix whose successive column vectors are $T(\mathbf{e}_1), T(\mathbf{e}_2), \ldots, T(\mathbf{e}_n)$.*

Example 19

Find the standard matrix for the transformation $T: R^3 \to R^4$ defined by

$$T\left(\begin{bmatrix} x_1 \\ x_2 \\ x_3 \end{bmatrix} \right) = \begin{bmatrix} x_1 + x_2 \\ x_1 - x_2 \\ x_3 \\ x_1 \end{bmatrix}$$

Solution.

$$T(\mathbf{e}_1) = T\left(\begin{bmatrix} 1 \\ 0 \\ 0 \end{bmatrix} \right) = \begin{bmatrix} 1 \\ 1 \\ 0 \\ 1 \end{bmatrix} \qquad T(\mathbf{e}_2) = T\left(\begin{bmatrix} 0 \\ 1 \\ 0 \end{bmatrix} \right) = \begin{bmatrix} 1 \\ -1 \\ 0 \\ 0 \end{bmatrix} \qquad T(\mathbf{e}_3) = T\left(\begin{bmatrix} 0 \\ 0 \\ 1 \end{bmatrix} \right) = \begin{bmatrix} 0 \\ 0 \\ 1 \\ 0 \end{bmatrix}$$

Using $T(\mathbf{e}_1)$, $T(\mathbf{e}_2)$, and $T(\mathbf{e}_3)$ as column vectors, we obtain

$$A = \begin{bmatrix} 1 & 1 & 0 \\ 1 & -1 & 0 \\ 0 & 0 & 1 \\ 1 & 0 & 0 \end{bmatrix}$$

As a check, observe that

$$A \begin{bmatrix} x_1 \\ x_2 \\ x_3 \end{bmatrix} = \begin{bmatrix} x_1 + x_2 \\ x_1 - x_2 \\ x_3 \\ x_1 \end{bmatrix}$$

which agrees with the given formula for T. ▲

Here is an interesting question to consider. Suppose that we *start* with an $m \times n$ matrix A and *define* $T:R^n \to R^m$ to be multiplication by A. According to Theorem 5, the linear transformation T is also multiplication by the standard matrix for T. Thus, T is multiplication by both A and $[T(\mathbf{e}_1) \,|\, T(\mathbf{e}_2) \,|\, \cdots \,|\, T(\mathbf{e}_n)]$. How are these two matrices related? The next example answers this question.

Example 20

Let $T:R^n \to R^m$ be multiplication by

$$A = \begin{bmatrix} a_{11} & a_{12} & \cdots & a_{1n} \\ a_{21} & a_{22} & \cdots & a_{2n} \\ \vdots & \vdots & & \vdots \\ a_{m1} & a_{m2} & \cdots & a_{mn} \end{bmatrix}$$

Find the standard matrix for T.

Solution. The vectors $T(\mathbf{e}_1)$, $T(\mathbf{e}_2)$, $\ldots$, $T(\mathbf{e}_n)$ are the successive column vectors of A. For example,

$$T(\mathbf{e}_1) = A\mathbf{e}_1 = \begin{bmatrix} a_{11} & a_{12} & \cdots & a_{1n} \\ a_{21} & a_{22} & \cdots & a_{2n} \\ \vdots & \vdots & & \vdots \\ a_{m1} & a_{m2} & \cdots & a_{mn} \end{bmatrix} \begin{bmatrix} 1 \\ 0 \\ 0 \\ \vdots \\ 0 \end{bmatrix} = \begin{bmatrix} a_{11} \\ a_{21} \\ \vdots \\ a_{m1} \end{bmatrix}$$

Thus, the standard matrix for T is

$$[T(\mathbf{e}_1) \,|\, T(\mathbf{e}_2) \,|\, \cdots \,|\, T(\mathbf{e}_n)] = A$$

In summary, *the standard matrix for a matrix transformation is the matrix itself.* ▲

Example 20 suggests a new way of thinking about matrices. An arbitrary $m \times n$ matrix A can be viewed as the standard matrix for the linear transformation that maps the standard basis for R^n into the column vectors of A. Thus,

$$A = \begin{bmatrix} 1 & -2 & 1 \\ 3 & 4 & 6 \end{bmatrix}$$

is the standard matrix for the linear transformation of R^3 to R^2 that maps

$$\mathbf{e}_1 = \begin{bmatrix} 1 \\ 0 \\ 0 \end{bmatrix} \qquad \mathbf{e}_2 = \begin{bmatrix} 0 \\ 1 \\ 0 \end{bmatrix} \qquad \mathbf{e}_3 = \begin{bmatrix} 0 \\ 0 \\ 1 \end{bmatrix}$$

into

$$\begin{bmatrix} 1 \\ 3 \end{bmatrix} \qquad \begin{bmatrix} -2 \\ 4 \end{bmatrix} \qquad \begin{bmatrix} 1 \\ 6 \end{bmatrix}$$

respectively.

In the rest of this section we shall study the geometric properties of *plane linear transformations*, that is, linear transformations from R^2 to R^2. If $T:R^2 \to R^2$ is such a transformation and

$$A = \begin{bmatrix} a & b \\ c & d \end{bmatrix}$$

is the standard matrix for T, then

$$T\left(\begin{bmatrix} x \\ y \end{bmatrix}\right) = \begin{bmatrix} a & b \\ c & d \end{bmatrix} \begin{bmatrix} x \\ y \end{bmatrix} = \begin{bmatrix} ax + by \\ cx + dy \end{bmatrix} \tag{5.9}$$

There are two equally good geometric interpretations of this formula. We may view the entries in the matrices

$$\begin{bmatrix} x \\ y \end{bmatrix} \qquad \text{and} \qquad \begin{bmatrix} ax + by \\ cx + dy \end{bmatrix}$$

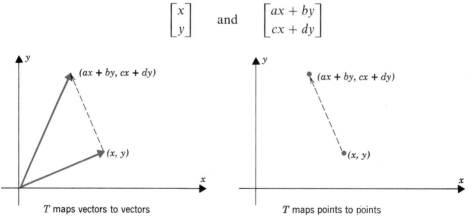

T maps vectors to vectors   T maps points to points

Figure 5.5

either as components of vectors or coordinates of points. With the first interpretation, T maps arrows to arrows, and with the second, points to points (Figure 5.5). The choice is a matter of taste. In the discussion to follow, we shall view plane linear transformations as mapping points to points.

Example 21

Let $T:R^2 \rightarrow R^2$ be the linear transformation that maps each point into its symmetric image about the y-axis (Figure 5.6). Find the standard matrix for T.

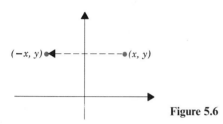

Figure 5.6

Solution.

$$T(\mathbf{e}_1) = T\left(\begin{bmatrix} 1 \\ 0 \end{bmatrix}\right) = \begin{bmatrix} -1 \\ 0 \end{bmatrix} \qquad T(\mathbf{e}_2) = T\left(\begin{bmatrix} 0 \\ 1 \end{bmatrix}\right) = \begin{bmatrix} 0 \\ 1 \end{bmatrix}$$

Using $T(\mathbf{e}_1)$ and $T(\mathbf{e}_2)$ as column vectors we obtain the standard matrix

$$A = \begin{bmatrix} -1 & 0 \\ 0 & 1 \end{bmatrix}$$

As a check,

$$\begin{bmatrix} -1 & 0 \\ 0 & 1 \end{bmatrix} \begin{bmatrix} x \\ y \end{bmatrix} = \begin{bmatrix} -x \\ y \end{bmatrix}$$

so that multiplication by A maps the point (x, y) into its symmetric image $(-x, y)$ about the y-axis. ▲

We will now focus our attention on five types of plane linear transformations that have special importance: rotations, reflections, expansions, compressions, and shears.

Rotations. If $T:R^2 \rightarrow R^2$ rotates each point in the plane about the origin through an angle θ, then it follows from Example 2 of Section 5.1 that the standard matrix for T is

$$\begin{bmatrix} \cos \theta & -\sin \theta \\ \sin \theta & \cos \theta \end{bmatrix}$$

Reflections. A *reflection* about a line *l* through the origin is a transformation that maps each point in the plane into its mirror image about *l*. It can be shown that reflections are linear transformations. The most important cases are reflections about the coordinate axes and about the line $y = x$. Following the method of Example 21, the reader should be able to show that the standard matrices for these transformations are:

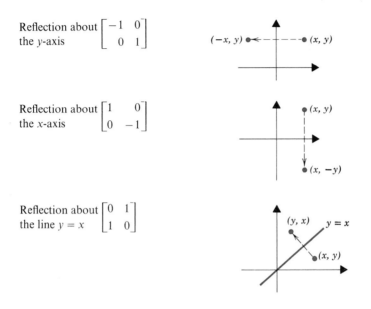

Reflection about $\begin{bmatrix} -1 & 0 \\ 0 & 1 \end{bmatrix}$
the y-axis

Reflection about $\begin{bmatrix} 1 & 0 \\ 0 & -1 \end{bmatrix}$
the x-axis

Reflection about $\begin{bmatrix} 0 & 1 \\ 1 & 0 \end{bmatrix}$
the line $y = x$

Expansions and Compressions. If the *x*-coordinate of each point in the plane is multiplied by a positive constant *k*, then the effect is to expand or compress each plane figure in the *x*-direction. If $0 < k < 1$, the result is a compression, and if $k > 1$, an expansion (Figure 5.7). We call such a transformation an *expansion* (or *compression*) *in the x-direction with factor k*. Similarly, if the *y*-coordinate of

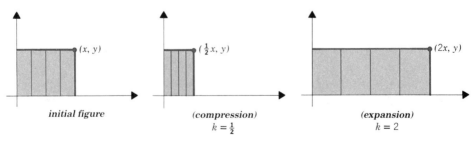

initial figure (compression) $k = \frac{1}{2}$ (expansion) $k = 2$

Figure 5.7

each point is multiplied by a positive constant k, we obtain an **expansion** (or **compression**) **in the y-direction with factor k**. It can be shown that expansions and compressions along the coordinate axes are linear transformations.

If $T:R^2 \rightarrow R^2$ is an expansion or compression in the x-direction with factor k, then

$$T(\mathbf{e}_1) = T\left(\begin{bmatrix} 1 \\ 0 \end{bmatrix}\right) = \begin{bmatrix} k \\ 0 \end{bmatrix} \qquad T(\mathbf{e}_2) = T\left(\begin{bmatrix} 0 \\ 1 \end{bmatrix}\right) = \begin{bmatrix} 0 \\ 1 \end{bmatrix}$$

so the standard matrix for T is

$$\begin{bmatrix} k & 0 \\ 0 & 1 \end{bmatrix}$$

Similarly, the standard matrix for an expansion or compression in the y-direction is

$$\begin{bmatrix} 1 & 0 \\ 0 & k \end{bmatrix}$$

Shears. A **shear in the x-direction with factor k** is a transformation that moves each point (x, y) parallel to the x-axis by an amount ky to the new position $(x + ky, y)$. Under such a transformation, points on the x-axis are unmoved since $y = 0$. However, as we progress away from the x-axis, the magnitude of y increases, so that points farther from the x-axis move a greater distance than those closer (Figure 5.8).

A **shear in the y-direction with factor k** is a transformation that moves each point (x, y) parallel to the y-axis by an amount kx to the new position $(x, y + kx)$. Under such a transformation, points on the y-axis remain fixed, and points farther from the y-axis move a greater distance than do those closer.

It can be shown that shears are linear transformations. If $T:R^2 \rightarrow R^2$ is a shear of factor k in the x-direction, then

$$T(\mathbf{e}_1) = T\left(\begin{bmatrix} 1 \\ 0 \end{bmatrix}\right) = \begin{bmatrix} 1 \\ 0 \end{bmatrix} \qquad T(\mathbf{e}_2) = T\left(\begin{bmatrix} 0 \\ 1 \end{bmatrix}\right) = \begin{bmatrix} k \\ 1 \end{bmatrix}$$

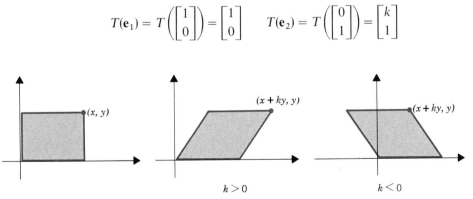

Figure 5.8

so the standard matrix for T is

$$\begin{bmatrix} 1 & k \\ 0 & 1 \end{bmatrix}$$

Similarly, the standard matrix for a shear in the y-direction of factor k is

$$\begin{bmatrix} 1 & 0 \\ k & 1 \end{bmatrix}$$

REMARK. Multiplication by the 2×2 identity matrix maps each point into itself. This is called the **identity transformation**. If desired, this transformation can be viewed as a rotation through $0°$, or as a shear along either axis with $k = 0$, or as a compression or expansion along either axis with factor $k = 1$.

If finitely many matrix transformations from R^n to R^n are performed in succession, then the same result can be achieved by a single matrix transformation. The following example illustrates this point.

Example 22

Suppose that the plane is rotated through an angle θ and then subjected to a shear of factor k in the x-direction. Find a single matrix transformation that produces the same effect as the two successive transformations.

Solution. Under the rotation, the point (x, y) is transformed into the point (x', y') with coordinates given by

$$\begin{bmatrix} x' \\ y' \end{bmatrix} = \begin{bmatrix} \cos \theta & -\sin \theta \\ \sin \theta & \cos \theta \end{bmatrix} \begin{bmatrix} x \\ y \end{bmatrix} \qquad (5.10)$$

Under the shear, the point (x', y') is then transformed into the point (x'', y'') with coordinates given by

$$\begin{bmatrix} x'' \\ y'' \end{bmatrix} = \begin{bmatrix} 1 & k \\ 0 & 1 \end{bmatrix} \begin{bmatrix} x' \\ y' \end{bmatrix} \qquad (5.11)$$

Substituting (5.10) in (5.11) yields

$$\begin{bmatrix} x'' \\ y'' \end{bmatrix} = \begin{bmatrix} 1 & k \\ 0 & 1 \end{bmatrix} \begin{bmatrix} \cos \theta & -\sin \theta \\ \sin \theta & \cos \theta \end{bmatrix} \begin{bmatrix} x \\ y \end{bmatrix}$$

or

$$\begin{bmatrix} x'' \\ y'' \end{bmatrix} = \begin{bmatrix} \cos \theta + k \sin \theta & -\sin \theta + k \cos \theta \\ \sin \theta & \cos \theta \end{bmatrix} \begin{bmatrix} x \\ y \end{bmatrix}$$

Thus, the rotation followed by the shear can be performed by the matrix transformation with matrix

$$\begin{bmatrix} \cos \theta + k \sin \theta & -\sin \theta + k \cos \theta \\ \sin \theta & \cos \theta \end{bmatrix}$$

In general, if the matrix transformations

$$T_1(\mathbf{x}) = A_1\mathbf{x}, \quad T_2(\mathbf{x}) = A_2\mathbf{x}, \dots, T_k(\mathbf{x}) = A_k\mathbf{x}$$

from R^n to R^n are performed in succession (first T_1, then T_2, etc.), then the same result is achieved by the single matrix transformation $T(x) = Ax$, where

$$A = A_k \cdots A_2 A_1 \tag{5.12}$$

Note that the order in which the transformations are performed is obtained by reading right to left in (5.12).

Example 23

(a) Find a matrix transformation from R^2 to R^2 that first shears by a factor of 2 in the x-direction and then reflects about $y = x$.
(b) Find a matrix transformation from R^2 to R^2 that first reflects about $y = x$ and then shears by a factor of 2 in the x-direction.

Solution (a). The standard matrix for the shear is

$$A_1 = \begin{bmatrix} 1 & 2 \\ 0 & 1 \end{bmatrix}$$

and for the reflection is

$$A_2 = \begin{bmatrix} 0 & 1 \\ 1 & 0 \end{bmatrix}$$

Thus, the standard matrix for the shear followed by the reflection is

$$A_2 A_1 = \begin{bmatrix} 0 & 1 \\ 1 & 0 \end{bmatrix}\begin{bmatrix} 1 & 2 \\ 0 & 1 \end{bmatrix} = \begin{bmatrix} 0 & 1 \\ 1 & 2 \end{bmatrix}$$

Solution (b). The reflection followed by the shear is represented by

$$A_1 A_2 = \begin{bmatrix} 1 & 2 \\ 0 & 1 \end{bmatrix}\begin{bmatrix} 0 & 1 \\ 1 & 0 \end{bmatrix} = \begin{bmatrix} 2 & 1 \\ 1 & 0 \end{bmatrix}$$

In the last example, note that $A_1 A_2 \neq A_2 A_1$, so that the effect of shearing and then reflecting is different from reflecting and then shearing. This is illustrated geometrically in Figure 5.9, where we show the effect of the transformations on a rectangle.

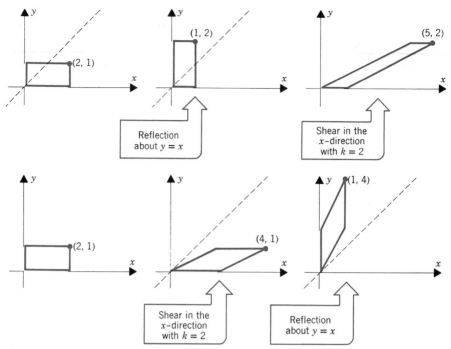

Figure 5.9

Example 24

Show that if $T: R^2 \to R^2$ is multiplication by an *elementary matrix*, then the transformation is one of the following:

(a) a shear along a coordinate axis
(b) a reflection about $y = x$
(c) a compression along a coordinate axis
(d) an expansion along a coordinate axis
(e) a reflection about a coordinate axis
(f) a compression or expansion along a coordinate axis followed by a reflection about a coordinate axis

Solution. Because a 2×2 elementary matrix results from performing a single elementary row operation on the 2×2 identity matrix, it must have one of the following forms (verify):

$$\begin{bmatrix} 1 & 0 \\ k & 1 \end{bmatrix} \quad \begin{bmatrix} 1 & k \\ 0 & 1 \end{bmatrix} \quad \begin{bmatrix} 0 & 1 \\ 1 & 0 \end{bmatrix} \quad \begin{bmatrix} k & 0 \\ 0 & 1 \end{bmatrix} \quad \begin{bmatrix} 1 & 0 \\ 0 & k \end{bmatrix}$$

The first two matrices represent shears along coordinate axes, and the third, a reflection about $y = x$. If $k > 0$, the last two matrices represent compressions or expansions along coordinate axes depending on whether $0 \le k \le 1$ or $k \ge 1$. If $k < 0$ and if we express k in the form $k = -k_1$ where $k_1 > 0$, then the last two matrices can be written as

$$\begin{bmatrix} k & 0 \\ 0 & 1 \end{bmatrix} = \begin{bmatrix} -k_1 & 0 \\ 0 & 1 \end{bmatrix} = \begin{bmatrix} -1 & 0 \\ 0 & 1 \end{bmatrix}\begin{bmatrix} k_1 & 0 \\ 0 & 1 \end{bmatrix} \tag{5.13}$$

$$\begin{bmatrix} 1 & 0 \\ 0 & k \end{bmatrix} = \begin{bmatrix} 1 & 0 \\ 0 & -k_1 \end{bmatrix} = \begin{bmatrix} 1 & 0 \\ 0 & -1 \end{bmatrix}\begin{bmatrix} 1 & 0 \\ 0 & k_1 \end{bmatrix} \tag{5.14}$$

Since $k_1 > 0$, the product in (5.13) represents a compression or expansion along the x-axis followed by a reflection about the y-axis, and (5.14) represents a compression or expansion along the y-axis followed by a reflection about the x-axis. In the case where $k = -1$, (5.13) and (5.14) are simply reflections about the y and x-axis, respectively. ▲

Let $T: R^2 \to R^2$ be multiplication by an *invertible* matrix A, and suppose that T maps the point (x, y) to the point (x', y'). Then

$$\begin{bmatrix} x' \\ y' \end{bmatrix} = A\begin{bmatrix} x \\ y \end{bmatrix}$$

and

$$\begin{bmatrix} x \\ y \end{bmatrix} = A^{-1}\begin{bmatrix} x' \\ y' \end{bmatrix}$$

It follows from these equations that if multiplication by A maps (x, y) to (x', y'), then multiplication by A^{-1} maps (x', y') back to its original position (x, y). For this reason, multiplication by A and multiplication by A^{-1} are said to be *inverse transformations*.

Example 25

If $T: R^2 \to R^2$ compresses the plane by a factor of $\frac{1}{2}$ in the y-direction, then it is intuitively clear that we must expand the plane by a factor of 2 in the y-direction to move each point back to its original position. This is indeed the case, for

$$A = \begin{bmatrix} 1 & 0 \\ 0 & \frac{1}{2} \end{bmatrix}$$

represents a compression of factor $\frac{1}{2}$ in the y-direction, and

$$A^{-1} = \begin{bmatrix} 1 & 0 \\ 0 & 2 \end{bmatrix}$$

is an expansion of factor 2 in the y-direction. ▲

Example 26

Multiplication by

$$A = \begin{bmatrix} \cos\theta & -\sin\theta \\ \sin\theta & \cos\theta \end{bmatrix}$$

rotates the points in the plane by an angle θ. To bring a point back to its original position, it must be rotated through an angle $-\theta$. This can be achieved by multiplying by the rotation matrix

$$\begin{bmatrix} \cos(-\theta) & -\sin(-\theta) \\ \sin(-\theta) & \cos(-\theta) \end{bmatrix}$$

Using the identities, $\cos(-\theta) = \cos\theta$ and $\sin(-\theta) = -\sin\theta$, this can be rewritten as

$$\begin{bmatrix} \cos\theta & \sin\theta \\ -\sin\theta & \cos\theta \end{bmatrix}$$

The reader should verify that this matrix is the inverse of A. ▲

We conclude this section with two theorems that provide some insight into the geometric properties of plane linear transformations.

Theorem 6. *If $T:R^2 \to R^2$ is multiplication by an invertible matrix A, then the geometric effect of T is the same as an appropriate succession of shears, compressions, expansions, and reflections.*

Proof. Since A is invertible, it can be reduced to the identity by a finite sequence of elementary row operations. An elementary row operation can be performed by multiplying on the left by an elementary matrix, and so there exist elementary matrices $E_1, E_2, \ldots, E_k$ such that

$$E_k \cdots E_2 E_1 A = I$$

Solving for A yields

$$A = E_1^{-1} E_2^{-1} \cdots E_k^{-1} I$$

or equivalently

$$A = E_1^{-1} E_2^{-1} \cdots E_k^{-1} \tag{5.15}$$

This equation expresses A as a product of elementary matrices (since the inverse of an elementary matrix is also elementary by Theorem 11 of Section 1.6.) The result now follows from Example 24. ▌

Example 27

Express

$$A = \begin{bmatrix} 1 & 2 \\ 3 & 4 \end{bmatrix}$$

as a product of elementary matrices, and then describe the geometric effect of multiplication by A in terms of shears, compressions, expansions, and reflections.

Solution. A can be reduced to I as follows:

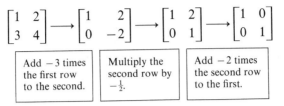

$$\begin{bmatrix} 1 & 2 \\ 3 & 4 \end{bmatrix} \longrightarrow \begin{bmatrix} 1 & 2 \\ 0 & -2 \end{bmatrix} \longrightarrow \begin{bmatrix} 1 & 2 \\ 0 & 1 \end{bmatrix} \longrightarrow \begin{bmatrix} 1 & 0 \\ 0 & 1 \end{bmatrix}$$

Add -3 times the first row to the second.	Multiply the second row by $-\frac{1}{2}$.	Add -2 times the second row to the first.

The three successive row operations can be performed by multiplying on the left successively by

$$E_1 = \begin{bmatrix} 1 & 0 \\ -3 & 1 \end{bmatrix} \qquad E_2 = \begin{bmatrix} 1 & 0 \\ 0 & -\frac{1}{2} \end{bmatrix} \qquad E_3 = \begin{bmatrix} 1 & -2 \\ 0 & 1 \end{bmatrix}$$

Inverting these matrices and using (5.15) yields

$$A = E_1^{-1} E_2^{-1} E_3^{-1} = \begin{bmatrix} 1 & 0 \\ 3 & 1 \end{bmatrix} \begin{bmatrix} 1 & 0 \\ 0 & -2 \end{bmatrix} \begin{bmatrix} 1 & 2 \\ 0 & 1 \end{bmatrix}$$

Reading from right to left and noting that

$$\begin{bmatrix} 1 & 0 \\ 0 & -2 \end{bmatrix} = \begin{bmatrix} 1 & 0 \\ 0 & -1 \end{bmatrix} \begin{bmatrix} 1 & 0 \\ 0 & 2 \end{bmatrix}$$

it follows that the effect of multiplying by A is equivalent to:

(1) shearing by a factor of 2 in the x-direction,
(2) then expanding by a factor of 2 in the y-direction,
(3) then reflecting about the x-axis,
(4) then shearing by a factor of 3 in the y-direction.

The proofs for parts of the following theorem are discussed in the exercises.

Theorem 7. *If* $T:R^2 \to R^2$ *is multiplication by an invertible matrix, then:*

(a) *The image of a straight line is a straight line.*
(b) *The image of a straight line through the origin is a straight line through the origin.*
(c) *The images of parallel straight lines are parallel straight lines.*
(d) *The image of the line segment joining points P and Q is the line segment joining the images of P and Q.*
(e) *The images of three points lie on a line if and only if the points themselves do.*

REMARK. It follows from parts (c), (d), and (e) that multiplication by an invertible 2×2 matrix A maps triangles into triangles and parallelograms into parallelograms.

Example 28

Sketch the image of the square with vertices $P_1(0, 0)$, $P_2(1, 0)$, $P_3(0, 1)$, and $P_4(1, 1)$ under multiplication by

$$A = \begin{bmatrix} -1 & 2 \\ 2 & -1 \end{bmatrix}$$

Solution. Since

$$\begin{bmatrix} -1 & 2 \\ 2 & -1 \end{bmatrix}\begin{bmatrix} 0 \\ 0 \end{bmatrix} = \begin{bmatrix} 0 \\ 0 \end{bmatrix} \qquad \begin{bmatrix} -1 & 2 \\ 2 & -1 \end{bmatrix}\begin{bmatrix} 1 \\ 0 \end{bmatrix} = \begin{bmatrix} -1 \\ 2 \end{bmatrix}$$

$$\begin{bmatrix} -1 & 2 \\ 2 & -1 \end{bmatrix}\begin{bmatrix} 0 \\ 1 \end{bmatrix} = \begin{bmatrix} 2 \\ -1 \end{bmatrix} \qquad \begin{bmatrix} -1 & 2 \\ 2 & -1 \end{bmatrix}\begin{bmatrix} 1 \\ 1 \end{bmatrix} = \begin{bmatrix} 1 \\ 1 \end{bmatrix}$$

the image is a parallelogram with vertices $(0, 0)$, $(-1, 2)$, $(2, -1)$, and $(1, 1)$ (Figure 5.10). ▲

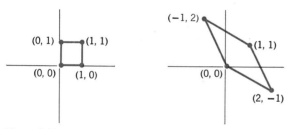

Figure 5.10

Example 29

According to Theorem 7, the invertible matrix

$$A = \begin{bmatrix} 3 & 1 \\ 2 & 1 \end{bmatrix}$$

maps the line $y = 2x + 1$ into another line. Find its equation.

Solution. Let (x, y) be a point on the line $y = 2x + 1$ and let (x', y') be its image under multiplication by A. Then

$$\begin{bmatrix} x' \\ y' \end{bmatrix} = \begin{bmatrix} 3 & 1 \\ 2 & 1 \end{bmatrix} \begin{bmatrix} x \\ y \end{bmatrix}$$

and

$$\begin{bmatrix} x \\ y \end{bmatrix} = \begin{bmatrix} 3 & 1 \\ 2 & 1 \end{bmatrix}^{-1} \begin{bmatrix} x' \\ y' \end{bmatrix} = \begin{bmatrix} 1 & -1 \\ -2 & 3 \end{bmatrix} \begin{bmatrix} x' \\ y' \end{bmatrix}$$

so

$$x = \quad x' - y'$$
$$y = -2x' + 3y'$$

Substituting in $y = 2x + 1$ yields

$$-2x' + 3y' = 2(x' - y') + 1$$

or equivalently

$$y' = \tfrac{4}{5}x' + \tfrac{1}{5}$$

Thus, (x', y') satisfies

$$y = \tfrac{4}{5}x + \tfrac{1}{5}$$

which is the equation we want. ▲

EXERCISE SET 5.3

1. Find the standard matrix of each of the following linear operators.

(a) $T\left(\begin{bmatrix} x_1 \\ x_2 \end{bmatrix}\right) = \begin{bmatrix} 2x_1 - x_2 \\ x_1 + x_2 \end{bmatrix}$

(b) $T\left(\begin{bmatrix} x_1 \\ x_2 \end{bmatrix}\right) = \begin{bmatrix} x_1 \\ x_2 \end{bmatrix}$

(c) $T\left(\begin{bmatrix} x_1 \\ x_2 \\ x_3 \end{bmatrix}\right) = \begin{bmatrix} x_1 + 2x_2 + x_3 \\ x_1 + 5x_2 \\ x_3 \end{bmatrix}$

(d) $T\left(\begin{bmatrix} x_1 \\ x_2 \\ x_3 \end{bmatrix}\right) = \begin{bmatrix} 4x_1 \\ 7x_2 \\ -8x_3 \end{bmatrix}$

2. Find the standard matrix of each of the following linear transformations.

(a) $T\left(\begin{bmatrix} x_1 \\ x_2 \end{bmatrix}\right) = \begin{bmatrix} x_2 \\ -x_1 \\ x_1 + 3x_2 \\ x_1 - x_2 \end{bmatrix}$

(b) $T\left(\begin{bmatrix} x_1 \\ x_2 \\ x_3 \\ x_4 \end{bmatrix}\right) = \begin{bmatrix} 7x_1 + 2x_2 - x_3 + x_4 \\ x_2 + x_3 \\ -x_1 \end{bmatrix}$

(c) $T\left(\begin{bmatrix} x_1 \\ x_2 \\ x_3 \end{bmatrix}\right) = \begin{bmatrix} 0 \\ 0 \\ 0 \\ 0 \\ 0 \end{bmatrix}$

(d) $T\left(\begin{bmatrix} x_1 \\ x_2 \\ x_3 \\ x_4 \end{bmatrix}\right) = \begin{bmatrix} x_4 \\ x_1 \\ x_3 \\ x_2 \\ x_1 - x_3 \end{bmatrix}$

3. Find the standard matrix for the plane linear transformation $T:R^2 \to R^2$ that maps a point (x, y) into (see Figure 5.11):
(a) its reflection about the line $y = -x$
(b) its reflection through the origin
(c) its orthogonal projection on the x-axis
(d) its orthogonal projection on the y-axis.

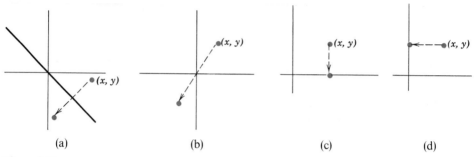

(a)   (b)   (c)   (d)

Figure 5.11

4. For each part of Exercise 3, use the matrix you have obtained to compute $T(2, 1)$. Check your answers geometrically by plotting the points $(2, 1)$ and $T(2, 1)$.

5. Find the standard matrix for the linear operator $T:R^3 \to R^3$ which maps a point (x, y, z) into:
(a) its reflection through the xy-plane
(b) its reflection through the xz-plane
(c) its reflection through the yz-plane.

6. For each part of Exercise 5, use the matrix you have obtained to compute $T(1, 1, 1)$. Check your answers geometrically by sketching the vectors $(1, 1, 1)$ and $T(1, 1, 1)$.

7. Find the standard matrix for the linear operator $T:R^3 \to R^3$ which
(a) rotates each vector $90°$ counterclockwise about the z-axis (looking along the positive z-axis toward the origin)
(b) rotates each vector $90°$ counterclockwise about the x-axis (looking along the positive x-axis toward the origin)
(c) rotates each vector $90°$ counterclockwise about the y-axis (looking along the positive y-axis toward the origin).

8. Sketch the image of the rectangle with vertices $(0, 0)$, $(1, 0)$, $(1, 2)$, and $(0, 2)$ under:
(a) a reflection about the x-axis
(b) a reflection about the y-axis
(c) a compression of factor $k = \frac{1}{4}$ in the y-direction
(d) an expansion of factor $k = 2$ in the x-direction
(e) a shear of factor $k = 3$ in the x-direction
(f) a shear of factor $k = -2$ in the y-direction.

9. Sketch the image of the square with vertices $(0, 0)$, $(1, 0)$, $(0, 1)$, and $(1, 1)$ under multiplication by

$$A = \begin{bmatrix} -3 & 0 \\ 0 & 1 \end{bmatrix}$$

10. Find the matrix that rotates a point (x, y) about the origin through:
(a) 45°   (b) 90°   (c) 180°   (d) 270°   (e) −30°

11. Find the matrix that shears by:
(a) a factor of $k = 4$ in the y-direction
(b) a factor of $k = -2$ in the x-direction.

12. Find the matrix that compresses or expands by:
(a) a factor of $\frac{1}{3}$ in the y-direction
(b) a factor of 6 in the x-direction.

13. In each part, describe the geometric effect of multiplication by the given matrix.

(a) $\begin{bmatrix} 3 & 0 \\ 0 & 1 \end{bmatrix}$ (b) $\begin{bmatrix} 1 & 0 \\ 0 & -5 \end{bmatrix}$ (c) $\begin{bmatrix} 1 & 4 \\ 0 & 1 \end{bmatrix}$

14. Express the matrix as a product of elementary matrices, and then describe the effect of multiplication by the given matrix in terms of compressions, expansions, reflections, and shears.

(a) $\begin{bmatrix} 2 & 0 \\ 0 & 3 \end{bmatrix}$ (b) $\begin{bmatrix} 1 & 4 \\ 2 & 9 \end{bmatrix}$

(c) $\begin{bmatrix} 0 & -2 \\ 4 & 0 \end{bmatrix}$ (d) $\begin{bmatrix} 1 & -3 \\ 4 & 6 \end{bmatrix}$

15. In each part, find a single matrix that performs the indicated succession of operations
(a) compresses by a factor of $\frac{1}{2}$ in the x-direction, then expands by a factor of 5 in the y-direction
(b) expands by a factor of 5 in the y-direction, then shears by a factor of 2 in the y-direction
(c) reflects about $y = x$, then rotates through an angle of 180°.

16. In each part, find a single matrix that performs the indicated succession of operations:
(a) reflects about the y-axis, then expands by a factor of 5 in the x-direction, and then reflects about $y = x$
(b) rotates through 30°, then shears by a factor of -2 in the y-direction, and then expands by a factor of 3 in the y-direction.

17. By matrix inversion, show the following:
(a) The inverse transformation for a reflection about $y = x$ is a reflection about $y = x$.
(b) The inverse transformation for a compression along an axis is an expansion along that axis.
(c) The inverse transformation for a reflection about an axis is a reflection about that axis.
(d) The inverse transformation for a shear along an axis is a shear along that axis.

18. Find the equation of the image of the line $y = -4x + 3$ under multiplication by

$$A = \begin{bmatrix} 4 & -3 \\ 3 & -2 \end{bmatrix}$$

19. In parts (a) through (e) find the equation of the image of the line $y = 2x$ under:
(a) a shear of factor 3 in the x-direction
(b) a compression of factor $\frac{1}{2}$ in the y-direction

(c) a reflection about $y = x$
(d) a reflection about the y-axis
(e) a rotation of $60°$

20. Let l be the line through the origin that makes an angle ϕ with the positive x-axis. Show that

$$\begin{bmatrix} \cos 2\phi & \sin 2\phi \\ \sin 2\phi & -\cos 2\phi \end{bmatrix}$$

is the standard matrix for the linear transformation that maps each point in the plane into its reflection about l. (**Hint.** The transformation can be performed by a rotation through the angle $-\phi$ to align l with the x-axis then a reflection about the x-axis and then a rotation through the angle ϕ to place l back in its original position.)

21. Find the matrix for a shear in the x-direction which transforms the triangle with vertices $(0, 0)$, $(2, 1)$, and $(3, 0)$ into a right triangle with the right angle at the origin.

22. (a) Show that multiplication by

$$A = \begin{bmatrix} 3 & 1 \\ 6 & 2 \end{bmatrix}$$

maps every point in the plane onto the line $y = 2x$.
(b) It follows from (a) that the noncollinear points $(1, 0)$ $(0, 1)$ $(-1, 0)$ are mapped on a line. Does this violate part (e) of Theorem 7?

23. Prove part (a) of Theorem 7. (**Hint.** A line in the plane has an equation $Ax + By + C = 0$, where A and B are not both zero. Use the method of Example 29 to show that the image of this line under multiplication by the invertible matrix

$$\begin{bmatrix} a & b \\ c & d \end{bmatrix}$$

has the equation $A'x + B'y + C = 0$, where $A' = (dA - cB)/(ad - bc)$ and $B' = (-bA + aB)/(ad - bc)$. Then show that A' and B' are not both zero to conclude that the image is a line.)

24. Use the hint in Exercise 23 to prove parts (b) and (c) of Theorem 7.

5.4 MATRICES OF LINEAR TRANSFORMATIONS

In this section we show that if V and W are finite-dimensional vector spaces (not necessarily R^n and R^m), then with a little ingenuity any linear transformation $T:V \to W$ can be regarded as a matrix transformation. The basic idea is to work with coordinate matrices of the vectors rather than with the vectors themselves.

Suppose that V is an n-dimensional vector space and W an m-dimensional vector space. If we choose bases B and B' for V and W, respectively, then for each $\mathbf{x}$ in V, the coordinate matrix $[\mathbf{x}]_B$ will be a vector in R^n, and the coordinate matrix $[T(\mathbf{x})]_{B'}$ will be a vector in R^m. Thus, in the process of mapping $\mathbf{x}$ into $T(\mathbf{x})$, the linear transformation T "generates" a mapping from R^n into R^m by sending $[\mathbf{x}]_B$ into $[T(\mathbf{x})]_{B'}$. It can be shown that this generated mapping is always a linear transformation. As such it can be carried out by using the standard matrix A for this transformation; that is,

$$A[\mathbf{x}]_B = [T(\mathbf{x})]_{B'} \tag{5.16}$$

To find a matrix A satisfying this equation, suppose that V is an n-dimensional space with basis $B = \{\mathbf{u}_1, \mathbf{u}_2, \ldots, \mathbf{u}_n\}$, and W is an m-dimensional space with basis $B' = \{\mathbf{v}_1, \mathbf{v}_2, \ldots, \mathbf{v}_m\}$. We are looking for an $m \times n$ matrix

$$A = \begin{bmatrix} a_{11} & a_{12} & \cdots & a_{1n} \\ a_{21} & a_{22} & \cdots & a_{2n} \\ \vdots & \vdots & & \vdots \\ a_{m1} & a_{m2} & \cdots & a_{mn} \end{bmatrix}$$

such that (5.16) holds for all vectors $\mathbf{x}$ in V. In particular, we want this equation to hold for the basis vectors $\mathbf{u}_1, \mathbf{u}_2, \ldots, \mathbf{u}_n$, that is,

$$A[\mathbf{u}_1]_B = [T(\mathbf{u}_1)]_{B'}, \quad A[\mathbf{u}_2]_B = [T(\mathbf{u}_2)]_{B'}, \ldots, \quad A[\mathbf{u}_n]_B = [T(\mathbf{u}_n)]_{B'} \tag{5.17}$$

But

$$[\mathbf{u}_1]_B = \begin{bmatrix} 1 \\ 0 \\ 0 \\ \vdots \\ 0 \end{bmatrix}, \quad [\mathbf{u}_2]_B = \begin{bmatrix} 0 \\ 1 \\ 0 \\ \vdots \\ 0 \end{bmatrix}, \ldots, \quad [\mathbf{u}_n]_B = \begin{bmatrix} 0 \\ 0 \\ 0 \\ \vdots \\ 1 \end{bmatrix}$$

so

$$A[\mathbf{u}_1]_B = \begin{bmatrix} a_{11} & a_{12} & \cdots & a_{1n} \\ a_{21} & a_{22} & \cdots & a_{2n} \\ \vdots & \vdots & & \vdots \\ a_{m1} & a_{m2} & \cdots & a_{mn} \end{bmatrix} \begin{bmatrix} 1 \\ 0 \\ 0 \\ \vdots \\ 0 \end{bmatrix} = \begin{bmatrix} a_{11} \\ a_{21} \\ \vdots \\ a_{m1} \end{bmatrix}$$

$$A[\mathbf{u}_2]_B = \begin{bmatrix} a_{11} & a_{12} & \cdots & a_{1n} \\ a_{21} & a_{22} & \cdots & a_{2n} \\ \vdots & \vdots & & \vdots \\ a_{m1} & a_{m2} & \cdots & a_{mn} \end{bmatrix} \begin{bmatrix} 0 \\ 1 \\ 0 \\ \vdots \\ 0 \end{bmatrix} = \begin{bmatrix} a_{12} \\ a_{22} \\ \vdots \\ a_{m2} \end{bmatrix}$$

$$\vdots$$

$$A[\mathbf{u}_n]_B = \begin{bmatrix} a_{11} & a_{12} & \cdots & a_{1n} \\ a_{21} & a_{22} & \cdots & a_{2n} \\ \vdots & \vdots & & \vdots \\ a_{m1} & a_{m2} & \cdots & a_{mn} \end{bmatrix} \begin{bmatrix} 0 \\ 0 \\ 0 \\ \vdots \\ 1 \end{bmatrix} = \begin{bmatrix} a_{1n} \\ a_{2n} \\ \vdots \\ a_{mn} \end{bmatrix}$$

Substituting these results into (5.17) yields

$$\begin{bmatrix} a_{11} \\ a_{21} \\ \vdots \\ a_{m1} \end{bmatrix} = [T(\mathbf{u}_1)]_{B'}, \quad \begin{bmatrix} a_{12} \\ a_{22} \\ \vdots \\ a_{m2} \end{bmatrix} = [T(\mathbf{u}_2)]_{B'}, \dots, \quad \begin{bmatrix} a_{1n} \\ a_{2n} \\ \vdots \\ a_{mn} \end{bmatrix} = [T(\mathbf{u}_n)]_{B'}$$

which shows that the successive columns of A are the coordinate matrices of

$$T(\mathbf{u}_1), \quad T(\mathbf{u}_2), \dots, \quad T(\mathbf{u}_n)$$

with respect to the basis B'. The unique matrix A obtained in this way is called the **matrix for T with respect to the bases B and B'**. Symbolically, we can denote this matrix by

$$A = \begin{bmatrix} \text{matrix for } T \text{ with} \\ \text{respect to the} \\ \text{bases } B \text{ and } B' \end{bmatrix} = \left[[T(\mathbf{u}_1)]_{B'} \,\middle|\, [T(\mathbf{u}_2)]_{B'} \,\middle|\, \cdots \,\middle|\, [T(\mathbf{u}_n)]_{B'} \right]$$

The matrix A is commonly denoted by the symbol

$$[T]_{B,B'}$$

so that the foregoing formula can also be written as

$$[T]_{B,B'} = \begin{bmatrix} \text{matrix for } T \text{ with} \\ \text{respect to the} \\ \text{bases } B \text{ and } B' \end{bmatrix} = \left[[T(\mathbf{u}_1)]_{B'} \,\middle|\, [T(\mathbf{u}_2)]_{B'} \,\middle|\, \cdots \,\middle|\, [T(\mathbf{u}_n)]_{B'} \right] \quad (5.18a)$$

where $B = \{\mathbf{u}_1, \mathbf{u}_2, \dots, \mathbf{u}_n\}$.

In the special case where $V = W$ (so that $T: V \to V$ is a linear operator) it is usual to take $B = B'$ when constructing a matrix for T. If this is done, then the resulting matrix is called the **matrix for T with respect to the basis B**. For simplicity, we shall write

$$[T]_B$$

rather than $[T]_{B,B}$. Thus, for a linear operator T, we have

$$[T]_B = \begin{bmatrix} \text{matrix for } T \text{ with} \\ \text{respect to the} \\ \text{basis } B \end{bmatrix} = \left[[T(\mathbf{u}_1)]_B \,\middle|\, [T(\mathbf{u}_2)]_B \,\middle|\, \cdots \,\middle|\, [T(\mathbf{u}_n)]_B \right] \quad (5.18b)$$

where $B = \{\mathbf{u}_1, \mathbf{u}_2, \dots, \mathbf{u}_n\}$.

Example 30

Let $T: P_1 \to P_2$ be the linear transformation defined by

$$T(p(x)) = xp(x)$$

Find the matrix for T with respect to the bases.

$$B = \{\mathbf{u}_1, \mathbf{u}_2\} \qquad \text{and} \qquad B' = \{\mathbf{v}_1, \mathbf{v}_2, \mathbf{v}_3\}$$

where

$$\mathbf{u}_1 = 1, \quad \mathbf{u}_2 = x; \quad \mathbf{v}_1 = 1, \quad \mathbf{v}_2 = x, \quad \mathbf{v}_3 = x^2$$

Solution. From the formula for T we obtain

$$T(\mathbf{u}_1) = T(1) = (x)(1) = x$$
$$T(\mathbf{u}_2) = T(x) = (x)(x) = x^2$$

By inspection, we can determine the coordinate matrices for $T(\mathbf{u}_1)$ and $T(\mathbf{u}_2)$ relative to B'; they are

$$[T(\mathbf{u}_1)]_{B'} = \begin{bmatrix} 0 \\ 1 \\ 0 \end{bmatrix}, \qquad [T(\mathbf{u}_2)]_{B'} = \begin{bmatrix} 0 \\ 0 \\ 1 \end{bmatrix}$$

Thus, the matrix for T with respect to B and B' is

$$[T]_{B,B'} = \left[[T(\mathbf{u}_1)]_{B'} \mid [T(\mathbf{u}_2)]_{B'} \right] = \begin{bmatrix} 0 & 0 \\ 1 & 0 \\ 0 & 1 \end{bmatrix}$$

Example 31

If $B = \{\mathbf{u}_1, \mathbf{u}_2, \dots, \mathbf{u}_n\}$ is any basis for a finite-dimensional vector space V and $I: V \to V$ is the identity operator on V, then $I(\mathbf{u}_1) = \mathbf{u}_1, I(\mathbf{u}_2) = \mathbf{u}_2, \dots, I(\mathbf{u}_n) = \mathbf{u}_n$. Therefore,

$$[I(\mathbf{u}_1)]_B = \begin{bmatrix} 1 \\ 0 \\ 0 \\ \vdots \\ 0 \end{bmatrix}, \quad [I(\mathbf{u}_2)]_B = \begin{bmatrix} 0 \\ 1 \\ 0 \\ \vdots \\ 0 \end{bmatrix}, \dots, \quad [I(\mathbf{u}_n)]_B = \begin{bmatrix} 0 \\ 0 \\ 0 \\ \vdots \\ 1 \end{bmatrix}$$

Thus,

$$[I]_B = \begin{bmatrix} 1 & 0 & \cdots & 0 \\ 0 & 1 & \cdots & 0 \\ 0 & 0 & \cdots & 0 \\ \vdots & \vdots & & \vdots \\ 0 & 0 & \cdots & 1 \end{bmatrix} = I_n$$

Consequently, the matrix of the identity operator with respect to any basis is the $n \times n$ identity matrix. ▲

Example 32

If $T: R^n \to R^m$ is a linear transformation and if B and B' are the standard bases for R^n and R^m, respectively, then the matrix for T with respect to B and B' is just the standard matrix for T discussed in the previous section (Exercise 13). ▲

Example 33

Let $T: R^2 \to R^2$ be the linear operator defined by

$$T\left(\begin{bmatrix} x_1 \\ x_2 \end{bmatrix}\right) = \begin{bmatrix} x_1 + x_2 \\ -2x_1 + 4x_2 \end{bmatrix}$$

Find the matrix for T with respect to the basis $B = \{u_1, u_2\}$ where

(a) $u_1 = \begin{bmatrix} 1 \\ 0 \end{bmatrix}$, $\quad u_2 = \begin{bmatrix} 0 \\ 1 \end{bmatrix}$ $\qquad$ (the standard basis)

(b) $u_1 = \begin{bmatrix} 1 \\ 1 \end{bmatrix}$, $\quad u_2 = \begin{bmatrix} 1 \\ 2 \end{bmatrix}$

Solution (a). Since B is the standard basis for R^2, it follows from Example 32 that $[T]_B$ is the standard matrix for T. But

$$T(u_1) = \begin{bmatrix} 1 \\ -2 \end{bmatrix} \quad \text{and} \quad T(u_2) = \begin{bmatrix} 1 \\ 4 \end{bmatrix}$$

so

$$[T]_B = \begin{bmatrix} \text{standard} \\ \text{matrix} \\ \text{for } T \end{bmatrix} = [T(u_1) \,|\, T(u_2)] = \begin{bmatrix} 1 & 1 \\ -2 & 4 \end{bmatrix}$$

Because B is the standard basis, it follows that $T(u_1) = [T(u_1)]_B$ and $T(u_2) = [T(u_2)]_B$, so that the same matrix would result if we were to use formula (5.18b).

Solution (b). From the definition of T,

$$T(u_1) = \begin{bmatrix} 2 \\ 2 \end{bmatrix} = 2u_1 \quad \text{and} \quad T(u_2) = \begin{bmatrix} 3 \\ 6 \end{bmatrix} = 3u_2$$

Therefore,

$$[T(\mathbf{u}_1)]_B = \begin{bmatrix} 2 \\ 0 \end{bmatrix} \quad \text{and} \quad [T(\mathbf{u}_2)]_B = \begin{bmatrix} 0 \\ 3 \end{bmatrix}$$

Consequently,

$$[T]_B = \left[[T(\mathbf{u}_1)]_B \mid [T(\mathbf{u}_2)]_B \right] = \begin{bmatrix} 2 & 0 \\ 0 & 3 \end{bmatrix}$$

REMARK. Observe that the basis in part (b) of the last example produced a simpler matrix for T than did the standard basis in part (a). As we shall see later, one of the most important problems in linear algebra is to find a basis for a vector space that produces the "simplest" possible matrix for a given linear operator on that space.

Example 34

Let $T:R^2 \to R^3$ be the linear transformation defined by

$$T\left(\begin{bmatrix} x_1 \\ x_2 \end{bmatrix} \right) = \begin{bmatrix} x_2 \\ -5x_1 + 13x_2 \\ -7x_1 + 16x_2 \end{bmatrix}$$

Find the matrix for T with respect to the bases $B = \{\mathbf{u}_1, \mathbf{u}_2\}$ for R^2 and $B' = \{\mathbf{v}_1, \mathbf{v}_2, \mathbf{v}_3\}$ for R^3, where

$$\mathbf{u}_1 = \begin{bmatrix} 3 \\ 1 \end{bmatrix}, \quad \mathbf{u}_2 = \begin{bmatrix} 5 \\ 2 \end{bmatrix}; \quad \mathbf{v}_1 = \begin{bmatrix} 1 \\ 0 \\ -1 \end{bmatrix}, \quad \mathbf{v}_2 = \begin{bmatrix} -1 \\ 2 \\ 2 \end{bmatrix}, \quad \mathbf{v}_3 = \begin{bmatrix} 0 \\ 1 \\ 2 \end{bmatrix}$$

Solution. From the formula for T,

$$T(\mathbf{u}_1) = \begin{bmatrix} 1 \\ -2 \\ -5 \end{bmatrix}, \quad T(\mathbf{u}_2) = \begin{bmatrix} 2 \\ 1 \\ -3 \end{bmatrix}$$

Expressing these vectors as linear combinations of $\mathbf{v}_1$, $\mathbf{v}_2$, and $\mathbf{v}_3$ we obtain (verify):

$$T(\mathbf{u}_1) = \mathbf{v}_1 - 2\mathbf{v}_3, \quad T(\mathbf{u}_2) = 3\mathbf{v}_1 + \mathbf{v}_2 - \mathbf{v}_3$$

Thus,

$$[T(\mathbf{u}_1)]_{B'} = \begin{bmatrix} 1 \\ 0 \\ -2 \end{bmatrix}, \quad [T(\mathbf{u}_2)]_{B'} = \begin{bmatrix} 3 \\ 1 \\ -1 \end{bmatrix}$$

so

$$[T]_{B,B'} = \left[[T(\mathbf{u}_1)]_{B'} \ \ [T(\mathbf{u}_2)]_{B'} \right] = \begin{bmatrix} 1 & 3 \\ 0 & 1 \\ -2 & -1 \end{bmatrix}$$

If $T:V \rightarrow W$ is a linear transformation, then with the notation of (5.18a), formula (5.16) can be written as

$$[T]_{B,B'}[\mathbf{x}]_B = [T(\mathbf{x})]_{B'} \qquad (5.19a)$$

and if $T:V \rightarrow V$ is a linear operator, then from (5.18b) and (5.16)

$$[T]_B[\mathbf{x}]_B = [T(\mathbf{x})]_B \qquad (5.19b)$$

Phrased informally, these formulas state that *the matrix for T times the coordinate matrix for* **x** *is the coordinate matrix for T(***x***).*

If T is a linear transformation, then as shown in Figure 5.12, the matrix $[T]_{B,B'}$ can be used to calculate $T(\mathbf{x})$ in three steps by the following *indirect* procedure:

(a) Compute the coordinate matrix $[\mathbf{x}]_B$.
(b) Multiply $[\mathbf{x}]_B$ on the left by $[T]_{B,B'}$ to produce $[T(\mathbf{x})]_{B'}$.
(c) Reconstruct $T(\mathbf{x})$ from its coordinate matrix $[T(\mathbf{x})]_{B'}$.

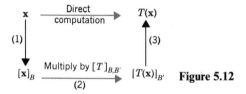

Figure 5.12

There are two major reasons why this indirect procedure is important, one quite practical and the other theoretical:

1. This procedure makes it possible to carry out linear transformations on a computer using matrix multiplication.

2. The procedure shows that by working with coordinate vectors, all linear transformations on finite-dimensional vector spaces can be represented as matrix transformations. Thus, answers to theoretical questions about general linear transformations on finite-dimensional vector spaces can often be obtained by studying just the matrix transformations. Such matters are considered in detail in more advanced linear algebra courses.

Example 35

Let $T:P_1 \to P_2$, B, and B' be as in Example 30, and let

$$\mathbf{x} = 1 - 2x$$

Use the matrix $[T]_{B,B'}$ obtained in Example 30 to compute $T(\mathbf{x})$ by the indirect procedure in Figure 5.12.

Solution. By inspection, the coordinate matrix of $\mathbf{x}$ with respect to B is

$$[\mathbf{x}]_B = \begin{bmatrix} 1 \\ -2 \end{bmatrix}$$

Therefore, from (5.19a),

$$[T(\mathbf{x})]_{B'} = [T]_{B,B'}[\mathbf{x}]_B = \begin{bmatrix} 0 & 0 \\ 1 & 0 \\ 0 & 1 \end{bmatrix} \begin{bmatrix} 1 \\ -2 \end{bmatrix} = \begin{bmatrix} 0 \\ 1 \\ -2 \end{bmatrix}$$

Thus,

$$T(\mathbf{x}) = 0\mathbf{v}_1 + 1\mathbf{v}_2 - 2\mathbf{v}_3 = 0(1) + 1(x) - 2(x^2) = x - 2x^2 \quad \blacktriangle$$

Example 36

Let $T:R^2 \to R^2$ be the linear operator defined by

$$T\left(\begin{bmatrix} x_1 \\ x_2 \end{bmatrix}\right) = \begin{bmatrix} x_1 + x_2 \\ -2x_1 + 4x_2 \end{bmatrix}$$

and $B = \{\mathbf{u}_1, \mathbf{u}_2\}$ the basis for R^2 with vectors

$$\mathbf{u}_1 = \begin{bmatrix} 1 \\ 1 \end{bmatrix} \quad \text{and} \quad \mathbf{u}_2 = \begin{bmatrix} 1 \\ 2 \end{bmatrix}$$

Use the indirect procedure in Figure 5.12 (with $B' = B$ and $[T]_{B,B'} = [T]_B$) to find $T(\mathbf{x})$, where

$$\mathbf{x} = \begin{bmatrix} 3 \\ 9 \end{bmatrix}$$

Solution. The coordinate matrix for $\mathbf{x}$ with respect to B is (verify)

$$[\mathbf{x}]_B = \begin{bmatrix} -3 \\ 6 \end{bmatrix}$$

Therefore, using formula (5.19b) and the matrix $[T]_B$ found in part (b) of Example 33, we obtain

$$[T(\mathbf{x})]_B = [T]_B[\mathbf{x}]_B = \begin{bmatrix} 2 & 0 \\ 0 & 3 \end{bmatrix} \begin{bmatrix} -3 \\ 6 \end{bmatrix} = \begin{bmatrix} -6 \\ 18 \end{bmatrix}$$

Thus,

$$T(\mathbf{x}) = -6\begin{bmatrix} 1 \\ 1 \end{bmatrix} + 18\begin{bmatrix} 1 \\ 2 \end{bmatrix} = \begin{bmatrix} 12 \\ 30 \end{bmatrix}$$

The reader should check this result by substituting $\mathbf{x}$ directly into the formula for T. ▲

REMARK. The last example is intended as an exercise to help you understand the concepts, not as a practical procedure for computing $T(\mathbf{x})$. Obviously, it is easier to calculate $T(\mathbf{x})$ directly rather than using the indirect procedure.

EXERCISE SET 5.4

1. Let $T:P_2 \to P_1$ be the linear transformation defined by

$$T(a_0 + a_1 x + a_2 x^2) = (a_0 + a_1) - (2a_1 + 3a_2)x$$

Find the matrix of T with respect to the standard bases for P_2 and P_1.

2. Let $T:R^2 \to R^3$ be defined by

$$T\left(\begin{bmatrix} x_1 \\ x_2 \end{bmatrix}\right) = \begin{bmatrix} x_1 + 2x_2 \\ -x_1 \\ 0 \end{bmatrix}$$

(a) Find the matrix of T with respect to the bases $B = \{\mathbf{u}_1, \mathbf{u}_2\}$ and $B' = \{\mathbf{v}_1, \mathbf{v}_2, \mathbf{v}_3\}$, where

$$\mathbf{u}_1 = \begin{bmatrix} 1 \\ 3 \end{bmatrix} \quad \mathbf{u}_2 = \begin{bmatrix} -2 \\ 4 \end{bmatrix} \quad \mathbf{v}_1 = \begin{bmatrix} 1 \\ 1 \\ 1 \end{bmatrix} \quad \mathbf{v}_2 = \begin{bmatrix} 2 \\ 2 \\ 0 \end{bmatrix} \quad \mathbf{v}_3 = \begin{bmatrix} 3 \\ 0 \\ 0 \end{bmatrix}$$

(b) Use the matrix obtained in (a) to compute

$$T\left(\begin{bmatrix} 8 \\ 3 \end{bmatrix}\right)$$

3. Let $T:R^3 \to R^3$ be defined by

$$T\left(\begin{bmatrix} x_1 \\ x_2 \\ x_3 \end{bmatrix}\right) = \begin{bmatrix} x_1 - x_2 \\ x_2 - x_1 \\ x_1 - x_3 \end{bmatrix}$$

(a) Find the matrix of T with respect to the basis $B = \{\mathbf{v}_1, \mathbf{v}_2, \mathbf{v}_3\}$, where

$$\mathbf{v}_1 = \begin{bmatrix} 1 \\ 0 \\ 1 \end{bmatrix} \quad \mathbf{v}_2 = \begin{bmatrix} 0 \\ 1 \\ 1 \end{bmatrix} \quad \mathbf{v}_3 = \begin{bmatrix} 1 \\ 1 \\ 0 \end{bmatrix}$$

(b) Use the matrix obtained in (a) to compute

$$T\left(\begin{bmatrix} 2 \\ 0 \\ 0 \end{bmatrix}\right)$$

4. Let $T:P_2 \to P_4$ be the linear transformation defined by $T(p(x)) = x^2p(x)$.
   (a) Find the matrix for T with respect to the bases $B = \{\mathbf{p}_1, \mathbf{p}_2, \mathbf{p}_3\}$ and B', where $\mathbf{p}_1 = 1 + x^2$, $\mathbf{p}_2 = 1 + 2x + 3x^2$, $\mathbf{p}_3 = 4 + 5x + x^2$, and B' is the standard basis for P_4.
   (b) Use the matrix obtained in (a) to compute $T(-3 + 5x - 2x^2)$.

5. Let $\mathbf{v}_1 = \begin{bmatrix} 1 \\ 3 \end{bmatrix}$ and $\mathbf{v}_2 = \begin{bmatrix} -1 \\ 4 \end{bmatrix}$, and let

$$A = \begin{bmatrix} 1 & 3 \\ -2 & 5 \end{bmatrix}$$

be the matrix for $T:R^2 \to R^2$ with respect to the basis $B = \{\mathbf{v}_1, \mathbf{v}_2\}$.
   (a) Find $[T(\mathbf{v}_1)]_B$ and $[T(\mathbf{v}_2)]_B$.
   (b) Find $T(\mathbf{v}_1)$ and $T(\mathbf{v}_2)$.
   (c) Find a formula for $T\left(\begin{bmatrix} x_1 \\ x_2 \end{bmatrix}\right)$.
   (d) Use the formula obtained in (c) to compute $T\left(\begin{bmatrix} 1 \\ 1 \end{bmatrix}\right)$.

6. Let $A = \begin{bmatrix} 3 & -2 & 1 & 0 \\ 1 & 6 & 2 & 1 \\ -3 & 0 & 7 & 1 \end{bmatrix}$ be the matrix of $T:R^4 \to R^3$ with respect to the bases

$B = \{\mathbf{v}_1, \mathbf{v}_2, \mathbf{v}_3, \mathbf{v}_4\}$ and $B' = \{\mathbf{w}_1, \mathbf{w}_2, \mathbf{w}_3\}$, where

$$\mathbf{v}_1 = \begin{bmatrix} 0 \\ 1 \\ 1 \\ 1 \end{bmatrix} \quad \mathbf{v}_2 = \begin{bmatrix} 2 \\ 1 \\ -1 \\ -1 \end{bmatrix} \quad \mathbf{v}_3 = \begin{bmatrix} 1 \\ 4 \\ -1 \\ 2 \end{bmatrix} \quad \mathbf{v}_4 = \begin{bmatrix} 6 \\ 9 \\ 4 \\ 2 \end{bmatrix}$$

$$\mathbf{w}_1 = \begin{bmatrix} 0 \\ 8 \\ 8 \end{bmatrix} \quad \mathbf{w}_2 = \begin{bmatrix} -7 \\ 8 \\ 1 \end{bmatrix} \quad \mathbf{w}_3 = \begin{bmatrix} -6 \\ 9 \\ 1 \end{bmatrix}$$

   (a) Find $[T(\mathbf{v}_1)]_{B'}$, $[T(\mathbf{v}_2)]_{B'}$, $[T(\mathbf{v}_3)]_{B'}$, and $[T(\mathbf{v}_4)]_{B'}$.
   (b) Find $T(\mathbf{v}_1)$, $T(\mathbf{v}_2)$, $T(\mathbf{v}_3)$, and $T(\mathbf{v}_4)$.
   (c) Find a formula for $T\left(\begin{bmatrix} x_1 \\ x_2 \\ x_3 \\ x_4 \end{bmatrix}\right)$.
   (d) Use the formula obtained in (c) to compute $T\left(\begin{bmatrix} 2 \\ 2 \\ 0 \\ 0 \end{bmatrix}\right)$.

7. Let $A = \begin{bmatrix} 1 & 3 & -1 \\ 2 & 0 & 5 \\ 6 & -2 & 4 \end{bmatrix}$ be the matrix of $T: P_2 \rightarrow P_2$ with respect to the basis $B =$

$\{v_1, v_2, v_3\}$, where $v_1 = 3x + 3x^2$, $v_2 = -1 + 3x + 2x^2$, $v_3 = 3 + 7x + 2x^2$.

(a) Find $[T(v_1)]_B$, $[T(v_2)]_B$, and $[T(v_3)]_B$.
(b) Find $T(v_1)$, $T(v_2)$, and $T(v_3)$.
(c) Find a formula for $T(a_0 + a_1x + a_2x^2)$.
(d) Use the formula obtained in (c) to compute $T(1 + x^2)$.

8. Show that if $T: V \rightarrow W$ is the zero transformation (Example 3), then the matrix of T with respect to any bases for V and W is a zero matrix.

9. Show that if $T: V \rightarrow V$ is a contraction or a dilation of V (Example 5), then the matrix of T with respect to any basis for V is a diagonal matrix.

10. Let $B = \{v_1, v_2, v_3, v_4\}$ be a basis for a vector space V. Find the matrix with respect to B of the linear operator $T: V \rightarrow V$ defined by $T(v_1) = v_2$, $T(v_2) = v_3$, $T(v_3) = v_4$, $T(v_4) = v_1$.

11. **(For readers who have studied calculus.)** Let $D: P_2 \rightarrow P_2$ be the differentiation operator $D(p) = p'$. In parts (a) and (b) find the matrix of D with respect to the basis $B = \{p_1, p_2, p_3\}$.
(a) $p_1 = 1$, $p_2 = x$, $p_3 = x^2$
(b) $p_1 = 2$, $p_2 = 2 - 3x$, $p_3 = 2 - 3x + 8x^2$
(c) Use the matrix in part (a) to compute $D(6 - 6x + 24x^2)$.
(d) Repeat the directions of part (c) for the matrix in part (b).

12. **(For readers who have studied calculus.)** In each part, $B = (f_1, f_2, f_3)$ is a basis for a subspace V of the vector space of real-valued functions defined on the real line. Find the matrix with respect to B of the differentiation operator $D: V \rightarrow V$.
(a) $f_1 = 1$, $f_2 = \sin x$, $f_3 = \cos x$
(b) $f_1 = 1$, $f_2 = e^x$, $f_3 = e^{2x}$
(c) $f_1 = e^{2x}$, $f_2 = xe^{2x}$, $f_3 = x^2e^{2x}$

13. Prove: If B and B' are the standard bases for R^n and R^m, respectively, then the matrix for a linear transformation $T: R^n \rightarrow R^m$ with respect to the bases B and B' is the standard matrix for T.

5.5 SIMILARITY

The matrix of a linear operator $T: V \rightarrow V$ depends on the basis selected for V. One of the fundamental problems of linear algebra is to choose a basis for V that makes the matrix for T as simple as possible. Often this problem is attacked by first finding a matrix for T relative to some "simple" basis like a standard basis. Usually, this choice does not yield the simplest matrix for T, so that one then looks for a way to change the basis in order to simplify the matrix. In order to address a problem like this, we must know how a change of basis affects the matrix of a linear operator.

The following theorem explains this.

Theorem 8. *Let* $T:V \to V$ *be a linear operator on a finite-dimensional vector space V. If A is the matrix of T with respect to a basis B, and A' is the matrix of T with respect to a basis B', then*

$$A' = P^{-1}AP \qquad (5.20)$$

where P is the transition matrix from B' to B.

Proof. To establish this theorem, it is convenient to describe the relationship

$$A\mathbf{u} = \mathbf{v}$$

pictorially by writing

$$\mathbf{u} \xrightarrow{\ A\ } \mathbf{v}$$

Since A is the matrix of T with respect to B, and A' is the matrix of T with respect to B', the following relationships hold for all $\mathbf{x}$ in V.

$$A[\mathbf{x}]_B = [T(\mathbf{x})]_B$$

and

$$A'[\mathbf{x}]_{B'} = [T(\mathbf{x})]_{B'}$$

These can be written

$$[\mathbf{x}]_B \xrightarrow{\ A\ } [T(\mathbf{x})]_B$$

and $\qquad\qquad\qquad\qquad\qquad\qquad\qquad\qquad\qquad\qquad\qquad$ (5.21)

$$[\mathbf{x}]_{B'} \xrightarrow{\ A'\ } [T(\mathbf{x})]_{B'}$$

To see how the matrices A and A' are related, let P be the transition matrix from the B' basis to the B basis, so that P^{-1} is the transition matrix from B to B'. Thus,

$$P[\mathbf{x}]_{B'} = [\mathbf{x}]_B$$

and

$$P^{-1}[T(\mathbf{x})]_B = [T(\mathbf{x})]_{B'}$$

which can be written as

$$[\mathbf{x}]_{B'} \xrightarrow{\ P\ } [\mathbf{x}]_B$$

and $\qquad\qquad\qquad\qquad\qquad\qquad\qquad\qquad\qquad\qquad\qquad$ (5.22)

$$[T(\mathbf{x})]_B \xrightarrow{\ P^{-1}\ } [T(\mathbf{x})]_{B'}$$

For compactness, relationships (5.21) and (5.22) can be linked together in a single figure as follows:

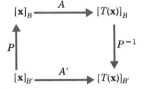

This figure illustrates that there are two ways to obtain the matrix $[T(\mathbf{x})]_{B'}$ from the matrix $[\mathbf{x}]_{B'}$. We can take the bottom path across the figure; that is,

$$A'[\mathbf{x}]_{B'} = [T(\mathbf{x})]_{B'} \qquad (5.23)$$

or we can go up the left side, across the top, and down the right side; that is,

$$P^{-1}AP[\mathbf{x}]_{B'} = [T(\mathbf{x})]_{B'} \qquad (5.24)$$

It follows from (5.23) and (5.24) that

$$P^{-1}AP[\mathbf{x}]_{B'} = A'[\mathbf{x}]_{B'} \qquad (5.25)$$

for all $\mathbf{x}$ in V. It follows from (5.25) and part (b) of Exercise 11 that

$$P^{-1}AP = A'$$

This proves Theorem 8. ▮

Warning. *When applying Theorem 8, it is easy to forget whether P is the transition matrix from B to B' (incorrect) or from B' to B (correct). It may help to call B the old basis, B' the new basis, A the old matrix, and A' the new matrix. Since P is the transition matrix from B' to B, P⁻¹ is the transition matrix from B to B'. Thus, (5.20) can be expressed as*:

$$new\ matrix = P^{-1}(old\ matrix)P$$

where P is the transition matrix from the new basis to the old basis.

For a more notational way to remember this formula, we can use the notation $[T]_B$ and $[T]_{B'}$ introduced in the last section. With this notation (5.20) can be expressed as:

$$[T]_{B'} = P^{-1}[T]_{B}P$$
$$\qquad\qquad\uparrow\qquad\qquad\uparrow$$
$$\qquad B\ to\ B'\qquad B'\ to\ B$$

Example 37

Let $T:R^2 \to R^2$ be defined by

$$T\left(\begin{bmatrix} x_1 \\ x_2 \end{bmatrix}\right) = \begin{bmatrix} x_1 + x_2 \\ -2x_1 + 4x_2 \end{bmatrix}$$

Find the standard matrix for T, that is, the matrix for T relative to the basis $B = \{\mathbf{e}_1, \mathbf{e}_2\}$, where

$$\mathbf{e}_1 = \begin{bmatrix} 1 \\ 0 \end{bmatrix} \qquad \mathbf{e}_2 = \begin{bmatrix} 0 \\ 1 \end{bmatrix}$$

and then use Theorem 8 to transform this matrix into the matrix for T relative to the basis $B' = \{\mathbf{u}_1', \mathbf{u}_2'\}$, where

$$\mathbf{u}_1' = \begin{bmatrix} 1 \\ 1 \end{bmatrix} \qquad \text{and} \qquad \mathbf{u}_2' = \begin{bmatrix} 1 \\ 2 \end{bmatrix}$$

Solution. In part (a) of Example 33 we found the matrix for T relative to the standard basis B to be

$$[T]_B = \begin{bmatrix} 1 & 1 \\ -2 & 4 \end{bmatrix}$$

Next we need the transition matrix from B' to B. For this transition matrix we need the coordinate matrices for the B' basis vectors relative to the basis B. By inspection,

$$\mathbf{u}_1' = \mathbf{e}_1 + \mathbf{e}_2$$
$$\mathbf{u}_2' = \mathbf{e}_1 + 2\mathbf{e}_2$$

so that

$$[\mathbf{u}_1']_B = \begin{bmatrix} 1 \\ 1 \end{bmatrix} \qquad \text{and} \qquad [\mathbf{u}_2']_B = \begin{bmatrix} 1 \\ 2 \end{bmatrix}$$

Thus, the transition matrix from B' to B is

$$P = \begin{bmatrix} 1 & 1 \\ 1 & 2 \end{bmatrix}$$

The reader can check that

$$P^{-1} = \begin{bmatrix} 2 & -1 \\ -1 & 1 \end{bmatrix}$$

so that by Theorem 8 the matrix of T relative to the basis B' is

$$P^{-1}[T]_B P = \begin{bmatrix} 2 & -1 \\ -1 & 1 \end{bmatrix} \begin{bmatrix} 1 & 1 \\ -2 & 4 \end{bmatrix} \begin{bmatrix} 1 & 1 \\ 1 & 2 \end{bmatrix} = \begin{bmatrix} 2 & 0 \\ 0 & 3 \end{bmatrix}$$

This agrees with the result obtained in part (b) of Example 33. ▲

This example illustrates that the standard basis for a vector space does not necessarily produce the simplest matrix for a linear operator; we saw that the standard matrix

$$A = \begin{bmatrix} 1 & 1 \\ -2 & 4 \end{bmatrix}$$

was not as simple in structure as the matrix

$$\begin{bmatrix} 2 & 0 \\ 0 & 3 \end{bmatrix} \tag{5.26}$$

relative to the basis B'. Matrix (5.26) is an example of a *diagonal matrix*, that is, a square matrix all of whose *non*diagonal entries are zeros. Diagonal matrices have many desirable properties. For example, the kth power of a diagonal matrix

$$D = \begin{bmatrix} d_1 & 0 & \cdots & 0 \\ 0 & d_2 & \cdots & 0 \\ \vdots & \vdots & & \vdots \\ 0 & 0 & \cdots & d_n \end{bmatrix}$$

is

$$D^k = \begin{bmatrix} d_1{}^k & 0 & \cdots & 0 \\ 0 & d_2{}^k & \cdots & 0 \\ \vdots & \vdots & & \vdots \\ 0 & 0 & \cdots & d_n{}^k \end{bmatrix}$$

Thus, to raise a diagonal matrix to the kth power, we need only raise each diagonal entry to the kth power. For a nondiagonal matrix there is much more computation involved in obtaining the kth power. Diagonal matrices also have other useful properties. In the next chapter we shall discuss the problem of finding bases that produce diagonal matrices for linear operators.

Theorem 8 motivates the following definition.

Definition. If A and B are square matrices, we say that **B is similar to A** if there is an invertible matrix P such that $B = P^{-1}AP$.

Note that the equation $B = P^{-1}AP$ can be rewritten as

$$A = PBP^{-1} \qquad \text{or} \qquad A = (P^{-1})^{-1}BP^{-1}$$

Letting $Q = P^{-1}$ yields

$$A = Q^{-1}BQ$$

which says that A is similar to B. Therefore, B is similar to A if and only if A is similar to B; consequently, we shall usually say simply that *A and B are similar*.

In this terminology, Theorem 8 asserts that *two matrices representing the same linear operator $T: V \rightarrow V$ with respect to different bases are similar.*

EXERCISE SET 5.5

In Exercises 1–7 find the matrix of T with respect to B, and use Theorem 8 to compute the matrix of T with respect to B'.

1. $T:R^2 \to R^2$ is defined by

$$T\left(\begin{bmatrix} x_1 \\ x_2 \end{bmatrix}\right) = \begin{bmatrix} x_1 - 2x_2 \\ -x_2 \end{bmatrix}$$

$B = \{\mathbf{u}_1, \mathbf{u}_2\}$ and $B' = \{\mathbf{v}_1, \mathbf{v}_2\}$, where

$$\mathbf{u}_1 = \begin{bmatrix} 1 \\ 0 \end{bmatrix} \quad \mathbf{u}_2 = \begin{bmatrix} 0 \\ 1 \end{bmatrix} \quad \mathbf{v}_1 = \begin{bmatrix} 2 \\ 1 \end{bmatrix} \quad \text{and} \quad \mathbf{v}_2 = \begin{bmatrix} -3 \\ 4 \end{bmatrix}$$

2. $T:R^2 \to R^2$ is defined by

$$T\left(\begin{bmatrix} x_1 \\ x_2 \end{bmatrix}\right) = \begin{bmatrix} x_1 + 7x_2 \\ 3x_1 - 4x_2 \end{bmatrix}$$

$B = \{\mathbf{u}_1, \mathbf{u}_2\}$ and $B' = \{\mathbf{v}_1, \mathbf{v}_2\}$, where

$$\mathbf{u}_1 = \begin{bmatrix} 2 \\ 2 \end{bmatrix} \quad \mathbf{u}_2 = \begin{bmatrix} 4 \\ -1 \end{bmatrix} \quad \mathbf{v}_1 = \begin{bmatrix} 1 \\ 3 \end{bmatrix} \quad \mathbf{v}_2 = \begin{bmatrix} -1 \\ -1 \end{bmatrix}$$

3. $T:R^2 \to R^2$ is the rotation about the origin through $45°$; B and B' are the bases in Exercise 1.

4. $T:R^3 \to R^3$ is defined by

$$T\left(\begin{bmatrix} x_1 \\ x_2 \\ x_3 \end{bmatrix}\right) = \begin{bmatrix} x_1 + 2x_2 - x_3 \\ -x_2 \\ x_1 + 7x_3 \end{bmatrix}$$

B is the standard basis for R^3 and $B' = \{\mathbf{v}_1, \mathbf{v}_2, \mathbf{v}_3\}$, where

$$\mathbf{v}_1 = \begin{bmatrix} 1 \\ 0 \\ 0 \end{bmatrix}, \quad \mathbf{v}_2 = \begin{bmatrix} 1 \\ 1 \\ 0 \end{bmatrix}, \quad \text{and} \quad \mathbf{v}_3 = \begin{bmatrix} 1 \\ 1 \\ 1 \end{bmatrix}$$

5. $T:R^3 \to R^3$ is the orthogonal projection on the xy-plane; B and B' are as in Exercise 4.

6. $T:R^2 \to R^2$ is defined by $T(\mathbf{x}) = 5\mathbf{x}$; B and B' are the bases in Exercise 2.

7. $T:P_1 \to P_1$ is defined by $T(a_0 + a_1x) = a_0 + a_1(x + 1)$; $B = \{\mathbf{p}_1, \mathbf{p}_2\}$ and $B' = \{\mathbf{q}_1, \mathbf{q}_2\}$, where $\mathbf{p}_1 = 6 + 3x$, $\mathbf{p}_2 = 10 + 2x$, $\mathbf{q}_1 = 2$, $\mathbf{q}_2 = 3 + 2x$.

8. Prove that if A and B are similar matrices, then $\det(A) = \det(B)$.

9. Prove that similar matrices have the same rank.

10. Prove that if A and B are similar matrices, then A^2 and B^2 are also similar. More generally, prove that A^k and B^k are similar, where k is any positive integer.

11. Let C and D be $m \times n$ matrices. Prove:
   (a) If $C\mathbf{x} = D\mathbf{x}$ for all $\mathbf{x}$ in R^n, then $C = D$.
   (b) If $B = \{\mathbf{v}_1, \mathbf{v}_2, \ldots, \mathbf{v}_n\}$ is a basis for a vector space V and $C[\mathbf{x}]_B = D[\mathbf{x}]_B$ for all $\mathbf{x}$ in V, then $C = D$.

SUPPLEMENTARY EXERCISES

1. Let A be an $n \times n$ matrix, B a nonzero $n \times 1$ matrix, and $\mathbf{x}$ a vector in R^n expressed in matrix notation. Is $T(\mathbf{x}) = A\mathbf{x} + B$ a linear operator on R^n? Justify your answer.

2. Let

$$A = \begin{bmatrix} \cos\theta & -\sin\theta \\ \sin\theta & \cos\theta \end{bmatrix}$$

(a) Show that

$$A^2 = \begin{bmatrix} \cos 2\theta & -\sin 2\theta \\ \sin 2\theta & \cos 2\theta \end{bmatrix} \quad \text{and} \quad A^3 = \begin{bmatrix} \cos 3\theta & -\sin 3\theta \\ \sin 3\theta & \cos 3\theta \end{bmatrix}$$

(b) Guess the form of the matrix A^n for any positive integer n.
(c) By considering the geometric effect of $T: R^2 \to R^2$, where T is multiplication by A, obtain the result in (b) geometrically.

3. Let $\mathbf{v}_0$ be a fixed vector in an inner product space V, and let $T: V \to V$ be defined by $T(\mathbf{v}) = \langle \mathbf{v}, \mathbf{v}_0 \rangle \mathbf{v}_0$. Show that T is a linear operator on V.

4. Let $\mathbf{v}_1, \mathbf{v}_2, \ldots, \mathbf{v}_m$ be fixed vectors in R^n and let $T: R^n \to R^m$ be defined by $T(\mathbf{x}) = (\mathbf{x} \cdot \mathbf{v}_1, \mathbf{x} \cdot \mathbf{v}_2, \ldots, \mathbf{x} \cdot \mathbf{v}_m)$, where $\mathbf{x} \cdot \mathbf{v}_i$ is the Euclidean inner product on R^n.
(a) Show that T is a linear transformation.
(b) Show that the matrix with row vectors $\mathbf{v}_1, \mathbf{v}_2, \ldots, \mathbf{v}_m$ is the standard matrix for T.

5. Let $\{\mathbf{e}_1, \mathbf{e}_2, \mathbf{e}_3, \mathbf{e}_4\}$ be the standard basis for R^4 and $T: R^4 \to R^3$ the linear transformation for which

$$\begin{aligned} T(\mathbf{e}_1) &= (1, 2, 1) & T(\mathbf{e}_2) &= (0, 1, 0) \\ T(\mathbf{e}_3) &= (1, 3, 0) & T(\mathbf{e}_4) &= (1, 1, 1) \end{aligned}$$

(a) Find bases for the range and kernel of T.
(b) Find the rank and nullity of T.

6. Let $A\mathbf{x} = \mathbf{0}$ be a homogeneous linear system with n unknowns, and let r be the rank of A. Find the dimension of the solution space if:
(a) $n = 5, r = 3$ (b) $n = 4, r = 4$

7. (a) Let l be the line in the xy-plane that passes through the origin and makes an angle ϕ with the positive x-axis; and let $P_0(x_0, y_0)$ be the orthogonal projection of a point $P(x, y)$ onto l (see Figure 5.13).

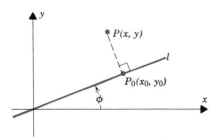

Figure 5.13

This projection can be accomplished by rotating the plane through the angle $-\phi$ to align l with the x-axis, then projecting the image of P under the rotation onto the x-axis, and then rotating the plane through the angle ϕ to return l to its original position. Use this observation to show that

$$\begin{bmatrix} x_0 \\ y_0 \end{bmatrix} = \begin{bmatrix} \cos^2 \phi & \sin \phi \cos \phi \\ \sin \phi \cos \phi & \sin^2 \phi \end{bmatrix} \begin{bmatrix} x \\ y \end{bmatrix}$$

(b) Use the result in (a) to find the orthogonal projection of the point $(1, 5)$ onto the line through the origin that makes an angle $\phi = 30°$ with the positive x-axis.

8. Let V and W be vector spaces, T, T_1, and T_2 linear transformations from V to W, and k a scalar. Define new transformations, $T_1 + T_2$ and kT, by the formulas

$$(T_1 + T_2)(\mathbf{x}) = T_1(\mathbf{x}) + T_2(\mathbf{x})$$
$$(kT)(\mathbf{x}) = k[T(\mathbf{x})]$$

(a) Show that $(T_1 + T_2): V \to W$ and $kT: V \to W$ are linear transformations.
(b) Show that the set of all linear transformations from V to W with the operations in (a) forms a vector space.

9. Let A and B be similar matrices. Prove:
(a) A^t and B^t are similar.
(b) A is invertible if and only if B is invertible.
(c) If A and B are invertible, then A^{-1} and B^{-1} are similar.

10. **(Fredholm Alternative Theorem.)** Let $T: V \to V$ be a linear operator on an n-dimensional vector space. Prove that exactly one of the following statements holds:
(i) The equation $T(\mathbf{x}) = \mathbf{b}$ has a solution for all vectors $\mathbf{b}$ in V.
(ii) Nullity of $T > 0$.

11. Let $T: M_{22} \to M_{22}$ be the linear operator defined by

$$T(X) = \begin{bmatrix} 1 & 1 \\ 0 & 0 \end{bmatrix} X + X \begin{bmatrix} 0 & 0 \\ 1 & 1 \end{bmatrix}$$

Find the rank and nullity of T.

12. Prove: If A and B are similar matrices, and if B and C are similar matrices, then A and C are similar matrices.

13. Let $T: M_{22} \to M_{22}$ be the linear operator defined by $T(M) = M^t$. Find the matrix for T with respect to the standard basis for M_{22}.

14. Let $B = \{\mathbf{u}_1, \mathbf{u}_2, \mathbf{u}_3\}$ and $B' = \{\mathbf{v}_1, \mathbf{v}_2, \mathbf{v}_3\}$ be bases for a vector space V, and let

$$P = \begin{bmatrix} 2 & -1 & 3 \\ 1 & 1 & 4 \\ 0 & 1 & 2 \end{bmatrix}$$

be the transition matrix from B' to B.

(a) Express $\mathbf{v}_1$, $\mathbf{v}_2$, $\mathbf{v}_3$ as linear combinations of $\mathbf{u}_1$, $\mathbf{u}_2$, $\mathbf{u}_3$.

(b) Express $\mathbf{u}_1$, $\mathbf{u}_2$, $\mathbf{u}_3$ as linear combinations of $\mathbf{v}_1$, $\mathbf{v}_2$, $\mathbf{v}_3$.

15. Let $B = \{\mathbf{u}_1, \mathbf{u}_2, \mathbf{u}_3\}$ be a basis for a vector space V and $T: V \rightarrow V$ a linear operator such that

$$[T]_B = \begin{bmatrix} -3 & 4 & 7 \\ 1 & 0 & -2 \\ 0 & 1 & 0 \end{bmatrix}$$

Find $[T]_{B'}$, where $B' = \{\mathbf{v}_1, \mathbf{v}_2, \mathbf{v}_3\}$ is the basis for V defined by

$$\mathbf{v}_1 = \mathbf{u}_1, \qquad \mathbf{v}_2 = \mathbf{u}_1 + \mathbf{u}_2, \qquad \mathbf{v}_3 = \mathbf{u}_1 + \mathbf{u}_2 + \mathbf{u}_3.$$

16. Show that the matrices

$$\begin{bmatrix} 1 & 1 \\ -1 & 4 \end{bmatrix} \quad \text{and} \quad \begin{bmatrix} 2 & 1 \\ 1 & 3 \end{bmatrix}$$

are similar, but

$$\begin{bmatrix} 3 & 1 \\ -6 & -2 \end{bmatrix} \quad \text{and} \quad \begin{bmatrix} -1 & 2 \\ 1 & 0 \end{bmatrix}$$

are not.

17. Suppose that $T: V \rightarrow V$ is a linear operator and B is a basis for V such that for any vector $\mathbf{x}$ in V

$$[T(\mathbf{x})]_B = \begin{bmatrix} x_1 - x_2 + x_3 \\ x_2 \\ x_1 - x_3 \end{bmatrix} \quad \text{if} \quad [\mathbf{x}]_B = \begin{bmatrix} x_1 \\ x_2 \\ x_3 \end{bmatrix}$$

Find $[T]_B$.

18. In each part find the standard matrix for the linear operator $T: R^3 \rightarrow R^3$ described by Figure 5.14.

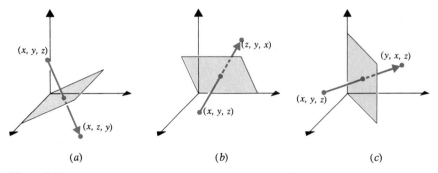

(a)   (b)   (c)

Figure 5.14

19. In R^3 the **shear in the xy-direction with factor k** is the linear transformation that moves each point (x, y, z) parallel to the xy-plane by an amount kz to the new position $(x + kz, y + kz, z)$. (See Figure 5.15.)

(a) Find the standard matrix for the shear in the xy-direction with factor k.

(b) How would you define the shear in the xz-direction with factor k and the shear in the yz-direction with factor k? Find the standard matrix for each of these linear transformations.

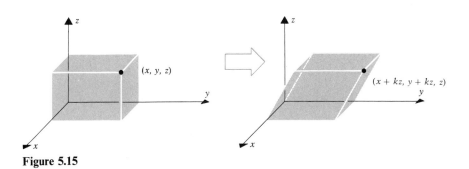

Figure 5.15

20. (For readers who have studied calculus.) Let $D: P_n \rightarrow P_n$ be the differentiation operator $D(\mathbf{p}) = \mathbf{p}'$. Show that the matrix for D with respect to the basis $B = \{1, x, x^2, \ldots, x^n\}$ is

$$\begin{bmatrix} 0 & 1 & 0 & 0 & \cdots & 0 \\ 0 & 0 & 2 & 0 & \cdots & 0 \\ 0 & 0 & 0 & 3 & \cdots & 0 \\ \vdots & \vdots & \vdots & \vdots & & \vdots \\ 0 & 0 & 0 & 0 & \cdots & n \\ 0 & 0 & 0 & 0 & \ldots & 0 \end{bmatrix}$$

21. (For readers who have studied calculus.) It can be shown that for any real number c, the vectors

$$1, \ x - c, \ \frac{(x - c)^2}{2!}, \ldots, \frac{(x - c)^n}{n!}$$

form a basis for P_n. Find the matrix for the differentiation operator of Exercise 20 with respect to this basis.

22. (For readers who have studied calculus.) Let $J: P_n \rightarrow P_{n+1}$ be the integration transformation defined by

$$J(\mathbf{p}) = \int (a_0 + a_1 x + \cdots + a_n x^n)\, dx = a_0 x + \frac{a_1}{2} x^2 + \cdots + \frac{a_n}{n+1} x^{n+1}$$

where $\mathbf{p} = a_0 + a_1 x + \cdots + a_n x^n$. Find the matrix for T with respect to the standard bases for P_n and P_{n+1}.

CHAPTER SIX

Eigenvalues, Eigenvectors

6.1 EIGENVALUES AND EIGENVECTORS

If A is an $n \times n$ matrix, there is often no obvious geometric relationship between a vector $\mathbf{x}$ and its image $A\mathbf{x}$ under multiplication by A (Figure 6.1a). However, frequently there are some nonzero vectors that A maps into scalar multiples of themselves (Figure 6.1b). Such vectors play an important role in analyses of linear transformations and arise naturally in the study of vibrations, electrical systems, genetics, chemical reactions, quantum mechanics, mechanical stress, economics, and geometry. In this section we shall show how to find these vectors, and in later sections we shall touch on some of their applications.

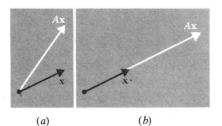

| (a) | (b) | **Figure 6.1** |

Definition. If A is an $n \times n$ matrix, then a nonzero vector $\mathbf{x}$ in R^n is called an *eigenvector* of A if $A\mathbf{x}$ is a scalar multiple of $\mathbf{x}$; that is,

$$A\mathbf{x} = \lambda \mathbf{x}$$

for some scalar λ. The scalar λ is called an *eigenvalue* of A and $\mathbf{x}$ is said to be an eigenvector *corresponding* to λ.

REMARK. The word "eigenvector" is a mixture of German and English. The German prefix "eigen" can be translated as "proper" or "characteristic"; hence eigenvalues are also called **proper values** or **characteristic values**. In the older literature they are sometimes called **latent roots**.

Example 1

The vector $\mathbf{x} = \begin{bmatrix} 1 \\ 2 \end{bmatrix}$ is an eigenvector of

$$A = \begin{bmatrix} 3 & 0 \\ 8 & -1 \end{bmatrix}$$

corresponding to the eigenvalue $\lambda = 3$ since

$$A\mathbf{x} = \begin{bmatrix} 3 & 0 \\ 8 & -1 \end{bmatrix}\begin{bmatrix} 1 \\ 2 \end{bmatrix} = \begin{bmatrix} 3 \\ 6 \end{bmatrix} = 3\mathbf{x}$$

Eigenvalues and eigenvectors have a useful geometric interpretation in R^2 and R^3. If λ is an eigenvalue of A corresponding to $\mathbf{x}$, then $A\mathbf{x} = \lambda\mathbf{x}$, so that multiplication by A dilates $\mathbf{x}$, contracts $\mathbf{x}$, or reverses the direction of $\mathbf{x}$, depending on the value of λ (Figure 6.2).

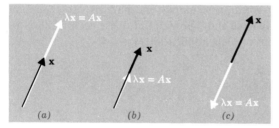

Figure 6.2 (*a*) Dilation $\lambda > 1$. (*b*) Contraction $0 < \lambda < 1$. (*c*) Reversal of direction $\lambda < 0$.

To find the eigenvalues of an $n \times n$ matrix A we rewrite $A\mathbf{x} = \lambda\mathbf{x}$ as

$$A\mathbf{x} = \lambda I\mathbf{x}$$

or equivalently

$$(\lambda I - A)\mathbf{x} = \mathbf{0} \tag{6.1}$$

For λ to be an eigenvalue, there must be a nonzero solution of this equation. However, by Theorem 15 of Section 4.6, equation (6.1) will have a nonzero solution if and only if

$$\det(\lambda I - A) = 0$$

This is called the ***characteristic equation*** of A; the scalars satisfying this equation are the eigenvalues of A. When expanded, the determinant $\det(\lambda I - A)$ is a polynomial in λ called the ***characteristic polynomial*** of A.

It can be shown (Exercise 15) that if A is an $n \times n$ matrix, then the characteristic polynomial of A has degree n and the coefficient of λ^n is 1. Thus, the characteristic polynomial of an $n \times n$ matrix has the form

$$\det(\lambda I - A) = \lambda^n + c_1\lambda^{n-1} + \cdots + c_n$$

Example 2

Find the eigenvalues of the matrix

$$A = \begin{bmatrix} 3 & 2 \\ -1 & 0 \end{bmatrix}$$

Solution. Since

$$\lambda I - A = \lambda \begin{bmatrix} 1 & 0 \\ 0 & 1 \end{bmatrix} - \begin{bmatrix} 3 & 2 \\ -1 & 0 \end{bmatrix} = \begin{bmatrix} \lambda - 3 & -2 \\ 1 & \lambda \end{bmatrix}$$

the characteristic polynomial of A is

$$\det(\lambda I - A) = \det \begin{bmatrix} \lambda - 3 & -2 \\ 1 & \lambda \end{bmatrix} = \lambda^2 - 3\lambda + 2$$

and the characteristic equation of A is

$$\lambda^2 - 3\lambda + 2 = 0$$

The solutions of this equation are $\lambda = 1$ and $\lambda = 2$; these are the eigenvalues of A. ▲

Example 3

Find the eigenvalues of the matrix

$$A = \begin{bmatrix} -2 & -1 \\ 5 & 2 \end{bmatrix}$$

Solution. Proceeding as in Example 2,

$$\det(\lambda I - A) = \det \begin{bmatrix} \lambda + 2 & 1 \\ -5 & \lambda - 2 \end{bmatrix} = \lambda^2 + 1$$

The eigenvalues of A must therefore satisfy the quadratic equation $\lambda^2 + 1 = 0$. Since the only solutions to this equation are the imaginary numbers $\lambda = i$ and

$\lambda = -i$, and since we are assuming in this section that all our scalars are real numbers, A has no real eigenvalues.* ▲

Example 4

Find the eigenvalues of

$$A = \begin{bmatrix} 0 & 1 & 0 \\ 0 & 0 & 1 \\ 4 & -17 & 8 \end{bmatrix}$$

Solution. As in the preceding examples

$$\det(\lambda I - A) = \det \begin{bmatrix} \lambda & -1 & 0 \\ 0 & \lambda & -1 \\ -4 & 17 & \lambda - 8 \end{bmatrix} = \lambda^3 - 8\lambda^2 + 17\lambda - 4$$

The eigenvalues of A must therefore satisfy the cubic equation

$$\lambda^3 - 8\lambda^2 + 17\lambda - 4 = 0 \qquad (6.2)$$

To solve this equation, we shall begin by searching for integer solutions. This task can be greatly simplified by exploiting the fact that all integer solutions (if there are any) to a polynomial equation with integer coefficients

$$\lambda^n + c_1\lambda^{n-1} + \cdots + c_n = 0$$

must be divisors of the constant term, c_n. Thus, the only possible integer solutions of (6.2) are the divisors of -4, that is, $\pm 1, \pm 2, \pm 4$. Successively substituting these values in (6.2) shows that $\lambda = 4$ is an integer solution. As a consequence, $\lambda - 4$ must be a factor of the left side of (6.2). Dividing $\lambda - 4$ into $\lambda^3 - 8\lambda^2 + 17\lambda - 4$ shows that (6.2) can be rewritten as

$$(\lambda - 4)(\lambda^2 - 4\lambda + 1) = 0$$

Thus, the remaining solutions of (6.2) satisfy the quadratic equation

$$\lambda^2 - 4\lambda + 1 = 0$$

which can be solved by the quadratic formula. Thus, the eigenvalues of A are

$$\lambda = 4 \qquad \lambda = 2 + \sqrt{3} \qquad \text{and} \qquad \lambda = 2 - \sqrt{3} \quad ▲$$

---

* As we pointed out in Section 4.2, there are some applications that require complex scalars and complex vector spaces. In such cases, matrices are allowed to have complex entries and complex eigenvalues. This will be discussed in Chapter Nine; however, until then we will allow real eigenvalues only, and all matrices will have real entries.

REMARK. In practical problems, the matrix A is often so large that computing the characteristic equation is not practical. As a result, various approximation methods are used to obtain eigenvalues; some of these are discussed in Chapter Eight.

The following theorem summarizes our results so far.

Theorem 1. *If A is an $n \times n$ matrix then the following are equivalent:*

(a) *λ is an eigenvalue of A.*
(b) *The system of equations $(\lambda I - A)\mathbf{x} = \mathbf{0}$ has nontrivial solutions.*
(c) *There is a nonzero vector $\mathbf{x}$ in R^n such that $A\mathbf{x} = \lambda\mathbf{x}$.*
(d) *λ is a real solution of the characteristic equation $\det(\lambda I - A) = 0$.*

Now that we know how to find eigenvalues, we turn to the problem of finding eigenvectors. The eigenvectors of A corresponding to an eigenvalue λ are the nonzero vectors $\mathbf{x}$ that satisfy $A\mathbf{x} = \lambda\mathbf{x}$. Equivalently, the eigenvectors corresponding to λ are the nonzero vectors in the solution space of $(\lambda I - A)\mathbf{x} = \mathbf{0}$. We call this solution space the *eigenspace* of A corresponding to λ.

Example 5

Find bases for the eigenspaces of

$$A = \begin{bmatrix} 3 & -2 & 0 \\ -2 & 3 & 0 \\ 0 & 0 & 5 \end{bmatrix}$$

Solution. The characteristic equation of A is $(\lambda - 1)(\lambda - 5)^2 = 0$ (verify), so that the eigenvalues of A are $\lambda = 1$ and $\lambda = 5$. Thus, there are two eigenspaces of A.
  By definition,

$$\mathbf{x} = \begin{bmatrix} x_1 \\ x_2 \\ x_3 \end{bmatrix}$$

is an eigenvector of A corresponding to λ if and only if $\mathbf{x}$ is a nontrivial solution of $(\lambda I - A)\mathbf{x} = \mathbf{0}$, that is, of

$$\begin{bmatrix} \lambda - 3 & 2 & 0 \\ 2 & \lambda - 3 & 0 \\ 0 & 0 & \lambda - 5 \end{bmatrix} \begin{bmatrix} x_1 \\ x_2 \\ x_3 \end{bmatrix} = \begin{bmatrix} 0 \\ 0 \\ 0 \end{bmatrix} \tag{6.3}$$

If $\lambda = 5$, then (6.3) becomes

$$\begin{bmatrix} 2 & 2 & 0 \\ 2 & 2 & 0 \\ 0 & 0 & 0 \end{bmatrix} \begin{bmatrix} x_1 \\ x_2 \\ x_3 \end{bmatrix} = \begin{bmatrix} 0 \\ 0 \\ 0 \end{bmatrix}$$

Solving this system yields (verify)

$$x_1 = -s \qquad x_2 = s \qquad x_3 = t$$

Thus, the eigenvectors of A corresponding to $\lambda = 5$ are the nonzero vectors of the form

$$\mathbf{x} = \begin{bmatrix} -s \\ s \\ t \end{bmatrix} = \begin{bmatrix} -s \\ s \\ 0 \end{bmatrix} + \begin{bmatrix} 0 \\ 0 \\ t \end{bmatrix} = s \begin{bmatrix} -1 \\ 1 \\ 0 \end{bmatrix} + t \begin{bmatrix} 0 \\ 0 \\ 1 \end{bmatrix}$$

Since

$$\begin{bmatrix} -1 \\ 1 \\ 0 \end{bmatrix} \qquad \text{and} \qquad \begin{bmatrix} 0 \\ 0 \\ 1 \end{bmatrix}$$

are linearly independent, they form a basis for the eigenspace corresponding to $\lambda = 5$.

If $\lambda = 1$, then (6.3) becomes

$$\begin{bmatrix} -2 & 2 & 0 \\ 2 & -2 & 0 \\ 0 & 0 & -4 \end{bmatrix} \begin{bmatrix} x_1 \\ x_2 \\ x_3 \end{bmatrix} = \begin{bmatrix} 0 \\ 0 \\ 0 \end{bmatrix}$$

Solving this system yields (verify)

$$x_1 = t \qquad x_2 = t \qquad x_3 = 0$$

Thus, the eigenvectors corresponding to $\lambda = 1$ are the nonzero vectors of the form

$$\mathbf{x} = \begin{bmatrix} t \\ t \\ 0 \end{bmatrix} = t \begin{bmatrix} 1 \\ 1 \\ 0 \end{bmatrix}$$

so that

$$\begin{bmatrix} 1 \\ 1 \\ 0 \end{bmatrix}$$

is a basis for the eigenspace corresponding to $\lambda = 1$.

OPTIONAL

Eigenvectors and eigenvalues can be defined for linear operators as well as matrices. A scalar λ is called an *eigenvalue* of a linear operator $T:V \to V$ if there is a nonzero vector x in V such that $Tx = \lambda x$. The vector x is called an *eigenvector* of T corresponding to λ. Equivalently, the eigenvectors of T corresponding to λ are the nonzero vectors in the kernel of $\lambda I - T$ (Exercise 20). This kernel is called the *eigenspace* of T corresponding to λ.

It can be shown that if V is a finite-dimensional vector space and A is the matrix of T with respect to *any* basis B, then:

1. The eigenvalues of T are the eigenvalues of the matrix A.
2. A vector x is an eigenvector of T corresponding to λ if and only if its coordinate matrix $[x]_B$ is an eigenvector of A corresponding to λ.

We leave the proofs for the exercises.

Example 6

Find the eigenvalues and bases for the eigenspaces of the linear operator $T:P_2 \to P_2$ defined by

$$T(a + bx + cx^2) = (3a - 2b) + (-2a + 3b)x + (5c)x^2$$

Solution. The matrix of T with respect to the standard basis $B = \{1, x, x^2\}$ is

$$A = \begin{bmatrix} 3 & -2 & 0 \\ -2 & 3 & 0 \\ 0 & 0 & 5 \end{bmatrix}$$

The eigenvalues of T are the eigenvalues of A; namely, $\lambda = 1$ and $\lambda = 5$ (Example 5). Also from Example 5 the eigenspace of A corresponding to $\lambda = 5$ has the basis $\{u_1, u_2\}$ and that corresponding to $\lambda = 1$ has the basis $\{u_3\}$, where

$$u_1 = \begin{bmatrix} -1 \\ 1 \\ 0 \end{bmatrix} \qquad u_2 = \begin{bmatrix} 0 \\ 0 \\ 1 \end{bmatrix} \qquad u_3 = \begin{bmatrix} 1 \\ 1 \\ 0 \end{bmatrix}$$

These matrices are the coordinate matrices with respect to B of

$$p_1 = -1 + x \qquad p_2 = x^2 \qquad p_3 = 1 + x$$

Thus, $\{-1 + x, x^2\}$ is a basis for the eigenspace of T corresponding to $\lambda = 5$, and $\{1 + x\}$ is a basis for the eigenspace corresponding to $\lambda = 1$. ▲

EXERCISE SET 6.1

1. Find the characteristic equations of the following matrices:

(a) $\begin{bmatrix} 3 & 0 \\ 8 & -1 \end{bmatrix}$ (b) $\begin{bmatrix} 10 & -9 \\ 4 & -2 \end{bmatrix}$ (c) $\begin{bmatrix} 0 & 3 \\ 4 & 0 \end{bmatrix}$

(d) $\begin{bmatrix} -2 & -7 \\ 1 & 2 \end{bmatrix}$ (e) $\begin{bmatrix} 0 & 0 \\ 0 & 0 \end{bmatrix}$ (f) $\begin{bmatrix} 1 & 0 \\ 0 & 1 \end{bmatrix}$

2. Find the eigenvalues of the matrices in Exercise 1.

3. Find bases for the eigenspaces of the matrices in Exercise 1.

4. In each part of Exercise 1, let $T:R^2 \to R^2$ be multiplication by the given matrix. In each case sketch the lines through the origin in R^2 that are mapped into themselves under T.

5. Find the characteristic equations of the following matrices:

(a) $\begin{bmatrix} 4 & 0 & 1 \\ -2 & 1 & 0 \\ -2 & 0 & 1 \end{bmatrix}$ (b) $\begin{bmatrix} 3 & 0 & -5 \\ \frac{1}{5} & -1 & 0 \\ 1 & 1 & -2 \end{bmatrix}$ (c) $\begin{bmatrix} -2 & 0 & 1 \\ -6 & -2 & 0 \\ 19 & 5 & -4 \end{bmatrix}$

(d) $\begin{bmatrix} -1 & 0 & 1 \\ -1 & 3 & 0 \\ -4 & 13 & -1 \end{bmatrix}$ (e) $\begin{bmatrix} 5 & 0 & 1 \\ 1 & 1 & 0 \\ -7 & 1 & 0 \end{bmatrix}$ (f) $\begin{bmatrix} 5 & 6 & 2 \\ 0 & -1 & -8 \\ 1 & 0 & -2 \end{bmatrix}$

6. Find the eigenvalues of the matrices in Exercise 5.

7. Find bases for the eigenspaces of the matrices in Exercise 5.

8. Find the characteristic equations of the following matrices:

(a) $\begin{bmatrix} 0 & 0 & 2 & 0 \\ 1 & 0 & 1 & 0 \\ 0 & 1 & -2 & 0 \\ 0 & 0 & 0 & 1 \end{bmatrix}$ (b) $\begin{bmatrix} 10 & -9 & 0 & 0 \\ 4 & -2 & 0 & 0 \\ 0 & 0 & -2 & -7 \\ 0 & 0 & 1 & 2 \end{bmatrix}$

9. Find the eigenvalues of the matrices in Exercise 8.

10. Find bases for the eigenspaces of the matrices in Exercise 8.

11. **(For readers of the optional material.)** Let $T:P_2 \to P_2$ be defined by

$$T(a_0 + a_1 x + a_2 x^2) = (5a_0 + 6a_1 + 2a_2) - (a_1 + 8a_2)x + (a_0 - 2a_2)x^2.$$

(a) Find the eigenvalues of T.
(b) Find bases for the eigenspaces of T.

12. **(For readers of the optional material.)** Let $T:M_{22} \to M_{22}$ be defined by

$$T\left(\begin{bmatrix} a & b \\ c & d \end{bmatrix} \right) = \begin{bmatrix} 2c & a+c \\ b-2c & d \end{bmatrix}$$

(a) Find the eigenvalues of T.

(b) Find bases for the eigenspaces of T.

13. Prove that $\lambda = 0$ is an eigenvalue of a matrix A if and only if A is not invertible.

14. Prove that the constant term in the characteristic polynomial of an $n \times n$ matrix A is $(-1)^n \det(A)$. (**Hint.** The constant term is the value of the characteristic polynomial when $\lambda = 0$.)

15. Let A be an $n \times n$ matrix.

    (a) Prove that the characteristic polynomial of A has degree n.

    (b) Prove that the coefficient of λ^n in the characteristic polynomial is 1.

16. The *trace* of a square matrix A is the sum of the elements on the main diagonal. Show that the characteristic equation of a 2×2 matrix A is $\lambda^2 - \text{tr}(A)\lambda + \det(A) = 0$, where $\text{tr}(A)$ is the trace of A.

17. Prove that the eigenvalues of a triangular matrix are the entries on the main diagonal.

18. Show that if λ is an eigenvalue of A, then λ^2 is an eigenvalue of A^2; more generally, show that λ^n is an eigenvalue of A^n if n is a positive integer.

19. Use the results of Exercises 17 and 18 to find the eigenvalues of A^9, where

$$A = \begin{bmatrix} 1 & 3 & 7 & 11 \\ 0 & -1 & 3 & 8 \\ 0 & 0 & -2 & 4 \\ 0 & 0 & 0 & 2 \end{bmatrix}$$

20. (**For readers of the optional material.**) Let λ be an eigenvalue of a linear operator $T:V \rightarrow V$. Prove that the eigenvectors of T corresponding to λ are the nonzero vectors in the kernel of $\lambda I - T$.

6.2 DIAGONALIZATION

In this section, we shall be concerned with the following problem:

The Diagonalization Problem. *Given a linear operator $T:V \rightarrow V$ on a finite-dimensional vector space, does there exist a basis for V with respect to which the matrix for T is diagonal?*

If A is the matrix for $T:V \rightarrow V$ with respect to some arbitrary basis, then this problem is equivalent to asking if there is a change of basis such that the new matrix for T is diagonal. By Theorem 8 in Section 5.5, the new matrix for T will be $P^{-1}AP$ where P is the appropriate transition matrix. Thus, we are led to the following matrix form of the diagonalization problem.

Matrix Form of the Diagonalization Problem. *Given a square matrix A, does there exist an invertible matrix P such that $P^{-1}AP$ is diagonal?*

This problem suggests the following definition.

Definition. A square matrix A is called **diagonalizable** if there is an invertible matrix P such that $P^{-1}AP$ is diagonal; the matrix P is said to **diagonalize** A.

The following theorem is the basic tool in the study of diagonalizability; its proof reveals the technique for diagonalizing a matrix.

Theorem 2. *If A is an $n \times n$ matrix, then the following are equivalent.*

(a) *A is diagonalizable.*
(b) *A has n linearly independent eigenvectors.*

Proof $(a) \Rightarrow (b)$. Since A is assumed diagonalizable, there is an invertible matrix

$$P = \begin{bmatrix} p_{11} & p_{12} & \cdots & p_{1n} \\ p_{21} & p_{22} & \cdots & p_{2n} \\ \vdots & \vdots & & \vdots \\ p_{n1} & p_{n2} & \cdots & p_{nn} \end{bmatrix}$$

such that $P^{-1}AP$ is diagonal, say $P^{-1}AP = D$, where

$$D = \begin{bmatrix} \lambda_1 & 0 & \cdots & 0 \\ 0 & \lambda_2 & \cdots & 0 \\ \vdots & \vdots & & \vdots \\ 0 & 0 & \cdots & \lambda_n \end{bmatrix}$$

Therefore, $AP = PD$; that is,

$$AP = \begin{bmatrix} p_{11} & p_{12} & \cdots & p_{1n} \\ p_{21} & p_{22} & \cdots & p_{2n} \\ \vdots & \vdots & & \vdots \\ p_{n1} & p_{n2} & \cdots & p_{nn} \end{bmatrix} \begin{bmatrix} \lambda_1 & 0 & \cdots & 0 \\ 0 & \lambda_2 & \cdots & 0 \\ \vdots & \vdots & & \vdots \\ 0 & 0 & \cdots & \lambda_n \end{bmatrix} = \begin{bmatrix} \lambda_1 p_{11} & \lambda_2 p_{12} & \cdots & \lambda_n p_{1n} \\ \lambda_1 p_{21} & \lambda_2 p_{22} & \cdots & \lambda_n p_{2n} \\ \vdots & \vdots & & \vdots \\ \lambda_1 p_{n1} & \lambda_2 p_{n2} & \cdots & \lambda_n p_{nn} \end{bmatrix}$$

$$(6.4)$$

If we now let $\mathbf{p}_1, \mathbf{p}_2, \ldots, \mathbf{p}_n$ denote the column vectors of P, then from (6.4) the successive columns of AP are $\lambda_1 \mathbf{p}_1, \lambda_2 \mathbf{p}_2, \ldots, \lambda_n \mathbf{p}_n$. However, from Example 18 of

Section 1.4 the successive columns of AP are $A\mathbf{p}_1, A\mathbf{p}_2, \ldots, A\mathbf{p}_n$. Thus, we must have

$$A\mathbf{p}_1 = \lambda_1\mathbf{p}_1, \ A\mathbf{p}_2 = \lambda_2\mathbf{p}_2, \ldots, A\mathbf{p}_n = \lambda_n\mathbf{p}_n \tag{6.5}$$

Since P is invertible, its column vectors are all nonzero; thus, by (6.5), $\lambda_1, \lambda_2, \ldots, \lambda_n$ are eigenvalues of A, and $\mathbf{p}_1, \mathbf{p}_2, \ldots, \mathbf{p}_n$ are corresponding eigenvectors. Since P is invertible, it follows from Theorem 15 in Section 4.6 that $\mathbf{p}_1, \mathbf{p}_2, \ldots, \mathbf{p}_n$ are linearly independent. Thus, A has n linearly independent eigenvectors.

$(b) \Rightarrow (a)$. Assume that A has n linearly independent eigenvectors, $\mathbf{p}_1, \mathbf{p}_2, \ldots, \mathbf{p}_n$, with corresponding eigenvalues $\lambda_1, \lambda_2, \ldots, \lambda_n$, and let

$$P = \begin{bmatrix} p_{11} & p_{12} & \cdots & p_{1n} \\ p_{21} & p_{22} & \cdots & p_{2n} \\ \vdots & \vdots & & \vdots \\ p_{n1} & p_{n2} & \cdots & p_{nn} \end{bmatrix}$$

be the matrix whose column vectors are $\mathbf{p}_1, \mathbf{p}_2, \ldots, \mathbf{p}_n$. By Example 17 in Section 1.4, the columns of the product AP are

$$A\mathbf{p}_1, A\mathbf{p}_2, \ldots, A\mathbf{p}_n$$

But

$$A\mathbf{p}_1 = \lambda_1\mathbf{p}_1, \ A\mathbf{p}_2 = \lambda_2\mathbf{p}_2, \ldots, A\mathbf{p}_n = \lambda_n\mathbf{p}_n$$

so that

$$AP = \begin{bmatrix} \lambda_1 p_{11} & \lambda_2 p_{12} & \cdots & \lambda_n p_{1n} \\ \lambda_1 p_{21} & \lambda_2 p_{22} & \cdots & \lambda_n p_{2n} \\ \vdots & \vdots & & \vdots \\ \lambda_1 p_{n1} & \lambda_2 p_{n2} & \cdots & \lambda_n p_{nn} \end{bmatrix} = \begin{bmatrix} p_{11} & p_{12} & \cdots & p_{1n} \\ p_{21} & p_{22} & \cdots & p_{2n} \\ \vdots & \vdots & & \vdots \\ p_{n1} & p_{n2} & \cdots & p_{nn} \end{bmatrix} \begin{bmatrix} \lambda_1 & 0 & \cdots & 0 \\ 0 & \lambda_2 & \cdots & 0 \\ \vdots & \vdots & & \vdots \\ 0 & 0 & \cdots & \lambda_n \end{bmatrix}$$

$$= PD \tag{6.6}$$

where D is the diagonal matrix having the eigenvalues $\lambda_1, \lambda_2, \ldots, \lambda_n$ on the main diagonal. Since the column vectors of P are linearly independent, P is invertible; thus, (6.6) can be rewritten as $P^{-1}AP = D$; that is, A is diagonalizable. ∎

From this proof we obtain the following procedure for diagonalizing a diagonalizable $n \times n$ matrix A.

Step 1. Find n linearly independent eigenvectors of A, say, $\mathbf{p}_1, \mathbf{p}_2, \ldots, \mathbf{p}_n$.

Step 2. Form the matrix P having $\mathbf{p}_1, \mathbf{p}_2, \ldots, \mathbf{p}_n$ as its column vectors.

Step 3. The matrix $P^{-1}AP$ will then be diagonal with $\lambda_1, \lambda_2, \ldots, \lambda_n$ as its successive diagonal entries, where λ_i is the eigenvalue corresponding to $\mathbf{p}_i$, $i = 1, 2, \ldots, n$.

Example 7

Find a matrix P that diagonalizes

$$A = \begin{bmatrix} 3 & -2 & 0 \\ -2 & 3 & 0 \\ 0 & 0 & 5 \end{bmatrix}$$

Solution. From Example 5 the eigenvalues of A are $\lambda = 1$ and $\lambda = 5$. Also from that example the vectors

$$\mathbf{p}_1 = \begin{bmatrix} -1 \\ 1 \\ 0 \end{bmatrix} \quad \text{and} \quad \mathbf{p}_2 = \begin{bmatrix} 0 \\ 0 \\ 1 \end{bmatrix}$$

form a basis for the eigenspace corresponding to $\lambda = 5$, and

$$\mathbf{p}_3 = \begin{bmatrix} 1 \\ 1 \\ 0 \end{bmatrix}$$

is a basis for the eigenspace corresponding to $\lambda = 1$. It is easy to check that $\{\mathbf{p}_1, \mathbf{p}_2, \mathbf{p}_3\}$ is linearly independent, so that

$$P = \begin{bmatrix} -1 & 0 & 1 \\ 1 & 0 & 1 \\ 0 & 1 & 0 \end{bmatrix}$$

diagonalizes A. As a check, the reader should verify that

$$P^{-1}AP = \begin{bmatrix} -\frac{1}{2} & \frac{1}{2} & 0 \\ 0 & 0 & 1 \\ \frac{1}{2} & \frac{1}{2} & 0 \end{bmatrix} \begin{bmatrix} 3 & -2 & 0 \\ -2 & 3 & 0 \\ 0 & 0 & 5 \end{bmatrix} \begin{bmatrix} -1 & 0 & 1 \\ 1 & 0 & 1 \\ 0 & 1 & 0 \end{bmatrix} = \begin{bmatrix} 5 & 0 & 0 \\ 0 & 5 & 0 \\ 0 & 0 & 1 \end{bmatrix}$$

There is no preferred order for the columns of P. Since the ith diagonal entry of $P^{-1}AP$ is an eigenvalue for the ith column vector of P, changing the order of the columns of P just changes the order of the eigenvalues on the diagonal of $P^{-1}AP$. Thus, had we written

$$P = \begin{bmatrix} -1 & 1 & 0 \\ 1 & 1 & 0 \\ 0 & 0 & 1 \end{bmatrix}$$

in the last example, we would have obtained

$$P^{-1}AP = \begin{bmatrix} 5 & 0 & 0 \\ 0 & 1 & 0 \\ 0 & 0 & 5 \end{bmatrix}$$

Example 8

The characteristic equation of

$$A = \begin{bmatrix} -3 & 2 \\ -2 & 1 \end{bmatrix}$$

is

$$\det(\lambda I - A) = \det \begin{bmatrix} \lambda + 3 & -2 \\ 2 & \lambda - 1 \end{bmatrix} = (\lambda + 1)^2 = 0$$

Thus, $\lambda = -1$ is the only eigenvalue of A; the eigenvectors corresponding to $\lambda = -1$ are the solutions of $(-I - A)\mathbf{x} = \mathbf{0}$, that is, of

$$2x_1 - 2x_2 = 0$$
$$2x_1 - 2x_2 = 0$$

The solutions of this system are $x_1 = t$, $x_2 = t$ (verify); hence the eigenspace consists of all vectors of the form

$$\begin{bmatrix} t \\ t \end{bmatrix} = t \begin{bmatrix} 1 \\ 1 \end{bmatrix}$$

Since this space is 1-dimensional, A does not have two linearly independent eigenvectors and is therefore not diagonalizable.

Example 9

Let $T:R^3 \to R^3$ be the linear operator given by

$$T\left(\begin{bmatrix} x_1 \\ x_2 \\ x_3 \end{bmatrix} \right) = \begin{bmatrix} 3x_1 - 2x_2 \\ -2x_1 + 3x_2 \\ 5x_3 \end{bmatrix}$$

Find a basis for R^3 relative to which the matrix of T is diagonal.

Solution. If $B = \{\mathbf{e}_1, \mathbf{e}_2, \mathbf{e}_3\}$ denotes the standard basis for R^3, then

$$T(\mathbf{e}_1) = T\left(\begin{bmatrix} 1 \\ 0 \\ 0 \end{bmatrix} \right) = \begin{bmatrix} 3 \\ -2 \\ 0 \end{bmatrix} \qquad T(\mathbf{e}_2) = T\left(\begin{bmatrix} 0 \\ 1 \\ 0 \end{bmatrix} \right) = \begin{bmatrix} -2 \\ 3 \\ 0 \end{bmatrix}$$

$$T(\mathbf{e}_3) = T\left(\begin{bmatrix} 0 \\ 0 \\ 1 \end{bmatrix} \right) = \begin{bmatrix} 0 \\ 0 \\ 5 \end{bmatrix}$$

so that the standard matrix for T is

$$A = \begin{bmatrix} 3 & -2 & 0 \\ -2 & 3 & 0 \\ 0 & 0 & 5 \end{bmatrix}$$

We now want to change from the standard basis to a new basis $B' = \{\mathbf{u}'_1, \mathbf{u}'_2, \mathbf{u}'_3\}$ in order to obtain a diagonal matrix A' for T. If we let P be the transition matrix from the unknown basis B' to the standard basis B, then by Theorem 8 of Section 5.5, A and A' will be related by

$$A' = P^{-1}AP$$

In other words, the transition matrix P diagonalizes A. We found this matrix in Example 7. From our work in that example

$$P = \begin{bmatrix} -1 & 0 & 1 \\ 1 & 0 & 1 \\ 0 & 1 & 0 \end{bmatrix} \quad \text{and} \quad A' = \begin{bmatrix} 5 & 0 & 0 \\ 0 & 5 & 0 \\ 0 & 0 & 1 \end{bmatrix}$$

Since P represents the transition matrix from the basis $B' = \{\mathbf{u}'_1, \mathbf{u}'_2, \mathbf{u}'_3\}$ to the standard basis $B = \{\mathbf{e}_1, \mathbf{e}_2, \mathbf{e}_3\}$, the columns of P are $[\mathbf{u}'_1]_B$, $[\mathbf{u}'_2]_B$, and $[\mathbf{u}'_3]_B$, so that

$$[\mathbf{u}'_1]_B = \begin{bmatrix} -1 \\ 1 \\ 0 \end{bmatrix}, \quad [\mathbf{u}'_2]_B = \begin{bmatrix} 0 \\ 0 \\ 1 \end{bmatrix}, \quad [\mathbf{u}'_3]_B = \begin{bmatrix} 1 \\ 1 \\ 0 \end{bmatrix}$$

Thus,

$$\mathbf{u}'_1 = (-1)\mathbf{e}_1 + (1)\mathbf{e}_2 + (0)\mathbf{e}_3 = \begin{bmatrix} -1 \\ 1 \\ 0 \end{bmatrix}$$

$$\mathbf{u}'_2 = (0)\mathbf{e}_1 + (0)\mathbf{e}_2 + (1)\mathbf{e}_3 = \begin{bmatrix} 0 \\ 0 \\ 1 \end{bmatrix}$$

$$\mathbf{u}'_3 = (1)\mathbf{e}_1 + (1)\mathbf{e}_2 + (0)\mathbf{e}_3 = \begin{bmatrix} 1 \\ 1 \\ 0 \end{bmatrix}$$

are the basis vectors that produce the diagonal matrix A' for T.

In many applications it is not important to actually compute a transition matrix P that diagonalizes a matrix A. Rather, one is concerned only with knowing whether A is diagonalizable, and if so, what the diagonal matrix is. Often, this information can be ascertained directly from the eigenvalues without going through the work of computing eigenvectors. To see why this is so we will need the following theorem, whose proof is deferred to the end of the section.

Theorem 3. *If $\mathbf{v}_1, \mathbf{v}_2, \ldots, \mathbf{v}_k$ are eigenvectors of A corresponding to distinct eigenvalues $\lambda_1, \lambda_2, \ldots, \lambda_k$, then $\{\mathbf{v}_1, \mathbf{v}_2, \ldots, \mathbf{v}_k\}$ is a linearly independent set.*

As a consequence of this theorem, we obtain the following useful result.

Theorem 4. *If an n × n matrix A has n distinct eigenvalues, then A is diagonalizable.*

Proof. If $\mathbf{v}_1, \mathbf{v}_2, \ldots, \mathbf{v}_n$ are eigenvectors corresponding to the distinct eigenvalues $\lambda_1, \lambda_2, \ldots, \lambda_n$, then by Theorem 3, $\mathbf{v}_1, \mathbf{v}_2, \ldots, \mathbf{v}_n$ are linearly independent. Thus, A is diagonalizable by Theorem 2.

Example 10

We saw in Example 4 that

$$A = \begin{bmatrix} 0 & 1 & 0 \\ 0 & 0 & 1 \\ 4 & -17 & 8 \end{bmatrix}$$

has 3 distinct eigenvalues, $\lambda = 4$, $\lambda = 2 + \sqrt{3}$, $\lambda = 2 - \sqrt{3}$. Therefore, A is diagonalizable. Further,

$$P^{-1}AP = \begin{bmatrix} 4 & 0 & 0 \\ 0 & 2 + \sqrt{3} & 0 \\ 0 & 0 & 2 - \sqrt{3} \end{bmatrix}$$

for some invertible matrix P. If desired, the matrix P can be found using the method shown in Example 7.

REMARK. Theorem 3 is a special case of a more general result: Suppose that $\lambda_1, \lambda_2, \ldots, \lambda_k$ are distinct eigenvalues and we choose a linearly independent set in each of the corresponding eigenspaces. If we then merge all these vectors into a single set, the result is still a linearly independent set. For example, if we choose three linearly independent vectors from one eigenspace and two linearly independent eigenvectors from another eigenspace, then the five vectors together form a linearly independent set. We omit the proof.

Example 11

The converse of Theorem 4 is false; that is, an $n \times n$ matrix A may be diagonalizable even if it does not have n distinct eigenvalues. For example, if

$$A = \begin{bmatrix} 3 & 0 \\ 0 & 3 \end{bmatrix}$$

then the characteristic equation of A is

$$\det(\lambda I - A) = (\lambda - 3)^2 = 0$$

so that $\lambda = 3$ is the only distinct eigenvalue of A. Yet A is obviously diagonalizable since with $P = I$,

$$P^{-1}AP = I^{-1}AI = A = \begin{bmatrix} 3 & 0 \\ 0 & 3 \end{bmatrix}$$

If A is an $n \times n$ matrix with fewer than n distinct eigenvalues, then there are theorems studied in more advanced courses which can be used to determine whether A is diagonalizable. However, these theorems do not provide a simple *computational* procedure for making this determination; all one can really do is perform the necessary calculations to see if A has n linearly independent eigenvectors.

OPTIONAL

We conclude this section with a proof of Theorem 3.

Proof. Let $\mathbf{v}_1, \mathbf{v}_2, \ldots, \mathbf{v}_k$ be eigenvectors of A corresponding to distinct eigenvalues $\lambda_1, \lambda_2, \ldots, \lambda_k$. We shall assume that $\mathbf{v}_1, \mathbf{v}_2, \ldots, \mathbf{v}_k$ are linearly dependent and obtain a contradiction. We can then conclude that $\mathbf{v}_1, \mathbf{v}_2, \ldots, \mathbf{v}_k$ are linearly independent.

Since an eigenvector is by definition nonzero, $\{\mathbf{v}_1\}$ is linearly independent. Let r be the largest integer such that $\{\mathbf{v}_1, \mathbf{v}_2, \ldots, \mathbf{v}_r\}$ is linearly independent. Since we are assuming that $\{\mathbf{v}_1, \mathbf{v}_2, \ldots, \mathbf{v}_k\}$ is linearly dependent, r satisfies $1 \leq r < k$. Moreover, by definition of r, $\{\mathbf{v}_1, \mathbf{v}_2, \ldots, \mathbf{v}_{r+1}\}$ is linearly dependent. Thus, there are scalars $c_1, c_2, \ldots, c_{r+1}$, not all zero, such that

$$c_1\mathbf{v}_1 + c_2\mathbf{v}_2 + \cdots + c_{r+1}\mathbf{v}_{r+1} = \mathbf{0} \tag{6.7}$$

Multiplying both sides of (6.7) by A and using

$$A\mathbf{v}_1 = \lambda_1\mathbf{v}_1, A\mathbf{v}_2 = \lambda_2\mathbf{v}_2, \ldots, A\mathbf{v}_{r+1} = \lambda_{r+1}\mathbf{v}_{r+1}$$

we obtain

$$c_1\lambda_1\mathbf{v}_1 + c_2\lambda_2\mathbf{v}_2 + \cdots + c_{r+1}\lambda_{r+1}\mathbf{v}_{r+1} = \mathbf{0} \tag{6.8}$$

Multiplying both sides of (6.7) by λ_{r+1} and subtracting the resulting equation from (6.8) yields

$$c_1(\lambda_1 - \lambda_{r+1})\mathbf{v}_1 + c_2(\lambda_2 - \lambda_{r+1})\mathbf{v}_2 + \cdots + c_r(\lambda_r - \lambda_{r+1})\mathbf{v}_r = \mathbf{0}$$

Since $\{\mathbf{v}_1, \mathbf{v}_2, \ldots, \mathbf{v}_r\}$ is linearly independent, this equation implies that

$$c_1(\lambda_1 - \lambda_{r+1}) = c_2(\lambda_2 - \lambda_{r+1}) = \cdots = c_r(\lambda_r - \lambda_{r+1}) = 0$$

and since $\lambda_1, \lambda_2, \ldots, \lambda_{r+1}$ are distinct, it follows that

$$c_1 = c_2 = \cdots = c_r = 0 \tag{6.9}$$

Substituting these values in (6.7) yields

$$c_{r+1}\mathbf{v}_{r+1} = \mathbf{0}$$

Since the eigenvector $\mathbf{v}_{r+1}$ is nonzero, it follows that

$$c_{r+1} = 0 \tag{6.10}$$

Equations (6.9) and (6.10) contradict the fact that $c_1, c_2, \ldots, c_{r+1}$ are not all zero; this completes the proof.

EXERCISE SET 6.2

Show that the matrices in Exercises 1–4 are not diagonalizable.

1. $\begin{bmatrix} 2 & 0 \\ 1 & 2 \end{bmatrix}$
2. $\begin{bmatrix} 2 & -3 \\ 1 & -1 \end{bmatrix}$
3. $\begin{bmatrix} 3 & 0 & 0 \\ 0 & 2 & 0 \\ 0 & 1 & 2 \end{bmatrix}$
4. $\begin{bmatrix} -1 & 0 & 1 \\ -1 & 3 & 0 \\ -4 & 13 & -1 \end{bmatrix}$

In Exercises 5–8 find a matrix P that diagonalizes A, and determine $P^{-1}AP$.

5. $A = \begin{bmatrix} -14 & 12 \\ -20 & 17 \end{bmatrix}$
6. $A = \begin{bmatrix} 1 & 0 \\ 6 & -1 \end{bmatrix}$

7. $A = \begin{bmatrix} 1 & 0 & 0 \\ 0 & 1 & 1 \\ 0 & 1 & 1 \end{bmatrix}$
8. $A = \begin{bmatrix} 2 & 0 & -2 \\ 0 & 3 & 0 \\ 0 & 0 & 3 \end{bmatrix}$

In Exercises 9–14 determine if A is diagonalizable. If so, find a matrix P that diagonalizes A, and determine $P^{-1}AP$.

9. $A = \begin{bmatrix} 19 & -9 & -6 \\ 25 & -11 & -9 \\ 17 & -9 & -4 \end{bmatrix}$
10. $A = \begin{bmatrix} -1 & 4 & -2 \\ -3 & 4 & 0 \\ -3 & 1 & 3 \end{bmatrix}$

11. $A = \begin{bmatrix} 5 & 0 & 0 \\ 1 & 5 & 0 \\ 0 & 1 & 5 \end{bmatrix}$
12. $A = \begin{bmatrix} 0 & 0 & 0 \\ 0 & 0 & 0 \\ 3 & 0 & 1 \end{bmatrix}$

13. $A = \begin{bmatrix} -2 & 0 & 0 & 0 \\ 0 & -2 & 0 & 0 \\ 0 & 0 & 3 & 0 \\ 0 & 0 & 1 & 3 \end{bmatrix}$
14. $A = \begin{bmatrix} -2 & 0 & 0 & 0 \\ 0 & -2 & 5 & -5 \\ 0 & 0 & 3 & 0 \\ 0 & 0 & 0 & 3 \end{bmatrix}$

15. Let $T:R^2 \to R^2$ be the linear operator given by

$$T\left(\begin{bmatrix} x_1 \\ x_2 \end{bmatrix}\right) = \begin{bmatrix} 3x_1 + 4x_2 \\ 2x_1 + x_2 \end{bmatrix}$$

Find a basis for R^2 relative to which the matrix of T is diagonal.

16. Let $T: R^3 \to R^3$ be the linear operator given by

$$T\left(\begin{bmatrix} x_1 \\ x_2 \\ x_3 \end{bmatrix}\right) = \begin{bmatrix} 2x_1 - x_2 - x_3 \\ x_1 \quad - x_3 \\ -x_1 + x_2 + 2x_3 \end{bmatrix}$$

Find a basis for R^3 relative to which the matrix of T is diagonal.

17. Let $T: P_1 \to P_1$ be the linear operator defined by

$$T(a_0 + a_1 x) = a_0 + (6a_0 - a_1)x$$

Find a basis for P_1 with respect to which the matrix for T is diagonal.

18. Let A be an $n \times n$ matrix and P an invertible $n \times n$ matrix. Show:
(a) $(P^{-1}AP)^2 = P^{-1}A^2 P$.
(b) $(P^{-1}AP)^k = P^{-1}A^k P$ (k a positive integer).

19. Use Exercise 18 to help compute A^{10}, where

$$A = \begin{bmatrix} 1 & 0 \\ -1 & 2 \end{bmatrix}$$

(*Hint.* Find a matrix P that diagonalizes A and compute $(P^{-1}AP)^{10}$.)

20. Let

$$A = \begin{bmatrix} a & b \\ c & d \end{bmatrix}$$

Show:
(a) A is diagonalizable if $(a - d)^2 + 4bc > 0$.
(b) A is not diagonalizable if $(a - d)^2 + 4bc < 0$.

6.3 ORTHOGONAL DIAGONALIZATION; SYMMETRIC MATRICES

If $T: V \to V$ is a linear operator on an *inner product space*, then the diagonalization problem can be posed in a different way. Instead of simply looking for any basis that produces a diagonal matrix for T, we can look for an *orthonormal* basis that produces a diagonal matrix for T. More precisely, we shall be concerned with the following problem.

The Orthogonal Diagonalization Problem. *Given a linear operator $T: V \to V$ on a finite-dimensional inner product space, does there exist an orthonormal basis for V with respect to which the matrix for T is diagonal?*

If A is the matrix for $T: V \to V$ with respect to some orthonormal basis, then this problem is equivalent to asking if there is a change of basis to a new orthonormal basis such that the new matrix for T is diagonal. By Theorem 32 in Section 4.10, the transition matrix for this change of basis will be orthogonal. Thus, we are led to the following matrix form of the orthogonal diagonalization problem.

Matrix Form of the Orthogonal Diagonalization Problem. *Given a square matrix A, does there exist an orthogonal matrix P such that $P^{-1}AP(=P^tAP)$ is diagonal?*

This problem suggests the following definition.

Definition. A square matrix A is called ***orthogonally diagonalizable*** if there is an orthogonal matrix P such that $P^{-1}AP$ $(= P^tAP)$ is diagonal; the matrix P is said to ***orthogonally diagonalize*** A.

We have two questions to consider:

1. Which matrices are orthogonally diagonalizable?
2. How do we find an orthogonal matrix to carry out the diagonalization?

To help us answer the first question we will need the following definition.

Definition. A square matrix A is called ***symmetric*** if $A = A^t$.

Example 12

If

$$A = \begin{bmatrix} 1 & 4 & 5 \\ 4 & -3 & 0 \\ 5 & 0 & 7 \end{bmatrix}$$

then

$$A^t = \begin{bmatrix} 1 & 4 & 5 \\ 4 & -3 & 0 \\ 5 & 0 & 7 \end{bmatrix} = A$$

so A is symmetric.

It is easy to recognize symmetric matrices by inspection: the entries on the main diagonal are arbitrary, but "mirror images" of entries across the main diagonal are equal (Figure 6.3).

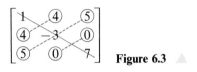

Figure 6.3

The next theorem is the main tool for determining whether a matrix is orthogonally diagonalizable. In this theorem and for the remainder of this section, *orthogonal* will mean orthogonal with respect to the Euclidean inner product on R^n.

Theorem 5. *If A is an n × n matrix, then the following are equivalent.*

(*a*) *A is orthogonally diagonalizable.*
(*b*) *A has an orthonormal set of n eigenvectors.*
(*c*) *A is symmetric.*

Proof $(a) \Rightarrow (b)$. Since A is orthogonally diagonalizable, there is an orthogonal matrix P such that $P^{-1}AP$ is diagonal. As shown in the proof of Theorem 2, the n column vectors of P are eigenvectors of A. Since P is orthogonal, these column vectors are orthonormal (see Theorem 33 of Section 4.10) so that A has n orthonormal eigenvectors.

$(b) \Rightarrow (a)$. Assume that A has an orthonormal set of n eigenvectors $\{\mathbf{p}_1, \mathbf{p}_2, \ldots, \mathbf{p}_n\}$. As shown in the proof of Theorem 2, the matrix P with these eigenvectors as columns diagonalizes A. Since these eigenvectors are orthonormal, P is orthogonal and thus orthogonally diagonalizes A.

$(a) \Rightarrow (c)$. In the proof that $(a) \Rightarrow (b)$ we showed that an orthogonally diagonalizable $n \times n$ matrix A is orthogonally diagonalized by an $n \times n$ matrix P whose columns form an orthonormal set of eigenvectors of A. Let D be the diagonal matrix

$$D = P^{-1}AP$$

Thus,

$$A = PDP^{-1}$$

or, since P is orthogonal,

$$A = PDP^t$$

Therefore,

$$A^t = (PDP^t)^t = PD^tP^t = PDP^t = A$$

which shows that A is symmetric.

$(c) \Rightarrow (a)$. The proof of this part is beyond the scope of this text and will be omitted. ▮

We now turn to the problem of finding an orthogonal matrix P to diagonalize a symmetric matrix. The key is the following theorem, whose proof is given at the end of the section.

> **Theorem 6.** *If A is a symmetric matrix, then eigenvectors from different eigenspaces are orthogonal.*

As a consequence of this theorem we obtain the following procedure for orthogonally diagonalizing a symmetric matrix.

Step 1. Find a basis for each eigenspace of A.

Step 2. Apply the Gram-Schmidt process to each of these bases to obtain an orthonormal basis for each eigenspace.

Step 3. Form the matrix P whose columns are the basis vectors constructed in Step 2; this matrix orthogonally diagonalizes A.

The justification of this procedure should be clear. Theorem 6 assures that eigenvectors from *distinct* eigenspaces are orthogonal, while the application of the Gram-Schmidt process assures that the eigenvectors obtained within the *same* eigenspace are orthonormal. Thus, the *entire* set of eigenvectors obtained by this procedure is orthonormal.

Example 13

Find an orthogonal matrix P that diagonalizes

$$A = \begin{bmatrix} 4 & 2 & 2 \\ 2 & 4 & 2 \\ 2 & 2 & 4 \end{bmatrix}$$

Solution. The characteristic equation of A is

$$\det(\lambda I - A) = \det \begin{bmatrix} \lambda - 4 & -2 & -2 \\ -2 & \lambda - 4 & -2 \\ -2 & -2 & \lambda - 4 \end{bmatrix} = (\lambda - 2)^2(\lambda - 8) = 0$$

Thus, the eigenvalues of A are $\lambda = 2$ and $\lambda = 8$. By the method used in Example 5, it can be shown that

$$\mathbf{u}_1 = \begin{bmatrix} -1 \\ 1 \\ 0 \end{bmatrix} \quad \text{and} \quad \mathbf{u}_2 = \begin{bmatrix} -1 \\ 0 \\ 1 \end{bmatrix}$$

form a basis for the eigenspace corresponding to $\lambda = 2$. Applying the Gram-Schmidt process to $\{\mathbf{u}_1, \mathbf{u}_2\}$ yields the following orthonormal eigenvectors (verify):

$$\mathbf{v}_1 = \begin{bmatrix} -\dfrac{1}{\sqrt{2}} \\ \dfrac{1}{\sqrt{2}} \\ 0 \end{bmatrix} \quad \text{and} \quad \mathbf{v}_2 = \begin{bmatrix} -\dfrac{1}{\sqrt{6}} \\ -\dfrac{1}{\sqrt{6}} \\ \dfrac{2}{\sqrt{6}} \end{bmatrix}$$

The eigenspace corresponding to $\lambda = 8$ has

$$\mathbf{u}_3 = \begin{bmatrix} 1 \\ 1 \\ 1 \end{bmatrix}$$

as a basis. Applying the Gram-Schmidt process to $\{\mathbf{u}_3\}$ yields

$$\mathbf{v}_3 = \begin{bmatrix} \dfrac{1}{\sqrt{3}} \\ \dfrac{1}{\sqrt{3}} \\ \dfrac{1}{\sqrt{3}} \end{bmatrix}$$

Finally, using $\mathbf{v}_1$, $\mathbf{v}_2$, and $\mathbf{v}_3$ as column vectors we obtain

$$P = \begin{bmatrix} -\dfrac{1}{\sqrt{2}} & -\dfrac{1}{\sqrt{6}} & \dfrac{1}{\sqrt{3}} \\ \dfrac{1}{\sqrt{2}} & -\dfrac{1}{\sqrt{6}} & \dfrac{1}{\sqrt{3}} \\ 0 & \dfrac{2}{\sqrt{6}} & \dfrac{1}{\sqrt{3}} \end{bmatrix}$$

which orthogonally diagonalizes A. (As a check, the reader may wish to verify that $P^t A P$ is a diagonal matrix.) ▲

We conclude this section with two important properties of symmetric matrices.

Theorem 7.

(a) *The characteristic equation of a symmetric matrix A has only real roots.*

(b) *If an eigenvalue λ of a symmetric matrix A is repeated k times as a root of the characteristic equation, then the eigenspace corresponding to λ is k-dimensional.*

The proof of the first result appears at the end of Section 9.6; the proof of the second will be omitted.

Example 14

The characteristic equation of the symmetric matrix

$$A = \begin{bmatrix} 3 & 1 & 0 & 0 & 0 \\ 1 & 3 & 0 & 0 & 0 \\ 0 & 0 & 2 & 1 & 1 \\ 0 & 0 & 1 & 2 & 1 \\ 0 & 0 & 1 & 1 & 2 \end{bmatrix}$$

is

$$(\lambda - 4)^2(\lambda - 1)^2(\lambda - 2) = 0$$

so that the eigenvalues are $\lambda = 4$, $\lambda = 1$, and $\lambda = 2$, where $\lambda = 4$ and $\lambda = 1$ are repeated twice and $\lambda = 2$ occurs once. Thus, the eigenspaces corresponding to $\lambda = 4$ and $\lambda = 1$ are 2-dimensional, and the eigenspace corresponding to $\lambda = 1$ is 1-dimensional. ▲

REMARK. We remind the reader that we have assumed to this point that all our matrices have real entries. Indeed, we shall see in Section 9.6 (Exercise 13) that part (a) of Theorem 7 is false for matrices with complex entries.

OPTIONAL

Proof of Theorem 6. Let λ_1 and λ_2 be two different eigenvalues of the $n \times n$ symmetric matrix A, and let

$$\mathbf{v}_1 = \begin{bmatrix} v_1 \\ v_2 \\ \vdots \\ v_n \end{bmatrix} \quad \text{and} \quad \mathbf{v}_2 = \begin{bmatrix} v'_1 \\ v'_2 \\ \vdots \\ v'_n \end{bmatrix}$$

be the corresponding eigenvectors. We want to show that

$$\mathbf{v}_1 \cdot \mathbf{v}_2 = v_1 v'_1 + v_2 v'_2 + \cdots + v_n v'_n = 0$$

Since $\mathbf{v}_1{}^t\mathbf{v}_2$ is a 1×1 matrix with $\mathbf{v}_1 \cdot \mathbf{v}_2$ as its only entry, we can complete the proof by showing that $\mathbf{v}_1{}^t\mathbf{v}_2 = 0$.

Since $\mathbf{v}_1$ and $\mathbf{v}_2$ are eigenvectors corresponding to the eigenvalues λ_1 and λ_2, we have

$$A\mathbf{v}_1 = \lambda_1\mathbf{v}_1 \tag{6.11}$$

$$A\mathbf{v}_2 = \lambda_2\mathbf{v}_2 \tag{6.12}$$

From (6.11)

$$(A\mathbf{v}_1)^t = (\lambda_1\mathbf{v}_1)^t$$

or

$$\mathbf{v}_1{}^tA^t = \lambda_1\mathbf{v}_1{}^t$$

or, since A is symmetric,

$$\mathbf{v}_1{}^tA = \lambda_1\mathbf{v}_1{}^t$$

Multiplying both sides of this equation on the right by $\mathbf{v}_2$ yields

$$\mathbf{v}_1{}^tA\mathbf{v}_2 = \lambda_1\mathbf{v}_1{}^t\mathbf{v}_2 \tag{6.13}$$

and multiplying both sides of (6.12) on the left by $\mathbf{v}_1{}^t$ yields

$$\mathbf{v}_1{}^tA\mathbf{v}_2 = \lambda_2\mathbf{v}_1{}^t\mathbf{v}_2 \tag{6.14}$$

Thus, from (6.13) and (6.14)

$$\lambda_1\mathbf{v}_1{}^t\mathbf{v}_2 = \lambda_2\mathbf{v}_1{}^t\mathbf{v}_2$$

or

$$(\lambda_1 - \lambda_2)\mathbf{v}_1{}^t\mathbf{v}_2 = 0$$

But $\lambda_1 \neq \lambda_2$, so $\mathbf{v}_1{}^t\mathbf{v}_2 = 0$, which is what we wanted to prove.

EXERCISE SET 6.3

1. Use part (*b*) of Theorem 7 to find the dimensions of the eigenspaces of the following symmetric matrices.

(a) $\begin{bmatrix} 1 & 1 \\ 1 & 1 \end{bmatrix}$
(b) $\begin{bmatrix} 1 & -4 & 2 \\ -4 & 1 & -2 \\ 2 & -2 & -2 \end{bmatrix}$

(c) $\begin{bmatrix} 1 & 1 & 1 \\ 1 & 1 & 1 \\ 1 & 1 & 1 \end{bmatrix}$
(d) $\begin{bmatrix} 6 & 0 & 0 \\ 0 & 3 & 3 \\ 0 & 3 & 3 \end{bmatrix}$

(e) $\begin{bmatrix} 4 & 4 & 0 & 0 \\ 4 & 4 & 0 & 0 \\ 0 & 0 & 0 & 0 \\ 0 & 0 & 0 & 0 \end{bmatrix}$
(f) $\begin{bmatrix} \frac{10}{3} & -\frac{4}{3} & 0 & -\frac{4}{3} \\ -\frac{4}{3} & -\frac{5}{3} & 0 & \frac{1}{3} \\ 0 & 0 & -2 & 0 \\ -\frac{4}{3} & \frac{1}{3} & 0 & -\frac{5}{3} \end{bmatrix}$

In Exercises 2–9 find a matrix P that orthogonally diagonalizes A, and determine $P^{-1}AP$.

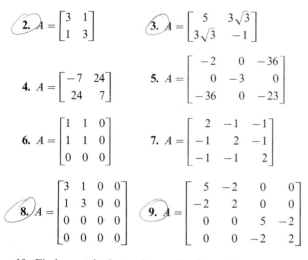

2. $A = \begin{bmatrix} 3 & 1 \\ 1 & 3 \end{bmatrix}$

3. $A = \begin{bmatrix} 5 & 3\sqrt{3} \\ 3\sqrt{3} & -1 \end{bmatrix}$

4. $A = \begin{bmatrix} -7 & 24 \\ 24 & 7 \end{bmatrix}$

5. $A = \begin{bmatrix} -2 & 0 & -36 \\ 0 & -3 & 0 \\ -36 & 0 & -23 \end{bmatrix}$

6. $A = \begin{bmatrix} 1 & 1 & 0 \\ 1 & 1 & 0 \\ 0 & 0 & 0 \end{bmatrix}$

7. $A = \begin{bmatrix} 2 & -1 & -1 \\ -1 & 2 & -1 \\ -1 & -1 & 2 \end{bmatrix}$

8. $A = \begin{bmatrix} 3 & 1 & 0 & 0 \\ 1 & 3 & 0 & 0 \\ 0 & 0 & 0 & 0 \\ 0 & 0 & 0 & 0 \end{bmatrix}$

9. $A = \begin{bmatrix} 5 & -2 & 0 & 0 \\ -2 & 2 & 0 & 0 \\ 0 & 0 & 5 & -2 \\ 0 & 0 & -2 & 2 \end{bmatrix}$

10. Find a matrix that orthogonally diagonalizes

$$\begin{bmatrix} a & b \\ b & a \end{bmatrix}$$

where $b \neq 0$.

11. Two $n \times n$ matrices, A and B, are called **orthogonally similar** if there is an orthogonal matrix P such that $B = P^t A P$. Show that if A is symmetric and A and B are orthogonally similar, then B is symmetric.

12. Prove Theorem 6 for 2×2 symmetric matrices.

13. Prove Theorem $7a$ for 2×2 symmetric matrices.

SUPPLEMENTARY EXERCISES

1. (a) Show that if $0 < \theta < \pi$, then

$$A = \begin{bmatrix} \cos\theta & -\sin\theta \\ \sin\theta & \cos\theta \end{bmatrix}$$

has no eigenvalues and consequently no eigenvectors.

(b) By considering the linear transformation of multiplication by A, give a geometric explanation of the result in (a).

2. Find the eigenvalues of

$$A = \begin{bmatrix} 0 & 1 & 0 \\ 0 & 0 & 1 \\ k^3 & -3k^2 & 3k \end{bmatrix}$$

3. (a) Show that if D is a diagonal matrix with nonnegative entries on the main diagonal, then there is a matrix S such that $S^2 = D$.

(b) Show that if A is a diagonalizable matrix with nonnegative eigenvalues, then there is a matrix S such that $S^2 = A$.

(c) Find a matrix S such that $S^2 = A$, if

$$A = \begin{bmatrix} 1 & 3 & 1 \\ 0 & 4 & 5 \\ 0 & 0 & 9 \end{bmatrix}$$

4. Prove: If A is a square matrix, then A and A^t have the same eigenvalues.

5. Prove: If A is a square matrix and $p(\lambda) = \det(\lambda I - A)$ is the characteristic polynomial of A, then the coefficient of λ^{n-1} in $p(\lambda)$ is the negative of the trace of A (see Exercise 15 of Section 6.1 for terminology).

6. Prove: If $b \neq 0$, then

$$A = \begin{bmatrix} a & b \\ 0 & a \end{bmatrix}$$

is not diagonalizable.

7. In advanced linear algebra, one proves the **Cayley-Hamilton Theorem**, which states that a square matrix A satisfies its characteristic equation; that is, if

$$c_0 + c_1\lambda + c_2\lambda^2 + \cdots + c_{n-1}\lambda^{n-1} + \lambda^n = 0$$

is the characteristic equation, then

$$c_0 I + c_1 A + c_2 A^2 + \cdots + c_{n-1}A^{n-1} + A^n = 0$$

Verify this result for:

(a) $A = \begin{bmatrix} 3 & 6 \\ 1 & 2 \end{bmatrix}$ (b) $A = \begin{bmatrix} 0 & 1 & 0 \\ 0 & 0 & 1 \\ 1 & -3 & 3 \end{bmatrix}$

Exercises 8–10 use the Cayley-Hamilton Theorem, stated in Exercise 7.

8. Use Exercise 16 of Section 6.1 to prove the Cayley-Hamilton Theorem for 2×2 matrices.

9. The Cayley-Hamilton Theorem provides an efficient method for calculating powers of a matrix. For example, if A is a 2×2 matrix with characteristic equation

$$c_0 + c_1\lambda + \lambda^2 = 0$$

then $c_0 I + c_1 A + A^2 = 0$, so

$$A^2 = -c_1 A - c_0 I$$

Multiplying through by A yields $A^3 = -c_1 A^2 - c_0 A$, which expresses A^3 in terms of A^2 and A, and multiplying through by A^2 yields $A^4 = -c_1 A^3 - c_0 A^2$, which expresses A^4 in terms of A^3 and A^2. Continuing in this way, we can calculate successive powers of

A simply by expressing them in terms of lower powers. Use this procedure to calculate

$$A^2, \quad A^3, \quad A^4, \quad \text{and} \quad A^5$$

for

$$A = \begin{bmatrix} 3 & 6 \\ 1 & 2 \end{bmatrix}$$

10. Use the method of the preceding exercise to calculate A^3 and A^4 if

$$A = \begin{bmatrix} 0 & 1 & 0 \\ 0 & 0 & 1 \\ 1 & -3 & 3 \end{bmatrix}$$

11. Show that

$$A = \begin{bmatrix} a & b \\ c & d \end{bmatrix}$$

has
(a) two distinct eigenvalues if $(a - d)^2 + 4bc > 0$
(b) one eigenvalue if $(a - d)^2 + 4bc = 0$
(c) no eigenvalues if $(a - d)^2 + 4bc < 0$.

12. (a) It was shown in Exercise 15 of Section 6.1 that if A is an $n \times n$ matrix, then the coefficient of λ^n in the characteristic polynomial of A is 1. (A polynomial with this property is called *monic*.) Show that the matrix

$$\begin{bmatrix} 0 & 0 & 0 & \cdots & 0 & -c_0 \\ 1 & 0 & 0 & \cdots & 0 & -c_1 \\ 0 & 1 & 0 & \cdots & 0 & -c_2 \\ \vdots & \vdots & \vdots & & \vdots & \vdots \\ 0 & 0 & 0 & \cdots & 1 & -c_{n-1} \end{bmatrix}$$

has characteristic polynomial $p(\lambda) = c_0 + c_1\lambda + \cdots + c_{n-1}\lambda^{n-1} + \lambda^n$. This shows that every monic polynomial is the characteristic polynomial of some matrix. The matrix in this example is called the *companion matrix* of $p(\lambda)$. (*Hint.* Evaluate all determinants in the problem by adding a multiple of the second row to the first to introduce a zero at the top of the first column, and then expanding by cofactors along the first column.)
(b) Find a matrix with characteristic polynomial $p(\lambda) = 1 - 2\lambda + \lambda^2 + 3\lambda^3 + \lambda^4$.

13. Prove: If a, b, c, and d are integers such that $a + b = c + d$, then

$$A = \begin{bmatrix} a & b \\ c & d \end{bmatrix}$$

has integer eigenvalues, namely, $x_1 = a + b$ and $x_2 = a - c$.

14. Prove: If A is an $n \times n$ matrix where n is odd, then A has at least one real eigenvalue.

15. Prove: If λ is an eigenvalue of an invertible matrix A and $\mathbf{x}$ is a corresponding eigenvector, then $1/\lambda$ is an eigenvalue of A^{-1} and $\mathbf{x}$ is a corresponding eigenvector.

16. Prove: If λ is an eigenvalue of A, $\mathbf{x}$ is a corresponding eigenvector, and s is a scalar, then $\lambda - s$ is an eigenvalue of $A - sI$ and $\mathbf{x}$ is a corresponding eigenvector.

17. Find the eigenvalues and bases for the eigenspaces of

$$A = \begin{bmatrix} -2 & 2 & 3 \\ -2 & 3 & 2 \\ -4 & 2 & 5 \end{bmatrix}$$

Then use Exercises 15 and 16 to find the eigenvalues and bases for the eigenspaces of
(a) A^{-1} (b) $A - 3I$ (c) $A + 2I$

18. In each part find as many linearly independent eigenvectors as you can by inspection (by visualizing the geometric effect of the transformation on R^2). For each of your eigenvectors find the corresponding eigenvalue by inspection, then check your results by computing the eigenvalues and bases for the eigenspaces from the standard matrix for the transformation (see Section 5.3 for these matrices).

(a) reflection about the x-axis
(b) reflection about the y-axis
(c) reflection about $y = x$
(d) shear in the x-direction with factor k
(e) shear in the y-direction with factor k
(f) rotation through the angle θ.

CHAPTER SEVEN

Applications

7.1 APPLICATION TO DIFFERENTIAL EQUATIONS

Many laws of physics, chemistry, biology, and economics are described in terms of *differential equations*, that is, equations involving functions and their derivatives. The purpose of this section is to illustrate one way in which linear algebra can be applied to solve certain systems of differential equations. The scope of this section is narrow, but it should serve to convince the reader that linear algebra has solid applications.

One of the simplest differential equations is

$$y' = ay \tag{7.1}$$

where $y = f(x)$ is an unknown function to be determined, $y' = dy/dx$ is its derivative, and a is a constant. Like most differential equations, (7.1) has infinitely many solutions; they are the functions of the form

$$y = ce^{ax} \tag{7.2}$$

where c is an arbitrary constant. Each function of this form is a solution of $y' = ay$ since

$$y' = cae^{ax} = ay$$

Conversely, every solution of $y' = ay$ must be a function of the form ce^{ax} (Exercise 7), so that (7.2) describes all solutions of $y' = ay$. We call (7.2) the ***general solution*** of $y' = ay$.

Sometimes the physical problem that generates a differential equation imposes some added condition that enables us to isolate one ***particular solution*** from the general solution. For example, if we require that the solution of $y' = ay$ satisfy the added condition

$$y(0) = 3 \tag{7.3}$$

that is, $y = 3$ when $x = 0$, then on substituting these values in the general solution $y = ce^{ax}$ we obtain a value for c, namely

$$3 = ce^0 = c$$

Thus,

$$y = 3e^{ax}$$

is the only solution of $y' = ay$ that satisfies the added condition. A condition, such as (7.3), which specifies the value of the solution at a point is called an ***initial condition***, and the problem of solving a differential equation subject to an initial condition is called an ***initial-value problem***.

In this section we will be concerned with solving systems of differential equations having the form

$$
\begin{aligned}
y_1' &= a_{11}y_1 + a_{12}y_2 + \cdots + a_{1n}y_n \\
y_2' &= a_{21}y_1 + a_{22}y_2 + \cdots + a_{2n}y_n \\
&\;\;\vdots \qquad \vdots \qquad \vdots \qquad\qquad \vdots \\
y_n' &= a_{n1}y_1 + a_{n2}y_2 + \cdots + a_{nn}y_n
\end{aligned}
\tag{7.4}
$$

where $y_1 = f_1(x)$, $y_2 = f_2(x)$, ..., $y_n = f_n(x)$ are functions to be determined, and the a_{ij}'s are constants. In matrix notation (7.4) can be written

$$
\begin{bmatrix} y_1' \\ y_2' \\ \vdots \\ y_n' \end{bmatrix} =
\begin{bmatrix}
a_{11} & a_{12} & \cdots & a_{1n} \\
a_{21} & a_{22} & \cdots & a_{2n} \\
\vdots & \vdots & & \vdots \\
a_{n1} & a_{n2} & \cdots & a_{nn}
\end{bmatrix}
\begin{bmatrix} y_1 \\ y_2 \\ \vdots \\ y_n \end{bmatrix}
$$

or more briefly

$$Y' = AY$$

Example 1

(a) Write the following system in matrix form:

$$
\begin{aligned}
y_1' &= 3y_1 \\
y_2' &= -2y_2 \\
y_3' &= 5y_3
\end{aligned}
$$

(b) Solve the system.
(c) Find a solution of the system which satisfies the initial conditions $y_1(0) = 1$, $y_2(0) = 4$, and $y_3(0) = -2$.

Solution (a).

$$
\begin{bmatrix} y_1' \\ y_2' \\ y_3' \end{bmatrix} =
\begin{bmatrix}
3 & 0 & 0 \\
0 & -2 & 0 \\
0 & 0 & 5
\end{bmatrix}
\begin{bmatrix} y_1 \\ y_2 \\ y_3 \end{bmatrix}
\tag{7.5}
$$

or

$$Y' = \begin{bmatrix} 3 & 0 & 0 \\ 0 & -2 & 0 \\ 0 & 0 & 5 \end{bmatrix} Y$$

(b) Because each equation involves only one unknown function, we can solve the equations individually. From (7.2), we obtain

$$y_1 = c_1 e^{3x}$$
$$y_2 = c_2 e^{-2x}$$
$$y_3 = c_3 e^{5x}$$

or in matrix notation

$$Y = \begin{bmatrix} y_1 \\ y_2 \\ y_3 \end{bmatrix} = \begin{bmatrix} c_1 e^{3x} \\ c_2 e^{-2x} \\ c_3 e^{5x} \end{bmatrix}$$

(c) From the given initial conditions, we obtain

$$1 = y_1(0) = c_1 e^0 = c_1$$
$$4 = y_2(0) = c_2 e^0 = c_2$$
$$-2 = y_3(0) = c_3 e^0 = c_3$$

so that the solution satisfying the initial conditions is

$$y_1 = e^{3x}, \ y_2 = 4e^{-2x}, \ y_3 = -2e^{5x}$$

or, in matrix notation,

$$Y = \begin{bmatrix} y_1 \\ y_2 \\ y_3 \end{bmatrix} = \begin{bmatrix} e^{3x} \\ 4e^{-2x} \\ -2e^{5x} \end{bmatrix}$$

The system in the foregoing example is easy to solve because each equation involves only one unknown function, and this is the case because the matrix of coefficients (7.5) for the system is diagonal. But how do we handle a system

$$Y' = AY$$

in which the matrix A is not diagonal? The idea is simple: try to make a substitution for Y that will yield a new system with a diagonal coefficient matrix; solve this new simpler system and then use this solution to determine the solution of the original system.

The kind of substitution we have in mind is

$$y_1 = p_{11}u_1 + p_{12}u_2 + \cdots + p_{1n}u_n$$
$$y_2 = p_{21}u_1 + p_{22}u_2 + \cdots + p_{2n}u_n$$
$$\vdots \quad \vdots \quad \vdots \quad \vdots$$
$$y_n = p_{n1}u_1 + p_{n2}u_2 + \cdots + p_{nn}u_n$$

(7.6)

or, in matrix notation,

$$\begin{bmatrix} y_1 \\ y_2 \\ \vdots \\ y_n \end{bmatrix} = \begin{bmatrix} p_{11} & p_{12} & \cdots & p_{1n} \\ p_{21} & p_{22} & \cdots & p_{2n} \\ \vdots & \vdots & & \vdots \\ p_{n1} & p_{n2} & \cdots & p_{nn} \end{bmatrix} \begin{bmatrix} u_1 \\ u_2 \\ \vdots \\ u_n \end{bmatrix}$$

or more briefly

$$Y = PU$$

In this substitution the p_{ij}'s are constants to be determined in such a way that the new system involving the unknown functions $u_1, u_2, \ldots, u_n$ has a diagonal coefficient matrix. We leave it as an exercise for the reader to differentiate each equation in (7.6) and deduce

$$Y' = PU'$$

If we make the substitutions $Y = PU$ and $Y' = PU'$ in the original system

$$Y' = AY$$

and if we assume P to be invertible, we obtain

$$PU' = A(PU)$$

or

$$U' = (P^{-1}AP)U$$

or

$$U' = DU$$

where $D = P^{-1}AP$. The choice for P is now clear; if we want the new coefficient matrix D to be diagonal, we must choose P to be a matrix that diagonalizes A.

This suggests the following procedure for solving a system

$$Y' = AY$$

with a diagonalizable coefficient matrix A.

Step 1. Find a matrix P that diagonalizes A.

Step 2. Make the substitutions $Y = PU$ and $Y' = PU'$ to obtain a new "diagonal system" $U' = DU$, where $D = P^{-1}AP$.

Step 3. Solve $U' = DU$.

Step 4. Determine Y from the equation $Y = PU$.

Example 2
(a) Solve

$$y_1' = y_1 + y_2$$
$$y_2' = 4y_1 - 2y_2$$

(b) Find the solution that satisfies the initial conditions $y_1(0) = 1$, $y_2(0) = 6$.

Solution (a). The coefficient matrix for the system is

$$A = \begin{bmatrix} 1 & 1 \\ 4 & -2 \end{bmatrix}$$

As discussed in Section 6.2, A will be diagonalized by any matrix P whose columns are linearly independent eigenvectors of A. Since

$$\det(\lambda I - A) = \begin{vmatrix} \lambda - 1 & -1 \\ -4 & \lambda + 2 \end{vmatrix} = \lambda^2 + \lambda - 6 = (\lambda + 3)(\lambda - 2)$$

the eigenvalues of A are $\lambda = 2$, $\lambda = -3$. By definition,

$$\mathbf{x} = \begin{bmatrix} x_1 \\ x_2 \end{bmatrix}$$

is an eigenvector of A corresponding to λ if and only if $\mathbf{x}$ is a nontrivial solution of $(\lambda I - A)\mathbf{x} = 0$, that is, of

$$\begin{bmatrix} \lambda - 1 & -1 \\ -4 & \lambda + 2 \end{bmatrix}\begin{bmatrix} x_1 \\ x_2 \end{bmatrix} = \begin{bmatrix} 0 \\ 0 \end{bmatrix}$$

If $\lambda = 2$, this system becomes

$$\begin{bmatrix} 1 & -1 \\ -4 & 4 \end{bmatrix}\begin{bmatrix} x_1 \\ x_2 \end{bmatrix} = \begin{bmatrix} 0 \\ 0 \end{bmatrix}$$

Solving this system yields

$$x_1 = t, \qquad x_2 = t$$

so that

$$\begin{bmatrix} x_1 \\ x_2 \end{bmatrix} = \begin{bmatrix} t \\ t \end{bmatrix} = t\begin{bmatrix} 1 \\ 1 \end{bmatrix}$$

Thus,

$$\mathbf{p}_1 = \begin{bmatrix} 1 \\ 1 \end{bmatrix}$$

is a basis for the eigenspace corresponding to $\lambda = 2$. Similarly, the reader can show that

$$\mathbf{p}_2 = \begin{bmatrix} -\frac{1}{4} \\ 1 \end{bmatrix}$$

is a basis for the eigenspace corresponding to $\lambda = -3$. Thus,

$$P = \begin{bmatrix} 1 & -\frac{1}{4} \\ 1 & 1 \end{bmatrix}$$

diagonalizes A, and

$$D = P^{-1}AP = \begin{bmatrix} 2 & 0 \\ 0 & -3 \end{bmatrix}$$

Therefore, the substitution

$$Y = PU \quad \text{and} \quad Y' = PU'$$

yields the new "diagonal system"

$$U' = DU = \begin{bmatrix} 2 & 0 \\ 0 & -3 \end{bmatrix} U \quad \text{or} \quad \begin{matrix} u'_1 = 2u_1 \\ u'_2 = -3u_2 \end{matrix}$$

From (7.2) the solution of this system is

$$\begin{matrix} u_1 = c_1 e^{2x} \\ u_2 = c_2 e^{-3x} \end{matrix} \quad \text{or} \quad U = \begin{bmatrix} c_1 e^{2x} \\ c_2 e^{-3x} \end{bmatrix}$$

so that the equation $Y = PU$ yields as the solution for Y

$$Y = \begin{bmatrix} y_1 \\ y_2 \end{bmatrix} = \begin{bmatrix} 1 & -\frac{1}{4} \\ 1 & 1 \end{bmatrix} \begin{bmatrix} c_1 e^{2x} \\ c_2 e^{-3x} \end{bmatrix} = \begin{bmatrix} c_1 e^{2x} - \frac{1}{4} c_2 e^{-3x} \\ c_1 e^{2x} + c_2 e^{-3x} \end{bmatrix}$$

or

$$\begin{matrix} y_1 = c_1 e^{2x} - \frac{1}{4} c_2 e^{-3x} \\ y_2 = c_1 e^{2x} + c_2 e^{-3x} \end{matrix} \tag{7.7}$$

(b) If we substitute the given initial conditions in (7.7), we obtain

$$c_1 - \tfrac{1}{4} c_2 = 1$$
$$c_1 + c_2 = 6$$

Solving this system we obtain

$$c_1 = 2, \quad c_2 = 4$$

so that from (7.7) the solution satisfying the initial conditions is

$$y_1 = 2e^{2x} - e^{-3x}$$
$$y_2 = 2e^{2x} + 4e^{-3x} \quad \blacktriangle$$

We have assumed in this section that the coefficient matrix of $Y' = AY$ is diagonalizable. If this is not the case, other methods must be used to solve the system. Such methods are discussed in more advanced texts.

EXERCISE SET 7.1

1. (a) Solve the system

$$y'_1 = y_1 + 4y_2$$
$$y'_2 = 2y_1 + 3y_2$$

(b) Find the solution that satisfies the initial conditions $y_1(0) = 0$, $y_2(0) = 0$.

2. (a) Solve the system

$$y'_1 = y_1 + 3y_2$$
$$y'_2 = 4y_1 + 5y_2$$

(b) Find the solution that satisfies the conditions $y_1(0) = 2$, $y_2'(0) = 1$.

3. (a) Solve the system

$$
\begin{aligned}
y_1' &= 4y_1 + y_3 \\
y_2' &= -2y_1 + y_2 \\
y_3' &= -2y_1 + y_3
\end{aligned}
$$

(b) Find the solution that satisfies the initial conditions $y_1(0) = -1$, $y_2(0) = 1$, $y_3(0) = 0$.

4. Solve the system

$$
\begin{aligned}
y_1' &= 4y_1 + 2y_2 + 2y_3 \\
y_2' &= 2y_1 + 4y_2 + 2y_3 \\
y_3' &= 2y_1 + 2y_2 + 4y_3
\end{aligned}
$$

5. Solve the differential equation $y'' - y' - 6y = 0$. (*Hint.* Let $y_1 = y$, $y_2 = y'$ and then show

$$
\begin{aligned}
y_1' &= y_2 \\
y_2' &= y'' = y' + 6y = y_1' + 6y_1 = 6y_1 + y_2
\end{aligned}
$$

6. Solve the differential equation $y''' - 6y'' + 11y' - 6y = 0$. (*Hint.* Let $y_1 = y$, $y_2 = y'$, $y_3 = y''$ and then show

$$
\begin{aligned}
y_1' &= y_2 \\
y_2' &= y_3 \\
y_3' &= 6y_1 - 11y_2 + 6y_3
\end{aligned}
$$

7. Prove: Every solution of $y' = ay$ has the form $y = ce^{ax}$. (*Hint.* Let $y = f(x)$ be a solution and show that $f(x)e^{-ax}$ is constant.)

8. Prove: If A is diagonalizable and

$$
Y = \begin{bmatrix} y_1 \\ y_2 \\ \vdots \\ y_n \end{bmatrix}
$$

satisfies $Y' = AY$, then each y_i is a linear combination of $e^{\lambda_1 x}, e^{\lambda_2 x}, \ldots, e^{\lambda_n x}$, where $\lambda_1, \lambda_2, \ldots, \lambda_n$ are eigenvalues of A.

7.2 APPROXIMATION PROBLEMS; FOURIER SERIES

In many applications one is concerned with finding the best possible approximation over an interval to a function f by another function in some specified class; for example:

(a) Find the best possible approximation to e^x over $[0, 1]$ by a polynomial of the form $a_0 + a_1 x + a_2 x^2$.

(b) Find the best possible approximation to $\sin \pi x$ over $[-1, 1]$ by a function of the form $a_0 + a_1 e^x + a_2 e^{2x} + a_3 e^{3x}$.

(c) Find the best possible approximation to $|x|$ over $[0, 2\pi]$ by a function of the form $a_0 + a_1 \sin x + a_2 \sin 2x + b_1 \cos x + b_2 \cos 2x$.

Observe that in each of these examples the approximating functions were drawn from some subspace of the vector space $C[a, b]$ (continuous functions on $[a, b]$). In the first example it was the subspace of $C[0, 1]$ spanned by 1, x, and x^2; in the second example it was the subspace of $C[-1, 1]$ spanned by 1, e^x, e^{2x}, and e^{3x}; and in the third example it was the subspace of $C[0, 2\pi]$ spanned by 1, $\sin x$, $\sin 2x$, $\cos x$, and $\cos 2x$. Thus, each of these examples states a problem of the following form.

Approximation Problem. Find the best possible approximation over $[a, b]$ to a given function f using only approximations from a specified subspace W of $C[a, b]$.

To solve this problem we must make the term "best possible approximation over $[a, b]$" more precise. Intuitively, the best possible approximation over $[a, b]$ is one that produces the smallest error. But what do we mean by "error"? If we were concerned only with approximating $f(x)$ at a single point x_0, then the error at x_0 by an approximation $g(x)$ would be simply

$$\text{error} = |f(x_0) - g(x_0)|$$

sometimes called the **deviation** between f and g at x_0 (Figure 7.1). However, we are concerned with approximation over an entire interval $[a, b]$, not at a single point.

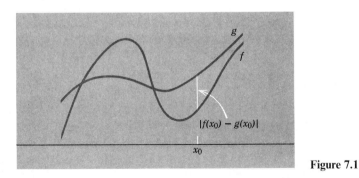

Figure 7.1

Consequently, in one part of the interval an approximation $g_1(x)$ may have smaller deviations from f than an approximation $g_2(x)$, and in another part of the interval it might be the other way around. How does one decide which is the better overall approximation? What we need is some way of measuring the overall error in an approximation $g(x)$. One possible measure of overall error is obtained by integrating the deviation $|f(x) - g(x)|$ over the entire interval; that is,

$$\text{error} = \int_a^b |f(x) - g(x)| \, dx \tag{7.8}$$

Geometrically, (7.8) is the area between the graphs of $f(x)$ and $g(x)$ over the interval $[a, b]$ (Figure 7.2); the greater the area, the greater the overall error.

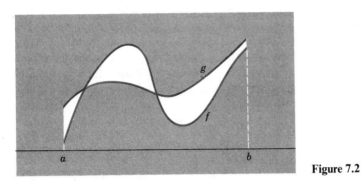

Figure 7.2

While (7.8) is natural and appealing geometrically, the occurrence of an absolute value sign is sufficiently bothersome in computations that most mathematicians and scientists generally favor the following alternative measure of error, called the *mean square error*.

$$\text{mean square error} = \int_a^b [f(x) - g(x)]^2 \, dx$$

Mean square error has the added advantage that it allows us to bring to bear the theory of inner product spaces in problems of approximation. To see how, consider the inner product

$$\langle \mathbf{f}, \mathbf{g} \rangle = \int_a^b f(x) g(x) \, dx \tag{7.9}$$

on the vector space $C[a, b]$. With this inner product

$$\|\mathbf{f} - \mathbf{g}\|^2 = \langle \mathbf{f} - \mathbf{g}, \mathbf{f} - \mathbf{g} \rangle = \int_a^b [f(x) - g(x)]^2 \, dx$$

which states that the mean square error that results from approximating $\mathbf{f}$ by $\mathbf{g}$ over $[a, b]$ is the square of the distance between $\mathbf{f}$ and $\mathbf{g}$ when these functions are viewed as vectors in $C[a, b]$ with the inner product (7.9). Thus, an approximation g from a subspace W of $C[a, b]$ minimizes the mean square error if and only if it minimizes $\|\mathbf{f} - \mathbf{g}\|^2$, or equivalently if and only if it minimizes $\|\mathbf{f} - \mathbf{g}\|$. In short, the approximation $\mathbf{g}$ in W that minimizes the mean square error is the vector $\mathbf{g}$ in W *closest* to $\mathbf{f}$ using inner product (7.9). But we already know what the vector $\mathbf{g}$ is; it is the orthogonal projection of $\mathbf{f}$ on the subspace W (Theorem 28, Section 4.9). See Figure 7.3. In summary, we have the following result.

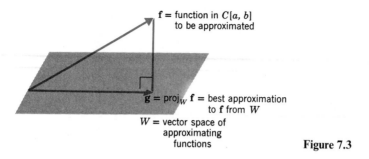

f = function in $C[a, b]$
to be approximated

g = proj_W **f** = best approximation
to **f** from W

W = vector space of
approximating
functions

Figure 7.3

Solution of the Approximation Problem. If **f** is a continuous function on $[a, b]$ and W is a finite dimensional subspace of $C[a, b]$, then the function **g** in W that minimizes the mean square error

$$\int_a^b [f(x) - g(x)]^2 \, dx$$

is **g** = proj_W **f**, the orthogonal projection of **f** on W, relative to inner product (7.9). The function **g** = proj_W **f** is called the *least squares approximation* to **f** from W.

Fourier Series. A function of the form

$$t(x) = c_0 + c_1 \cos x + c_2 \cos 2x + \cdots + c_n \cos nx$$
$$+ d_1 \sin x + d_2 \sin 2x + \cdots + d_n \sin nx \tag{7.10}$$

is called a *trigonometric polynomial*; if c_n and d_n are not both zero, then $t(x)$ is said to have *order n*.

Example 3

$$t(x) = 2 + \cos x - 3 \cos 2x + 7 \sin 4x$$

is a trigonometric polynomial with

$$c_0 = 2, c_1 = 1, c_2 = -3, d_1 = 0, d_2 = 0, d_3 = 0, d_4 = 7$$

The order of $t(x)$ is 4. ▲

It is evident from (7.10) that the trigonometric polynomials of order n or less are the various possible linear combinations of

$$1, \quad \cos x, \quad \cos 2x, \ldots, \cos nx, \quad \sin x, \quad \sin 2x, \ldots, \sin nx \tag{7.11}$$

Thus, the trigonometric polynomials of order n or less form a subspace W of the vector space of continuous functions; it is the subspace spanned by the $2n + 1$ functions listed in (7.11). It can be shown that these functions are linearly independent and consequently form a basis for W.

Let us consider the problem of approximating a continuous function $f(x)$ over the interval $[0, 2\pi]$ by a trigonometric polynomial of order n or less. As noted earlier, the least squares approximation to **f** from W is the orthogonal projection of **f** on W. To find this orthogonal projection, we must find an orthonormal basis $\mathbf{g}_0, \mathbf{g}_1, \ldots, \mathbf{g}_{2n}$ for W, after which we can compute the orthogonal projection on W from the formula

$$\text{proj}_W \, \mathbf{f} = \langle \mathbf{f}, \mathbf{g}_0 \rangle \mathbf{g}_0 + \langle \mathbf{f}, \mathbf{g}_1 \rangle \mathbf{g}_1 + \cdots + \langle \mathbf{f}, \mathbf{g}_{2n} \rangle \mathbf{g}_{2n} \qquad (7.12)$$

(see formula (4.30) of Section 4.9). An orthonormal basis for W can be obtained by applying the Gram-Schmidt process to the basis (7.11), using the inner product

$$\langle \mathbf{u}, \mathbf{v} \rangle = \int_0^{2\pi} u(x)v(x)\,dx$$

This yields (Exercise 6) the orthonormal basis

$$\mathbf{g}_0 = \frac{1}{\sqrt{2\pi}}, \, \mathbf{g}_1 = \frac{1}{\sqrt{\pi}}\cos x, \ldots, \mathbf{g}_n = \frac{1}{\sqrt{\pi}}\cos nx,$$

$$\mathbf{g}_{n+1} = \frac{1}{\sqrt{\pi}}\sin x, \ldots, \mathbf{g}_{2n} = \frac{1}{\sqrt{\pi}}\sin nx \qquad (7.13)$$

If we introduce the notation

$$a_0 = \frac{2}{\sqrt{2\pi}}\langle \mathbf{f}, \mathbf{g}_0 \rangle, \, a_1 = \frac{1}{\sqrt{\pi}}\langle \mathbf{f}, \mathbf{g}_1 \rangle, \ldots, a_n = \frac{1}{\sqrt{\pi}}\langle \mathbf{f}, \mathbf{g}_n \rangle$$

$$b_1 = \frac{1}{\sqrt{\pi}}\langle \mathbf{f}, \mathbf{g}_{n+1} \rangle, \ldots, b_n = \frac{1}{\sqrt{\pi}}\langle \mathbf{f}, \mathbf{g}_{2n} \rangle$$

then on substituting (7.13) in (7.12) we obtain

$$\text{proj}_W \, \mathbf{f} = \frac{a_0}{2} + [a_1 \cos x + \cdots + a_n \cos nx] + [b_1 \sin x + \cdots + b_n \sin nx]$$

where

$$a_0 = \frac{2}{\sqrt{2\pi}}\langle \mathbf{f}, \mathbf{g}_0 \rangle = \frac{2}{\sqrt{2\pi}}\int_0^{2\pi} f(x)\frac{1}{\sqrt{2\pi}}\,dx = \frac{1}{\pi}\int_0^{2\pi} f(x)\,dx$$

$$a_1 = \frac{1}{\sqrt{\pi}}\langle \mathbf{f}, \mathbf{g}_1 \rangle = \frac{1}{\sqrt{\pi}}\int_0^{2\pi} f(x)\frac{1}{\sqrt{\pi}}\cos x\,dx = \frac{1}{\pi}\int_0^{2\pi} f(x)\cos x\,dx$$

$$\vdots$$

$$a_n = \frac{1}{\sqrt{\pi}}\langle \mathbf{f}, \mathbf{g}_n \rangle = \frac{1}{\sqrt{\pi}}\int_0^{2\pi} f(x)\frac{1}{\sqrt{\pi}}\cos nx\,dx = \frac{1}{\pi}\int_0^{2\pi} f(x)\cos nx\,dx$$

$$b_1 = \frac{1}{\sqrt{\pi}}\langle \mathbf{f}, \mathbf{g}_{n+1} \rangle = \frac{1}{\sqrt{\pi}}\int_0^{2\pi} f(x)\frac{1}{\sqrt{\pi}}\sin x\,dx = \frac{1}{\pi}\int_0^{2\pi} f(x)\sin x\,dx$$

$$\vdots$$

$$b_n = \frac{1}{\sqrt{\pi}}\langle \mathbf{f}, \mathbf{g}_{2n} \rangle = \frac{1}{\sqrt{\pi}}\int_0^{2\pi} f(x)\frac{1}{\sqrt{\pi}}\sin nx\,dx = \frac{1}{\pi}\int_0^{2\pi} f(x)\sin nx\,dx$$

In short

$$a_k = \frac{1}{\pi} \int_0^{2\pi} f(x)\cos kx \, dx, \qquad b_k = \frac{1}{\pi} \int_0^{2\pi} f(x)\sin kx \, dx$$

The numbers $a_0, a_1, \ldots, a_n, b_1, \ldots, b_n$ are called the **Fourier* coefficients** of **f**.

Example 4

Find the least squares approximation of $f(x) = x$ on $[0, 2\pi]$ by
(a) a trigonometric polynomial of order 2 or less
(b) a trigonometric polynomial of order n or less.

Solution.

$$a_0 = \frac{1}{\pi} \int_0^{2\pi} f(x) \, dx = \frac{1}{\pi} \int_0^{2\pi} x \, dx = 2\pi$$

For $k = 1, 2, \ldots$ integration by parts yields (verify):

$$a_k = \frac{1}{\pi} \int_0^{2\pi} f(x)\cos kx \, dx = \frac{1}{\pi} \int_0^{2\pi} x \cos kx \, dx = 0$$

$$b_k = \frac{1}{\pi} \int_0^{2\pi} f(x)\sin kx \, dx = \frac{1}{\pi} \int_0^{2\pi} x \sin kx \, dx = -\frac{2}{k} \qquad (7.14)$$

Thus, the least squares approximation to x on $[0, 2\pi]$ by a trigonometric polynomial of order 2 or less is

$$x \simeq \frac{a_0}{2} + a_1 \cos x + a_2 \cos 2x + b_1 \sin x + b_2 \sin 2x$$

or from (7.14)

$$x \simeq \pi - 2 \sin x - \sin 2x$$

(b) The least squares approximation to x on $[0, 2\pi]$ by a trigonometric polynomial of order n or less is

$$x \simeq \frac{a_0}{2} + [a_1 \cos x + \cdots + a_n \cos nx] + [b_1 \sin x + \cdots + b_n \sin nx]$$

---

* *Jean Baptiste Joseph Fourier (1768–1830)* was a French mathematician and physicist who discovered the Fourier series and related ideas while working on problems of heat diffusion. This discovery is one of the most influential in the history of mathematics; it is the cornerstone of many fields of mathematical research and a basic tool in many branches of engineering. Fourier, a political activist during the French revolution, spent time in jail for his defense of many victims during the Terror. He later became a favorite of Napoleon and was named both a baron and a count.

or from (7.14)

$$x \simeq \pi - 2\left(\sin x + \frac{\sin 2x}{2} + \frac{\sin 3x}{3} + \cdots + \frac{\sin nx}{n} \right)$$

It is natural to expect that the mean square error will diminish as the number of terms in the least squares approximation

$$f(x) \simeq \frac{a_0}{2} + \sum_{k=1}^{n} (a_k \cos kx + b_k \sin kx)$$

increases. It can be proved that the mean square error approaches zero as $n \to +\infty$; this is denoted by writing

$$f(x) = \frac{a_0}{2} + \sum_{k=1}^{\infty} (a_k \cos kx + b_k \sin kx)$$

The right side of this equation is called the **Fourier series** for f. Such series are of major importance in engineering, science, and mathematics.

EXERCISE SET 7.2

1. Find the least squares approximation of $f(x) = 1 + x$ over the interval $[0, 2\pi]$ by
   (a) a trigonometric polynomial of order 2 or less
   (b) a trigonometric polynomial of order n or less.

2. Find the least squares approximation of $f(x) = x^2$ over the interval $[0, 2\pi]$ by
   (a) a trigonometric polynomial of order 3 or less
   (b) a trigonometric polynomial of order n or less.

3. (a) Find the least squares approximation of x over the interval $[0, 1]$ by a function of the form $a + be^x$.
   (b) Find the mean square error of the approximation.

4. (a) Find the least squares approximation of e^x over the interval $[0, 1]$ by a polynomial of the form $a_0 + a_1 x$.
   (b) Find the mean square error of the approximation.

5. (a) Find the least squares approximation of $\sin \pi x$ over the interval $[-1, 1]$ by a polynomial of the form $a_0 + a_1 x + a_2 x^2$.
   (b) Find the mean square error of the approximation.

6. Use the Gram-Schmidt process to obtain the orthonormal basis (7.13) from the basis (7.11).

7. Carry out the integrations in (7.14).

8. Find the Fourier series of $f(x) = \pi - x$.

7.3 QUADRATIC FORMS

At the very beginning of this text we defined a linear equation in the n variables $x_1, x_2, \ldots, x_n$ to be one that can be expressed in the form

$$a_1 x_1 + a_2 x_2 + \cdots + a_n x_n = b$$

The left side of this equation

$$a_1 x_1 + a_2 x_2 + \cdots + a_n x_n$$

is a function of n variables, called a ***linear form***. In a linear form all variables occur to the first power, and there are no products of variables in the expression. In this section we shall study functions, called ***quadratic forms***, in which the terms are squares of variables or products of two variables. Quadratic forms arise in a variety of applications, including geometry, vibrations of mechanical systems, statistics, and electrical engineering.

A quadratic form in two variables, x and y, is defined to be an expression that can be written as

$$ax^2 + 2bxy + cy^2 \tag{7.15}$$

Example 5

The following are quadratic forms in x and y:

$$2x^2 + 6xy - 7y^2 \qquad (a = 2,\ b = 3,\ c = -7)$$
$$4x^2 - 5y^2 \qquad (a = 4,\ b = 0,\ c = -5)$$
$$xy \qquad (a = 0,\ b = 1/2,\ c = 0)$$

If we agree to omit the brackets on 1×1 matrices, then (7.15) can be written in matrix form as

$$ax^2 + 2bxy + cy^2 = [x \quad y] \begin{bmatrix} a & b \\ b & c \end{bmatrix} \begin{bmatrix} x \\ y \end{bmatrix} \tag{7.16}$$

(Verify by multiplying the matrices.) Note that the 2×2 matrix in (7.16) is symmetric, the diagonal entries are the coefficients of the squared terms, and the entries off the main diagonal are each half the coefficient of the product term xy.

Example 6

$$2x^2 + 6xy - 7y^2 = [x \quad y] \begin{bmatrix} 2 & 3 \\ 3 & -7 \end{bmatrix} \begin{bmatrix} x \\ y \end{bmatrix}$$

$$4x^2 - 5y^2 = [x \quad y] \begin{bmatrix} 4 & 0 \\ 0 & -5 \end{bmatrix} \begin{bmatrix} x \\ y \end{bmatrix}$$

$$xy = [x \quad y] \begin{bmatrix} 0 & \frac{1}{2} \\ \frac{1}{2} & 0 \end{bmatrix} \begin{bmatrix} x \\ y \end{bmatrix}$$

Quadratic forms are not limited to two variables. A general quadratic form is defined as follows.

Definition. A *quadratic form* in $x_1, x_2, \ldots, x_n$ is an expression that can be written as

$$\begin{bmatrix} x_1 & x_2 & \cdots & x_n \end{bmatrix} A \begin{bmatrix} x_1 \\ x_2 \\ \vdots \\ x_n \end{bmatrix} \tag{7.17}$$

where A is a symmetric $n \times n$ matrix.

If we let

$$\mathbf{x} = \begin{bmatrix} x_1 \\ x_2 \\ \vdots \\ x_n \end{bmatrix}$$

then (7.17) can be written more compactly as

$$\mathbf{x}^t A \mathbf{x} \tag{7.18}$$

Moreover, it can be shown that if the matrices in (7.18) are multiplied out, the resulting expression has the form

$$\mathbf{x}^t A \mathbf{x} = a_{11} x_1^2 + a_{22} x_2^2 + \cdots + a_{nn} x_n^2 + \sum_{i \neq j} a_{ij} x_i x_j$$

where

$$\sum_{i \neq j} a_{ij} x_i x_j$$

denotes a sum of terms of the form $a_{ij} x_i x_j$, where x_i and x_j are different variables. These are called the *cross-product terms* of the quadratic form.

Symmetric matrices are useful but not essential for representing quadratic forms in matrix notation. For example, for the quadratic form $2x^2 + 6xy - 7y^2$ in Example 6, we might split the coefficient of the cross-product term into $5 + 1$ or $4 + 2$ and write

$$2x^2 + 6xy - 7y^2 = \begin{bmatrix} x & y \end{bmatrix} \begin{bmatrix} 2 & 5 \\ 1 & -7 \end{bmatrix} \begin{bmatrix} x \\ y \end{bmatrix}$$

or

$$2x^2 + 6xy - 7y^2 = \begin{bmatrix} x & y \end{bmatrix} \begin{bmatrix} 2 & 4 \\ 2 & -7 \end{bmatrix} \begin{bmatrix} x \\ y \end{bmatrix}$$

However, symmetric matrices generally produce the simplest results, so we will always use them. Thus, when we denote a quadratic form by $\mathbf{x}^t A \mathbf{x}$ it will be understood that A is symmetric, even if this is not stated explicitly.

REMARK. If we use the fact that A is symmetric, that is, $A = A^t$, then (7.18) can be expressed in terms of the Euclidean inner product by writing

$$\mathbf{x}^t A \mathbf{x} = \mathbf{x}^t A^t \mathbf{x} = (A\mathbf{x})^t \mathbf{x} = \langle \mathbf{x}, A\mathbf{x} \rangle \tag{7.19}$$

Example 7

The following is a quadratic form in x_1, x_2, and x_3:

$$x_1^2 + 7x_2^2 - 3x_3^2 + 4x_1x_2 - 2x_1x_3 + 6x_2x_3 = [x_1 \ \ x_2 \ \ x_3] \begin{bmatrix} 1 & 2 & -1 \\ 2 & 7 & 3 \\ -1 & 3 & -3 \end{bmatrix} \begin{bmatrix} x_1 \\ x_2 \\ x_3 \end{bmatrix}$$

Note that the coefficients of the squared terms appear on the main diagonal of the 3×3 matrix, and the coefficients of the cross-product terms are each split in half and appear in the off-diagonal positions as follows:

Coefficient of	Positions in Matrix A	
x_1x_2	a_{12} and	a_{21}
x_1x_3	a_{13} and	a_{31}
x_2x_3	a_{23} and	a_{32}

The study of quadratic forms is an extensive topic that we can only touch on in this section. The following are some of the important mathematical problems relating to quadratic forms.

1. Find the maximum and minimum values of the quadratic form $\mathbf{x}^t A \mathbf{x}$ if $\mathbf{x}$ is constrained so that

$$\|\mathbf{x}\| = (x_1^2 + x_2^2 + \cdots + x_n^2)^{1/2} = 1$$

2. What conditions must A satisfy in order for a quadratic form to satisfy the inequality $\mathbf{x}^t A \mathbf{x} > 0$ for all $\mathbf{x} \neq \mathbf{0}$?

3. If $\mathbf{x}^t A \mathbf{x}$ is a quadratic form in two or three variables and c is a constant, what does the graph of the equation $\mathbf{x}^t A \mathbf{x} = c$ look like?

4. If P is an orthogonal matrix, the change of variables $\mathbf{x} = P\mathbf{y}$ converts the quadratic form $\mathbf{x}^t A \mathbf{x}$ to $(P\mathbf{y})^t A(P\mathbf{y}) = \mathbf{y}^t (P^t A P)\mathbf{y}$. But $P^t A P$ is a symmetric matrix if A is (verify), so $\mathbf{y}^t (P^t A P)\mathbf{y}$ is a new quadratic form in the variables of $\mathbf{y}$. It is important to know if P can be chosen so that this new quadratic form has no cross-product terms.

In this section we shall study the first two problems, and in the following sections we shall study the last two. The following theorem provides a solution to the first problem. The proof is deferred to the end of this section.

Theorem 1. *Let A be a symmetric $n \times n$ matrix whose eigenvalues in decreasing size order are $\lambda_1 \geq \lambda_2 \geq \cdots \geq \lambda_n$. If $\mathbf{x}$ is constrained so that $\|\mathbf{x}\| = 1$ relative to the Euclidean inner product on R^n, then*

(a) $\lambda_1 \geq \mathbf{x}^t A \mathbf{x} \geq \lambda_n$
(b) $\mathbf{x}^t A \mathbf{x} = \lambda_n$ *if $\mathbf{x}$ is an eigenvector of A corresponding to λ_n and $\mathbf{x}^t A \mathbf{x} = \lambda_1$ if $\mathbf{x}$ is an eigenvector of A corresponding to λ_1.*

It follows from this theorem that subject to the constraint

$$\|\mathbf{x}\| = (x_1^2 + x_2^2 + \cdots + x_n^2)^{1/2} = 1$$

the quadratic form $\mathbf{x}^t A \mathbf{x}$ has a maximum value of λ_1 (the largest eigenvalue) and a minimum value of λ_n (the smallest eigenvalue).

Example 8

Find the maximum and minimum values of the quadratic form

$$x_1^2 + x_2^2 + 4x_1 x_2$$

subject to the constraint $x_1^2 + x_2^2 = 1$, and determine values of x_1 and x_2 at which the maximum and minimum occur.

Solution. The quadratic form can be written as

$$x_1^2 + x_2^2 + 4x_1 x_2 = \mathbf{x}^t A \mathbf{x} = \begin{bmatrix} x_1 & x_2 \end{bmatrix} \begin{bmatrix} 1 & 2 \\ 2 & 1 \end{bmatrix} \begin{bmatrix} x_1 \\ x_2 \end{bmatrix}$$

The characteristic equation of A is

$$\det(\lambda I - A) = \det \begin{bmatrix} \lambda - 1 & -2 \\ -2 & \lambda - 1 \end{bmatrix} = \lambda^2 - 2\lambda - 3 = (\lambda - 3)(\lambda + 1) = 0$$

Thus, the eigenvalues of A are $\lambda = 3$ and $\lambda = -1$, which are the maximum and minimum values, respectively, of the quadratic form subject to the constraint. To find values of x_1 and x_2 at which these extreme values occur, we must find eigenvectors corresponding to these eigenvalues and then normalize these eigenvectors to satisfy the condition $x_1^2 + x_2^2 = 1$.

We leave it for the reader to show that bases for the eigenspaces are

$$\lambda = 3: \begin{bmatrix} 1 \\ 1 \end{bmatrix}, \qquad \lambda = -1: \begin{bmatrix} 1 \\ -1 \end{bmatrix}$$

Normalizing these eigenvectors yields

$$\begin{bmatrix} 1/\sqrt{2} \\ 1/\sqrt{2} \end{bmatrix}, \qquad \begin{bmatrix} 1/\sqrt{2} \\ -1/\sqrt{2} \end{bmatrix}$$

Thus, subject to the constraint $x_1^2 + x_2^2 = 1$, the maximum value of the quadratic form is $\lambda = 3$, which occurs if $x_1 = 1/\sqrt{2}$, $x_2 = 1/\sqrt{2}$; and the minimum value is $\lambda = -1$, which occurs if $x_1 = 1/\sqrt{2}$, $x_2 = -1/\sqrt{2}$. Moreover, alternative bases for the eigenspaces can be obtained by multiplying the basis vectors above by -1. Thus, the maximum value, $\lambda = 3$, also occurs if $x_1 = -1/\sqrt{2}$, $x_2 = -1/\sqrt{2}$, and the minimum value, $\lambda = -1$, also occurs if $x_1 = -1/\sqrt{2}$, $x_2 = 1/\sqrt{2}$. ▲

Definition. A quadratic form $x^t A x$ is called ***positive definite*** if $x^t A x > 0$ for all $x \neq 0$, and a symmetric matrix A is called a ***positive definite matrix*** if $x^t A x$ is a positive definite quadratic form.

The following theorem is the main result about positive definite matrices.

Theorem 2. *A symmetric matrix A is positive definite if and only if all the eigenvalues of A are positive.*

Proof. Assume that A is positive definite, and let λ be any eigenvalue of A. If x is an eigenvector of A corresponding to λ, then $x \neq 0$, so

$$0 < x^t A x = x^t \lambda x = \lambda x^t x = \lambda \|x\|^2 \qquad (7.20)$$

where $\|x\|$ is the Euclidean norm of x. Since $\|x\|^2 > 0$ it follows that $\lambda > 0$, which is what we wanted to show.

Conversely, assume that all eigenvalues of A are positive. We must show that $x^t A x > 0$ for all $x \neq 0$. But if $x \neq 0$, we can normalize x to obtain the vector $y = x/\|x\|$ with the property $\|y\| = 1$. It now follows from Theorem 1 that

$$y^t A y \geq \lambda_n > 0$$

where λ_n is the smallest eigenvalue of A. Thus,

$$y^t A y = \left(\frac{x}{\|x\|} \right)^t A \left(\frac{x}{\|x\|} \right) = \frac{1}{\|x\|^2} x^t A x > 0$$

Multiplying through by $\|x\|^2$ yields

$$x^t A x > 0$$

which is what we wanted to show. ∎

Example 9

In Example 13 of Section 6.3 we showed that the symmetric matrix

$$A = \begin{bmatrix} 4 & 2 & 2 \\ 2 & 4 & 2 \\ 2 & 2 & 4 \end{bmatrix}$$

has eigenvalues $\lambda = 2$ and $\lambda = 8$. Since these are positive, the matrix A is positive definite, and for all $\mathbf{x} \neq \mathbf{0}$

$$\mathbf{x}^t A \mathbf{x} = 4x_1^2 + 4x_2^2 + 4x_3^2 + 4x_1 x_2 + 4x_1 x_3 + 4x_2 x_3 > 0 \quad \triangle$$

Our next objective is to give a criterion that can be used to determine whether a symmetric matrix is positive definite without finding its eigenvalues. To do this it will be helpful to introduce some terminology. If

$$A = \begin{bmatrix} a_{11} & a_{12} & \cdots & a_{1n} \\ a_{21} & a_{22} & \cdots & a_{2n} \\ \vdots & \vdots & & \vdots \\ a_{n1} & a_{n2} & \cdots & a_{nn} \end{bmatrix}$$

is a square matrix, then the ***principal submatrices*** of A are the submatrices formed from the first r rows and r columns of A for $r = 1, 2, \ldots, n$. These submatrices are

$$A_1 = [a_{11}], \quad A_2 = \begin{bmatrix} a_{11} & a_{12} \\ a_{21} & a_{22} \end{bmatrix}, \quad A_3 = \begin{bmatrix} a_{11} & a_{12} & a_{13} \\ a_{21} & a_{22} & a_{23} \\ a_{31} & a_{32} & a_{33} \end{bmatrix}, \ldots,$$

$$A_n(=A) = \begin{bmatrix} a_{11} & a_{12} & \cdots & a_{1n} \\ a_{21} & a_{22} & \cdots & a_{2n} \\ \vdots & \vdots & & \vdots \\ a_{n1} & a_{n2} & \cdots & a_{nn} \end{bmatrix}$$

Theorem 3. *A symmetric matrix A is positive definite if and only if the determinant of every principal submatrix is positive.*

We omit the proof.

Example 10

The matrix

$$A = \begin{bmatrix} 2 & -1 & -3 \\ -1 & 2 & 4 \\ -3 & 4 & 9 \end{bmatrix}$$

is positive definite since

$$|2| = 2, \quad \begin{vmatrix} 2 & -1 \\ -1 & 2 \end{vmatrix} = 3, \quad \begin{vmatrix} 2 & -1 & -3 \\ -1 & 2 & 4 \\ -3 & 4 & 9 \end{vmatrix} = 1$$

all of which are positive. Thus, we are guaranteed that all eigenvalues of A are positive and $\mathbf{x}^t A \mathbf{x} > 0$ for all $\mathbf{x} \neq \mathbf{0}$. ▲

REMARK. A symmetric matrix A and the quadratic form $\mathbf{x}^t A \mathbf{x}$ are called

positive semidefinite	if $\mathbf{x}^t A \mathbf{x} \geq 0$ for all $\mathbf{x}$
negative definite	if $\mathbf{x}^t A \mathbf{x} < 0$ for $\mathbf{x} \neq \mathbf{0}$
negative semidefinite	if $\mathbf{x}^t A \mathbf{x} \leq 0$ for all $\mathbf{x}$
indefinite	if $\mathbf{x}^t A \mathbf{x}$ has both positive and negative values

Theorems 2 and 3 can be modified in an obvious way to apply to matrices of the first three types. For example, a symmetric matrix A is positive semi-definite if and only if all of its eigenvalues are *nonnegative*. Also, A is positive semidefinite if and only if all its principal submatrices have *nonnegative* determinants.

OPTIONAL

Proof of Theorem 1(a). Since A is symmetric it follows from Theorem 5(*b*) of Section 6.3 that there is an orthonormal basis for R^n consisting of eigenvectors of A. Let $S = \{\mathbf{v}_1, \mathbf{v}_2, \ldots, \mathbf{v}_n\}$ be such a basis. If $\langle \, , \, \rangle$ denotes the Euclidean inner product, then it follows from Theorem 23 of Section 4.9 that for any $\mathbf{x}$ in R^n

$$\mathbf{x} = \langle \mathbf{x}, \mathbf{v}_1 \rangle \mathbf{v}_1 + \langle \mathbf{x}, \mathbf{v}_2 \rangle \mathbf{v}_2 + \cdots + \langle \mathbf{x}, \mathbf{v}_n \rangle \mathbf{v}_n$$

Thus,

$$\begin{aligned} A\mathbf{x} &= \langle \mathbf{x}, \mathbf{v}_1 \rangle A\mathbf{v}_1 + \langle \mathbf{x}, \mathbf{v}_2 \rangle A\mathbf{v}_2 + \ldots + \langle \mathbf{x}, \mathbf{v}_n \rangle A\mathbf{v}_n \\ &= \langle \mathbf{x}, \mathbf{v}_1 \rangle \lambda_1 \mathbf{v}_1 + \langle \mathbf{x}, \mathbf{v}_2 \rangle \lambda_2 \mathbf{v}_2 + \ldots + \langle \mathbf{x}, \mathbf{v}_n \rangle \lambda_n \mathbf{v}_n \\ &= \lambda_1 \langle \mathbf{x}, \mathbf{v}_1 \rangle \mathbf{v}_1 + \lambda_2 \langle \mathbf{x}, \mathbf{v}_2 \rangle \mathbf{v}_2 + \ldots + \lambda_n \langle \mathbf{x}, \mathbf{v}_n \rangle \mathbf{v}_n \end{aligned}$$

It follows that the coordinate vectors for $\mathbf{x}$ and $A\mathbf{x}$ relative to the basis S are

$$(\mathbf{x})_S = (\langle \mathbf{x}, \mathbf{v}_1 \rangle, \langle \mathbf{x}, \mathbf{v}_2 \rangle, \ldots, \langle \mathbf{x}, \mathbf{v}_n \rangle)$$
$$(A\mathbf{x})_S = (\lambda_1 \langle \mathbf{x}, \mathbf{v}_1 \rangle, \lambda_2 \langle \mathbf{x}, \mathbf{v}_2 \rangle, \ldots, \lambda_n \langle \mathbf{x}, \mathbf{v}_n \rangle)$$

Thus, from Theorem 30 in Section 4.10 and the fact that $\|\mathbf{x}\| = 1$, we obtain:

$$\|\mathbf{x}\|^2 = \langle \mathbf{x}, \mathbf{v}_1 \rangle^2 + \langle \mathbf{x}, \mathbf{v}_2 \rangle^2 + \cdots + \langle \mathbf{x}, \mathbf{v}_n \rangle^2 = 1$$
$$\langle \mathbf{x}, A\mathbf{x} \rangle = \lambda_1 \langle \mathbf{x}, \mathbf{v}_1 \rangle^2 + \lambda_2 \langle \mathbf{x}, \mathbf{v}_2 \rangle^2 + \cdots + \lambda_n \langle \mathbf{x}, \mathbf{v}_n \rangle^2$$

Using these two equations and formula (7.19), we can prove that $\mathbf{x}^t A \mathbf{x} \leq \lambda_1$ as follows:

$$\mathbf{x}^t A \mathbf{x} = \langle \mathbf{x}, A\mathbf{x} \rangle = \lambda_1 \langle \mathbf{x}, \mathbf{v}_1 \rangle^2 + \lambda_2 \langle \mathbf{x}, \mathbf{v}_2 \rangle^2 + \cdots + \lambda_n \langle \mathbf{x}, \mathbf{v}_n \rangle^2$$
$$\leq \lambda_1 \langle \mathbf{x}, \mathbf{v}_1 \rangle^2 + \lambda_1 \langle \mathbf{x}, \mathbf{v}_2 \rangle^2 + \cdots + \lambda_1 \langle \mathbf{x}, \mathbf{v}_n \rangle^2$$
$$= \lambda_1 (\langle \mathbf{x}, \mathbf{v}_1 \rangle^2 + \langle \mathbf{x}, \mathbf{v}_2 \rangle^2 + \cdots + \langle \mathbf{x}, \mathbf{v}_n \rangle^2)$$
$$= \lambda_1$$

The proof that $\lambda_n \leq \mathbf{x}^t A \mathbf{x}$ is similar and is left as an exercise.

(b) If $\mathbf{x}$ is an eigenvector of A corresponding to λ_1 and $\|\mathbf{x}\| = 1$, then

$$\mathbf{x}^t A \mathbf{x} = \langle \mathbf{x}, A\mathbf{x} \rangle = \langle \mathbf{x}, \lambda_1 \mathbf{x} \rangle = \lambda_1 \langle \mathbf{x}, \mathbf{x} \rangle = \lambda_1 \|\mathbf{x}\|^2 = \lambda_1$$

Similarly, $\mathbf{x}^t A \mathbf{x} = \lambda_n$ if $\|\mathbf{x}\| = 1$ and $\mathbf{x}$ is an eigenvector of A corresponding to λ_n.

EXERCISE SET 7.3

1. Which of the following are quadratic forms?
   (a) $x^2 - \sqrt{2}xy$
   (b) $5x_1^2 - 2x_2^3 + 4x_1 x_2$
   (c) $4x_1^2 - 3x_2^2 + x_3^2 - 5x_1 x_3$
   (d) $x_1^2 - 7x_2^2 + x_3^2 + 4x_1 x_2 x_3$
   (e) $x_1 x_2 - 3x_1 x_3 + 2x_2 x_3$
   (f) $x_1^2 - 6x_2^2 + x_1 - 5x_2$
   (g) $(x_1 - 3x_2)^2$
   (h) $(x_1 - x_3)^2 + 2(x_1 + 4x_2)^2$

2. Express the following quadratic forms in the matrix notation $\mathbf{x}^t A \mathbf{x}$, where A is a symmetric matrix.
   (a) $3x_1^2 + 7x_2^2$
   (b) $4x_1^2 - 9x_2^2 - 6x_1 x_2$
   (c) $5x_1^2 + 5x_1 x_2$
   (d) $-7x_1 x_2$

3. Express the following quadratic forms in the matrix notation $\mathbf{x}^t A \mathbf{x}$, where A is a symmetric matrix.
   (a) $9x_1^2 - x_2^2 + 4x_3^2 + 6x_1 x_2 - 8x_1 x_3 + x_2 x_3$
   (b) $x_1^2 + x_2^2 - 3x_3^2 - 5x_1 x_2 + 9x_1 x_3$
   (c) $x_1 x_2 + x_1 x_3 + x_2 x_3$
   (d) $\sqrt{2}x_1^2 - \sqrt{3}x_3^2 + 2\sqrt{2}x_1 x_2 - 8\sqrt{3}x_1 x_3$
   (e) $x_1^2 + x_2^2 - x_3^2 - x_4^2 + 2x_1 x_2 - 10x_1 x_4 + 4x_3 x_4$

4. In each part find a formula for the quadratic form that does not use matrices.

   (a) $\begin{bmatrix} x & y \end{bmatrix} \begin{bmatrix} 2 & -3 \\ -3 & 5 \end{bmatrix} \begin{bmatrix} x \\ y \end{bmatrix}$
   (b) $\begin{bmatrix} x_1 & x_2 \end{bmatrix} \begin{bmatrix} 7 & \frac{5}{2} \\ \frac{5}{2} & 0 \end{bmatrix} \begin{bmatrix} x_1 \\ x_2 \end{bmatrix}$

   (c) $\begin{bmatrix} x & y & z \end{bmatrix} \begin{bmatrix} 1 & 0 & 0 \\ 0 & -3 & 0 \\ 0 & 0 & 5 \end{bmatrix} \begin{bmatrix} x \\ y \\ z \end{bmatrix}$

(d) $\begin{bmatrix} x_1 & x_2 & x_3 \end{bmatrix} \begin{bmatrix} -2 & \frac{7}{2} & \frac{1}{2} \\ \frac{7}{2} & 0 & 6 \\ \frac{1}{2} & 6 & 3 \end{bmatrix} \begin{bmatrix} x_1 \\ x_2 \\ x_3 \end{bmatrix}$

(e) $\begin{bmatrix} x_1 & x_2 & x_3 & x_4 \end{bmatrix} \begin{bmatrix} 0 & 1 & 1 & 1 \\ 1 & 0 & 1 & 1 \\ 1 & 1 & 0 & 1 \\ 1 & 1 & 1 & 0 \end{bmatrix} \begin{bmatrix} x_1 \\ x_2 \\ x_3 \\ x_4 \end{bmatrix}$

5. In each part find the maximum and minimum values of the quadratic form subject to the constraint $x_1^2 + x_2^2 = 1$, and determine the values of x_1 and x_2 at which the maximum and minimum occur.
   (a) $5x_1^2 - x_2^2$ (b) $7x_1^2 + 4x_2^2 + x_1 x_2$
   (c) $5x_1^2 + 2x_2^2 - x_1 x_2$ (d) $2x_1^2 + 2x_2^2 + 3x_1 x_2$

6. In each part find the maximum and minimum values of the quadratic form subject to the constraint $x_1^2 + x_2^2 + x_3^2 = 1$, and determine the values of x_1, x_2, and x_3 at which the maximum and minimum occur.
   (a) $x_1^2 + x_2^2 + 2x_3^2 - 2x_1 x_2 + 4x_1 x_3 + 4x_2 x_3$
   (b) $2x_1^2 + x_2^2 + x_3^2 + 2x_1 x_3 + 2x_1 x_2$
   (c) $3x_1^2 + 2x_2^2 + 3x_3^2 + 2x_1 x_3$

7. Use Theorem 2 to determine which of the following matrices are positive definite.

   (a) $\begin{bmatrix} 2 & 3 \\ 3 & 2 \end{bmatrix}$ (b) $\begin{bmatrix} 5 & -1 \\ -1 & 5 \end{bmatrix}$ (c) $\begin{bmatrix} 2 & -2 \\ -2 & -1 \end{bmatrix}$

8. Use Theorem 3 to determine which of the matrices in Exercise 7 are positive definite.

9. Use Theorem 2 to determine which of the following matrices are positive definite.

   (a) $\begin{bmatrix} 3 & -1 & 0 \\ -1 & 2 & -1 \\ 0 & -1 & 3 \end{bmatrix}$ (b) $\begin{bmatrix} 0 & 1 & 1 \\ 1 & 0 & 1 \\ 1 & 1 & 0 \end{bmatrix}$ (c) $\begin{bmatrix} 1 & 2 & 1 \\ 2 & 1 & 1 \\ 1 & 1 & 3 \end{bmatrix}$

10. Use Theorem 3 to determine which of the matrices in Exercise 9 are positive definite.

11. In each part classify the quadratic form as positive definite, positive semidefinite, negative definite, negative semidefinite, or indefinite.
    (a) $x_1^2 + x_2^2$ (b) $-x_1^2 - 3x_2^2$
    (c) $(x_1 - x_2)^2$ (d) $-(x_1 - x_2)^2$
    (e) $x_1^2 - x_2^2$ (f) $x_1 x_2$

12. In each part classify the matrix as positive definite, positive semidefinite, negative definite, negative semidefinite, or indefinite.

    (a) $\begin{bmatrix} 3 & 0 & 0 \\ 0 & -2 & 0 \\ 0 & 0 & 1 \end{bmatrix}$ (b) $\begin{bmatrix} -5 & 0 & 0 \\ 0 & 0 & 0 \\ 0 & 0 & 1 \end{bmatrix}$ (c) $\begin{bmatrix} 6 & 7 & 1 \\ 0 & 9 & 2 \\ 0 & 0 & 1 \end{bmatrix}$

$$
\text{(d)} \begin{bmatrix} -4 & 0 & 0 \\ 7 & -3 & 0 \\ 8 & 9 & -1 \end{bmatrix} \quad \text{(e)} \begin{bmatrix} 0 & 0 & 0 \\ 0 & 0 & 0 \\ 0 & 0 & 0 \end{bmatrix} \quad \text{(f)} \begin{bmatrix} 1 & 0 & 0 \\ 0 & 1 & 0 \\ 0 & 0 & 1 \end{bmatrix}
$$

13. Let $x^t A x$ be a quadratic form in $x_1, x_2, \ldots, x_n$ and define $T:R^n \to R$ by $T(x) = x^t A x$.
    (a) Show that $T(x + y) = T(x) + 2x^t Ay + T(y)$
    (b) Show that $T(kx) = k^2 T(x)$
    (c) Is T a linear transformation? Explain.

14. In each part find all values of k for which the quadratic form is positive definite.
    (a) $x_1^2 + kx_2^2 - 4x_1x_2$
    (b) $5x_1^2 + x_2^2 + kx_3^2 + 4x_1x_2 - 2x_1x_3 - 2x_2x_3$
    (c) $3x_1^2 + x_2^2 + 2x_3^2 + 2x_1x_3 + 2kx_2x_3$

15. Express the quadratic form $(c_1x_1 + c_2x_2 + \cdots + c_nx_n)^2$ in the matrix notation $x^t A x$, where A is symmetric.

16. Let $x = (x_1, x_2, \ldots, x_n)$. In statistics the quantity

$$
\bar{x} = \frac{1}{n}(x_1 + x_2 + \cdots + x_n)
$$

is called the **sample mean** of $x_1, x_2, \ldots, x_n$, and

$$
s_x^2 = \frac{1}{n-1}\left[(x_1 - \bar{x})^2 + (x_2 - \bar{x})^2 + \cdots + (x_n - \bar{x})^2\right]
$$

is called the **sample variance**.
    (a) Express the quadratic form s_x^2 in the matrix notation $x^t A x$, where A is symmetric.
    (b) Is s_x^2 a positive definite quadratic form? Explain.

17. Complete the proof of Theorem 1 by showing that $\lambda_n \le x^t A x$ if $\|x\| = 1$ and $\lambda_n = x^t A x$ if x is an eigenvector of A corresponding to λ_n.

7.4 DIAGONALIZING QUADRATIC FORMS; APPLICATION TO CONIC SECTIONS

In this section we shall show how to remove the cross-product terms from a quadratic form by changing variables, and we shall use our results to study the graphs of the conic sections.
    Let

$$
x^t A x = \begin{bmatrix} x_1 & x_2 & \cdots & x_n \end{bmatrix} \begin{bmatrix} a_{11} & a_{12} & \cdots & a_{1n} \\ a_{21} & a_{22} & \cdots & a_{2n} \\ \vdots & \vdots & & \vdots \\ a_{n1} & a_{n2} & \cdots & a_{nn} \end{bmatrix} \begin{bmatrix} x_1 \\ x_2 \\ \vdots \\ x_n \end{bmatrix} \tag{7.21}
$$

be a quadratic form, where A is a symmetric matrix. We know from Theorem 5 in Section 6.3 that there is an orthogonal matrix P that diagonalizes A; that is,

$$P^t A P = D = \begin{bmatrix} \lambda_1 & 0 & \cdots & 0 \\ 0 & \lambda_2 & \cdots & 0 \\ \vdots & \vdots & & \vdots \\ 0 & 0 & \cdots & \lambda_n \end{bmatrix}$$

where $\lambda_1, \lambda_2, \ldots, \lambda_n$ are the eigenvalues of A. If we let

$$\mathbf{y} = \begin{bmatrix} y_1 \\ y_2 \\ \vdots \\ y_n \end{bmatrix}$$

where $y_1, y_2, \ldots, y_n$ are new variables, and if we make the substitution $\mathbf{x} = P\mathbf{y}$ in (7.21), then we obtain

$$\mathbf{x}^t A \mathbf{x} = (P\mathbf{y})^t A P\mathbf{y} = \mathbf{y}^t P^t A P\mathbf{y} = \mathbf{y}^t D\mathbf{y}$$

But

$$\mathbf{y}^t D\mathbf{y} = \begin{bmatrix} y_1 & y_2 & \cdots & y_n \end{bmatrix} \begin{bmatrix} \lambda_1 & 0 & \cdots & 0 \\ 0 & \lambda_2 & \cdots & 0 \\ \vdots & \vdots & & \vdots \\ 0 & 0 & \cdots & \lambda_n \end{bmatrix} \begin{bmatrix} y_1 \\ y_2 \\ \vdots \\ y_n \end{bmatrix}$$

$$= \lambda_1 y_1^2 + \lambda_2 y_2^2 + \cdots + \lambda_n y_n^2$$

which is a quadratic form with no cross-product terms.

In summary, we have the following result.

Theorem 4. *Let $\mathbf{x}^t A \mathbf{x}$ be a quadratic form in the variables $x_1, x_2, \ldots, x_n$, where A is symmetric. If P orthogonally diagonalizes A, and if new variables $y_1, y_2, \ldots, y_n$ are defined by the equation $\mathbf{x} = P\mathbf{y}$, then substituting this equation in $\mathbf{x}^t A \mathbf{x}$ yields*

$$\mathbf{x}^t A \mathbf{x} = \mathbf{y}^t D\mathbf{y} = \lambda_1 y_1^2 + \lambda_2 y_2^2 + \cdots + \lambda_n y_n^2$$

where $\lambda_1, \lambda_2, \ldots, \lambda_n$ are the eigenvalues of A and $D = P^t A P$.

The matrix P in this theorem is said to **orthogonally diagonalize** the quadratic form or **reduce the quadratic form to a sum of squares**.

Example 11

Find a change of variables that reduces the quadratic form $x_1^2 - x_3^2 - 4x_1x_2 + 4x_2x_3$ to a sum of squares and express the quadratic form in terms of the new variables.

Solution. The quadratic form can be written as

$$
\begin{bmatrix} x_1 & x_2 & x_3 \end{bmatrix}
\begin{bmatrix} 1 & -2 & 0 \\ -2 & 0 & 2 \\ 0 & 2 & -1 \end{bmatrix}
\begin{bmatrix} x_1 \\ x_2 \\ x_3 \end{bmatrix}
$$

The characteristic equation of the 3×3 matrix is

$$
\begin{vmatrix} \lambda - 1 & 2 & 0 \\ 2 & \lambda & -2 \\ 0 & -2 & \lambda + 1 \end{vmatrix} = \lambda^3 - 9\lambda = \lambda(\lambda + 3)(\lambda - 3) = 0
$$

so the eigenvalues are $\lambda = 0$, $\lambda = -3$, $\lambda = 3$. We leave it for the reader to show that orthonormal bases for the three eigenspaces are

$$
\lambda = 0: \begin{bmatrix} \frac{2}{3} \\ \frac{1}{3} \\ \frac{2}{3} \end{bmatrix} \qquad
\lambda = -3: \begin{bmatrix} -\frac{1}{3} \\ -\frac{2}{3} \\ \frac{2}{3} \end{bmatrix} \qquad
\lambda = 3: \begin{bmatrix} -\frac{2}{3} \\ \frac{2}{3} \\ \frac{1}{3} \end{bmatrix}
$$

Thus, the substitution $\mathbf{x} = P\mathbf{y}$ that eliminates cross-product terms is

$$
\begin{bmatrix} x_1 \\ x_2 \\ x_3 \end{bmatrix} =
\begin{bmatrix} \frac{2}{3} & -\frac{1}{3} & -\frac{2}{3} \\ \frac{1}{3} & -\frac{2}{3} & \frac{2}{3} \\ \frac{2}{3} & \frac{2}{3} & \frac{1}{3} \end{bmatrix}
\begin{bmatrix} y_1 \\ y_2 \\ y_3 \end{bmatrix}
$$

or equivalently

$$
x_1 = \tfrac{2}{3}y_1 - \tfrac{1}{3}y_2 - \tfrac{2}{3}y_3
$$
$$
x_2 = \tfrac{1}{3}y_1 - \tfrac{2}{3}y_2 + \tfrac{2}{3}y_3
$$
$$
x_3 = \tfrac{2}{3}y_1 + \tfrac{2}{3}y_2 + \tfrac{1}{3}y_3
$$

The new quadratic form is

$$
\begin{bmatrix} y_1 & y_2 & y_3 \end{bmatrix}
\begin{bmatrix} 0 & 0 & 0 \\ 0 & -3 & 0 \\ 0 & 0 & 3 \end{bmatrix}
\begin{bmatrix} y_1 \\ y_2 \\ y_3 \end{bmatrix}
$$

or equivalently

$$
-3y_2^2 + 3y_3^2
$$

REMARK. There are other methods for eliminating the cross-product terms from a quadratic form, which we shall not discuss here. Two such methods, *Lagrange's reduction* and *Kronecker's reduction*, are discussed in more advanced books.

We shall now apply our work on quadratic forms to the study of equations of the form

$$
ax^2 + 2bxy + cy^2 + dx + ey + f = 0 \tag{7.22}
$$

where $a, b, \ldots, f$ are real numbers, and at least one of the numbers a, b, c is not zero. An equation of this type is called a **quadratic equation** in x and y, and

$$ax^2 + 2bxy + cy^2$$

is called the **associated quadratic form**.

Example 12

In the quadratic equation

$$3x^2 + 5xy - 7y^2 + 2x + 7 = 0$$

the constants in (7.22) are

$$a = 3 \qquad b = \tfrac{5}{2} \qquad c = -7 \qquad d = 2 \qquad e = 0 \qquad f = 7$$

Example 13

Quadratic Equation	Associated Quadratic Form
$3x^2 + 5xy - 7y^2 + 2x + 7 = 0$	$3x^2 + 5xy - 7y^2$
$4x^2 - 5y^2 + 8y + 9 = 0$	$4x^2 - 5y^2$
$xy + y = 0$	xy

Graphs of quadratic equations in x and y are called **conics** or **conic sections**. The most important conics are ellipses, circles, hyperbolas, and parabolas; these are called the **nondegenerate** conics. The remaining conics are called **degenerate** and include single points and pairs of lines (see Exercise 15).

A nondegenerate conic is said to be in **standard position** relative to the coordinate axes if its equation can be expressed in one of the forms given in Figure 7.4.

Example 14

The equation

$$\frac{x^2}{4} + \frac{y^2}{9} = 1 \text{ is of the form } \frac{x^2}{k^2} + \frac{y^2}{l^2} = 1 \text{ with } k = 2, l = 3$$

Thus, its graph is an ellipse in standard position intersecting the x-axis at $(-2, 0)$ and $(2, 0)$ and intersecting the y-axis at $(0, -3)$ and $(0, 3)$.

The equation $x^2 - 8y^2 = -16$ can be rewritten as $y^2/2 - x^2/16 = 1$, which is of the form $y^2/k^2 - x^2/l^2 = 1$ with $k = \sqrt{2}, l = 4$. Its graph is thus a hyperbola in standard position intersecting the y-axis at $(0, -\sqrt{2})$ and $(0, \sqrt{2})$.

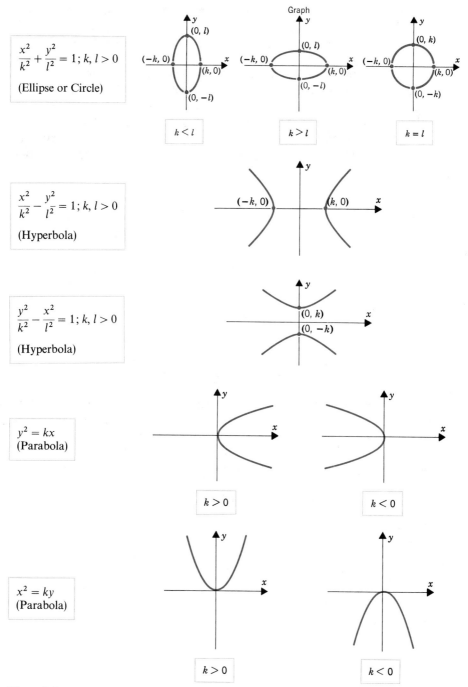

Figure 7.4

The equation $5x^2 + 2y = 0$ can be rewritten as $x^2 = -\frac{2}{5}y$, which is of the form $x^2 = ky$ with $k = -\frac{2}{5}$. Since $k < 0$, its graph is a parabola in standard position opening downward. ▲

Observe that no conic in standard position has an xy-term (that is, a cross-product term) in its equation; the presence of an xy-term in the equation of a nondegenerate conic indicates that the conic is rotated out of standard position (Fig. 7.5a). Also, no conic in standard position has both an x^2 and x term or both a y^2 and y term. If there is no cross-product term, the occurrence of either of these pairs in the equation of a nondegenerate conic indicates that the conic is translated out of standard position (Fig. 7.5b).

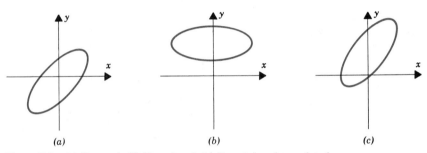

Figure 7.5   (a) Rotated. (b) Translated. (c) Rotated and translated.

One technique for identifying the graph of a nondegenerate conic that is not in standard position consists of rotating and translating the xy-coordinate axes to obtain an $x'y'$-coordinate system relative to which the conic is in standard position. Once this is done, the equation of the conic in the $x'y'$-system will have one of the forms given in Figure 7.4 and can then easily be identified.

Example 15

Since the quadratic equation

$$2x^2 + y^2 - 12x - 4y + 18 = 0$$

contains x^2, x, y^2, and y terms but no cross-product term, its graph is a conic that is translated out of standard position but not rotated. This conic can be brought into standard position by properly translating the coordinate axes. To do this, first collect x and y terms. This yields

$$(2x^2 - 12x) + (y^2 - 4y) + 18 = 0$$

or

$$2(x^2 - 6x) + (y^2 - 4y) = -18$$

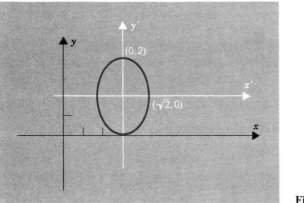

Figure 7.6

By completing the squares* on the two expressions in parentheses, we obtain

$$2(x^2 - 6x + 9) + (y^2 - 4y + 4) = -18 + 18 + 4$$

or

$$2(x - 3)^2 + (y - 2)^2 = 4 \qquad (7.23)$$

If we translate the coordinate axes by means of the translation equations

$$x' = x - 3 \qquad y' = y - 2$$

(Example 3 in Section 3.1), then (7.23) becomes

$$2x'^2 + y'^2 = 4$$

or

$$\frac{x'^2}{2} + \frac{y'^2}{4} = 1$$

which is the equation of an ellipse in standard position in the $x'y'$-system. This ellipse is sketched in Figure 7.6.

   We shall now show how to identify conics that are rotated out of standard position. If we omit the brackets on 1×1 matrices then (7.22) can be written in the matrix form

$$\begin{bmatrix} x & y \end{bmatrix} \begin{bmatrix} a & b \\ b & c \end{bmatrix} \begin{bmatrix} x \\ y \end{bmatrix} + \begin{bmatrix} d & e \end{bmatrix} \begin{bmatrix} x \\ y \end{bmatrix} + f = 0$$

---

* To complete the square on an expression of the form $x^2 + px$, add and subtract the constant $(p/2)^2$ to obtain

$$x^2 + px = x^2 + px + \left(\frac{p}{2}\right)^2 - \left(\frac{p}{2}\right)^2 = \left(x + \frac{p}{2}\right)^2 - \left(\frac{p}{2}\right)^2$$

or

$$\mathbf{x}^t A \mathbf{x} + K \mathbf{x} + f = 0$$

where

$$\mathbf{x} = \begin{bmatrix} x \\ y \end{bmatrix} \qquad A = \begin{bmatrix} a & b \\ b & c \end{bmatrix} \qquad K = [d \quad e]$$

Now consider a conic C whose equation in xy-coordinates is

$$\mathbf{x}^t A \mathbf{x} + K \mathbf{x} + f = 0 \tag{7.24}$$

We would like to rotate the xy-coordinate axes so that the equation of the conic in the new $x'y'$-coordinate system has no cross-product term. This can be done as follows.

Step 1. Find a matrix

$$P = \begin{bmatrix} p_{11} & p_{12} \\ p_{21} & p_{22} \end{bmatrix}$$

that orthogonally diagonalizes the quadratic form $\mathbf{x}^t A \mathbf{x}$.

Step 2. Interchange the columns of P, if necessary, to make $\det(P) = 1$. This assures that the orthogonal coordinate transformation

$$\mathbf{x} = P\mathbf{x}', \text{ that is, } \begin{bmatrix} x \\ y \end{bmatrix} = P \begin{bmatrix} x' \\ y' \end{bmatrix} \tag{7.25}$$

is a rotation (see the discussion preceding Example 76 in Section 4.10).

Step 3. To obtain the equation for C in the $x'y'$-system, substitute (7.25) into (7.24). This yields

$$(P\mathbf{x}')^t A (P\mathbf{x}') + K(P\mathbf{x}') + f = 0$$

or

$$\mathbf{x}'^t (P^t A P)\mathbf{x}' + (KP)\mathbf{x}' + f = 0 \tag{7.26}$$

Since P orthogonally diagonalizes A

$$P^t A P = \begin{bmatrix} \lambda_1 & 0 \\ 0 & \lambda_2 \end{bmatrix}$$

where λ_1 and λ_2 are eigenvalues of A. Thus, (7.26) can be rewritten as

$$[x' \quad y'] \begin{bmatrix} \lambda_1 & 0 \\ 0 & \lambda_2 \end{bmatrix} \begin{bmatrix} x' \\ y' \end{bmatrix} + [d \quad e] \begin{bmatrix} p_{11} & p_{12} \\ p_{21} & p_{22} \end{bmatrix} \begin{bmatrix} x' \\ y' \end{bmatrix} + f = 0$$

or

$$\lambda_1 x'^2 + \lambda_2 y'^2 + d'x' + e'y' + f = 0$$

(where $d' = dp_{11} + ep_{21}$ and $e' = dp_{12} + ep_{22}$). This equation has no cross-product term.

The following theorem summarizes this discussion.

Theorem 5 (*Principal Axes Theorem for R²*). *Let*

$$ax^2 + 2bxy + cy^2 + dx + ey + f = 0$$

be the equation of a conic C, and let

$$\mathbf{x}^t A \mathbf{x} = ax^2 + 2bxy + cy^2$$

be the associated quadratic form. Then the coordinate axes can be rotated so that the equation for C in the new x′y′-coordinate system has the form

$$\lambda_1 x'^2 + \lambda_2 y'^2 + d'x' + e'y' + f = 0$$

where λ_1 and λ_2 are the eigenvalues of A. The rotation can be accomplished by the substitution

$$\mathbf{x} = P\mathbf{x}'$$

where P orthogonally diagonalizes $\mathbf{x}^t A \mathbf{x}$ and det(P) = 1.

Example 16

Describe the conic C whose equation is $5x^2 - 4xy + 8y^2 - 36 = 0$.

Solution. The matrix form of this equation is

$$\mathbf{x}^t A \mathbf{x} - 36 = 0 \tag{7.27}$$

where

$$A = \begin{bmatrix} 5 & -2 \\ -2 & 8 \end{bmatrix}$$

The characteristic equation of A is

$$\det(\lambda I - A) = \det \begin{bmatrix} \lambda - 5 & 2 \\ 2 & \lambda - 8 \end{bmatrix} = (\lambda - 9)(\lambda - 4) = 0$$

so the eigenvalues of A are $\lambda = 4$ and $\lambda = 9$. We leave it for the reader to show that orthonormal bases for the eigenspaces are

$$\lambda = 4 \colon \begin{bmatrix} 2/\sqrt{5} \\ 1/\sqrt{5} \end{bmatrix} \qquad \lambda = 9 \colon \begin{bmatrix} -1/\sqrt{5} \\ 2/\sqrt{5} \end{bmatrix}$$

Thus,

$$P = \begin{bmatrix} 2/\sqrt{5} & -1/\sqrt{5} \\ 1/\sqrt{5} & 2/\sqrt{5} \end{bmatrix}$$

orthogonally diagonalizes $\mathbf{x}^t A \mathbf{x}$. Moreover, $\det(P) = 1$ so that the orthogonal co-ordinate transformation

$$\mathbf{x} = P\mathbf{x}' \tag{7.28}$$

is a rotation. Substituting (7.28) into (7.27) yields

$$(P\mathbf{x}')^t A(P\mathbf{x}') - 36 = 0$$

or

$$(\mathbf{x}')^t (P^t AP)\mathbf{x}' - 36 = 0$$

Since

$$P^t AP = \begin{bmatrix} 4 & 0 \\ 0 & 9 \end{bmatrix}$$

this equation can be written as

$$\begin{bmatrix} x' & y' \end{bmatrix} \begin{bmatrix} 4 & 0 \\ 0 & 9 \end{bmatrix} \begin{bmatrix} x' \\ y' \end{bmatrix} - 36 = 0$$

or

$$4x'^2 + 9y'^2 - 36 = 0$$

or

$$\frac{x'^2}{9} + \frac{y'^2}{4} = 1$$

which is the equation of the ellipse sketched in Figure 7.7.

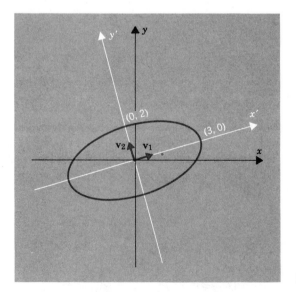

Figure 7.7

Example 17

Describe the conic C whose equation is

$$5x^2 - 4xy + 8y^2 + \frac{20}{\sqrt{5}}x - \frac{80}{\sqrt{5}}y + 4 = 0$$

Solution. The matrix form of this equation is

$$\mathbf{x}^t A \mathbf{x} + K \mathbf{x} + 4 = 0 \qquad (7.29)$$

where

$$A = \begin{bmatrix} 5 & -2 \\ -2 & 8 \end{bmatrix} \quad \text{and} \quad K = [20/\sqrt{5} \quad -80/\sqrt{5}]$$

As shown in Example 16

$$P = \begin{bmatrix} 2/\sqrt{5} & -1/\sqrt{5} \\ 1/\sqrt{5} & 2/\sqrt{5} \end{bmatrix}$$

orthogonally diagonalizes $\mathbf{x}^t A \mathbf{x}$. Substituting $\mathbf{x} = P\mathbf{x}'$ into (7.29) gives

$$(P\mathbf{x}')^t A(P\mathbf{x}') + K(P\mathbf{x}') + 4 = 0$$

or

$$(\mathbf{x}')^t (P^t A P)\mathbf{x}' + (KP)\mathbf{x}' + 4 = 0 \qquad (7.30)$$

Since

$$P^t A P = \begin{bmatrix} 4 & 0 \\ 0 & 9 \end{bmatrix} \quad \text{and} \quad KP = [20/\sqrt{5} \quad -80/\sqrt{5}]\begin{bmatrix} 2/\sqrt{5} & -1/\sqrt{5} \\ 1/\sqrt{5} & 2/\sqrt{5} \end{bmatrix} = [-8 \quad -36]$$

(7.30) can be written as

$$4x'^2 + 9y'^2 - 8x' - 36y' + 4 = 0 \qquad (7.31)$$

To bring the conic into standard position, the $x'y'$-axes must be translated. Proceeding as in Example 15, we rewrite (7.31) as

$$4(x'^2 - 2x') + 9(y'^2 - 4y') = -4$$

Completing the squares yields

$$4(x'^2 - 2x' + 1) + 9(y'^2 - 4y' + 4) = -4 + 4 + 36$$

or

$$4(x' - 1)^2 + 9(y' - 2)^2 = 36 \qquad (7.32)$$

If we translate the coordinate axes by means of the translation equations

$$x'' = x' - 1 \qquad y'' = y' - 2$$

then (7.32) becomes

$$4x''^2 + 9y''^2 = 36$$

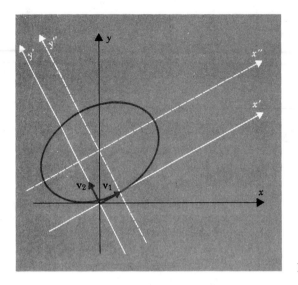

Figure 7.8

or

$$\frac{x''^2}{9} + \frac{y''^2}{4} = 1$$

which is the equation of the ellipse sketched in Figure 7.8. ▲

EXERCISE SET 7.4

1. In each part find a change of variables that reduces the quadratic form to a sum of squares, and express the quadratic form in terms of the new variables.
(a) $2x_1^2 + 2x_2^2 - 2x_1x_2$
(b) $5x_1^2 + 2x_2^2 + 4x_1x_2$
(c) $2x_1x_2$
(d) $-3x_1^2 + 5x_2^2 + 2x_1x_2$

2. In each part find a change of variables that reduces the quadratic form to a sum of squares, and express the quadratic form in terms of the new variables.
(a) $3x_1^2 + 4x_2^2 + 3x_3^2 + 4x_1x_2 - 4x_2x_3$
(b) $2x_1^2 + 5x_2^2 + 5x_3^2 + 4x_1x_2 - 4x_1x_3 - 8x_2x_3$
(c) $-5x_1^2 + x_2^2 - x_3^2 + 6x_1x_3 + 4x_1x_2$
(d) $2x_1x_3 + 6x_2x_3$

3. Find the quadratic forms associated with the following quadratic equations.
(a) $2x^2 - 3xy + 4y^2 - 7x + 2y + 7 = 0$
(b) $x^2 - xy + 5x + 8y - 3 = 0$

   (c) $5xy = 8$
   (d) $4x^2 - 2y^2 = 7$
   (e) $y^2 + 7x - 8y - 5 = 0$

4. Find the matrices of the quadratic forms in Exercise 3.

5. Express each of the quadratic equations in Exercise 3 in the matrix form
$$\mathbf{x}^t A \mathbf{x} + K \mathbf{x} + f = 0.$$

6. Name the following conics.
   (a) $2x^2 + 5y^2 = 20$ (b) $4x^2 + 9y^2 = 1$
   (c) $x^2 - y^2 - 8 = 0$ (d) $4y^2 - 5x^2 = 20$
   (e) $x^2 + y^2 - 25 = 0$ (f) $7y^2 - 2x = 0$
   (g) $-x^2 = 2y$ (h) $3x - 11y^2 = 0$
   (i) $y - x^2 = 0$ (j) $x^2 - 3 = -y^2$

7. In each part a translation will put the conic in standard position. Name the conic and give its equation in the translated coordinate system.
   (a) $9x^2 + 4y^2 - 36x - 24y + 36 = 0$
   (b) $x^2 - 16y^2 + 8x + 128y = 256$
   (c) $y^2 - 8x - 14y + 49 = 0$
   (d) $x^2 + y^2 + 6x - 10y + 18 = 0$
   (e) $2x^2 - 3y^2 + 6x + 20y = -41$
   (f) $x^2 + 10x + 7y = -32$

8. The following nondegenerate conics are rotated out of standard position. In each part rotate the coordinate axes to remove the xy-term. Name the conic and give its equation in the rotated coordinate system.
   (a) $2x^2 - 4xy - y^2 + 8 = 0$ (b) $x^2 + 2xy + y^2 + 8x + y = 0$
   (c) $5x^2 + 4xy + 5y^2 = 9$ (d) $11x^2 + 24xy + 4y^2 - 15 = 0$

In Exercises 9–14 translate and rotate the coordinate axes, if necessary, to put the conic in standard position. Name the conic and give its equation in the final coordinate system.

9. $9x^2 - 4xy + 6y^2 - 10x - 20y = 5$

10. $3x^2 - 8xy - 12y^2 - 30x - 64y = 0$

11. $2x^2 - 4xy - y^2 - 4x - 8y = -14$

12. $21x^2 + 6xy + 13y^2 - 114x + 34y + 73 = 0$

13. $x^2 - 6xy - 7y^2 + 10x + 2y + 9 = 0$

14. $4x^2 - 20xy + 25y^2 - 15x - 6y = 0$

15. The graph of a quadratic equation in x and y can, in certain cases, be a point, a line, or a pair of lines. These are called **degenerate** conics. It is also possible that the equation is not satisfied by any real values of x and y. In such cases the equation has no graph; it is said to represent an **imaginary conic**. Each of the following represents a degenerate or imaginary conic. Where possible, sketch the graph.
   (a) $x^2 - y^2 = 0$ (b) $x^2 + 3y^2 + 7 = 0$
   (c) $8x^2 + 7y^2 = 0$ (d) $x^2 - 2xy + y^2 = 0$
   (e) $9x^2 + 12xy + 4y^2 - 52 = 0$ (f) $x^2 + y^2 - 2x - 4y = -5$

7.5 QUADRATIC FORMS; APPLICATION TO QUADRIC SURFACES

In this section the techniques of the previous section are extended to quadratic equations in three variables.

An equation of the form

$$ax^2 + by^2 + cz^2 + 2dxy + 2exz + 2fyz + gx + hy + iz + j = 0 \qquad (7.33)$$

where $a, b, \ldots, f$ are not all zero, is called a *quadratic equation in x, y, and z*; the expression

$$ax^2 + by^2 + cz^2 + 2dxy + 2exz + 2fyz$$

is called the *associated quadratic form*.

Equation (7.33) can be written in the matrix form

$$\begin{bmatrix} x & y & z \end{bmatrix} \begin{bmatrix} a & d & e \\ d & b & f \\ e & f & c \end{bmatrix} \begin{bmatrix} x \\ y \\ z \end{bmatrix} + \begin{bmatrix} g & h & i \end{bmatrix} \begin{bmatrix} x \\ y \\ z \end{bmatrix} + j = 0$$

or

$$\mathbf{x}^t A \mathbf{x} + K \mathbf{x} + j = 0$$

where

$$\mathbf{x} = \begin{bmatrix} x \\ y \\ z \end{bmatrix} \qquad A = \begin{bmatrix} a & d & e \\ d & b & f \\ e & f & c \end{bmatrix} \qquad K = \begin{bmatrix} g & h & i \end{bmatrix}$$

Example 18

The quadratic form associated with the quadratic equation

$$3x^2 + 2y^2 - z^2 + 4xy + 3xz - 8yz + 7x + 2y + 3z - 7 = 0$$

is

$$3x^2 + 2y^2 - z^2 + 4xy + 3xz - 8yz \quad \blacktriangle$$

Graphs of quadratic equations in x, y, and z are called *quadrics* or *quadric surfaces*. We now give some examples of quadrics and their equations.

A quadric whose equation can be expressed in one of the forms in Figure 7.9 is said to be in *standard position* relative to the coordinate axes. The presence of one or more of the cross-product terms xy, xz, and yz in the equation of a quadric indicates that the quadric is rotated out of standard position; the presence of both x^2 and x terms, y^2 and y terms, or z^2 and z terms in a quadric with no cross-product term indicates the quadric is translated out of standard position.

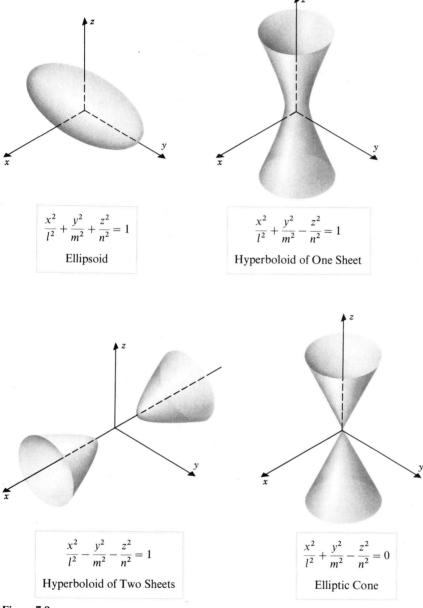

Figure 7.9

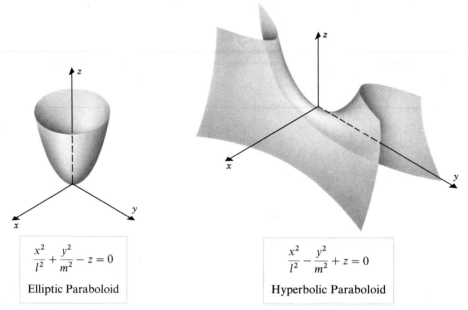

$$\frac{x^2}{l^2} + \frac{y^2}{m^2} - z = 0$$

Elliptic Paraboloid

$$\frac{x^2}{l^2} - \frac{y^2}{m^2} + z = 0$$

Hyperbolic Paraboloid

Figure 7.9

Example 19

Describe the quadric surface whose equation is

$$4x^2 + 36y^2 - 9z^2 - 16x - 216y + 304 = 0$$

Solution. Rearranging terms gives

$$4(x^2 - 4x) + 36(y^2 - 6y) - 9z^2 = -304$$

Completing the squares yields

$$4(x^2 - 4x + 4) + 36(y^2 - 6y + 9) - 9z^2 = -304 + 16 + 324$$

or

$$4(x - 2)^2 + 36(y - 3)^2 - 9z^2 = 36$$

or

$$\frac{(x - 2)^2}{9} + (y - 3)^2 - \frac{z^2}{4} = 1$$

Translating the axes by means of the translation equations

$$x' = x - 2 \qquad y' = y - 3 \qquad z' = z$$

yields

$$\frac{x'^2}{9} + y'^2 - \frac{z'^2}{4} = 1$$

which is the equation of a hyperboloid of one sheet. ▲

The procedure for identifying quadrics that are rotated out of standard position is similar to the procedure for conics. Let Q be a quadric surface whose equation in xyz-coordinates is

$$\mathbf{x}^t A \mathbf{x} + K \mathbf{x} + j = 0 \tag{7.34}$$

We want to rotate the xyz-coordinate axes so that the equation of the quadric in the new $x'y'z'$-coordinate system has no cross-product terms. This can be done as follows:

Step 1. Find a matrix P that orthogonally diagonalizes $\mathbf{x}^t A \mathbf{x}$.

Step 2. Interchange two columns of P, if necessary, to make $\det(P) = 1$. This assures that the orthogonal coordinate transformation

$$\mathbf{x} = P\mathbf{x}', \text{ that is, } \begin{bmatrix} x \\ y \\ z \end{bmatrix} = P \begin{bmatrix} x' \\ y' \\ z' \end{bmatrix} \tag{7.35}$$

is a rotation.

Step 3. Substitute (7.35) into (7.34). This will produce an equation for the quadric in $x'y'z'$-coordinates with no cross-product terms. (The proof is similar to that for conics and is left as an exercise.)

The following theorem summarizes this discussion.

Theorem 6 (*Principal Axes Theorem for R^3*). *Let*

$$ax^2 + by^2 + cz^2 + 2dxy + 2exz + 2fyz + gx + hy + iz + j = 0$$

be the equation of a quadric Q, and let

$$\mathbf{x}^t A \mathbf{x} = ax^2 + by^2 + cz^2 + 2dxy + 2exz + 2fyz$$

be the associated quadratic form. The coordinate axes can be rotated so that the equation of Q in the $x'y'z'$-coordinate system has the form

$$\lambda_1 x'^2 + \lambda_2 y'^2 + \lambda_3 z'^2 + g'x' + h'y' + i'z' + j = 0$$

where λ_1, λ_2, and λ_3 are the eigenvalues of A. The rotation can be accomplished by the substitution

$$\mathbf{x} = P\mathbf{x}'$$

where P orthogonally diagonalizes $\mathbf{x}^t A \mathbf{x}$ and $\det(P) = 1$.

Example 20

Describe the quadric surface whose equation is

$$4x^2 + 4y^2 + 4z^2 + 4xy + 4xz + 4yz - 3 = 0$$

Solution. The matrix form of the above quadratic equation is

$$\mathbf{x}^t A \mathbf{x} - 3 = 0 \tag{7.36}$$

where

$$A = \begin{bmatrix} 4 & 2 & 2 \\ 2 & 4 & 2 \\ 2 & 2 & 4 \end{bmatrix}$$

As shown in Example 13 of Section 6.3, the eigenvalues of A are $\lambda = 2$ and $\lambda = 8$, and A is orthogonally diagonalized by the matrix

$$P = \begin{bmatrix} -\dfrac{1}{\sqrt{2}} & -\dfrac{1}{\sqrt{6}} & \dfrac{1}{\sqrt{3}} \\ \dfrac{1}{\sqrt{2}} & -\dfrac{1}{\sqrt{6}} & \dfrac{1}{\sqrt{3}} \\ 0 & \dfrac{2}{\sqrt{6}} & \dfrac{1}{\sqrt{3}} \end{bmatrix}$$

where the first two column vectors in P are eigenvectors corresponding to $\lambda = 2$, and the third column vector is an eigenvector corresponding to $\lambda = 8$.

Since $\det(P) = 1$ (verify), the orthogonal coordinate transformation $\mathbf{x} = P\mathbf{x}'$ is a rotation. Substituting this expression in (7.36) yields

$$(P\mathbf{x}')^t A (P\mathbf{x}') - 3 = 0$$

or equivalently

$$(\mathbf{x}')^t (P^t A P)\mathbf{x}' - 3 = 0 \tag{7.37}$$

But

$$P^t A P = \begin{bmatrix} 2 & 0 & 0 \\ 0 & 2 & 0 \\ 0 & 0 & 8 \end{bmatrix}$$

so (7.37) becomes

$$\begin{bmatrix} x' & y' & z' \end{bmatrix} \begin{bmatrix} 2 & 0 & 0 \\ 0 & 2 & 0 \\ 0 & 0 & 8 \end{bmatrix} \begin{bmatrix} x' \\ y' \\ z' \end{bmatrix} - 3 = 0$$

or

$$2x'^2 + 2y'^2 + 8z'^2 = 3$$

This can be rewritten as

$$\frac{x'^2}{3/2} + \frac{y'^2}{3/2} + \frac{z'^2}{3/8} = 1$$

which is the equation of an ellipsoid.   ▲

EXERCISE SET 7.5

1. Find the quadratic forms associated with the following quadratic equations.
   (a) $x^2 + 2y^2 - z^2 + 4xy - 5yz + 7x + 2z = 3$
   (b) $3x^2 + 7z^2 + 2xy - 3xz + 4yz - 3x = 4$
   (c) $xy + xz + yz = 1$
   (d) $x^2 + y^2 - z^2 = 7$
   (e) $3z^2 + 3xz - 14y + 9 = 0$
   (f) $2z^2 + 2xz + y^2 + 2x - y + 3z = 0$

2. Find the matrices of the quadratic forms in Exercise 1.

3. Express each of the quadratic equations given in Exercise 1 in the matrix form $x^t Ax + Kx + j = 0$.

4. Name the following quadrics.
   (a) $36x^2 + 9y^2 + 4z^2 - 36 = 0$
   (b) $2x^2 + 6y^2 - 3z^2 = 18$
   (c) $6x^2 - 3y^2 - 2z^2 - 6 = 0$
   (d) $9x^2 + 4y^2 - z^2 = 0$
   (e) $16x^2 + y^2 = 16z$
   (f) $7x^2 - 3y^2 + z = 0$
   (g) $x^2 + y^2 + z^2 = 25$

5. In each part determine the translation equations that will put the quadric in standard position. Name the quadric.
   (a) $9x^2 + 36y^2 + 4z^2 - 18x - 144y - 24z + 153 = 0$
   (b) $6x^2 + 3y^2 - 2z^2 + 12x - 18y - 8z = -7$
   (c) $3x^2 - 3y^2 - z^2 + 42x + 144 = 0$
   (d) $4x^2 + 9y^2 - z^2 - 54y - 50z = 544$
   (e) $x^2 + 16y^2 + 2x - 32y - 16z - 15 = 0$
   (f) $7x^2 - 3y^2 + 126x + 72y + z + 135 = 0$
   (g) $x^2 + y^2 + z^2 - 2x + 4y - 6z = 11$

6. In each part find a rotation $x = Px'$ that removes the cross-product terms. Name the quadric and give its equation in the $x'y'z'$-system.
   (a) $2x^2 + 3y^2 + 23z^2 + 72xz + 150 = 0$
   (b) $4x^2 + 4y^2 + 4z^2 + 4xy + 4xz + 4yz - 5 = 0$
   (c) $144x^2 + 100y^2 + 81z^2 - 216xz - 540x - 720z = 0$
   (d) $2xy + z = 0$

In Exercises 7–10 translate and rotate the coordinate axes to put the quadric in standard position. Name the quadric and give its equation in the final coordinate system.

7. $2xy + 2xz + 2yz - 6x - 6y - 4z = -9$

8. $7x^2 + 7y^2 + 10z^2 - 2xy - 4xz + 4yz - 12x + 12y + 60z = 24$

9. $2xy - 6x + 10y + z - 31 = 0$

10. $2x^2 + 2y^2 + 5z^2 - 4xy - 2xz + 2yz + 10x - 26y - 2z = 0$

11. Prove Theorem 6.

CHAPTER EIGHT

Introduction to Numerical Methods of Linear Algebra

8.1 COMPARISON OF PROCEDURES FOR SOLVING LINEAR SYSTEMS

In this chapter we shall discuss some practical aspects of solving systems of linear equations, inverting matrices, and finding eigenvalues. Although we have previously discussed methods for performing these computations, those methods are not directly applicable to the computer solution of the large-scale problems that arise in real-world applications.

Since computers are limited in the number of decimal places they can carry, they round off or truncate most numerical quantities. For example, a computer designed to store eight decimal places might record $\frac{2}{3}$ either as .66666667 (rounded off) or .66666666 (truncated). In either case, an error is introduced that we shall call *roundoff error* or *rounding error*.

The main practical considerations in solving linear algebra problems on digital computers are minimizing the computer time (and thus cost) needed to obtain the solution, and minimizing inaccuracies due to roundoff errors. Thus, a good computer algorithm uses as few operations as possible and performs those operations in a way that minimizes the effect of roundoff errors.

In this text we have studied four methods for solving a linear system, $A\mathbf{x} = \mathbf{b}$, of n equations in n unknowns:

1. Gaussian elimination with back substitution
2. Gauss-Jordan elimination
3. Computing A^{-1}, then $\mathbf{x} = A^{-1}\mathbf{b}$
4. Cramer's Rule

To determine how these methods compare as computational tools, we need to know how many arithmetic operations each requires. On a large modern computer, typical execution times in microseconds (1 microsecond = 10^{-6} second) for the basic arithmetic operations are:

$$
\begin{aligned}
\text{Multiplication} &= 1.0 \text{ microseconds} \\
\text{Division} \quad &= 3.0 \text{ microseconds} \\
\text{Addition} \quad &= 0.5 \text{ microseconds} \\
\text{Subtraction} \quad &= 0.5 \text{ microseconds}
\end{aligned}
$$

In our analysis we shall group divisions and multiplications together (average execution time = 2.0 microseconds), and we shall also group additions and subtractions together (average execution time = 0.5 microseconds). We shall refer to either multiplications or divisions as "multiplications" and to additions or subtractions as "additions."

In Table 8.1 we list the number of operations required to solve a linear system $A\mathbf{x} = \mathbf{b}$ of n equations in n unknowns by each of the four methods discussed in the text, as well as the number of operations required to invert A or to compute its determinant by row reduction.

Note that the text methods of Gauss-Jordan elimination and Gaussian elimination have the same operation counts. It is not hard to see why this is so. Both methods begin by reducing the augmented matrix to row-echelon form. This is called the *forward phase* or *forward pass*. Then the solution is completed by back-substitution in Gaussian elimination and by continued reduction to reduced row-

TABLE 8.1 *Operation Counts for an Invertible n × n Matrix A*

Method	Number of Additions	Number of Multiplications
Solve $A\mathbf{x} = \mathbf{b}$ by Gauss-Jordan elimination	$\frac{1}{3}n^3 + \frac{1}{2}n^2 - \frac{5}{6}n$	$\frac{1}{3}n^3 + n^2 - \frac{1}{3}n$
Solve $A\mathbf{x} = \mathbf{b}$ by Gaussian elimination	$\frac{1}{3}n^3 + \frac{1}{2}n^2 - \frac{5}{6}n$	$\frac{1}{3}n^3 + n^2 - \frac{1}{3}n$
Find A^{-1} by reducing $[A \vert I]$ to $[I \vert A^{-1}]$	$n^3 - 2n^2 + n$	n^3
Solve $A\mathbf{x} = \mathbf{b}$ as $\mathbf{x} = A^{-1}\mathbf{b}$	$n^3 - n^2$	$n^3 + n^2$
Find $\det(A)$ by row reduction	$\frac{1}{3}n^3 - \frac{1}{2}n^2 + \frac{1}{6}n$	$\frac{1}{3}n^3 + \frac{2}{3}n - 1$
Solve $A\mathbf{x} = \mathbf{b}$ by Cramer's rule	$\frac{1}{3}n^4 - \frac{1}{6}n^3 - \frac{1}{3}n^2 + \frac{1}{6}n$	$\frac{1}{3}n^4 + \frac{1}{3}n^3 + \frac{2}{3}n^2 - \frac{1}{3}n - 1$

echelon form in Gauss-Jordan elimination. This is called the **backward phase** or **backward pass**. It turns out that the number of operations required for the backward phase is the same whether one uses back-substitution or continued reduction to reduce row-echelon form. Thus, the text methods of Gaussian elimination and Gauss-Jordan elimination have the same operation counts.

REMARK. There is a common variation of Gauss-Jordan elimination that is less efficient than the one presented in this text. In our method the augmented matrix is first reduced to reduced row-echelon form by introducing zeros *below* the leading ones. Then the reduction is completed by introducing zeros above the leading ones. An alternate procedure is to introduce zeros *above* and *below* a leading one as soon as it is obtained. This method requires

$$\frac{n^3}{2} - \frac{n}{2} \text{ additions } \quad \text{and} \quad \frac{n^3}{2} + n^2 - \frac{n}{2} \text{ multiplications}$$

both of which are larger than our values for all $n \geq 3$.

To illustrate how the results in Table 8.1 are computed, we shall derive the operation counts for Gauss-Jordan elimination. For this discussion we need the following formulas for the sum of the first n positive integers and the sum of the squares of the first n positive integers:

$$1 + 2 + 3 + \cdots + n = \frac{n(n + 1)}{2} \tag{8.1}$$

$$1^2 + 2^2 + 3^2 + \cdots + n^2 = \frac{n(n + 1)(2n + 1)}{6} \tag{8.2}$$

Derivations of these formulas are discussed in the exercises. We also need formulas for the sum of the first $n - 1$ positive integers and the sum of the squares of the first $n - 1$ positive integers. These can be obtained by substituting $n - 1$ for n in (8.1) and (8.2).

$$1 + 2 + 3 + \cdots + (n - 1) = \frac{(n - 1)n}{2} \tag{8.3}$$

$$1^2 + 2^2 + 3^2 + \cdots + (n - 1)^2 = \frac{(n - 1)n(2n - 1)}{6} \tag{8.4}$$

Operation Count for Gauss-Jordan Elimination. Let $A\mathbf{x} = \mathbf{b}$ be a system of n linear equations in n unknowns, and assume that A is invertible, so the system has a unique solution. Also assume, for simplicity, that no row interchanges are required to put the augmented matrix $[A \mid \mathbf{b}]$ in reduced row-echelon form. This assumption is justified by the fact that row interchanges are performed as bookkeeping operations on a computer and require much less time than an arithmetic operation.

Since no row interchanges are required, the first step in the Gauss-Jordan elimination process is to introduce a leading 1 in the first row by multiplying the elements in that row by the reciprocal of the leftmost entry in the row. We shall represent this step schematically as follows:

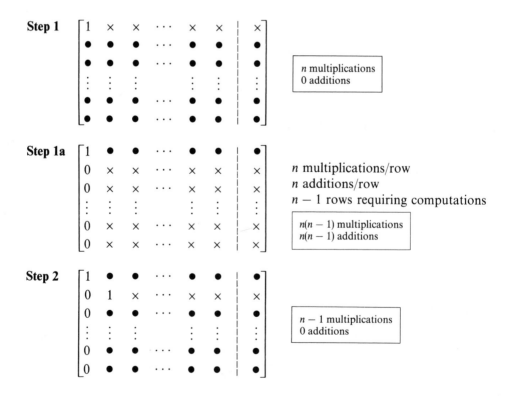

$$\begin{bmatrix} 1 & \times & \times & \cdots & \times & \times & | & \times \\ \bullet & \bullet & \bullet & \cdots & \bullet & \bullet & | & \bullet \\ \bullet & \bullet & \bullet & \cdots & \bullet & \bullet & | & \bullet \\ \vdots & \vdots & \vdots & & \vdots & \vdots & | & \vdots \\ \bullet & \bullet & \bullet & \cdots & \bullet & \bullet & | & \bullet \\ \bullet & \bullet & \bullet & \cdots & \bullet & \bullet & | & \bullet \end{bmatrix}$$

> × denotes a quantity to be computed
> • denotes a quantity that is not computed
> The matrix size is $n \times (n + 1)$

Note that the leading 1 is simply recorded and requires no computation; only the remaining n entries in the first row must be computed.

The following is a schematic description of the steps and the number of operations required to reduce $[A \,|\, \mathbf{b}]$ to row-echelon form.

Step 1
$$\begin{bmatrix} 1 & \times & \times & \cdots & \times & \times & | & \times \\ \bullet & \bullet & \bullet & \cdots & \bullet & \bullet & | & \bullet \\ \bullet & \bullet & \bullet & \cdots & \bullet & \bullet & | & \bullet \\ \vdots & \vdots & \vdots & & \vdots & \vdots & | & \vdots \\ \bullet & \bullet & \bullet & \cdots & \bullet & \bullet & | & \bullet \\ \bullet & \bullet & \bullet & \cdots & \bullet & \bullet & | & \bullet \end{bmatrix}$$

> n multiplications
> 0 additions

Step 1a
$$\begin{bmatrix} 1 & \bullet & \bullet & \cdots & \bullet & \bullet & | & \bullet \\ 0 & \times & \times & \cdots & \times & \times & | & \times \\ 0 & \times & \times & \cdots & \times & \times & | & \times \\ \vdots & \vdots & \vdots & & \vdots & \vdots & | & \vdots \\ 0 & \times & \times & \cdots & \times & \times & | & \times \\ 0 & \times & \times & \cdots & \times & \times & | & \times \end{bmatrix}$$

n multiplications/row
n additions/row
$n - 1$ rows requiring computations

> $n(n - 1)$ multiplications
> $n(n - 1)$ additions

Step 2
$$\begin{bmatrix} 1 & \bullet & \bullet & \cdots & \bullet & \bullet & | & \bullet \\ 0 & 1 & \times & \cdots & \times & \times & | & \times \\ 0 & \bullet & \bullet & \cdots & \bullet & \bullet & | & \bullet \\ \vdots & \vdots & \vdots & & \vdots & \vdots & | & \vdots \\ 0 & \bullet & \bullet & \cdots & \bullet & \bullet & | & \bullet \\ 0 & \bullet & \bullet & \cdots & \bullet & \bullet & | & \bullet \end{bmatrix}$$

> $n - 1$ multiplications
> 0 additions

Step 2a
$$\begin{bmatrix} 1 & \bullet & \bullet & \cdots & \bullet & \bullet & \vdots & \bullet \\ 0 & 1 & \bullet & \cdots & \bullet & \bullet & \vdots & \bullet \\ 0 & 0 & \times & \cdots & \times & \times & \vdots & \times \\ \vdots & \vdots & \vdots & & \vdots & \vdots & \vdots & \vdots \\ 0 & 0 & \times & \cdots & \times & \times & \vdots & \times \\ 0 & 0 & \times & \cdots & \times & \times & \vdots & \times \end{bmatrix}$$

$n - 1$ multiplications/row
$n - 1$ additions/row
$n - 2$ rows requiring computations

$(n-1)(n-2)$ multiplications
$(n-1)(n-2)$ additions

Step 3
$$\begin{bmatrix} 1 & \bullet & \bullet & \cdots & \bullet & \bullet & \vdots & \bullet \\ 0 & 1 & \bullet & \cdots & \bullet & \bullet & \vdots & \bullet \\ 0 & 0 & 1 & & \times & \times & \vdots & \times \\ \vdots & \vdots & \vdots & \cdots & \vdots & \vdots & \vdots & \vdots \\ 0 & 0 & \bullet & \cdots & \bullet & \bullet & \vdots & \bullet \\ 0 & 0 & \bullet & \cdots & \bullet & \bullet & \vdots & \bullet \end{bmatrix}$$

$n - 2$ multiplications
0 additions

Step 3a
$$\begin{bmatrix} 1 & \bullet & \bullet & \cdots & \bullet & \bullet & \vdots & \bullet \\ 0 & 1 & \bullet & \cdots & \bullet & \bullet & \vdots & \bullet \\ 0 & 0 & 1 & & \bullet & \bullet & \vdots & \bullet \\ \vdots & \vdots & \vdots & \cdots & \vdots & \vdots & \vdots & \vdots \\ 0 & 0 & 0 & \cdots & \times & \times & \vdots & \times \\ 0 & 0 & 0 & \cdots & \times & \times & \vdots & \times \end{bmatrix}$$

$\vdots$

$n - 2$ multiplications/row
$n - 2$ additions/row
$n - 3$ rows requiring computations

$(n-2)(n-3)$ multiplications
$(n-2)(n-3)$ additions

Step $(n - 1)$
$$\begin{bmatrix} 1 & \bullet & \bullet & \cdots & \bullet & \bullet & \vdots & \bullet \\ 0 & 1 & \bullet & \cdots & \bullet & \bullet & \vdots & \bullet \\ 0 & 0 & 1 & & \bullet & \bullet & \vdots & \bullet \\ \vdots & \vdots & \vdots & \cdots & \vdots & \vdots & \vdots & \vdots \\ 0 & 0 & 0 & \cdots & 1 & \times & \vdots & \times \\ 0 & 0 & 0 & \cdots & \bullet & \bullet & \vdots & \bullet \end{bmatrix}$$

2 multiplications
0 additions

Step $(n - 1)$a
$$\begin{bmatrix} 1 & \bullet & \bullet & \cdots & \bullet & \bullet & \vdots & \bullet \\ 0 & 1 & \bullet & \cdots & \bullet & \bullet & \vdots & \bullet \\ 0 & 0 & 1 & & \bullet & \bullet & \vdots & \bullet \\ \vdots & \vdots & \vdots & \cdots & \vdots & \vdots & \vdots & \vdots \\ 0 & 0 & 0 & \cdots & 1 & \bullet & \vdots & \bullet \\ 0 & 0 & 0 & \cdots & 0 & \times & \vdots & \times \end{bmatrix}$$

2 multiplications/row
2 additions/row
1 row requiring computations

2 multiplications
2 additions

Step n
$$\begin{bmatrix} 1 & \bullet & \bullet & \cdots & \bullet & \bullet & \vdots & \bullet \\ 0 & 1 & \bullet & \cdots & \bullet & \bullet & \vdots & \bullet \\ 0 & 0 & 1 & & \bullet & \bullet & \vdots & \bullet \\ \vdots & \vdots & \vdots & \cdots & \vdots & \vdots & \vdots & \vdots \\ 0 & 0 & 0 & \cdots & 1 & \bullet & \vdots & \bullet \\ 0 & 0 & 0 & \cdots & 0 & 1 & \vdots & \times \end{bmatrix}$$

1 multiplication
0 additions

Thus, the number of operations required to complete successive steps is as follows:

Steps 1 and 1a
Multiplications: $n + n(n - 1) = n^2$
Additions: $n(n - 1) = n^2 - n$

Steps 2 and 2a
Multiplications: $(n - 1) + (n - 1)(n - 2) = (n - 1)^2$
Additions: $(n - 1)(n - 2) = (n - 1)^2 - (n - 1)$

Steps 3 and 3a
Multiplications: $(n - 2) + (n - 2)(n - 3) = (n - 2)^2$
Additions: $(n - 2)(n - 3) = (n - 2)^2 - (n - 2)$

$$\vdots$$

Steps $(n - 1)$ and $(n - 1)$a
Multiplications: $4 (= 2^2)$
Additions: $2 (= 2^2 - 2)$

Step n
Multiplications: $1 (= 1^2)$
Additions: $0 (= 1^2 - 1)$

Therefore, the total number of operations required to reduce $[A\,|\,\mathbf{b}]$ to row-echelon form is:

Multiplications: $n^2 + (n - 1)^2 + (n - 2)^2 + \cdots + 1^2$

Additions:
$$[n^2 + (n - 1)^2 + (n - 2)^2 + \cdots + 1^2] - [n + (n - 1) + (n - 2) + \cdots + 1]$$

or, on applying formulas (8.1) and (8.2),

$$\text{Multiplications:} \quad \frac{n(n + 1)(2n + 1)}{6} = \frac{n^3}{3} + \frac{n^2}{2} + \frac{n}{6} \tag{8.5}$$

$$\text{Additions:} \quad \frac{n(n + 1)(2n + 1)}{6} - \frac{n(n + 1)}{2} = \frac{n^3}{3} - \frac{n}{3} \tag{8.6}$$

This completes the operation count for the forward phase. For the backward phase we must put the row-echelon form of $[A\,|\,\mathbf{b}]$ into reduced row-echelon form by introducing zeros above the leading 1's. The operations are as follows:

Step 1
$$\begin{bmatrix} 1 & \bullet & \bullet & \cdots & \bullet & 0 & | & \times \\ 0 & 1 & \bullet & \cdots & \bullet & 0 & | & \times \\ 0 & 0 & 1 & \cdots & \bullet & 0 & | & \times \\ \vdots & \vdots & \vdots & & \vdots & \vdots & | & \vdots \\ 0 & 0 & 0 & \cdots & 1 & 0 & | & \times \\ 0 & 0 & 0 & \cdots & 0 & 1 & | & \bullet \end{bmatrix}$$

$n - 1$ multiplications
$n - 1$ additions

Step 2
$$\left[\begin{array}{cccccc|c} 1 & \bullet & \bullet & \cdots & 0 & 0 & \times \\ 0 & 1 & \bullet & \cdots & 0 & 0 & \times \\ 0 & 0 & 1 & \cdots & 0 & 0 & \times \\ \vdots & \vdots & \vdots & & \vdots & \vdots & \vdots \\ 0 & 0 & 0 & \cdots & 1 & 0 & \bullet \\ 0 & 0 & 0 & \cdots & 0 & 1 & \bullet \end{array}\right]$$

$n - 2$ multiplications
$n - 2$ additions

$\vdots$

Step (n − 2)
$$\left[\begin{array}{cccccc|c} 1 & \bullet & 0 & \cdots & 0 & 0 & \times \\ 0 & 1 & 0 & \cdots & 0 & 0 & \times \\ 0 & 0 & 1 & \cdots & 0 & 0 & \bullet \\ \vdots & \vdots & \vdots & & \vdots & \vdots & \vdots \\ 0 & 0 & 0 & \cdots & 1 & 0 & \bullet \\ 0 & 0 & 0 & \cdots & 0 & 1 & \bullet \end{array}\right]$$

2 multiplications
2 additions

Step (n − 1)
$$\left[\begin{array}{cccccc|c} 1 & 0 & 0 & \cdots & 0 & 0 & \times \\ 0 & 1 & 0 & \cdots & 0 & 0 & \bullet \\ 0 & 0 & 1 & \cdots & 0 & 0 & \bullet \\ \vdots & \vdots & \vdots & & \vdots & \vdots & \vdots \\ 0 & 0 & 0 & \cdots & 1 & 0 & \bullet \\ 0 & 0 & 0 & \cdots & 0 & 1 & \bullet \end{array}\right]$$

1 multiplication
1 addition

Thus, the number of operations required for the backward phase is

Multiplications: $(n - 1) + (n - 2) + \cdots + 2 + 1$

Additions: $(n - 1) + (n - 2) + \cdots + 2 + 1$

or, on applying formulas (8.3) and (8.4),

$$\text{Multiplications:} \quad \frac{(n - 1)n}{2} = \frac{n^2}{2} - \frac{n}{2} \tag{8.7}$$

$$\text{Additions:} \quad \frac{(n - 1)n}{2} = \frac{n^2}{2} - \frac{n}{2} \tag{8.8}$$

Thus, from (8.5), (8.6), (8.7), and (8.8) the total operation count for Gauss-Jordan elimination is

$$\text{Multiplications:} \quad \left(\frac{n^3}{3} + \frac{n^2}{2} + \frac{n}{6}\right) + \left(\frac{n^2}{2} - \frac{n}{2}\right) = \frac{n^3}{3} + n^2 - \frac{n}{3} \tag{8.9}$$

$$\text{Additions:} \quad \left(\frac{n^3}{3} - \frac{n}{3}\right) + \left(\frac{n^2}{2} - \frac{n}{2}\right) = \frac{n^3}{3} + \frac{n^2}{2} - \frac{5n}{6} \tag{8.10}$$

In practical applications it is not uncommon to encounter linear systems with thousands of equations in thousands of unknowns. Thus, we shall be interested in

Table 8.1 for large values of n. It is a fact about polynomials that for large values of the variable, a polynomial can be approximated well by its term of highest degree; that is,

$$a_0 + a_1 x + \cdots + a_k x^k \approx a_k x^k \quad \text{for large } x$$

(see Exercise 12). Thus, for large values of n the operation counts in Table 8.1 can be approximated as shown in Table 8.2.

TABLE 8.2 *Approximate Operation Counts for an Invertible* $n \times n$ *Matrix with Large* n

Method	*Number of Additions*	*Number of Multiplications*
Solve $Ax = b$ by Gauss-Jordan elimination	$\approx \dfrac{n^3}{3}$	$\approx \dfrac{n^3}{3}$
Solve $Ax = b$ by Gaussian elimination	$\approx \dfrac{n^3}{3}$	$\approx \dfrac{n^3}{3}$
Find A^{-1} by reducing $[A\,|\,I]$ to $[I\,|\,A^{-1}]$	$\approx n^3$	$\approx n^3$
Solve $Ax = b$ as $x = A^{-1}b$	$\approx n^3$	$\approx n^3$
Find $\det(A)$ by row reduction	$\approx \dfrac{n^3}{3}$	$\approx \dfrac{n^3}{3}$
Solve $Ax = b$ by Cramer's rule	$\approx \dfrac{n^4}{3}$	$\approx \dfrac{n^4}{3}$

It follows from Table 8.2 that for large n the best methods for solving $Ax = b$ are Gaussian elimination and Gauss-Jordan elimination. The method of multiplying by A^{-1} is much worse than these (it requires three times as many operations), and the poorest of the four methods is Cramer's Rule.

REMARK. We observed in the remark following Table 8.1 that if Gauss-Jordan elimination is performed by introducing zeros above and below leading 1's as soon as they are obtained, then the operation count is

$$\frac{n^3}{2} - \frac{n}{2} \quad \text{additions} \quad \text{and} \quad \frac{n^3}{2} + \frac{n^2}{2} \quad \text{multiplications}$$

Thus, for large n this procedure requires $\approx n^3/2$ multiplications, which is 50% greater than the $n^3/3$ multiplications required by the text method. Similarly for additions.

It is reasonable to ask if it is possible to devise other methods for solving linear systems that might require significantly fewer than the $\approx n^3/3$ additions and multiplications needed in Gaussian elimination and Gauss-Jordan elimination. The answer is a qualified "yes." In recent years methods have been devised that require $\approx Cn^q$ multiplications, where q is slightly larger than 2.5. However, these methods have little practical value because the programming is complicated, the constant C is very large, and the number of additions required is excessive. In short, there is currently no practical method for solving general linear systems that significantly improves on the operation counts for Gaussian elimination and the text method of Gauss-Jordan elimination.

EXERCISE SET 8.1

1. Find the number of additions and multiplications required to compute AB if A is an $m \times n$ matrix and B is an $n \times p$ matrix.

2. Use the result in Exercise 1 to find the number of additions and multiplications required to compute A^k by direct multiplication if A is an $n \times n$ matrix.

3. Assuming A to be an $n \times n$ matrix, use the formulas in Table 8.1 to determine the number of operations required for the procedures in Table 8.3.

TABLE 8.3

	$n = 5$		$n = 10$		$n = 100$		$n = 1000$	
	+	×	+	×	+	×	+	×
Solve $A\mathbf{x} = \mathbf{b}$ by Gauss-Jordan elimination								
Solve $A\mathbf{x} = \mathbf{b}$ by Gaussian elimination								
Find A^{-1} by reducing $[A \mid I]$ to $[I \mid A^{-1}]$								
Solve $A\mathbf{x} = \mathbf{b}$ as $\mathbf{x} = A^{-1}\mathbf{b}$								
Find $\det(A)$ by row reduction								
Solve $A\mathbf{x} = \mathbf{b}$ by Cramer's rule								

380 / Introduction to Numerical Methods of Linear Algebra

4. Assuming a computer execution time of 2.0 microseconds for multiplications and 0.5 microseconds for additions, use the results in Exercise 3 to fill in the execution times in seconds for the procedures in Table 8.4.

TABLE 8.4

	$n = 5$	$n = 10$	$n = 100$	$n = 1000$
	Execution Time (sec)	Execution Time (sec)	Execution Time (sec)	Execution Time (sec)
Solve $Ax = b$ by Gauss-Jordan elimination				
Solve $Ax = b$ by Gaussian elimination				
Find A^{-1} by reducing $[A \mid I]$ to $[I \mid A^{-1}]$				
Solve $Ax = b$ as $x = A^{-1}b$				
Find $\det(A)$ by row reduction				
Solve $Ax = b$ by Cramer's rule				

5. Derive the formula

$$1 + 2 + 3 + \cdots + n = \frac{n(n + 1)}{2}$$

(*Hint.* Let $S_n = 1 + 2 + 3 + \cdots + n$. Write the terms of S_n in reverse order and add the two expressions for S_n.)

6. Use the result in Exercise 5 to show that

$$1 + 2 + 3 + \cdots + (n - 1) = \frac{(n - 1)n}{2}$$

7. Derive the formula

$$1^2 + 2^2 + 3^2 + \cdots + n^2 = \frac{n(n + 1)(2n + 1)}{6}$$

using the following steps:
(a) Show that $(k + 1)^3 - k^3 = 3k^2 + 3k + 1$
(b) Show that

$$[2^3 - 1^3] + [3^3 - 2^3] + [4^3 - 3^3] + \cdots + [(n + 1)^3 - n^3] = (n + 1)^3 - 1$$

(c) Apply (a) to each term on the left side of (b) to show that

$$(n + 1)^3 - 1 = 3[1^2 + 2^2 + 3^2 + \cdots + n^2] + 3[1 + 2 + 3 + \cdots + n] + n$$

(d) Solve the equation in (c) for $1^2 + 2^2 + 3^2 + \cdots + n^2$, use the result of Exercise 5, and then simplify.

8. Use the result in Exercise 7 to show that

$$1^2 + 2^2 + 3^2 + \cdots + (n - 1)^2 = \frac{(n - 1)n(2n - 1)}{6}$$

9. Let R be a row-echelon form of an invertible $n \times n$ matrix. Show that solving the linear system $Rx = b$ by back-substitution requires:

$$\frac{n^2}{2} - \frac{n}{2} \qquad \text{multiplications}$$

$$\frac{n^2}{2} - \frac{n}{2} \qquad \text{additions}$$

10. Show that to reduce an invertible $n \times n$ matrix to I_n by the text method requires:

$$\frac{n^3}{3} - \frac{n}{3} \qquad \text{multiplications}$$

$$\frac{n^3}{3} - \frac{n^2}{2} + \frac{n}{6} \qquad \text{additions}$$

(*Note.* Assume that no row interchanges are required.)

11. Consider the variation of Gauss-Jordan elimination in which zeros are introduced above and below a leading 1 as soon as it is obtained, and let A be an invertible $n \times n$ matrix. Show that to solve a linear system $Ax = b$ using this version of Gauss-Jordan elimination requires:

$$\frac{n^3}{2} + n^2 - \frac{n}{2} \qquad \text{multiplications}$$

$$\frac{n^3}{2} - \frac{n}{2} \qquad \text{additions}$$

(*Note.* Assume that no row interchanges are required.)

12. (**For readers who have studied calculus.**)
Show that if $p(x) = a_0 + a_1 x + \cdots + a_k x^k$, where $a_k \neq 0$, then

$$\lim_{x \to +\infty} \frac{p(x)}{a_k x^k} = 1$$

(*Note.* This result justifies the approximation $a_0 + a_1 x + \cdots + a_k x^k \approx a_k x^k$ for large x.)

8.2 *LU*-DECOMPOSITIONS

With Gaussian elimination and Gauss-Jordan elimination a linear system is solved by operating systematically on the augmented matrix. In this section we shall discuss a different approach, one based on factoring the coefficient matrix into a product of lower and upper triangular matrices. This method is well suited for digital computers and is the basis for many practical computer programs.*

We shall proceed in two stages. First we shall show how a linear system $A\mathbf{x} = \mathbf{b}$ can be readily solved once A is factored into a product of lower and upper triangular matrices. Then we shall show how to construct such factorizations.

If an $n \times n$ matrix A can be factored as

$$A = LU$$

where L is lower triangular and U is upper triangular, then the linear system $A\mathbf{x} = \mathbf{b}$ can be solved as follows:

Step 1. Rewrite the system $A\mathbf{x} = \mathbf{b}$ as

$$LU\mathbf{x} = \mathbf{b} \tag{8.11}$$

Step 2. Define a new $n \times 1$ matrix $\mathbf{y}$ by

$$U\mathbf{x} = \mathbf{y} \tag{8.12}$$

Step 3. Use (8.12) to rewrite (8.11) as $L\mathbf{y} = \mathbf{b}$ and solve this system for $\mathbf{y}$.

Step 4. Substitute $\mathbf{y}$ in (8.12) and solve for $\mathbf{x}$.

Although this procedure replaces the problem of solving the single system $A\mathbf{x} = \mathbf{b}$ by the problem of solving the two systems, $L\mathbf{y} = \mathbf{b}$ and $U\mathbf{x} = \mathbf{y}$, the latter systems are easy to solve because the coefficient matrices are triangular. The following example illustrates this procedure.

Example 1

Later in this section we will derive the factorization

$$\begin{bmatrix} 2 & 6 & 2 \\ -3 & -8 & 0 \\ 4 & 9 & 2 \end{bmatrix} = \begin{bmatrix} 2 & 0 & 0 \\ -3 & 1 & 0 \\ 4 & -3 & 7 \end{bmatrix} \begin{bmatrix} 1 & 3 & 1 \\ 0 & 1 & 3 \\ 0 & 0 & 1 \end{bmatrix}$$

---

* In 1979 an important library of machine-independent linear algebra programs, called LINPAK, was developed at Argonne National Laboratories. Many of the programs in that library are based on the methods discussed in this section.

Use this result and the method described above to solve the system

$$
\begin{bmatrix} 2 & 6 & 2 \\ -3 & -8 & 0 \\ 4 & 9 & 2 \end{bmatrix} \begin{bmatrix} x_1 \\ x_2 \\ x_3 \end{bmatrix} = \begin{bmatrix} 2 \\ 2 \\ 3 \end{bmatrix} \tag{8.13}
$$

Solution. Rewrite (8.13) as

$$
\begin{bmatrix} 2 & 0 & 0 \\ -3 & 1 & 0 \\ 4 & -3 & 7 \end{bmatrix} \begin{bmatrix} 1 & 3 & 1 \\ 0 & 1 & 3 \\ 0 & 0 & 1 \end{bmatrix} \begin{bmatrix} x_1 \\ x_2 \\ x_3 \end{bmatrix} = \begin{bmatrix} 2 \\ 2 \\ 3 \end{bmatrix} \tag{8.14}
$$

As specified in Step 2 above, define y_1, y_2, and y_3 by the equation

$$
\begin{bmatrix} 1 & 3 & 1 \\ 0 & 1 & 3 \\ 0 & 0 & 1 \end{bmatrix} \begin{bmatrix} x_1 \\ x_2 \\ x_3 \end{bmatrix} = \begin{bmatrix} y_1 \\ y_2 \\ y_3 \end{bmatrix} \tag{8.15}
$$

so (8.13) can be rewritten as

$$
\begin{bmatrix} 2 & 0 & 0 \\ -3 & 1 & 0 \\ 4 & -3 & 7 \end{bmatrix} \begin{bmatrix} y_1 \\ y_2 \\ y_3 \end{bmatrix} = \begin{bmatrix} 2 \\ 2 \\ 3 \end{bmatrix}
$$

or equivalently

$$
\begin{aligned}
2y_1 \quad\quad\quad\;\;\; &= 2 \\
-3y_1 + y_2 \quad\;\; &= 2 \\
4y_1 - 3y_2 + 7y_3 &= 3
\end{aligned}
$$

The procedure for solving this system is similar to back-substitution except that the equations are solved from the top down instead of the bottom up. This procedure, called *forward-substitution*, yields

$$
y_1 = 1, \quad y_2 = 5, \quad y_3 = 2
$$

(Verify.) Substituting these values in (8.15) yields the linear system

$$
\begin{bmatrix} 1 & 3 & 1 \\ 0 & 1 & 3 \\ 0 & 0 & 1 \end{bmatrix} \begin{bmatrix} x_1 \\ x_2 \\ x_3 \end{bmatrix} = \begin{bmatrix} 1 \\ 5 \\ 2 \end{bmatrix}
$$

or equivalently

$$
\begin{aligned}
x_1 + 3x_2 + x_3 &= 1 \\
x_2 + 3x_3 &= 5 \\
x_3 &= 2
\end{aligned}
$$

Solving this system by back-substitution yields the solution

$$
x_1 = 2, \quad x_2 = -1, \quad x_3 = 2
$$

(Verify.)

Now that we have seen how a linear system of n equations in n unknowns can be solved by factoring the coefficient matrix, we shall turn to the problem of constructing such factorizations. To motivate the method, suppose that an $n \times n$ matrix A has been reduced to a row-echelon form U by a sequence of elementary row operations. By Theorem 10 of Section 1.6 each of these operations can be accomplished by multiplying on the left by an appropriate elementary matrix. Thus, we can find elementary matrices $E_1, E_2, \ldots, E_k$ such that

$$E_k \cdots E_2 E_1 A = U \tag{8.16}$$

By Theorem 11 of Section 1.6, $E_1, E_2, \ldots, E_k$ are invertible, so we can multiply both sides of equation (8.16) on the left successively by

$$E_k^{-1}, \ldots, E_2^{-1}, E_1^{-1}$$

to obtain

$$A = E_1^{-1} E_2^{-1} \cdots E_k^{-1} U \tag{8.17}$$

In Exercise 15 we will help the reader to show that the matrix L defined by

$$L = E_1^{-1} E_2^{-1} \cdots E_k^{-1} \tag{8.18}$$

is lower triangular provided that *no row interchanges are used in reducing A to U*. Assuming this to be so, substituting (8.18) into (8.17) yields

$$A = LU$$

which is a factorization of A into a product of a lower triangular matrix and an upper triangular matrix.

The following theorem summarizes the above result.

Theorem 1. *If A is a square matrix that can be reduced to a row-echelon form U without using row interchanges, then A can be factored as $A = LU$ where L is a lower triangular matrix.*

Definition. A factorization of a square matrix A as $A = LU$, where L is lower triangular and U is upper triangular is called an *LU-decomposition* or *triangular decomposition* of A.

Example 2

Find an LU-decomposition of

$$A = \begin{bmatrix} 2 & 6 & 2 \\ -3 & -8 & 0 \\ 4 & 9 & 2 \end{bmatrix}$$

Solution. To obtain an *LU*-decomposition $A = LU$ we shall reduce A to a row-echelon form U, then calculate L from (8.18). The steps are as follows:

	Reduction to Row-Echelon Form	Elementary Matrix Corresponding to the Row Operation	Inverse of the Elementary Matrix
	$\begin{bmatrix} 2 & 6 & 2 \\ -3 & -8 & 0 \\ 4 & 9 & 2 \end{bmatrix}$		
Step 1		$E_1 = \begin{bmatrix} \frac{1}{2} & 0 & 0 \\ 0 & 1 & 0 \\ 0 & 0 & 1 \end{bmatrix}$	$E_1^{-1} = \begin{bmatrix} 2 & 0 & 0 \\ 0 & 1 & 0 \\ 0 & 0 & 1 \end{bmatrix}$
	$\begin{bmatrix} 1 & 3 & 1 \\ -3 & -8 & 0 \\ 4 & 9 & 2 \end{bmatrix}$		
Step 2		$E_2 = \begin{bmatrix} 1 & 0 & 0 \\ 3 & 1 & 0 \\ 0 & 0 & 1 \end{bmatrix}$	$E_2^{-1} = \begin{bmatrix} 1 & 0 & 0 \\ -3 & 1 & 0 \\ 0 & 0 & 1 \end{bmatrix}$
	$\begin{bmatrix} 1 & 3 & 1 \\ 0 & 1 & 3 \\ 4 & 9 & 2 \end{bmatrix}$		
Step 3		$E_3 = \begin{bmatrix} 1 & 0 & 0 \\ 0 & 1 & 0 \\ -4 & 0 & 1 \end{bmatrix}$	$E_3^{-1} = \begin{bmatrix} 1 & 0 & 0 \\ 0 & 1 & 0 \\ 4 & 0 & 1 \end{bmatrix}$
	$\begin{bmatrix} 1 & 3 & 1 \\ 0 & 1 & 3 \\ 0 & -3 & -2 \end{bmatrix}$		
Step 4		$E_4 = \begin{bmatrix} 1 & 0 & 0 \\ 0 & 1 & 0 \\ 0 & 3 & 1 \end{bmatrix}$	$E_4^{-1} = \begin{bmatrix} 1 & 0 & 0 \\ 0 & 1 & 0 \\ 0 & -3 & 1 \end{bmatrix}$
	$\begin{bmatrix} 1 & 3 & 1 \\ 0 & 1 & 3 \\ 0 & 0 & 7 \end{bmatrix}$		
Step 5		$E_5 = \begin{bmatrix} 1 & 0 & 0 \\ 0 & 1 & 0 \\ 0 & 0 & \frac{1}{7} \end{bmatrix}$	$E_5^{-1} = \begin{bmatrix} 1 & 0 & 0 \\ 0 & 1 & 0 \\ 0 & 0 & 7 \end{bmatrix}$
	$\begin{bmatrix} 1 & 3 & 1 \\ 0 & 1 & 3 \\ 0 & 0 & 1 \end{bmatrix}$		

Thus,

$$U = \begin{bmatrix} 1 & 3 & 1 \\ 0 & 1 & 3 \\ 0 & 0 & 1 \end{bmatrix}$$

and, from (8.18),

$$L = \begin{bmatrix} 2 & 0 & 0 \\ 0 & 1 & 0 \\ 0 & 0 & 1 \end{bmatrix} \begin{bmatrix} 1 & 0 & 0 \\ -3 & 1 & 0 \\ 0 & 0 & 1 \end{bmatrix} \begin{bmatrix} 1 & 0 & 0 \\ 0 & 1 & 0 \\ 4 & 0 & 1 \end{bmatrix} \begin{bmatrix} 1 & 0 & 0 \\ 0 & 1 & 0 \\ 0 & -3 & 1 \end{bmatrix} \begin{bmatrix} 1 & 0 & 0 \\ 0 & 1 & 0 \\ 0 & 0 & 7 \end{bmatrix}$$

$$= \begin{bmatrix} 2 & 0 & 0 \\ -3 & 1 & 0 \\ 4 & -3 & 7 \end{bmatrix}$$

so

$$\begin{bmatrix} 2 & 6 & 2 \\ -3 & -8 & 0 \\ 4 & 9 & 2 \end{bmatrix} = \begin{bmatrix} 2 & 0 & 0 \\ -3 & 1 & 0 \\ 4 & -3 & 7 \end{bmatrix} \begin{bmatrix} 1 & 3 & 1 \\ 0 & 1 & 3 \\ 0 & 0 & 1 \end{bmatrix}$$

is an *LU*-decomposition of *A*. ▲

As this example shows, most of the work in constructing an *LU*-decomposition is expended in the calculation of *L*. However, *all* this work can be eliminated by some careful bookkeeping of the operations used to reduce *A* to *U*. Because we are assuming that no row interchanges are required to reduce *A* to *U*, there are only two types of operations involved: multiplying a row by a nonzero constant, and adding a multiple of one row to another. The first operation is used to introduce the leading ones and the second to introduce zeros below the leading ones.

In Example 2, the multipliers needed to introduce the leading 1's in successive rows were as follows:

$\frac{1}{2}$ for the first row

1 for the second row

$\frac{1}{7}$ for the third row

Note that the successive diagonal entries in *L* were precisely the reciprocals of these multipliers (Figure 8.1).

$$L = \begin{bmatrix} ② & 0 & 0 \\ -3 & ① & 0 \\ 4 & -3 & ⑦ \end{bmatrix}$$

Figure 8.1

Next, observe that to introduce zeros below the leading 1 in the first row we used the following operations:

> add 3 times the first row to the second
> add -4 times the first row to the third

and to introduce the zero below the leading 1 in the second row, we used the operation:

> add 3 times the second row to the third

Now note that in each position below the main diagonal of L the entry is the *negative* of the multiplier in the operation that introduced the zero in that position in U.

$$L = \begin{bmatrix} 2 & 0 & 0 \\ -3 & 1 & 0 \\ 4 & -3 & 7 \end{bmatrix}$$

Figure 8.2

In summary, we have the following procedure for constructing an LU-decomposition of a square matrix A that can be reduced to row-echelon form without row interchanges.

Step 1. Reduce A to a row-echelon form U without using row interchanges, keeping track of the multipliers used to introduce the leading 1's and the multipliers used to introduce the zeros below the leading 1's.

Step 2. In each position along the main diagonal of L, place the reciprocal of the multiplier that introduced the leading 1 in that position in U.

Step 3. In each position below the main diagonal of L, place the negative of the multiplier used to introduce the zero in that position in U.

Step 4. Form the decomposition $A = LU$.

Example 3

Find an LU-decomposition of

$$A = \begin{bmatrix} 6 & -2 & 0 \\ 9 & -1 & 1 \\ 3 & 7 & 5 \end{bmatrix}$$

Solution. We begin by reducing A to row-echelon form, keeping track of all multipliers.

$$\begin{bmatrix} 6 & -2 & 0 \\ 9 & -1 & 1 \\ 3 & 7 & 5 \end{bmatrix}$$

$$\begin{bmatrix} \boxed{1} & -\frac{1}{3} & 0 \\ 9 & -1 & 1 \\ 3 & 7 & 5 \end{bmatrix} \longleftarrow \text{multiplier} = \frac{1}{6}$$

$$\begin{bmatrix} 1 & -\frac{1}{3} & 0 \\ \boxed{0} & 2 & 1 \\ \boxed{0} & 8 & 5 \end{bmatrix} \begin{matrix} \\ \longleftarrow \text{multiplier} = -9 \\ \longleftarrow \text{multiplier} = -3 \end{matrix}$$

$$\begin{bmatrix} 1 & -\frac{1}{3} & 0 \\ 0 & \boxed{1} & \frac{1}{2} \\ 0 & 8 & 5 \end{bmatrix} \longleftarrow \text{multiplier} = \frac{1}{2}$$

$$\begin{bmatrix} 1 & -\frac{1}{3} & 0 \\ 0 & 1 & \frac{1}{2} \\ 0 & \boxed{0} & 1 \end{bmatrix} \longleftarrow \text{multiplier} = -8$$

$$\begin{bmatrix} 1 & -\frac{1}{3} & 0 \\ 0 & 1 & \frac{1}{2} \\ 0 & 0 & \boxed{1} \end{bmatrix} \longleftarrow \text{multiplier} = 1$$

> No actual operation is performed here, since there is already a leading 1 in the third row.

Constructing L from the multipliers yields the LU-decomposition.

$$A = LU = \begin{bmatrix} 6 & 0 & 0 \\ 9 & 2 & 0 \\ 3 & 8 & 1 \end{bmatrix} \begin{bmatrix} 1 & -\frac{1}{3} & 0 \\ 0 & 1 & \frac{1}{2} \\ 0 & 0 & 1 \end{bmatrix}$$

We conclude this section by briefly discussing two fundamental questions about LU-decompositions:

1. Does every square matrix have an LU-decomposition?
2. Can a square matrix have more than one LU-decomposition?

We already know that if a square matrix A can be reduced to row-echelon form without using row interchanges, then A has an LU-decomposition. In general, if row interchanges are required to reduce A to row-echelon form, then there is no LU-decomposition of A. However, in such cases it is possible to factor A in the form

$$A = PLU$$

where L is lower triangular, U is upper triangular, and P is a matrix obtained by interchanging the rows of I_n appropriately (see Exercise 17).

In absence of additional restrictions, LU-decompositions are not unique. For example, if

$$A = LU = \begin{bmatrix} a_{11} & 0 & 0 \\ a_{21} & a_{22} & 0 \\ a_{31} & a_{32} & a_{33} \end{bmatrix} \begin{bmatrix} 1 & u_{12} & u_{13} \\ 0 & 1 & u_{23} \\ 0 & 0 & 1 \end{bmatrix}$$

and L has nonzero diagonal entries, then we can shift the diagonal entries from the left factor to the right factor by writing

$$A = \begin{bmatrix} 1 & 0 & 0 \\ \dfrac{a_{21}}{a_{11}} & 1 & 0 \\ \dfrac{a_{31}}{a_{11}} & \dfrac{a_{32}}{a_{22}} & 1 \end{bmatrix} \begin{bmatrix} a_{11} & 0 & 0 \\ 0 & a_{22} & 0 \\ 0 & 0 & a_{33} \end{bmatrix} \begin{bmatrix} 1 & u_{12} & u_{13} \\ 0 & 1 & u_{23} \\ 0 & 0 & 1 \end{bmatrix}$$

$$= \begin{bmatrix} 1 & 0 & 0 \\ \dfrac{a_{21}}{a_{11}} & 1 & 0 \\ \dfrac{a_{31}}{a_{11}} & \dfrac{a_{32}}{a_{22}} & 1 \end{bmatrix} \begin{bmatrix} a_{11} & a_{11}u_{12} & a_{11}u_{13} \\ 0 & a_{22} & a_{22}u_{23} \\ 0 & 0 & a_{33} \end{bmatrix}$$

which is another triangular decomposition of A.

EXERCISE SET 8.2

1. Use the method of Example 1 and the LU-decomposition

$$\begin{bmatrix} 3 & -6 \\ -2 & 5 \end{bmatrix} = \begin{bmatrix} 3 & 0 \\ -2 & 1 \end{bmatrix} \begin{bmatrix} 1 & -2 \\ 0 & 1 \end{bmatrix}$$

to solve the system

$$3x_1 - 6x_2 = 0$$
$$-2x_1 + 5x_2 = 1$$

2. Use the method of Example 1 and the LU-decomposition

$$\begin{bmatrix} 3 & -6 & -3 \\ 2 & 0 & 6 \\ -4 & 7 & 4 \end{bmatrix} = \begin{bmatrix} 3 & 0 & 0 \\ 2 & 4 & 0 \\ -4 & -1 & 2 \end{bmatrix} \begin{bmatrix} 1 & -2 & -1 \\ 0 & 1 & 2 \\ 0 & 0 & 1 \end{bmatrix}$$

to solve the system

$$3x_1 - 6x_2 - 3x_3 = -3$$
$$2x_1 \qquad + 6x_3 = -22$$
$$-4x_1 + 7x_2 + 4x_3 = 3$$

In Exercises 3–10 find an LU-decomposition of the coefficient matrix; then use the method of Example 1 to solve the system.

3. $\begin{bmatrix} 2 & 8 \\ -1 & -1 \end{bmatrix} \begin{bmatrix} x_1 \\ x_2 \end{bmatrix} = \begin{bmatrix} -2 \\ -2 \end{bmatrix}$

4. $\begin{bmatrix} -5 & -10 \\ 6 & 5 \end{bmatrix} \begin{bmatrix} x_1 \\ x_2 \end{bmatrix} = \begin{bmatrix} -10 \\ 19 \end{bmatrix}$

5. $\begin{bmatrix} 2 & -2 & -2 \\ 0 & -2 & 2 \\ -1 & 5 & 2 \end{bmatrix} \begin{bmatrix} x_1 \\ x_2 \\ x_3 \end{bmatrix} = \begin{bmatrix} -4 \\ -2 \\ 6 \end{bmatrix}$

6. $\begin{bmatrix} -3 & 12 & -6 \\ 1 & -2 & 2 \\ 0 & 1 & 1 \end{bmatrix} \begin{bmatrix} x_1 \\ x_2 \\ x_3 \end{bmatrix} = \begin{bmatrix} -33 \\ 7 \\ -1 \end{bmatrix}$

7. $\begin{bmatrix} 5 & 5 & 10 \\ -8 & -7 & -9 \\ 0 & 4 & 26 \end{bmatrix} \begin{bmatrix} x_1 \\ x_2 \\ x_3 \end{bmatrix} = \begin{bmatrix} 0 \\ 1 \\ 4 \end{bmatrix}$

8. $\begin{bmatrix} -1 & -3 & -4 \\ 3 & 10 & -10 \\ -2 & -4 & 11 \end{bmatrix} \begin{bmatrix} x_1 \\ x_2 \\ x_3 \end{bmatrix} = \begin{bmatrix} -6 \\ -3 \\ 9 \end{bmatrix}$

9. $\begin{bmatrix} -1 & 0 & 1 & 0 \\ 2 & 3 & -2 & 6 \\ 0 & -1 & 2 & 0 \\ 0 & 0 & 1 & 5 \end{bmatrix} \begin{bmatrix} x_1 \\ x_2 \\ x_3 \\ x_4 \end{bmatrix} = \begin{bmatrix} 5 \\ -1 \\ 3 \\ 7 \end{bmatrix}$

10. $\begin{bmatrix} 2 & -4 & 0 & 0 \\ 1 & 2 & 12 & 0 \\ 0 & -1 & -4 & -5 \\ 0 & 0 & 2 & 11 \end{bmatrix} \begin{bmatrix} x_1 \\ x_2 \\ x_3 \\ x_4 \end{bmatrix} = \begin{bmatrix} 8 \\ 0 \\ 1 \\ 0 \end{bmatrix}$

11. Let

$$A = \begin{bmatrix} 2 & 1 & -1 \\ -2 & -1 & 2 \\ 2 & 1 & 0 \end{bmatrix}$$

(a) Find an LU-decomposition of A.
(b) Express A in the form $A = L_1 D U_1$, where L_1 is lower triangular with 1's along the main diagonal, U_1 is upper triangular, and D is a diagonal matrix.
(c) Express A in the form $A = L_2 U_2$, where L_2 is lower triangular with 1's along the main diagonal and U_2 is upper triangular.

12. Show that the matrix

$$\begin{bmatrix} 0 & 1 \\ 1 & 0 \end{bmatrix}$$

has no LU-decomposition.

13. Let

$$A = \begin{bmatrix} a & b \\ c & d \end{bmatrix}$$

(a) Prove: If $a \neq 0$ then A has a unique LU-decomposition with 1's along the main diagonal of L.
(b) Find the LU-decomposition described in (a).

14. Let $A\mathbf{x} = \mathbf{b}$ be a linear system of n equations in n unknowns, and assume that A is an invertible matrix that can be reduced to row-echelon form without row interchanges. How many additions and multiplications are required to solve the system by the method of Example 1? [*Note.* Count subtractions as additions and divisions as multiplications.]

15. (a) Prove: If L_1 and L_2 are $n \times n$ lower triangular matrices, then so is L_1L_2.
    (b) The result in (a) is a special case of a general result that states that a product of finitely many lower triangular matrices is lower triangular. Use this fact to prove that the matrix L in (8.18) is lower triangular. [*Hint.* See Exercise 26 of Section 2.4.]

16. Use the result in Exercise 15(b) to prove that a product of finitely many upper triangular matrices is upper triangular. [*Hint.* Take transposes.]

17. Prove: If A is any $n \times n$ matrix, then A can be factored as $A = PLU$, where L is lower triangular, U is upper triangular, and P can be obtained by interchanging the rows of I_n appropriately. [*Hint.* Let U be a row-echelon form of A, and let all row interchanges required in the reduction of A to U be performed first.]

18. Factor

$$A = \begin{bmatrix} 3 & -1 & 0 \\ 3 & -1 & 1 \\ 0 & 2 & 1 \end{bmatrix}$$

as $A = PLU$, where P is obtained from I_3 by interchanging rows appropriately, L is lower triangular, and U is upper triangular.

8.3 THE GAUSS-SEIDEL AND JACOBI METHODS

Although Gaussian elimination (or the text version of Gauss-Jordan elimination) is generally the method of choice for solving a linear system of n equations in n unknowns, there are other approaches to solving linear systems, called *iterative* or *indirect methods*, which are better in certain situations. These methods start with an initial approximation to a solution and then generate a succession of better and better approximations that tend toward an exact solution. In this section we shall study two iterative methods and discuss situations in which they should be used.

We shall begin by discussing the simplest iterative method, which is called *Jacobi iteration* or the *method of simultaneous displacements*. This method applies to linear systems of n equations in n unknowns. Suppose that the system

$$\begin{aligned} a_{11}x_1 + a_{12}x_2 + \cdots + a_{1n}x_n &= b_1 \\ a_{21}x_1 + a_{22}x_2 + \cdots + a_{2n}x_n &= b_2 \\ \vdots \qquad \vdots \qquad\qquad \vdots \qquad \vdots \\ a_{n1}x_1 + a_{n2}x_2 + \cdots + a_{nn}x_n &= b_n \end{aligned} \qquad (8.19)$$

has exactly one solution and that the diagonal entries $a_{11}, a_{22}, \ldots, a_{nn}$ are nonzero.

To start, rewrite system (8.19) by solving the first equation for x_1 in terms of the remaining unknowns, solving the second equation for x_2 in terms of the remaining unknowns, solving the third equation for x_3 in terms of the remaining unknowns, and so on. This yields

$$x_1 = \frac{1}{a_{11}} (b_1 - a_{12}x_2 - a_{13}x_3 - \cdots - a_{1n}x_n)$$

$$x_2 = \frac{1}{a_{22}} (b_2 - a_{21}x_1 - a_{23}x_3 - \cdots - a_{2n}x_n)$$

$$\vdots$$

$$x_n = \frac{1}{a_{nn}} (b_n - a_{n1}x_1 - a_{n2}x_2 - \cdots - a_{nn-1}x_{n-1})$$

$$(8.20)$$

For example, the system

$$\begin{aligned} 20x_1 + \quad x_2 - \quad x_3 &= 17 \\ x_1 - 10x_2 + \quad x_3 &= 13 \\ -x_1 + \quad x_2 + 10x_3 &= 18 \end{aligned} \qquad (8.21)$$

would be rewritten

$$\begin{aligned} x_1 &= \tfrac{17}{20} - \tfrac{1}{20}x_2 + \tfrac{1}{20}x_3 \\ x_2 &= -\tfrac{13}{10} + \tfrac{1}{10}x_1 + \tfrac{1}{10}x_3 \\ x_3 &= \tfrac{18}{10} + \tfrac{1}{10}x_1 - \tfrac{1}{10}x_2 \end{aligned}$$

or

$$\begin{aligned} x_1 &= .850 - .05x_2 + .05x_3 \\ x_2 &= -1.3 + .1x_1 + .1x_3 \\ x_3 &= \quad 1.8 + .1x_1 - .1x_2 \end{aligned} \qquad (8.22)$$

If an approximation to the solution of (8.19) is known, and these approximate values are substituted into the right side of (8.20), it is often the case that the values of $x_1, x_2, \ldots, x_n$ that result on the left side form an even better approximation to the solution. This observation is the key to the Jacobi method.

To solve system (8.19) by Jacobi iteration, make an initial approximation to the solution. When no better choice is available use $x_1 = 0, x_2 = 0, x_3 = 0, \ldots$.

Substitute this initial approximation into the right side of (8.20) and use the values of $x_1, x_2, \ldots$ that result on the left side as a new approximation to the solution.

For example, to solve (8.21) by the Jacobi method, we would substitute the initial approximation $x_1 = 0$, $x_2 = 0$, $x_3 = 0$ into the right side of (8.22), and calculate the new approximation

$$x_1 = .850 \qquad x_2 = -1.3 \qquad x_3 = 1.8 \qquad (8.23)$$

To improve the approximation, we can repeat the substitution process. For example, in solving (8.21), we would substitute the approximation (8.23) into the right side of (8.22) to obtain the next approximation

$$x_1 = .850 - .05(-1.3) + .05(1.8) = 1.005$$
$$x_2 = -1.3 + .1(.850) + .1(1.8) = -1.035$$
$$x_3 = 1.8 + .1(.850) - .1(-1.3) = 2.015$$

In this way a succession of approximations can be generated that, under certain conditions, get closer and closer to the exact solution of the system. In Table 8.5 we have summarized the results obtained in solving system (8.21) by Jacobi iteration. All computations were rounded off to five significant digits. At the end of six substitutions (called *iterations*), the exact solution $x_1 = 1$, $x_2 = -1$, $x_3 = 2$ is accurately known to five significant digits.

TABLE 8.5

	Initial Approx- imation	First Approx- imation	Second Approx- imation	Third Approx- imation	Fourth Approx- imation	Fifth Approx- imation	Sixth Approx- imation
x_1	0	.850	1.005	1.0025	1.0001	.99997	1.0000
x_2	0	-1.3	-1.035	-.9980	-.99935	-.99999	-1.0000
x_3	0	1.8	2.015	2.004	2.0000	1.9999	2.0000

We now discuss a minor modification of the Jacobi method that often reduces the number of iterations needed to obtain a given degree of accuracy. The technique is called *Gauss-Seidel iteration* or the method of *successive displacements*.

In each iteration of the Jacobi method, the new approximation is obtained by substituting the previous approximation into the right side of (8.20) and solving for new values of $x_1, x_2, \ldots$. These new x-values are not all computed simultaneously; first x_1 is obtained from the top equation, then x_2 is obtained from the second equation, then x_3, and so on. Since the new x-values are generally closer to the exact solution, this suggests that better accuracy might be obtained by using the new x-values as soon as they are known. For example, consider system (8.21). In the first iteration of the Jacobi method, the initial approximation $x_1 = 0$, $x_2 = 0$, $x_3 = 0$ was substituted into each equation on the right side of (8.22) to obtain the new approximation

$$x_1 = .850 \qquad x_2 = -1.3 \qquad x_3 = 1.8 \qquad (8.24)$$

In the first iteration of the Gauss-Seidel method, the new approximation would be computed as follows. Substitute the initial approximation $x_1 = 0$, $x_2 = 0$, $x_3 = 0$ into the right side of the first equation in (8.22). This yields the new estimate $x_1 = .850$.

Use this new value of x_1 immediately by substituting

$$x_1 = .850 \qquad x_2 = 0 \qquad x_3 = 0$$

into the right side of the second equation in (8.22). This yields the new estimate $x_2 = -1.215$.

Use this new value of x_2 immediately by substituting

$$x_1 = .850 \qquad x_2 = -1.215 \qquad x_3 = 0$$

into the right side of the third equation in (8.22). This yields the new estimate $x_3 = 2.0065$.

Thus, at the end of the first iteration of the Gauss-Seidel method the new approximation is

$$x_1 = .850 \qquad x_2 = -1.215 \qquad x_3 = 2.0065 \qquad (8.25)$$

The computations for the second iteration would be carried out as follows.

Substituting (8.25) into the right side of the first equation in (8.22) and rounding off to five significant digits yields

$$x_1 = .850 - .05(-1.215) + .05(2.0065) = 1.0111$$

Substituting

$$x_1 = 1.0111 \qquad x_2 = -1.215 \qquad x_3 = 2.0065$$

into the right side of the second equation in (8.22) and rounding off to five significant digits yields

$$x_2 = -1.3 + .1(1.0111) + .1(2.0065) = -.99824$$

Substituting

$$x_1 = 1.0111 \qquad x_2 = -.99824 \qquad x_3 = 2.0065$$

into the right side of the third equation in (8.22) and rounding off to five significant digits yields

$$x_3 = 1.8 + .1(1.0111) - .1(-.99824) = 2.0009$$

Thus, at the end of the second iteration of the Gauss-Seidel method, the new approximation is

$$x_1 = 1.0111 \qquad x_2 = -.99824 \qquad x_3 = 2.0009$$

In Table 8.6 we have summarized the results obtained using four iterations of the Gauss-Seidel method to solve (8.21). All numbers were rounded off to five significant digits.

Comparing Tables 8.5 and 8.6, we see that the Gauss-Seidel method produces the solution of (8.21) (accurate to five significant digits) in four iterations, while six iterations are needed to attain the same accuracy with the Jacobi method.

Although it may seem surprising, it is possible to construct examples in which Jacobi iteration is better than Gauss-Seidel iteration. However, for most practical problems Gauss-Seidel iteration is a better choice.

TABLE 8.6

	Initial Approx- imation	First Approx- imation	Second Approx- imation	Third Approx- imation	Fourth Approx- imation
x_1	0	.850	1.0111	.99995	1.0000
x_2	0	-1.215	$-.99824$	$-.99992$	-1.0000
x_3	0	2.0065	2.0009	2.0000	2.0000

The Gauss-Seidel and Jacobi methods do not always work. In some cases, one or both of these methods can fail to produce a good approximation to the solution, regardless of the number of iterations performed. In such cases the approximations are said to *diverge*. However, if by performing sufficiently many iterations, the solution can be obtained to any desired degree of accuracy, the approximations are said to *converge*. We shall now discuss some conditions that ensure convergence.

A square matrix

$$A = \begin{bmatrix} a_{11} & a_{12} & \cdots & a_{1n} \\ a_{21} & a_{22} & \cdots & a_{2n} \\ \vdots & \vdots & & \vdots \\ a_{n1} & a_{n2} & \cdots & a_{nn} \end{bmatrix}$$

is called *strictly diagonally dominant* if the absolute value of each diagonal entry is greater than the sum of the absolute values of the remaining entries in the same row; that is,

$$|a_{11}| > |a_{12}| + |a_{13}| + \cdots + |a_{1n}|$$
$$|a_{22}| > |a_{21}| + |a_{23}| + \cdots + |a_{2n}|$$
$$\vdots \qquad \vdots \qquad \vdots \qquad \qquad \vdots$$
$$|a_{nn}| > |a_{n1}| + |a_{n2}| + \cdots + |a_{nn-1}|$$

Example 4

$$\begin{bmatrix} 7 & -2 & 3 \\ 4 & 1 & -6 \\ 5 & 12 & -4 \end{bmatrix}$$

is not strictly diagonally dominant since in the second row, $|1|$ is not greater than $|4| + |-6|$, and in the third row, $|-4|$ is not greater than $|5| + |12|$.

If the second and third rows are interchanged, however, the resulting matrix

$$\begin{bmatrix} 7 & -2 & 3 \\ 5 & 12 & -4 \\ 4 & 1 & -6 \end{bmatrix}$$

is strictly diagonally dominant since

$$|7| > |-2| + |3|$$
$$|12| > |5| + |-4|$$
$$|-6| > |4| + |1|$$

Two results on convergence of iterative methods are proved in some of the references at the end of this chapter.

1. if A is strictly diagonally dominant, then the Gauss-Seidel and Jacobi approximations to the solution of $A\mathbf{x} = \mathbf{b}$ both converge to the exact solution of the system for all choices of the initial approximation.
2. if A is a positive definite symmetric matrix, then the Gauss-Seidel approximations to the solution of $A\mathbf{x} = \mathbf{b}$ converge to the exact solution of the system for all choices of the initial approximation.

In practical problems we are concerned not only with the convergence of iterative methods but also with how fast they converge. For example, there are linear systems in which the Gauss-Seidel approximations converge, but the convergence is so slow that millions of iterations would be required to obtain any reasonable accuracy. Thus, numerical analysts have devised various methods to improve the convergence rate of Gauss-Seidel iteration. The most important of these methods is known as ***extrapolated Gauss-Seidel iteration*** or the ***method of successive over-relaxation*** (abbreviated as SOR in the literature.) Readers interested in this topic are referred to the references at the end of this chapter.

Faced with a linear system in practice, one must decide whether to solve the system by a direct method, such as Gaussian elimination, or by an indirect method such as Gauss-Seidel iteration. Here are some considerations that enter into making that choice:

1. When the coefficient matrix has a high proportion of zeros (such matrices are called *sparse*), iterative methods can be used to advantage because these zeros simplify the iteration equations, thereby reducing the amount of calculation. Gaussian elimination and Gauss-Jordan elimination generally destroy the sparseness by replacing zeros with nonzero entries, especially above the main diagonal.
2. It may happen that a good estimate of the solution is known. If this is used as a starting value in an iterative method, then there is a good chance of obtaining a satisfactory approximate solution with fewer computations than a direct method would require.
3. With clever programming, less computer memory is needed for iterative methods than direct methods. Thus, if memory space is a problem, iterative methods may be essential.
4. If the iterative methods diverge or the rate of convergence is too slow, direct methods may be essential.

EXERCISE SET 8.3

In Exercises 1–4 solve the systems by Jacobi iteration. Start with $x_1 = 0$, $x_2 = 0$. Use four iterations and round off the computations to three significant digits. Compare your results to the exact solutions.

1. $2x_1 + x_2 = 7$
$\quad x_1 - 2x_2 = 1$

2. $3x_1 - x_2 = 5$
$\quad 2x_1 + 3x_2 = -4$

3. $5x_1 - 2x_2 = -13$
$\quad x_1 + 7x_2 = -10$

4. $4x_1 + .1x_2 = .2$
$\quad .3x_1 + .7x_2 = 1.4$

In Exercises 5–8 solve the systems by Gauss-Seidel iteration. Start with $x_1 = 0$, $x_2 = 0$. Use three iterations and round off the computations to three significant digits. Compare your results to the exact solutions.

5. Solve the system in Exercise 1.

6. Solve the system in Exercise 2.

7. Solve the system in Exercise 3.

8. Solve the system in Exercise 4.

In Exercises 9 and 10 solve the systems by Jacobi iteration. Start with $x_1 = 0$, $x_2 = 0$, $x_3 = 0$. Use three iterations and round off the computations to three significant digits. Compare your results to the exact solutions.

9. $10x_1 + x_2 + 2x_3 = 3$
$\quad x_1 + 10x_2 - x_3 = \frac{3}{2}$
$\quad 2x_1 + x_2 + 10x_3 = -9$

10. $20x_1 - x_2 + x_3 = 20$
$\quad 2x_1 + 10x_2 - x_3 = 11$
$\quad x_1 + x_2 - 20x_3 = -18$

In Exercises 11 and 12 solve the systems by Gauss-Seidel iteration. Start with $x_1 = 0$, $x_2 = 0$, $x_3 = 0$. Use three iterations and round off the computations to three significant digits. Compare your results to the exact solutions.

11. Solve the system in Exercise 9.

12. Solve the system in Exercise 10.

13. Which of the following are strictly diagonally dominant?

(a) $\begin{bmatrix} 2 & 1 \\ -1 & 4 \end{bmatrix}$ (b) $\begin{bmatrix} 3 & -5 \\ 1 & 2 \end{bmatrix}$

(c) $\begin{bmatrix} 6 & 0 & 1 \\ 3 & 5 & 3 \\ 0 & 0 & 1 \end{bmatrix}$ (d) $\begin{bmatrix} 4 & 1 & 2 \\ 0 & 3 & 2 \\ 4 & 1 & -7 \end{bmatrix}$ (e) $\begin{bmatrix} 5 & 1 & 2 & 0 \\ 3 & -7 & 2 & 1 \\ 0 & 2 & 5 & 1 \\ 1 & 1 & 2 & -5 \end{bmatrix}$

14. Consider the system

$$x_1 + 3x_2 = 4$$
$$x_1 - x_2 = 0$$

(a) Show that the approximations obtained by Jacobi iteration diverge.

(b) Is the coefficient matrix

$$A = \begin{bmatrix} 1 & 3 \\ 1 & -1 \end{bmatrix}$$

strictly diagonally dominant?

15. Show that if one or more of the diagonal entries $a_{11}, a_{22}, \ldots, a_{nn}$ in (8.19) is zero, then it is possible to interchange equations and relabel the unknowns so that the diagonal entries in the resulting system are all nonzero.

8.4 PARTIAL PIVOTING; REDUCTION OF ROUNDOFF ERROR

We noted in Section 8.1 that minimizing roundoff error is a primary consideration when solving linear systems on a computer. In this section we shall discuss a procedure, called *partial pivoting* (or *pivotal condensation*), which is designed to minimize the cumulative effect of roundoff error in solving linear systems.

Most computer arithmetic is performed using **normalized floating point numbers**. This means the numbers are expressed in the form*

$$\pm M \times 10^k \tag{8.26}$$

where k is an integer and M is a fraction that satisfies

$$.1 \leq M < 1$$

The fraction M is called the **mantissa**.

Example 5

The following numbers are expressed in normalized floating point form.

$$73 = .73 \times 10^2$$
$$-.000152 = -.152 \times 10^{-3}$$
$$1{,}579 = .1579 \times 10^4$$
$$-1/4 = -.25 \times 10^0$$

The number of decimal places in the mantissa and the allowable size of the exponent k in (8.26) depend on the computer being used. For example, the INTEL 8087 coprocessor chip stores the equivalent of fifteen decimal digits in the mantissa and allows 10^k to range from 10^{-308} to 10^{308}. A computer that uses n decimal places in the mantissa is said to round off numbers to **n significant digits**.

---

* Most computers convert decimal numbers (base 10) to binary numbers (base 2). For simplicity, however, we shall think in terms of decimals.

Example 6

The following numbers are rounded off to three significant digits.

Number	Normalized Floating Point Form	Rounded Value
7/3	$.233 \times 10^1$	2.33
1,758	$.176 \times 10^4$	1,760
.0000092143	$.921 \times 10^{-5}$	.00000921
−.12	$-.120 \times 10^0$	−.12
13.850	$.138 \times 10^2$	13.8
−.08495	$-.850 \times 10^{-1}$	−.085

(If, as in the last two cases, the portion of decimal to be discarded in the rounding process is exactly half a unit, we shall adopt the convention of rounding so that the last retained digit is even. In practice, the treatment of this situation varies from computer to computer.) ▲

We shall now introduce a technique, called *partial pivoting* (or *pivotal condensation*), which can be used to reduce roundoff error in Gaussian elimination, Gauss-Jordan elimination, or any other computational procedure that uses elementary row operations. To explain the method we shall describe its steps and illustrate them by solving the system

$$3x_1 + 2x_2 - x_3 = 1$$
$$6x_1 + 6x_2 + 2x_3 = 12$$
$$3x_1 - 2x_2 + x_3 = 11$$

using Gaussian elimination with partial pivoting applied to the augmented matrix.

Step 1.   In the leftmost column find an entry that has the largest absolute value. This is called the *pivot entry*.

$$\begin{bmatrix} 3 & 2 & -1 & 1 \\ ⑥ & 6 & 2 & 12 \\ 3 & -2 & 1 & 11 \end{bmatrix}$$

Pivot entry

Leftmost column

Step 2.   Perform a row interchange, if necessary, to bring the pivot entry to the top of the column.

$$\begin{bmatrix} 6 & 6 & 2 & 12 \\ 3 & 2 & -1 & 1 \\ 3 & -2 & 1 & 11 \end{bmatrix}$$

The first and second rows of the previous matrix were interchanged.

Step 3. If the pivot entry is a, multiply the top row by $1/a$.

$$\begin{bmatrix} 1 & 1 & \frac{1}{3} & 2 \\ 3 & 2 & -1 & 1 \\ 3 & -2 & 1 & 11 \end{bmatrix}$$

> The first row of the previous matrix was multiplied by 1/6.

Step 4. Add suitable multiples of the top row to the rows below so that in the column located in Step 1, all the entries below the top become zeros.

$$\begin{bmatrix} 1 & 1 & \frac{1}{3} & 2 \\ 0 & -1 & -2 & -5 \\ 0 & -5 & 0 & 5 \end{bmatrix}$$

> -3 times the first row of the previous matrix was added to the second and third rows.

Step 5. Cover the top row in the matrix and begin again with Step 1, applied to the submatrix that remains. Continue in this way until the *entire* matrix is in row-echelon form.

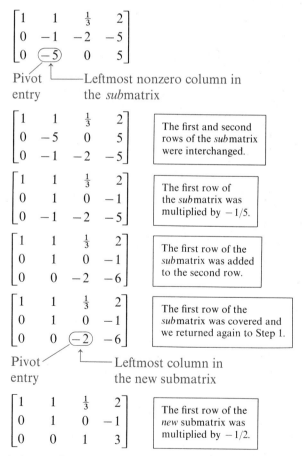

The entire matrix is now in row-echelon form.

Step 6.   Solve the corresponding system of equations by back-substitution. The corresponding system of equations is

$$x_1 + x_2 + \tfrac{1}{3}x_3 = 2$$
$$x_2 \phantom{+ \tfrac{1}{3}x_3} = -1$$
$$x_3 = 3$$

Solving by back-substitution yields

$$x_3 = 3 \qquad x_2 = -1 \qquad x_1 = 2$$

Since the above computations are exact, this example does not illustrate the effectiveness of partial pivoting in reducing rounding error; the next example does.

Example 7

Solve the following system by Gaussian elimination with partial pivoting. After each calculation round off the result to three significant digits.

$$.00044x_1 + .0003x_2 - .0001x_3 = .00046$$
$$4x_1 + x_2 + x_3 = 1.5 \qquad\qquad (8.27)$$
$$3x_1 - 9.2x_2 - .5x_3 = -8.2$$

Solution (with partial pivoting). The augmented matrix is

$$\begin{bmatrix} .00044 & .0003 & -.0001 & .00046 \\ 4 & 1 & 1 & 1.5 \\ 3 & -9.2 & -.5 & -8.2 \end{bmatrix}$$

To bring the pivot entry to the top of the first column, we interchange the first and second rows; this yields

$$\begin{bmatrix} 4 & 1 & 1 & 1.5 \\ .00044 & .0003 & -.0001 & .00046 \\ 3 & -9.2 & -.5 & -8.2 \end{bmatrix}$$

Dividing each entry in the first row by 4 yields

$$\begin{bmatrix} 1 & .25 & .25 & .375 \\ .00044 & .0003 & -.0001 & .00046 \\ 3 & -9.2 & -.5 & -8.2 \end{bmatrix}$$

Adding $-.00044$ times the first row to the second and -3 times the first row to the third yields (after rounding to three significant digits)

$$\begin{bmatrix} 1 & .25 & .25 & .375 \\ 0 & .000190 & -.00021 & .000295 \\ 0 & -9.95 & -1.25 & -9.32 \end{bmatrix}$$

Interchanging the second and third rows yields

$$\begin{bmatrix} 1 & .25 & .25 & .375 \\ 0 & -9.95 & -1.25 & -9.32 \\ 0 & .000190 & -.00021 & .000295 \end{bmatrix}$$

Dividing each entry in the second row by -9.95 yields

$$\begin{bmatrix} 1 & .25 & .25 & .375 \\ 0 & 1 & .126 & .937 \\ 0 & .000190 & -.00021 & .000295 \end{bmatrix}$$

Adding $-.000190$ times the second row to the third yields

$$\begin{bmatrix} 1 & .25 & .25 & .375 \\ 0 & 1 & .126 & .937 \\ 0 & 0 & -.000234 & .000117 \end{bmatrix}$$

Dividing each entry in the third row by $-.000234$ yields the row-echelon form

$$\begin{bmatrix} 1 & .25 & .25 & .375 \\ 0 & 1 & .126 & .937 \\ 0 & 0 & 1 & -.5 \end{bmatrix}$$

The corresponding system of equations is

$$x_1 + .25x_2 + .25x_3 = .375$$
$$x_2 + .126x_3 = .937$$
$$x_3 = -.5$$

Solving by back-substitution yields (to three significant digits)

$$x_1 = .250 \qquad x_2 = 1.00 \qquad x_3 = -.500 \qquad (8.28)$$

If (8.27) is solved by Gaussian elimination *without* partial pivoting and each calculation is rounded to three significant digits, one obtains (details omitted)

$$x_1 = .245 \qquad x_2 = 1.01 \qquad x_3 = -.492 \qquad (8.29)$$

Comparing (8.28) and (8.29) to the exact solution

$$x_1 = \tfrac{1}{4} \qquad x_2 = 1 \qquad x_3 = -\tfrac{1}{2}$$

shows that partial pivoting yields more accurate results. ▲

In spite of the fact that partial pivoting can reduce the cumulative effect of rounding error, there are certain systems of equations, called ***ill-conditioned*** systems, that are so extremely sensitive that even the slightest errors in the coefficients can result in major inaccuracies in the solution. For example, consider the system

$$
\begin{aligned}
x_1 + \quad x_2 &= -3 \\
x_1 + 1.016x_2 &= \quad 5
\end{aligned}
\tag{8.30}
$$

If we assume that this system is to be solved on a computer that rounds off to three significant digits, the computer will store this system as

$$
\begin{aligned}
x_1 + \quad x_2 &= -3 \\
x_1 + 1.02x_2 &= \quad 5
\end{aligned}
\tag{8.31}
$$

The exact solution to (8.30) is $x_1 = -503$, $x_2 = 500$, and the exact solution to (8.31) is $x_1 = -403$, $x_2 = 400$. Thus, a rounding error of only .004 in one coefficient of (8.30) results in a gross error in the solution.

There is little that can be done computationally to avoid large errors in the solutions of ill-conditioned linear systems. However, in physical problems, where ill-conditioned systems arise, it is sometimes possible to reformulate the problem giving rise to the system to avoid ill-conditioning. Some of the texts referenced at the end of this chapter explain how to recognize ill-conditioned systems.

In this section we have explained how to reduce roundoff error by partial pivoting, but have not explained why the method works. Suffice it to say that when a quantity has some error in it resulting from roundoff, division of that quantity by a number near zero amplifies the roundoff error—the closer the divisor is to zero, the greater the amplification. Partial pivoting is designed to keep all divisors in the reduction process as far from zero as possible.

Finally, we note that there is a procedure, called *complete pivoting*, which uses both row and column interchanges to reduce roundoff error. Although this method is more effective than partial pivoting in reducing roundoff error, it is so time consuming that its advantage is often lost. For most purposes partial pivoting is adequate.

EXERCISE SET 8.4

1. Express the following in normalized floating point form.
(a) $\frac{14}{5}$ (b) 3,452   (c) .000003879
(d) $-.135$ (e) 17.921   (f) $-.0863$

2. Round off the numbers in Exercise 1 to three significant digits.

3. Round off the numbers in Exercise 1 to two significant digits.

In Exercises 4–7 use Gaussian elimination with partial pivoting to solve the system exactly. Check your work by using Gaussian elimination without partial pivoting to solve the system.

4.
$$
\begin{aligned}
3x_1 + x_2 &= -2 \\
-5x_1 + x_2 &= 22
\end{aligned}
$$
     5.
$$
\begin{aligned}
x_1 + \quad x_2 + \quad x_3 &= 6 \\
2x_1 - \quad x_2 + 4x_3 &= 12 \\
-3x_1 + 2x_2 - \quad x_3 &= -4
\end{aligned}
$$

6. $2x_1 + 3x_2 - x_3 = 5$
$4x_1 + 4x_2 - 3x_3 = 3$
$2x_1 - 3x_2 + x_3 = -1$

7. $5x_1 + 6x_2 - x_3 + 2x_4 = -3$
$2x_1 - x_2 + x_3 + x_4 = 0$
$-8x_1 + x_2 + 2x_3 - x_4 = 3$
$5x_1 + 2x_2 + 3x_3 - x_4 = 4$

In Exercises 8–9 solve the system by Gaussian elimination with partial pivoting. Round off all calculations to three significant digits.

8. $.21x_1 + .33x_2 = .54$
$.70x_1 + .24x_2 = .94$

9. $.11x_1 - .13x_2 + .20x_3 = -.02$
$.10x_1 + .36x_2 + .45x_3 = .25$
$.50x_1 - .01x_2 + .30x_3 = -.70$

10. Solve

$$.0001x_1 + x_2 = 1$$

$$x_1 + x_2 = 2$$

by Gaussian elimination with and without partial pivoting. Round off all computations to three significant digits. Compare the results with the exact solution.

8.5 APPROXIMATING EIGENVALUES BY THE POWER METHOD

We learned earlier that the eigenvalues of a square matrix can be found by solving its characteristic equation. In practical problems, especially those involving large matrices, that procedure for finding eigenvalues has many computational difficulties, so other methods for finding eigenvalues are used. In this section we shall study a simple algorithm, called the *power method*, that produces an approximation to the eigenvalue with greatest absolute value and a corresponding eigenvector. In the next section we will discuss methods for approximating the remaining eigenvalues and eigenvectors.

Definition. An eigenvalue of a matrix A is called the *dominant eigenvalue* of A if its absolute value is larger than the absolute values of the remaining eigenvalues. An eigenvector corresponding to the dominant eigenvalue is called a *dominant eigenvector* of A.

Example 8

If a 4×4 matrix A has eigenvalues

$$\lambda_1 = -4 \qquad \lambda_2 = 3 \qquad \lambda_3 = -2 \qquad \lambda_4 = 2$$

then $\lambda_1 = -4$ is the dominant eigenvalue since

$$|-4| > |3| \qquad |-4| > |-2| \qquad \text{and} \qquad |-4| > |2|$$

Example 9

A 3×3 matrix A with eigenvalues

$$\lambda_1 = 7 \qquad \lambda_2 = -7 \qquad \lambda_3 = 2$$

has no dominant eigenvalue.

Let A be a diagonalizable $n \times n$ matrix with a dominant eigenvalue. We shall show at the end of this section that if $\mathbf{x}_0$ is an arbitrary nonzero vector in R^n, then the vector

$$A^p \mathbf{x}_0 \tag{8.32}$$

is usually a good approximation to a dominant eigenvector of A when the exponent p is large. The following example illustrates this idea.

Example 10

As shown in Example 2 of Chapter 6, the matrix

$$A = \begin{bmatrix} 3 & 2 \\ -1 & 0 \end{bmatrix}$$

has eigenvalues $\lambda_1 = 2$ and $\lambda_2 = 1$.

The eigenspace corresponding to the dominant eigenvalue $\lambda_1 = 2$ is the solution space of the system

$$(2I - A)\mathbf{x} = \mathbf{0}$$

That is,

$$\begin{bmatrix} -1 & -2 \\ 1 & 2 \end{bmatrix} \begin{bmatrix} x_1 \\ x_2 \end{bmatrix} = \begin{bmatrix} 0 \\ 0 \end{bmatrix}$$

Solving this system yields $x_1 = -2t$, $x_2 = t$. Thus, the eigenvectors corresponding to $\lambda_1 = 2$ are the nonzero vectors of the form

$$\mathbf{x} = \begin{bmatrix} -2t \\ t \end{bmatrix} \tag{8.33}$$

We now illustrate a procedure for using (8.32) to estimate a dominant eigenvector of A. To start, we arbitrarily select

$$\mathbf{x}_0 = \begin{bmatrix} 1 \\ 1 \end{bmatrix}$$

as an initial approximation to a dominant eigenvector. Repeatedly multiplying $\mathbf{x}_0$ by A yields

$$A\mathbf{x}_0 = \begin{bmatrix} 3 & 2 \\ -1 & 0 \end{bmatrix}\begin{bmatrix} 1 \\ 1 \end{bmatrix} = \begin{bmatrix} 5 \\ -1 \end{bmatrix}$$

$$A^2\mathbf{x}_0 = A(A\mathbf{x}_0) = \begin{bmatrix} 3 & 2 \\ -1 & 0 \end{bmatrix}\begin{bmatrix} 5 \\ -1 \end{bmatrix} = \begin{bmatrix} 13 \\ -5 \end{bmatrix} = 5\begin{bmatrix} 2.6 \\ -1 \end{bmatrix}$$

$$A^3\mathbf{x}_0 = A(A^2\mathbf{x}_0) = \begin{bmatrix} 3 & 2 \\ -1 & 0 \end{bmatrix}\begin{bmatrix} 13 \\ -5 \end{bmatrix} = \begin{bmatrix} 29 \\ -13 \end{bmatrix} \approx 13\begin{bmatrix} 2.23 \\ -1 \end{bmatrix}$$

$$A^4\mathbf{x}_0 = A(A^3\mathbf{x}_0) = \begin{bmatrix} 3 & 2 \\ -1 & 0 \end{bmatrix}\begin{bmatrix} 29 \\ -13 \end{bmatrix} = \begin{bmatrix} 61 \\ -29 \end{bmatrix} \approx 29\begin{bmatrix} 2.10 \\ -1 \end{bmatrix}$$

$$A^5\mathbf{x}_0 = A(A^4\mathbf{x}_0) = \begin{bmatrix} 3 & 2 \\ -1 & 0 \end{bmatrix}\begin{bmatrix} 61 \\ -29 \end{bmatrix} = \begin{bmatrix} 125 \\ -61 \end{bmatrix} \approx 61\begin{bmatrix} 2.05 \\ -1 \end{bmatrix}$$

$$A^6\mathbf{x}_0 = A(A^5\mathbf{x}_0) = \begin{bmatrix} 3 & 2 \\ -1 & 0 \end{bmatrix}\begin{bmatrix} 125 \\ -61 \end{bmatrix} = \begin{bmatrix} 253 \\ -125 \end{bmatrix} \approx 125\begin{bmatrix} 2.02 \\ -1 \end{bmatrix}$$

$$A^7\mathbf{x}_0 = A(A^6\mathbf{x}_0) = \begin{bmatrix} 3 & 2 \\ -1 & 0 \end{bmatrix}\begin{bmatrix} 253 \\ -125 \end{bmatrix} = \begin{bmatrix} 509 \\ -253 \end{bmatrix} \approx 253\begin{bmatrix} 2.01 \\ -1 \end{bmatrix}$$

It is evident from these calculations that the products are getting closer and closer to scalar multiples of

$$\begin{bmatrix} 2 \\ -1 \end{bmatrix}$$

which is the dominant eigenvector of A obtained by letting $t = -1$ in (8.33). Since a scalar multiple of a dominant eigenvector is also a dominant eigenvector, the above calculations are producing better and better approximations to a dominant eigenvector of A. ▲

We now show how to approximate the dominant eigenvalue once an approximation to a dominant eigenvector is known. Let λ be an eigenvalue of A and $\mathbf{x}$ a corresponding eigenvector. If $\langle\ ,\ \rangle$ denotes the Euclidean inner product, then

$$\frac{\langle \mathbf{x}, A\mathbf{x} \rangle}{\langle \mathbf{x}, \mathbf{x} \rangle} = \frac{\langle \mathbf{x}, \lambda\mathbf{x} \rangle}{\langle \mathbf{x}, \mathbf{x} \rangle} = \frac{\lambda\langle \mathbf{x}, \mathbf{x} \rangle}{\langle \mathbf{x}, \mathbf{x} \rangle} = \lambda$$

Thus, if $\tilde{\mathbf{x}}$ is an approximation to a dominant eigenvector, the dominant eigenvalue λ_1 can be approximated by

$$\lambda_1 \approx \frac{\langle \tilde{\mathbf{x}}, A\tilde{\mathbf{x}} \rangle}{\langle \tilde{\mathbf{x}}, \tilde{\mathbf{x}} \rangle} \tag{8.34}$$

The ratio in (8.34) is called the **Rayleigh quotient.***

---

* *John William Strutt Rayleigh* (1842–1919) was a British physicist awarded the Nobel prize in physics in 1904 for his part in the discovery of argon in 1894. His research ranged over almost the entire field of physics, including sound, wave theory, optics, color vision, electrodynamics, electromagnetism, scattering of light, viscosity, and photography.

Example 11

In Example 10 we obtained

$$\tilde{\mathbf{x}} = \begin{bmatrix} 509 \\ -253 \end{bmatrix}$$

as an approximation to a dominant eigenvector corresponding to the dominant eigenvalue $\lambda_1 = 2$. Substituting,

$$A\tilde{\mathbf{x}} = \begin{bmatrix} 3 & 2 \\ -1 & 0 \end{bmatrix} \begin{bmatrix} 509 \\ -253 \end{bmatrix} = \begin{bmatrix} 1021 \\ -509 \end{bmatrix}$$

in (8.34) we obtain

$$\lambda_1 \approx \frac{\langle \tilde{\mathbf{x}}, A\tilde{\mathbf{x}} \rangle}{\langle \tilde{\mathbf{x}}, \tilde{\mathbf{x}} \rangle} = \frac{(509)(1021) + (-253)(-509)}{(509)(509) + (-253)(-253)} \approx 2.007$$

which is a relatively good approximation to the dominant eigenvalue $\lambda_1 = 2$.

The technique illustrated in Examples 10 and 11 for approximating dominant eigenvalues and eigenvectors is often called the **power method** or **iteration method**. As illustrated in Example 10, the power method often generates vectors that have inconveniently large components. To remedy this problem it is usual to "scale down" the approximate eigenvector at each step so its components lie between $+1$ and -1. This can be achieved by multiplying the approximate eigenvector by the reciprocal of the component having largest absolute value. To illustrate, in the first step of Example 10, the approximation to a dominant eigenvector is

$$\begin{bmatrix} 5 \\ -1 \end{bmatrix}$$

The component with largest absolute value is 5; thus, the scaled-down eigenvector is

$$\frac{1}{5} \begin{bmatrix} 5 \\ -1 \end{bmatrix} = \begin{bmatrix} 1 \\ -.2 \end{bmatrix}$$

We now summarize the steps in the power method with scaling.

Step 0. Pick an arbitrary nonzero vector $\mathbf{x}_0$.

Step 1. Compute $A\mathbf{x}_0$ and scale down to obtain the first approximation to a dominant eigenvector. Call it $\mathbf{x}_1$.

Step 2. Compute $A\mathbf{x}_1$ and scale down to obtain the second approximation, $\mathbf{x}_2$.

Step 3. Compute $A\mathbf{x}_2$ and scale down to obtain the third approximation, $\mathbf{x}_3$.

Continuing in this way, a succession, $\mathbf{x}_0, \mathbf{x}_1, \mathbf{x}_2, \ldots$, of better and better approximations to a dominant eigenvector will be obtained.

Example 12

Use the power method with scaling to approximate a dominant eigenvector and the dominant eigenvalue of the matrix A in Example 10.

Solution. We arbitrarily select

$$\mathbf{x}_0 = \begin{bmatrix} 1 \\ 1 \end{bmatrix}$$

as an initial approximation. Multiplying $\mathbf{x}_0$ by A and scaling down yields

$$A\mathbf{x}_0 = \begin{bmatrix} 3 & 2 \\ -1 & 0 \end{bmatrix}\begin{bmatrix} 1 \\ 1 \end{bmatrix} = \begin{bmatrix} 5 \\ -1 \end{bmatrix} \qquad \mathbf{x}_1 = \frac{1}{5}\begin{bmatrix} 5 \\ -1 \end{bmatrix} = \begin{bmatrix} 1 \\ -.2 \end{bmatrix}$$

Multiplying $\mathbf{x}_1$ by A and scaling down yields

$$A\mathbf{x}_1 = \begin{bmatrix} 3 & 2 \\ -1 & 0 \end{bmatrix}\begin{bmatrix} 1 \\ -.2 \end{bmatrix} = \begin{bmatrix} 2.6 \\ -1 \end{bmatrix} \qquad \mathbf{x}_2 = \frac{1}{2.6}\begin{bmatrix} 2.6 \\ -1 \end{bmatrix} = \begin{bmatrix} 1 \\ -.385 \end{bmatrix}$$

From the Rayleigh quotient the first estimate of the dominant eigenvalue is

$$\lambda_1 \approx \frac{\langle \mathbf{x}_1, A\mathbf{x}_1 \rangle}{\langle \mathbf{x}_1, \mathbf{x}_1 \rangle} = \frac{(1)(2.6) + (-.2)(-1)}{(1)(1) + (-.2)(-.2)} = 2.692$$

Multiplying $\mathbf{x}_2$ by A and scaling down yields

$$A\mathbf{x}_2 = \begin{bmatrix} 3 & 2 \\ -1 & 0 \end{bmatrix}\begin{bmatrix} 1 \\ -.385 \end{bmatrix} = \begin{bmatrix} 2.23 \\ -1 \end{bmatrix} \qquad \mathbf{x}_3 = \frac{1}{2.23}\begin{bmatrix} 2.23 \\ -1 \end{bmatrix} \approx \begin{bmatrix} 1 \\ -.448 \end{bmatrix}$$

From the Rayleigh quotient the second estimate of the dominant eigenvalue is

$$\lambda_1 \approx \frac{\langle \mathbf{x}_2, A\mathbf{x}_2 \rangle}{\langle \mathbf{x}_2, \mathbf{x}_2 \rangle} = \frac{(1)(2.23) + (-.385)(-1)}{(1)(1) + (-.385)(-.385)} = 2.278$$

Multiplying $\mathbf{x}_3$ by A and scaling down gives

$$A\mathbf{x}_3 = \begin{bmatrix} 3 & 2 \\ -1 & 0 \end{bmatrix}\begin{bmatrix} 1 \\ -.448 \end{bmatrix} = \begin{bmatrix} 2.104 \\ -1 \end{bmatrix} \qquad \mathbf{x}_4 = \frac{1}{2.104}\begin{bmatrix} 2.104 \\ -1 \end{bmatrix} = \begin{bmatrix} 1 \\ -.475 \end{bmatrix}$$

The third estimate of the dominant eigenvalue is

$$\lambda_1 \approx \frac{\langle \mathbf{x}_3, A\mathbf{x}_3 \rangle}{\langle \mathbf{x}_3, \mathbf{x}_3 \rangle} = \frac{(1)(2.104) + (-.448)(-1)}{(1)(1) + (-.448)(-.448)} = 2.125$$

Continuing in this way, we generate a succession of approximations to a dominant eigenvector and the dominant eigenvalue. These values, together with results of further estimates, are tabulated in Table 8.7.

TABLE 8.7

Step i	0	1	2	3	4	5	6	7
$\mathbf{x}_i$ = the scaled down approximation to a dominant eigenvector	$\begin{bmatrix}1\\1\end{bmatrix}$	$\begin{bmatrix}1\\-.2\end{bmatrix}$	$\begin{bmatrix}1\\-.385\end{bmatrix}$	$\begin{bmatrix}1\\-.488\end{bmatrix}$	$\begin{bmatrix}1\\-.475\end{bmatrix}$	$\begin{bmatrix}1\\-.488\end{bmatrix}$	$\begin{bmatrix}1\\-.494\end{bmatrix}$	$\begin{bmatrix}1\\-.497\end{bmatrix}$
$A\mathbf{x}_i$		$\begin{bmatrix}5\\-1\end{bmatrix}$	$\begin{bmatrix}2.6\\-1\end{bmatrix}$	$\begin{bmatrix}2.23\\-1\end{bmatrix}$	$\begin{bmatrix}2.104\\-1\end{bmatrix}$	$\begin{bmatrix}2.050\\-1\end{bmatrix}$	$\begin{bmatrix}2.024\\-1\end{bmatrix}$	$\begin{bmatrix}2.012\\-1\end{bmatrix}$	—
Approximation to λ_1	—	2.692	2.278	2.125	2.060	2.029	2.014	—

There are no hard and fast rules for determining how many steps to use in the power method. We shall consider one possible procedure that is widely used. If $\tilde{q}$ denotes an approximation to a quantity q, then the *relative error* in the approximation is defined to be

$$\left|\frac{q - \tilde{q}}{q}\right| \tag{8.35}$$

and the *percentage error* in the approximation is defined to be

$$\left|\frac{q - \tilde{q}}{q}\right| \times 100\%$$

Example 13
If the exact value of a certain eigenvalue is $\lambda = 5$, and if $\tilde{\lambda} = 5.1$ is an approximation to λ, then the relative error is

$$\left|\frac{\lambda - \tilde{\lambda}}{\lambda}\right| = \left|\frac{5 - 5.1}{5}\right| = |-.02| = .02$$

and the percentage error is

$$(.02) \times 100\% = 2\%$$

In the power method one would ideally like to decide in advance the relative error E that can be tolerated in the eigenvalue, and then stop the computations once the relative error is less than E. Thus, if $\tilde{\lambda}(i)$ denotes the approximation to the dominant eigenvalue λ_1 at the ith step, the computations would be stopped once the condition

$$\left|\frac{\lambda_1 - \tilde{\lambda}(i)}{\lambda_1}\right| < E$$

is satisfied. Unfortunately, it is not possible to carry out this idea since the exact value λ_1 is unknown. To remedy this, it is usual to estimate λ_1 by $\tilde{\lambda}(i)$ and stop the computations at the ith step if

$$\left| \frac{\tilde{\lambda}(i) - \tilde{\lambda}(i-1)}{\tilde{\lambda}(i)} \right| < E \tag{8.36}$$

The quantity on the left side of (8.36) is called the **estimated relative error**. When multiplied by 100%, it is called the **estimated percentage error**.

Example 14

In Example 12, how many steps should be used to assure that the estimated percentage error in the dominant eigenvalue is less than 2%?

Solution. Let $\tilde{\lambda}(i)$ denote the approximation to the dominant eigenvalue at the ith step. Thus, from Table 8.7

$$\tilde{\lambda}(1) = 2.692, \qquad \tilde{\lambda}(2) = 2.278, \qquad \tilde{\lambda}(3) = 2.125, \quad \text{etc.}$$

From (8.36) the estimated relative error after two steps is

$$\left| \frac{\tilde{\lambda}(2) - \tilde{\lambda}(1)}{\tilde{\lambda}(2)} \right| = \left| \frac{2.278 - 2.692}{2.278} \right| \approx |-.182| = .182$$

so the estimated percentage error after two steps is 18.2%. The estimated relative error after three steps is

$$\left| \frac{\tilde{\lambda}(3) - \tilde{\lambda}(2)}{\tilde{\lambda}(3)} \right| = \left| \frac{2.125 - 2.278}{2.125} \right| \approx |-.072| = .072$$

and the estimated percentage error is 7.2%. The remaining percentage errors are listed in Table 8.8. From this table we see that the estimated percentage error is less than 2% at the end of the fifth step. ▲

TABLE 8.8

i = step number	2	3	4	5	6
$\tilde{\lambda}(i)$	2.278	2.125	2.060	2.029	2.014
Estimated relative error after i steps	.182	.072	.032	.015	.007
Estimated percentage error after i steps	18.2%	7.2%	3.2%	1.5%	.7%

OPTIONAL

We conclude this section with a proof that the power method works when A is a diagonalizable matrix with a dominant eigenvalue.

Let A be a diagonalizable $n \times n$ matrix. By Theorem 2 in Section 6.2, A has n linearly independent eigenvectors $\mathbf{v}_1, \mathbf{v}_2, \ldots, \mathbf{v}_n$. Let $\lambda_1, \lambda_2, \ldots, \lambda_n$ be the corresponding eigenvalues, and assume

$$|\lambda_1| > |\lambda_2| \geq \cdots \geq |\lambda_n| \tag{8.37}$$

By Theorem 11a in Section 4.5 the eigenvectors $\mathbf{v}_1, \mathbf{v}_2, \ldots, \mathbf{v}_n$ form a basis for R^n; thus, an arbitrary vector $\mathbf{x}_0$ in R^n can be expressed in the form

$$\mathbf{x}_0 = k_1\mathbf{v}_1 + k_2\mathbf{v}_2 + \cdots + k_n\mathbf{v}_n \tag{8.38}$$

Multiplying both sides on the left by A gives

$$\begin{aligned} A\mathbf{x}_0 &= A(k_1\mathbf{v}_1 + k_2\mathbf{v}_2 + \cdots + k_n\mathbf{v}_n) \\ &= k_1(A\mathbf{v}_1) + k_2(A\mathbf{v}_2) + \cdots + k_n(A\mathbf{v}_n) \\ &= k_1\lambda_1\mathbf{v}_1 + k_2\lambda_2\mathbf{v}_2 + \cdots + k_n\lambda_n\mathbf{v}_n \end{aligned}$$

Multiplying by A again gives

$$\begin{aligned} A^2\mathbf{x}_0 &= A(k_1\lambda_1\mathbf{v}_1 + k_2\lambda_2\mathbf{v}_2 + \cdots + k_n\lambda_n\mathbf{v}_n) \\ &= k_1\lambda_1(A\mathbf{v}_1) + k_2\lambda_2(A\mathbf{v}_2) + \cdots + k_n\lambda_n(A\mathbf{v}_n) \\ &= k_1\lambda_1^2\mathbf{v}_1 + k_2\lambda_2^2\mathbf{v}_2 + \cdots + k_n\lambda_n^2\mathbf{v}_n \end{aligned}$$

Continuing, we would obtain after p multiplications by A

$$A^p\mathbf{x}_0 = k_1\lambda_1^p\mathbf{v}_1 + k_2\lambda_2^p\mathbf{v}_2 + \cdots + k_n\lambda_n^p\mathbf{v}_n \tag{8.39}$$

Since $\lambda_1 \neq 0$ (see (8.37)), (8.39) can be rewritten as

$$A^p\mathbf{x}_0 = \lambda_1^p\left(k_1\mathbf{v}_1 + k_2\left(\frac{\lambda_2}{\lambda_1}\right)^p\mathbf{v}_2 + \cdots + k_n\left(\frac{\lambda_n}{\lambda_1}\right)^p\mathbf{v}_n\right) \tag{8.40}$$

It follows from (8.37) that

$$\frac{\lambda_2}{\lambda_1}, \ldots, \frac{\lambda_n}{\lambda_1}$$

are all less than one in absolute value; thus, $(\lambda_2/\lambda_1)^p, \ldots, (\lambda_n/\lambda_1)^p$ get steadily closer to zero as p increases, and from (8.40), the approximation

$$A^p\mathbf{x}_0 \approx \lambda_1^p k_1\mathbf{v}_1 \tag{8.41}$$

gets better and better.

If $k_1 \neq 0$,* then $\lambda_1^p k_1 \mathbf{v}_1$ is a nonzero scalar multiple of the dominant eigenvector $\mathbf{v}_1$; thus, $\lambda_1^p k_1 \mathbf{v}_1$ is also a dominant eigenvector. Therefore, by (8.41), $A^p \mathbf{x}_0$ becomes a better and better estimate of a dominant eigenvector as p is increased.

EXERCISE SET 8.5

1. Find the dominant eigenvalue (if it exists).

(a) $\begin{bmatrix} -1 & 4 \\ 1 & -1 \end{bmatrix}$ (b) $\begin{bmatrix} 0 & 1 \\ 4 & 0 \end{bmatrix}$ (c) $\begin{bmatrix} 4 & 2 & 1 \\ 0 & -5 & 3 \\ 0 & 0 & 6 \end{bmatrix}$ (d) $\begin{bmatrix} 1 & -12 & 0 \\ 1 & 0 & 0 \\ 0 & 0 & 3 \end{bmatrix}$

2. Let

$$A = \begin{bmatrix} 3 & 4 \\ 1 & 3 \end{bmatrix}$$

(a) Use the power method with scaling to approximate a dominant eigenvector of A. Start with

$$\mathbf{x}_0 = \begin{bmatrix} 1 \\ 1 \end{bmatrix}$$

Round off all computations to three significant digits, and stop after three iterations (that is, three multiplications by A).

(b) Use the result of part (a) and the Rayleigh quotient to approximate the dominant eigenvalue of A.

(c) Find the exact values of the dominant eigenvector and eigenvalue.

(d) Find the percentage error in the approximation of the dominant eigenvalue.

Exercises 3 and 4 follow the directions given in Exercise 2.

3. $A = \begin{bmatrix} 5 & 4 \\ 3 & 4 \end{bmatrix}$ 4. $A = \begin{bmatrix} -3 & 2 \\ 2 & 0 \end{bmatrix}$

5. Let

$$A = \begin{bmatrix} 18 & 17 \\ 2 & 3 \end{bmatrix}$$

(a) Use the power method with scaling to approximate the dominant eigenvalue and a dominant eigenvector of A. Start with

$$\mathbf{x}_0 = \begin{bmatrix} 1 \\ 1 \end{bmatrix}$$

Round off all computations to three significant digits, and stop when the estimated percentage error in the dominant eigenvalue is less than 2%.

(b) Find the exact values of the dominant eigenvalue and eigenvector.

---

* One cannot usually tell by inspection of the $\mathbf{x}_0$ selected whether $k_1 \neq 0$. If by accident $k_1 = 0$, the power method still works in practical problems, since computer roundoff errors generally build up to make k_1 small but nonzero. This is one instance where errors help to obtain correct results!

6. Repeat the directions of Exercise 5 with

$$A = \begin{bmatrix} -5 & 5 \\ 6 & -4 \end{bmatrix}$$

7. Let

$$A = \begin{bmatrix} 2 & 1 & 0 \\ 1 & 2 & 0 \\ 0 & 0 & 10 \end{bmatrix}$$

(a) Use the power method with scaling to approximate a dominant eigenvector of A. Start with

$$\mathbf{x}_0 = \begin{bmatrix} 1 \\ 1 \\ 1 \end{bmatrix}$$

Round off all computations to three significant digits, and stop after three iterations.

(b) Use the result of part (a) and the Rayleigh quotient to approximate the dominant eigenvalue of A.

(c) Find the exact values for the dominant eigenvalue and eigenvector.

(d) Find the percentage error in the approximation of the dominant eigenvalue.

8.6 APPROXIMATING NONDOMINANT EIGENVALUES: DEFLATION AND INVERSE POWER METHODS

In many applications only the dominant eigenvalue and eigenvector of a matrix are needed, in which case the power method alone can be tried. However, if additional eigenvalues and eigenvectors are needed, then other methods are required. In this section we shall briefly discuss some ways in which the power method can be adapted to find nondominant eigenvalues and eigenvectors.

We shall need the following theorem, which is proved in the exercises (Exercise 6).

Theorem 2. *Let A be a symmetric $n \times n$ matrix with eigenvalues $\lambda_1, \lambda_2, \ldots, \lambda_n$. If $\mathbf{v}_1$ is an eigenvector corresponding to λ_1, and $\|\mathbf{v}_1\| = 1$, then:*

(a) *The matrix $B = A - \lambda_1 \mathbf{v}_1 \mathbf{v}_1{}^t$ has eigenvalues $0, \lambda_2, \ldots, \lambda_n$.*

(b) *If $\mathbf{v}$ is an eigenvector of B corresponding to a nonzero eigenvalue in the set $\{\lambda_2, \ldots, \lambda_n\}$, then $\mathbf{v}$ is also an eigenvector of A corresponding to this eigenvalue.*

REMARK. In this theorem the eigenvalues $\lambda_1, \lambda_2, \ldots, \lambda_n$ need not be distinct. Also, we are assuming that $\mathbf{v}_1$ is expressed as an $n \times 1$ matrix so $\mathbf{v}_1\mathbf{v}_1{}^t$ is an $n \times n$ matrix.

Example 15

It was shown in Example 5 of Section 6.1 that

$$A = \begin{bmatrix} 3 & -2 & 0 \\ -2 & 3 & 0 \\ 0 & 0 & 5 \end{bmatrix}$$

has eigenvalues $\lambda_1 = 5$, $\lambda_2 = 5$, $\lambda_3 = 1$, and

$$\mathbf{v} = \begin{bmatrix} -1 \\ 1 \\ 0 \end{bmatrix}$$

is an eigenvector corresponding to $\lambda_1 = 5$. Normalizing $\mathbf{v}$ yields

$$\mathbf{v}_1 = \frac{1}{\sqrt{2}} \begin{bmatrix} -1 \\ 1 \\ 0 \end{bmatrix} = \begin{bmatrix} -\dfrac{1}{\sqrt{2}} \\ \dfrac{1}{\sqrt{2}} \\ 0 \end{bmatrix}$$

which is an eigenvector of norm 1 corresponding to $\lambda_1 = 5$.
By Theorem 2 the matrix

$$B = A - \lambda_1\mathbf{v}_1\mathbf{v}_1{}^t = \begin{bmatrix} 3 & -2 & 0 \\ -2 & 3 & 0 \\ 0 & 0 & 5 \end{bmatrix} - 5 \begin{bmatrix} -\dfrac{1}{\sqrt{2}} \\ \dfrac{1}{\sqrt{2}} \\ 0 \end{bmatrix} \begin{bmatrix} -\dfrac{1}{\sqrt{2}} & \dfrac{1}{\sqrt{2}} & 0 \end{bmatrix}$$

$$= \begin{bmatrix} 3 & -2 & 0 \\ -2 & 3 & 0 \\ 0 & 0 & 5 \end{bmatrix} - 5 \begin{bmatrix} \frac{1}{2} & -\frac{1}{2} & 0 \\ -\frac{1}{2} & \frac{1}{2} & 0 \\ 0 & 0 & 0 \end{bmatrix} = \begin{bmatrix} \frac{1}{2} & \frac{1}{2} & 0 \\ \frac{1}{2} & \frac{1}{2} & 0 \\ 0 & 0 & 5 \end{bmatrix}$$

should have eigenvalues $\lambda = 0$, 5, and 1. As a check, the characteristic equation of B is

$$\det(\lambda I - B) = \det \begin{bmatrix} \lambda - \frac{1}{2} & -\frac{1}{2} & 0 \\ -\frac{1}{2} & \lambda - \frac{1}{2} & 0 \\ 0 & 0 & \lambda - 5 \end{bmatrix} = \lambda(\lambda - 5)(\lambda - 1) = 0$$

Hence, the eigenvalues of B are $\lambda = 0$, $\lambda = 5$, $\lambda = 1$ as predicted by Theorem 2.

The eigenspace of B corresponding to $\lambda = 5$ is the solution space of the system

$$(5I - B)\mathbf{x} = \mathbf{0}$$

That is,

$$\begin{bmatrix} \frac{9}{2} & -\frac{1}{2} & 0 \\ -\frac{1}{2} & \frac{9}{2} & 0 \\ 0 & 0 & 0 \end{bmatrix} \begin{bmatrix} x_1 \\ x_2 \\ x_3 \end{bmatrix} = \begin{bmatrix} 0 \\ 0 \\ 0 \end{bmatrix}$$

Solving this system yields $x_1 = 0$, $x_2 = 0$, $x_3 = t$. Thus, the eigenvectors of B corresponding to $\lambda = 5$ are the nonzero vectors of the form

$$\mathbf{x} = \begin{bmatrix} 0 \\ 0 \\ t \end{bmatrix}$$

As predicted by part (*b*) of Theorem 2, these are also eigenvectors of A corresponding to $\lambda = 5$, since

$$A \begin{bmatrix} 0 \\ 0 \\ t \end{bmatrix} = \begin{bmatrix} 3 & -2 & 0 \\ -2 & 3 & 0 \\ 0 & 0 & 5 \end{bmatrix} \begin{bmatrix} 0 \\ 0 \\ t \end{bmatrix} = \begin{bmatrix} 0 \\ 0 \\ 5t \end{bmatrix}$$

That is,

$$A \begin{bmatrix} 0 \\ 0 \\ t \end{bmatrix} = 5 \begin{bmatrix} 0 \\ 0 \\ t \end{bmatrix}$$

Similarly, the eigenvectors of B corresponding to $\lambda = 1$ are also eigenvectors of A corresponding to $\lambda = 1$. ▲

Theorem 2, to a limited extent, makes it possible to determine the nondominant eigenvalues and eigenvectors of a *symmetric* $n \times n$ matrix A. To see how, assume the eigenvalues of A can be ordered according to the size of their absolute values as follows.

$$|\lambda_1| > |\lambda_2| > |\lambda_3| \geq \cdots \geq |\lambda_n| \tag{8.42}$$

Suppose the dominant eigenvalue and a dominant eigenvector of A have been obtained by the power method. By normalizing the dominant eigenvector, we can obtain a dominant eigenvector $\mathbf{v}_1$ having norm one. By Theorem 2 the eigenvalues of $B = A - \lambda_1 \mathbf{v}_1 \mathbf{v}_1{}^t$ will be $0, \lambda_2, \lambda_3, \ldots, \lambda_n$. From (8.42) these eigenvalues will be ordered according to their absolute values as follows:

$$|\lambda_2| > |\lambda_3| \geq \cdots \geq |\lambda_n| \geq 0$$

Thus, λ_2 is the dominant eigenvalue of B. Now, by applying the power method to B, we can approximate the eigenvalue λ_2 and a corresponding eigenvector. This technique for approximating the eigenvalue with the second largest absolute value is called **deflation**.

Unfortunately, there are practical limitations to the deflation method. Since λ_1 and $\mathbf{v}_1$ are only approximated in the power method, an error is introduced into B when deflation is used. If the deflation process is applied again, the next matrix has additional errors introduced through the approximation of λ_2 and $\mathbf{v}_2$. As the process continues, this compounding of errors can affect the accuracy of the results so seriously that the approximations to the eigenvalues with smaller absolute values are worthless. In practice, one should avoid finding more than two or three eigenvalues by deflation.

There is a way of using the power method to approximate the eigenvalue of smallest absolute value when the matrix is invertible. It uses the fact that inverting a matrix takes reciprocals of the eigenvalues but leaves the corresponding eigenvectors unchanged. More precisely, if

$$\lambda_1, \lambda_2, \ldots, \lambda_n$$

are the eigenvalues of an invertible matrix A, then

$$\frac{1}{\lambda_1}, \frac{1}{\lambda_2}, \ldots, \frac{1}{\lambda_n}$$

are the eigenvalues of A^{-1}. Moreover, $\mathbf{x}$ is an eigenvector of A corresponding to λ if and only if $\mathbf{x}$ is an eigenvector of A^{-1} corresponding to $1/\lambda$ (see Exercise 5). It follows from this result that if the eigenvalues of A can be ordered according to the size of their absolute values as follows:

$$|\lambda_1| \geq |\lambda_2| \geq \cdots \geq |\lambda_{n-1}| > |\lambda_n|$$

then $1/\lambda_n$ will be the dominant eigenvalue of A^{-1}, which can be approximated by the power method. Once obtained, the reciprocal of the dominant eigenvalue of A^{-1} can be taken to find the eigenvalue of A with smallest absolute value. This procedure is called the ***inverse power method***.

Example 16

In Example 10 we noted that the eigenvalues of

$$A = \begin{bmatrix} 3 & 2 \\ -1 & 0 \end{bmatrix}$$

are $\lambda_1 = 2$ and $\lambda_2 = 1$, and we used the power method to approximate the dominant eigenvalue ($\lambda_1 = 2$) and a corresponding eigenvector. Now we shall use the inverse power method to approximate the eigenvalue with smallest absolute value ($\lambda_2 = 1$) and a corresponding eigenvector.

If we start with the initial vector

$$\mathbf{x}_0 = \begin{bmatrix} 1 \\ 1 \end{bmatrix}$$

and multiply repeatedly by

$$A^{-1} = \begin{bmatrix} 0 & -1 \\ \frac{1}{2} & \frac{3}{2} \end{bmatrix} = \begin{bmatrix} 0 & -1 \\ .5 & 1.5 \end{bmatrix}$$

the power method yields the results in Table 8.9. Thus, after six iterations with the inverse power method, the approximations to λ_2 and a corresponding eigenvector are

$$\lambda_2 \approx .999, \qquad \mathbf{x} \approx \begin{bmatrix} -.989 \\ 1 \end{bmatrix}$$

The exact values are

$$\lambda_2 = 1, \qquad \mathbf{x} = \begin{bmatrix} -1 \\ 1 \end{bmatrix}$$

When the ratio $|\lambda_2/\lambda_1|$ is close to one, the power method has a slow rate of convergence; that is, many steps are needed to obtain a reasonable degree of accuracy. Readers interested in studying techniques for "speeding up" this rate of convergence and learning more about the numerical methods of linear algebra may want to consult the references that follow the exercise set.

TABLE 8.9

Step i	0	1	2	3	4	5	6
$\mathbf{x}_i$ = the scaled-down approximation to a dominant eigenvector of A^{-1}	$\begin{bmatrix} 1 \\ 1 \end{bmatrix}$	$\begin{bmatrix} -.5 \\ 1 \end{bmatrix}$	$\begin{bmatrix} -.8 \\ 1 \end{bmatrix}$	$\begin{bmatrix} -.909 \\ 1 \end{bmatrix}$	$\begin{bmatrix} -.956 \\ 1 \end{bmatrix}$	$\begin{bmatrix} -.978 \\ 1 \end{bmatrix}$	$\begin{bmatrix} -.989 \\ 1 \end{bmatrix}$
$A^{-1}\mathbf{x}_i$	$\begin{bmatrix} -1 \\ 2 \end{bmatrix}$	$\begin{bmatrix} -1 \\ 1.25 \end{bmatrix}$	$\begin{bmatrix} -1 \\ 1.1 \end{bmatrix}$	$\begin{bmatrix} -1 \\ 1.046 \end{bmatrix}$	$\begin{bmatrix} -1 \\ 1.022 \end{bmatrix}$	$\begin{bmatrix} -1 \\ 1.011 \end{bmatrix}$	$\begin{bmatrix} -1 \\ 1.006 \end{bmatrix}$
Approximation to $1/\lambda_2$	—	1.4	1.159	1.070	1.033	1.016	1.001
Approximation to λ_2	—	.714	.863	.935	.968	.984	.999

EXERCISE SET 8.6

1. Let

$$A = \begin{bmatrix} 6 & 2 \\ 2 & 3 \end{bmatrix}$$

(a) Use the power method to approximate a dominant eigenvector. Start with

$$\mathbf{x}_0 = \begin{bmatrix} 1 \\ 1 \end{bmatrix}$$

Round off all computations to three significant digits, and stop after three iterations (three multiplications by A).

(b) Use the result of part (a) and the Rayleigh quotient to approximate the dominant eigenvalue of A.

(c) Use deflation to approximate the remaining eigenvalue and a corresponding eigenvector; that is, apply the power method to

$$B = A - \tilde{\lambda}_1 \tilde{\mathbf{v}}_1 \tilde{\mathbf{v}}_1^t$$

where $\tilde{\mathbf{v}}_1$ and $\tilde{\lambda}_1$ are the approximations obtained in parts (a) and (b). Start with

$$\mathbf{x}_0 = \begin{bmatrix} 1 \\ 1 \end{bmatrix}$$

Round off all computations to three significant digits and stop after three iterations.

(d) Find the exact values of the eigenvalues and eigenvectors.

2. Follow the directions given in Exercise 1 with

$$A = \begin{bmatrix} 10 & 4 \\ 4 & 4 \end{bmatrix}$$

3. Use the power method and the inverse power method to approximate both eigenvalues and corresponding eigenvectors of

$$A = \begin{bmatrix} -9 & 11 \\ -1 & 3 \end{bmatrix}$$

Start with

$$\mathbf{x}_0 = \begin{bmatrix} 1 \\ 0 \end{bmatrix}$$

Round off all computations to three significant digits and stop after five iterations.

4. Follow the directions of Exercise 3 with

$$A = \begin{bmatrix} -3 & 2 \\ 2 & -3 \end{bmatrix}$$

5. Let A be an invertible matrix.
   (a) Prove that $1/\lambda$ is an eigenvalue of A^{-1} if and only if λ is an eigenvalue of A.
   (b) Prove that $\mathbf{x}$ is an eigenvector of A corresponding to the eigenvalue λ if and only if $\mathbf{x}$ is an eigenvector of A^{-1} corresponding to $1/\lambda$.

6. In this exercise we will help the reader to prove Theorem 2. Observe first that since A is a symmetric $n \times n$ matrix, there exists an orthonormal set of eigenvectors $\mathbf{v}_1, \mathbf{v}_2, \ldots, \mathbf{v}_n$ corresponding to the eigenvalues $\lambda_1, \lambda_2, \ldots, \lambda_n$, respectively (by Theorem 5 of Section 6.3).
   (a) Prove that $B\mathbf{v}_1 = \mathbf{0}$ and $B\mathbf{v}_i = \lambda_i \mathbf{v}_i$ for $i = 2, \ldots, n$. This shows that $0, \lambda_2, \ldots, \lambda_n$ are eigenvalues of B, which proves part (a) of Theorem 2.
   (b) Prove that B is a symmetric matrix.
   (c) Let $\mathbf{v}$ be an eigenvector of B corresponding to a nonzero eigenvalue λ in the set $\{\lambda_1, \lambda_2, \ldots, \lambda_n\}$. As shown in part (a) above, the eigenvector $\mathbf{v}_1$ corresponds to the zero eigenvalue of B, so $\mathbf{v}_1$ and $\mathbf{v}$ lie in distinct eigenspaces of B. Use this fact and part (b) of this exercise to prove that $\mathbf{v}$ is an eigenvector of A corresponding to λ. This proves part (b) of Theorem 2.

REFERENCES

FADDEEV, D. K., and FADDEEVA, V. N. (1960). Computational Methods of Linear Algebra (Russian ed.), Freeman, San Francisco.

FORSYTHE, G., and MOLER, C. B. (1967). *Computer Solution of Linear Algebraic Systems*, Prentice-Hall, Englewood Cliffs, New Jersey.

GOLUB, G. H., and VAN LOAN, C. F. (1983). *Matrix Computations*, Johns Hopkins, Baltimore.

HOUSEHOLDER, A. S. (1964). *The Theory of Matrices in Numerical Analysis*, Ginn (Blaisdell), Boston.

ISSACSON, E., and KELLER, H. B. (1966). *Analysis of Numerical Methods*, Wiley, New York.

MARON, M. (1982). *Numerical Analysis; A Practical Approach*, Macmillan, New York.

RICE, J. R. (1981). *Matrix Computations and Mathematical Software*, McGraw-Hill, New York.

STEWART, G. W. (1973). *Introduction to Matrix Computations*, Academic Press, New York.

VARGA, R. S. (1962). *Matrix Iterative Analysis*, Prentice-Hall, Englewood Cliffs, New Jersey.

WILKINSON, J. H. (1963). *Rounding Errors in Algebraic Processes*, Prentice-Hall, Englewood Cliffs, New Jersey.

WILKINSON, J. M., and REINSCH, C. (eds.) (1971). *Handbook for Automatic Computation, Vol. II, Linear Algebra*, Springer-Verlag, New York.

YOUNG, D. M. (1971). *Iterative Solution of Large Linear Systems*, Academic Press, New York.

CHAPTER NINE

Complex Vector Spaces

9.1 COMPLEX NUMBERS

Up to now we have considered only vector spaces for which the scalars are real numbers. However, for many important applications of vectors it is desirable to allow the scalars to be complex numbers. A vector space that allows complex scalars is called a *complex vector space*, and one that allows real scalars only is called a *real vector space*. One advantage of allowing complex scalars is that all matrices with scalar entries have eigenvalues, which is not true if only real scalars are allowed. For example, the matrix

$$A = \begin{bmatrix} -2 & -1 \\ 5 & 2 \end{bmatrix}$$

has characteristic polynomial

$$\det(\lambda I - A) = \det \begin{bmatrix} \lambda + 2 & 1 \\ -5 & \lambda - 2 \end{bmatrix} = \lambda^2 + 1$$

so the characteristic equation, $\lambda^2 + 1 = 0$, has no real solutions and therefore no real eigenvalues.

In the first three sections of this chapter we will review some of the basic properties of complex numbers, and in subsequent sections we will discuss complex vector spaces.

Since $x^2 \geq 0$ for every real number x, the equation

$$x^2 = -1$$

has no real solutions. To deal with this problem, mathematicians of the eighteenth century introduced the "imaginary" number,

$$i = \sqrt{-1}$$

which they assumed had the property

$$i^2 = (\sqrt{-1})^2 = -1$$

but which otherwise could be treated as a real number. Expressions of the form

$$a + bi$$

where a and b are real numbers were called "complex numbers," and these were manipulated according to the standard rules of arithmetic with the added property that $i^2 = -1$.

By the beginning of the nineteenth century it was recognized that a complex number,

$$a + bi$$

could be regarded as an alternative symbol for the ordered pair

$$(a, b)$$

of real numbers, and that operations of addition, subtraction, multiplication, and division could be defined on these ordered pairs so that the familiar laws of arithmetic hold and $i^2 = -1$. This is the approach we will follow.

Definition. A *complex number* is an ordered pair of real numbers, denoted either by (a, b) or $a + bi$.

Example 1

Some examples of complex numbers in both notations are:

Ordered Pair	Equivalent Notation
$(3, 4)$	$3 + 4i$
$(-1, 2)$	$-1 + 2i$
$(0, 1)$	$0 + i$
$(2, 0)$	$2 + 0i$
$(4, -2)$	$4 + (-2)i$

For simplicity, the last three complex numbers would usually be abbreviated as:

$$0 + i = i \qquad 2 + 0i = 2 \qquad 4 + (-2)i = 4 - 2i$$

Geometrically, a complex number can be viewed either as a point or a vector in the xy-plane (Figure 9.1).

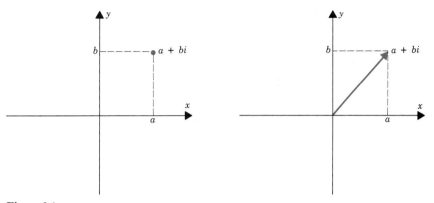

Figure 9.1

Example 2

Some complex numbers are shown as points in Figure 9.2*a* and as vectors in Figure 9.2*b*.

Sometimes it is convenient to use a single letter, such as *z*, to denote a complex number. Thus, we might write   .

$$z = a + bi$$

The real number *a* is called the ***real part of z*** and the real number *b* the ***imaginary part of z***. These numbers are denoted by Re(*z*) and Im(*z*), respectively. Thus,

$$\text{Re}(4 - 3i) = 4 \qquad \text{and} \qquad \text{Im}(4 - 3i) = -3$$

When complex numbers are represented geometrically in an *xy*-coordinate system, the *x*-axis is called the ***real axis***, the *y*-axis the ***imaginary axis***, and the plane is called the ***complex plane*** (Figure 9.3).

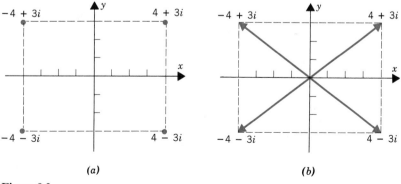

(*a*)   (*b*)

Figure 9.2

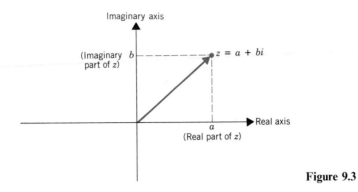

Figure 9.3

Just as two vectors in R^2 are defined to be equal if they have the same components, so we define two complex numbers to be equal if their real parts are equal and their imaginary parts are equal:

Definition. Two complex numbers, $a + bi$ and $c + di$, are defined to be *equal*, written

$$a + bi = c + di$$

if $a = c$ and $b = d$.

If $b = 0$, then the complex number $a + bi$ reduces to $a + 0i$, which we write simply as a. Thus, for any real number a,

$$a = a + 0i$$

so that the real numbers can be regarded as complex numbers with an imaginary part of zero. Geometrically, the real numbers correspond to points on the real axis. If $a = 0$, then $a + bi$ reduces to $0 + bi$, which we usually write as bi. These complex numbers, which correspond to points on the imaginary axis, are called *pure imaginary numbers*.

Just as vectors in R^2 are added by adding corresponding components, so complex numbers are added by adding their real parts and adding their imaginary parts:

$$(a + bi) + (c + di) = (a + c) + (b + d)i$$

The operations of subtraction and multiplication by a *real* number are also similar to the corresponding vector operations in R^2:

$$(a + bi) - (c + di) = (a - c) + (b - d)i$$

$$k(a + bi) = (ka) + (kb)i, \qquad k \text{ real}$$

Because the operations of addition, subtraction, and multiplication of a complex number by a real number parallel the corresponding operations for vectors in R^2, the familiar geometric interpretations of these operations hold for complex numbers (Figure 9.4).

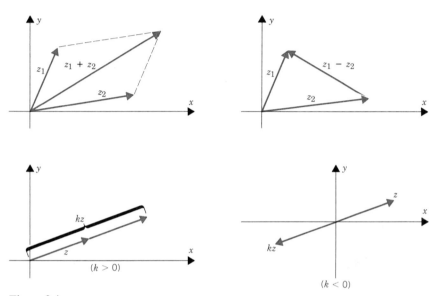

Figure 9.4

Because $(-1)z + z = 0$ (verify), we denote $(-1)z$ as $-z$ and call it the ***negative of z***.

Example 3

If $z_1 = 4 - 5i$ and $z_2 = -1 + 6i$, find $z_1 + z_2$, $z_1 - z_2$, $3z_1$, and $-z_2$

Solution.

$$z_1 + z_2 = (4 - 5i) + (-1 + 6i) = (4 - 1) + (-5 + 6)i = 3 + i$$
$$z_1 - z_2 = (4 - 5i) - (-1 + 6i) = (4 + 1) + (-5 - 6)i = 5 - 11i$$
$$3z_1 = 3(4 - 5i) = 12 - 15i$$
$$-z_2 = (-1)z_2 = (-1)(-1 + 6i) = 1 - 6i$$

So far, there has been a parallel between complex numbers and vectors in R^2. However, we now define multiplication of complex numbers, an operation with no vector analog in R^2. To motivate the definition, we expand the product

$$(a + bi)(c + di)$$

following the usual rules of algebra, but treating i^2 as -1. This yields

$$(a + bi)(c + di) = ac + bdi^2 + adi + bci$$
$$= (ac - bd) + (ad + bc)i$$

which suggests the following *definition*:

$$(a + bi)(c + di) = (ac - bd) + (ad + bc)i \qquad (9.1)$$

Example 4

$$(3 + 2i)(4 + 5i) = (3 \cdot 4 - 2 \cdot 5) + (3 \cdot 5 + 2 \cdot 4)i$$
$$= 2 + 23i$$
$$(4 - i)(2 - 3i) = [4 \cdot 2 - (-1)(-3)] + [(4)(-3) + (-1)(2)]i$$
$$= 5 - 14i$$
$$i^2 = (0 + i)(0 + i) = (0 \cdot 0 - 1 \cdot 1) + (0 \cdot 1 + 1 \cdot 0)i = -1 \quad \blacktriangle$$

We leave it as an exercise to verify the following rules of complex arithmetic:

$$z_1 + z_2 = z_2 + z_1$$
$$z_1 z_2 = z_2 z_1$$
$$z_1 + (z_2 + z_3) = (z_1 + z_2) + z_3$$
$$z_1(z_2 z_3) = (z_1 z_2)z_3$$
$$z_1(z_2 + z_3) = z_1 z_2 + z_1 z_3$$
$$0 + z = z$$
$$z + (-z) = 0$$
$$1 \cdot z = z$$

These rules make it possible to multiply complex numbers without using Formula (9.1) directly. Following the procedure used to motivate this formula, we can simply multiply each term of $a + bi$ with each term of $c + di$, set $i^2 = -1$, and simplify.

Example 5

$$(3 + 2i)(4 + i) = 12 + 3i + 8i + 2i^2 = 12 + 11i - 2 = 10 + 11i$$
$$(5 - \tfrac{1}{2}i)(2 + 3i) = 10 + 15i - i - \tfrac{3}{2}i^2 = 10 + 14i + \tfrac{3}{2} = \tfrac{23}{2} + 14i$$
$$i(1 + i)(1 - 2i) = i(1 - 2i + i - 2i^2) = i(3 - i) = 3i - i^2 = 1 + 3i \quad \blacktriangle$$

REMARK. Unlike the real numbers, there is no size ordering for the complex numbers. Thus, the order symbols $<$, $\leq$, $>$, and $\geq$ are not used with complex numbers.

Now that we have defined addition, subtraction, and multiplication of complex numbers, it is possible to add, subtract, and multiply matrices with complex entries and multiply a matrix by a complex number. Without going into detail, we note that the matrix operations and terminology discussed in Section 1.4 carry over without change to matrices with complex entries.

Example 6

If

$$A = \begin{bmatrix} 1 & -i \\ 1+i & 4-i \end{bmatrix} \quad \text{and} \quad B = \begin{bmatrix} i & 1-i \\ 2-3i & 4 \end{bmatrix}$$

then

$$A + B = \begin{bmatrix} 1+i & 1-2i \\ 3-2i & 8-i \end{bmatrix} \quad A - B = \begin{bmatrix} 1-i & -1 \\ -1+4i & -i \end{bmatrix}$$

$$iA = \begin{bmatrix} i & -i^2 \\ i+i^2 & 4i-i^2 \end{bmatrix} = \begin{bmatrix} i & 1 \\ -1+i & 1+4i \end{bmatrix}$$

$$AB = \begin{bmatrix} 1 & -i \\ 1+i & 4-i \end{bmatrix}\begin{bmatrix} i & 1-i \\ 2-3i & 4 \end{bmatrix}$$

$$= \begin{bmatrix} 1 \cdot i + (-i) \cdot (2-3i) & 1 \cdot (1-i) + (-i) \cdot 4 \\ (1+i) \cdot i + (4-i) \cdot (2-3i) & (1+i) \cdot (1-i) + (4-i) \cdot 4 \end{bmatrix}$$

$$= \begin{bmatrix} -3-i & 1-5i \\ 4-13i & 18-4i \end{bmatrix} \quad \blacktriangle$$

EXERCISE SET 9.1

1. In each part plot the point and sketch the vector that corresponds to the given complex number.
   (a) $2 + 3i$ (b) -4 (c) $-3 - 2i$ (d) $-5i$

2. Express each complex number in Exercise 1 as an ordered pair of real numbers.

3. In each part use the given information to find the real numbers x and y.
   (a) $x - iy = -2 + 3i$ (b) $(x + y) + (x - y)i = 3 + i$

4. Given that $z_1 = 1 - 2i$ and $z_2 = 4 + 5i$, find
   (a) $z_1 + z_2$ (b) $z_1 - z_2$ (c) $4z_1$
   (d) $-z_2$ (e) $3z_1 + 4z_2$ (f) $\frac{1}{2}z_1 - \frac{3}{2}z_2$

5. In each part solve for z.
(a) $z + (1 - i) = 3 + 2i$ (b) $-5z = 5 + 10i$ (c) $(i - z) + (2z - 3i) = -2 + 7i$

6. In each part sketch the vectors z_1, z_2, $z_1 + z_2$, and $z_1 - z_2$.
(a) $z_1 = 3 + i, z_2 = 1 + 4i$ (b) $z_1 = -2 + 2i, z_2 = 4 + 5i$

7. In each part sketch the vectors z and kz.
(a) $z = 1 + i, k = 2$ (b) $z = -3 - 4i, k = -2$ (c) $z = 4 + 6i, k = \frac{1}{2}$

8. In each part find real numbers k_1 and k_2 that satisfy the equation.
(a) $k_1 i + k_2(1 + i) = 3 - 2i$ (b) $k_1(2 + 3i) + k_2(1 - 4i) = 7 + 5i$

9. In each part find $z_1 z_2$, z_1^2, and z_2^2.
(a) $z_1 = 3i, z_2 = 1 - i$
(b) $z_1 = 4 + 6i, z_2 = 2 - 3i$
(c) $z_1 = \frac{1}{3}(2 + 4i), z_2 = \frac{1}{2}(1 - 5i)$

10. Given that $z_1 = 2 - 5i$ and $z_2 = -1 - i$, find
(a) $z_1 - z_1 z_2$ (b) $(z_1 + 3z_2)^2$ (c) $[z_1 + (1 + z_2)]^2$ (d) $iz_2 - z_1^2$

In Exercises 11–18 perform the calculations and express the result in the form $a + bi$.

11. $(1 + 2i)(4 - 6i)^2$

12. $(2 - i)(3 + i)(4 - 2i)$

13. $(1 - 3i)^3$

14. $i(1 + 7i) - 3i(4 + 2i)$

15. $[(2 + i)(\frac{1}{2} + \frac{3}{4}i)]^2$

16. $(\sqrt{2} + i) - i\sqrt{2}(1 + \sqrt{2}i)$

17. $(1 + i + i^2 + i^3)^{100}$

18. $(3 - 2i)^2 - (3 + 2i)^2$

19. Let

$$A = \begin{bmatrix} 1 & i \\ -i & 3 \end{bmatrix} \qquad B = \begin{bmatrix} 2 & 2 + i \\ 3 - i & 4 \end{bmatrix}$$

Find
(a) $A + 3iB$ (b) BA (c) AB (d) $B^2 - A^2$

20. Let

$$A = \begin{bmatrix} 3 + 2i & 0 \\ -i & 2 \\ 1 + i & 1 - i \end{bmatrix} \qquad B = \begin{bmatrix} -i & 2 \\ 0 & i \end{bmatrix} \qquad C = \begin{bmatrix} -1 - i & 0 & -i \\ 3 & 2i & -5 \end{bmatrix}$$

Find
(a) $A(BC)$ (b) $(BC)A$ (c) $(CA)B^2$ (d) $(1 + i)(AB) + (3 - 4i)A$

21. Show that
(a) $\text{Im}(iz) = \text{Re}(z)$ (b) $\text{Re}(iz) = -\text{Im}(z)$

22. In each part solve the equation by the quadratic formula and check your results by substituting the solutions into the given equation.
(a) $z^2 + 2z + 2 = 0$ (b) $z^2 - z + 1 = 0$

23. (a) Show that if n is a positive integer, then the only possible values for i^n are 1, -1, i, and $-i$.

(b) Find i^{2509}. (**Hint.** The value of i^n can be determined from the remainder when n is divided by 4.)

24. Prove: If $z_1 z_2 = 0$, then $z_1 = 0$ or $z_2 = 0$.

25. Use the result of Exercise 24 to prove: If $zz_1 = zz_2$ and $z \neq 0$, then $z_1 = z_2$.

26. Prove that for all complex numbers z_1, z_2, and z_3
   (a) $z_1 + z_2 = z_2 + z_1$ (b) $z_1 + (z_2 + z_3) = (z_1 + z_2) + z_3$

27. Prove that for all complex numbers z_1, z_2, and z_3
   (a) $z_1 z_2 = z_2 z_1$ (b) $z_1(z_2 z_3) = (z_1 z_2) z_3$

28. Prove that $z_1(z_2 + z_3) = z_1 z_2 + z_1 z_3$ for all complex numbers z_1, z_2, and z_3.

29. In quantum mechanics the **Dirac*** **matrices** are:

$$\beta = \begin{bmatrix} 1 & 0 & 0 & 0 \\ 0 & 1 & 0 & 0 \\ 0 & 0 & -1 & 0 \\ 0 & 0 & 0 & -1 \end{bmatrix} \qquad \alpha_x = \begin{bmatrix} 0 & 0 & 0 & 1 \\ 0 & 0 & 1 & 0 \\ 0 & 1 & 0 & 0 \\ 1 & 0 & 0 & 0 \end{bmatrix}$$

$$\alpha_y = \begin{bmatrix} 0 & 0 & 0 & -i \\ 0 & 0 & i & 0 \\ 0 & -i & 0 & 0 \\ i & 0 & 0 & 0 \end{bmatrix} \qquad \alpha_z = \begin{bmatrix} 0 & 0 & 1 & 0 \\ 0 & 0 & 0 & -1 \\ 1 & 0 & 0 & 0 \\ 0 & -1 & 0 & 0 \end{bmatrix}$$

(a) Prove that $\beta^2 = \alpha_x^2 = \alpha_y^2 = \alpha_z^2 = I$.

(b) Two matrices A and B are called **anticommutative** if $AB = -BA$. Prove that any two Dirac matrices are anticommutative.

9.2   MODULUS, COMPLEX CONJUGATE, DIVISION

Our main objective in this section is to define the division of complex numbers. However, it will be helpful to begin with some preliminary ideas.

If $z = a + bi$ is any complex number, then the **conjugate of** z denoted by $\bar{z}$ (read "z bar"), is defined by

$$\bar{z} = a - bi$$

In words, $\bar{z}$ is obtained by reversing the sign of the imaginary part of z. Geometrically, $\bar{z}$ is the reflection of z about the real axis (Figure 9.5).

---

* *Paul Adrien Maurice Dirac (1902–1984)* was a British theoretical physicist who devised a new form of quantum mechanics and a theory that predicted electron spin and the existence of a fundamental atomic particle called a positron. In 1933 he received the Nobel Prize for Physics and in 1939, the medal of the Royal Society.

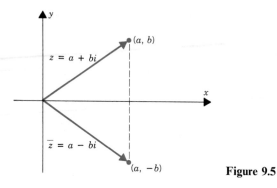

Figure 9.5

Example 7

$$z = 3 + 2i \qquad \bar{z} = 3 - 2i$$
$$z = -4 - 2i \qquad \bar{z} = -4 + 2i$$
$$z = i \qquad \bar{z} = -i$$
$$z = 4 \qquad z = 4$$

REMARK. The last line in Example 7 illustrates the fact that a real number is the same as its conjugate. More precisely, it can be shown (Exercise 22) that $z = \bar{z}$ if and only if z is a real number.

If a complex number z is viewed as a vector in R^2, then the norm or length of the vector is called the modulus (or *absolute value*) of z. More precisely:

Definition. The ***modulus*** of a complex number $z = a + bi$, denoted by $|z|$, is defined by

$$|z| = \sqrt{a^2 + b^2} \qquad (9.2)$$

If $b = 0$, then $z = a$ is a real number, and

$$|z| = \sqrt{a^2 + 0^2} = \sqrt{a^2} = |a|$$

so the modulus of a real number is simply its absolute value. Thus, the modulus of z is also called the ***absolute value*** of z.

Example 8

Find $|z|$ if $z = 3 - 4i$.

Solution. From (9.2) with $a = 3$ and $b = -4$, $|z| = \sqrt{(3)^2 + (-4)^2} = \sqrt{25} = 5$.

The following theorem establishes a basic relationship between $\bar{z}$ and $|z|$.

Theorem 1. *For any complex number z,*

$$z\bar{z} = |z|^2$$

Proof. If $z = a + bi$, then

$$z\bar{z} = (a + bi)(a - bi) = a^2 - abi + bai - b^2i^2$$
$$= a^2 + b^2 = |z|^2 \quad \blacksquare$$

We now turn to the division of complex numbers. Our objective is to define division as the inverse of multiplication. Thus, if $z_2 \neq 0$, then our definition of $z = z_1/z_2$ should be such that

$$z_1 = z_2 z \tag{9.3}$$

Our procedure will be to prove that (9.3) has a unique solution for z if $z_2 \neq 0$, and then define z_1/z_2 to be this value of z.

Theorem 2. *If $z_2 \neq 0$, then equation (9.3) has a unique solution, which is*

$$z = \frac{1}{|z_2|^2} z_1 \bar{z}_2 \tag{9.4}$$

Proof. Let $z = x + iy$, $z_1 = x_1 + iy_1$, and $z_2 = x_2 + iy_2$. Then (9.3) can be written as

$$x_1 + iy_1 = (x_2 + iy_2)(x + iy)$$

or

$$x_1 + iy_1 = (x_2 x - y_2 y) + i(y_2 x + x_2 y)$$

or, on equating real and imaginary parts,

$$x_2 x - y_2 y = x_1$$
$$y_2 x + x_2 y = y_1$$

or

$$\begin{bmatrix} x_2 & -y_2 \\ y_2 & x_2 \end{bmatrix} \begin{bmatrix} x \\ y \end{bmatrix} = \begin{bmatrix} x_1 \\ y_1 \end{bmatrix} \tag{9.5}$$

Since $z_2 = x_2 + iy_2 \neq 0$, it follows that x_2 and y_2 are not both zero, so

$$\begin{vmatrix} x_2 & -y_2 \\ y_2 & x_2 \end{vmatrix} = x_2^2 + y_2^2 \neq 0$$

Thus, by Cramer's Rule (Theorem 10 of Section 2.4), system (9.5) has the unique solution

$$x = \frac{\begin{vmatrix} x_1 & -y_2 \\ y_1 & x_2 \end{vmatrix}}{\begin{vmatrix} x_2 & -y_2 \\ y_2 & x_2 \end{vmatrix}} = \frac{x_1 x_2 + y_1 y_2}{x_2^2 + y_2^2} = \frac{x_1 x_2 + y_1 y_2}{|z_2|^2}$$

$$y = \frac{\begin{vmatrix} x_2 & x_1 \\ y_2 & y_1 \end{vmatrix}}{\begin{vmatrix} x_2 & -y_2 \\ y_2 & x_2 \end{vmatrix}} = \frac{y_1 x_2 - x_1 y_2}{x_2^2 + y_2^2} = \frac{y_1 x_2 - x_1 y_2}{|z_2|^2}$$

Thus,

$$z = x + iy = \frac{1}{|z_2|^2}(x_1 x_2 + y_1 y_2) + i(y_1 x_2 - x_1 y_2)$$

$$= \frac{1}{|z_2|^2}(x_1 + iy_1)(x_2 - iy_2)$$

$$= \frac{1}{|z_2|^2} z_1 \bar{z}_2 \quad \blacksquare$$

Thus, for $z_2 \neq 0$ we define

$$\frac{z_1}{z_2} = \frac{1}{|z_2|^2} z_1 \bar{z}_2 \qquad (9.6)$$

REMARK. To remember this formula, multiply numerator and denominator of z_1/z_2 by $\bar{z}_2$:

$$\frac{z_1}{z_2} = \frac{z_1 \bar{z}_2}{z_2 \bar{z}_2} = \frac{z_1 \bar{z}_2}{|z_2|^2} = \frac{1}{|z_2|^2} z_1 \bar{z}_2$$

Example 9

Express

$$\frac{3 + 4i}{1 - 2i}$$

in the form $a + bi$.

Solution. From (9.6) with $z_1 = 3 + 4i$ and $z_2 = 1 - 2i$,

$$\frac{3 + 4i}{1 - 2i} = \frac{1}{|1 - 2i|^2}(3 + 4i)(\overline{1 - 2i})$$

$$= \frac{1}{5}(3 + 4i)(1 + 2i)$$

$$= \frac{1}{5}(-5 + 10i)$$

$$= -1 + 2i$$

Alternative Solution. As in the remark above, multiply numerator and denominator by the conjugate of the denominator:

$$\frac{3 + 4i}{1 - 2i} = \frac{3 + 4i}{1 - 2i} \cdot \frac{1 + 2i}{1 + 2i} = \frac{-5 + 10i}{5} = -1 + 2i \; \blacktriangle$$

Systems of linear equations with complex coefficients arise in various applications. Without going into detail we note that all the results about linear systems studied in Chapters 1 to 3 carry over without change to systems with complex coefficients.

Example 10

Use Cramer's Rule to solve

$$ix + 2y = 1 - 2i$$
$$4x - iy = -1 + 3i$$

Solution.

$$x = \frac{\begin{vmatrix} 1 - 2i & 2 \\ -1 + 3i & -i \end{vmatrix}}{\begin{vmatrix} i & 2 \\ 4 & -i \end{vmatrix}} = \frac{(-i)(1 - 2i) - 2(-1 + 3i)}{i(-i) - 2(4)} = \frac{-7i}{-7} = i$$

$$y = \frac{\begin{vmatrix} i & 1 - 2i \\ 4 & -1 + 3i \end{vmatrix}}{\begin{vmatrix} i & 2 \\ 4 & -i \end{vmatrix}} = \frac{(i)(-1 + 3i) - 4(1 - 2i)}{i(-i) - 2(4)} = \frac{-7 + 7i}{-7} = 1 - i$$

Thus, the solution is $x = i$, $y = 1 - i$. $\blacktriangle$

We conclude this section by listing some properties of the complex conjugate that will be useful in later sections.

Theorem 3. *For any complex numbers z, z_1, and z_2*

(a) $\overline{z_1 + z_2} = \bar{z}_1 + \bar{z}_2$

(b) $\overline{z_1 - z_2} = \bar{z}_1 - \bar{z}_2$

(c) $\overline{z_1 z_2} = \bar{z}_1 \bar{z}_2$

(d) $\overline{(z_1/z_2)} = \bar{z}_1/\bar{z}_2$

(e) $\bar{\bar{z}} = z$

We prove (a) and leave the rest as exercises.

Proof (a): Let $z_1 = a_1 + b_1 i$ and $z_2 = a_2 + b_2 i$; then

$$\overline{z_1 + z_2} = \overline{(a_1 + a_2) + (b_1 + b_2)i}$$
$$= (a_1 + a_2) - (b_1 + b_2)i$$
$$= (a_1 - b_1 i) + (a_2 - b_2 i)$$
$$= \bar{z}_1 + \bar{z}_2 \quad \blacksquare$$

REMARK. It is possible to extend part (a) of Theorem 3 to n terms and part (c) to n factors. More precisely,

$$\overline{z_1 + z_2 + \cdots + z_n} = \bar{z}_1 + \bar{z}_2 + \cdots + \bar{z}_n$$
$$\overline{z_1 z_2 \cdots z_n} = \bar{z}_1 \bar{z}_2 \cdots \bar{z}_n$$

EXERCISE SET 9.2

1. In each part find $\bar{z}$.
   (a) $z = 2 + 7i$ (b) $z = -3 - 5i$ (c) $z = 5i$
   (d) $z = -i$ (e) $z = -9$ (f) $z = 0$

2. In each part find $|z|$.
   (a) $z = i$ (b) $z = -7i$ (c) $z = -3 - 4i$
   (d) $z = 1 + i$ (e) $z = -8$ (f) $z = 0$

3. Verify that $z\bar{z} = |z|^2$ for
   (a) $z = 2 - 4i$ (b) $z = -3 + 5i$ (c) $z = \sqrt{2} - \sqrt{2}i$

4. Given that $z_1 = 1 - 5i$ and $z_2 = 3 + 4i$, find
   (a) z_1/z_2 (b) $\bar{z}_1/z_2$ (c) $z_1/\bar{z}_2$ (d) $\overline{(z_1/z_2)}$ (e) $z_1/|z_2|$ (f) $|z_1/z_2|$

5. In each part find $1/z$.

   (a) $z = i$ (b) $z = 1 - 5i$ (c) $z = \dfrac{-i}{7}$

6. Given that $z_1 = 1 + i$ and $z_2 = 1 - 2i$, find

(a) $z_1 - \left(\dfrac{z_1}{z_2}\right)$ (b) $\dfrac{z_1 - 1}{z_2}$ (c) $z_1^2 - \left(\dfrac{iz_1}{z_2}\right)$ (d) $\dfrac{z_1}{iz_2}$

In Exercises 7–14 perform the calculations and express the result in the form $a + bi$.

7. $\dfrac{i}{1 + i}$ **8.** $\dfrac{2}{(1 - i)(3 + i)}$

9. $\dfrac{1}{(3 + 4i)^2}$ **10.** $\dfrac{2 + i}{i(-3 + 4i)}$

11. $\dfrac{\sqrt{3} + i}{(1 - i)(\sqrt{3} - i)}$ **12.** $\dfrac{1}{i(3 - 2i)(1 + i)}$

13. $\dfrac{i}{(1 - i)(1 - 2i)(1 + 2i)}$ **14.** $\dfrac{1 - 2i}{3 + 4i} - \dfrac{2 + i}{5i}$

15. In each part solve for z.
  (a) $iz = 2 - i$ (b) $(4 - 3i)\bar{z} = i$

16. Use Theorem 3 to prove the following identities:

(a) $\overline{z + 5i} = z - 5i$ (b) $\overline{iz} = -i\bar{z}$ (c) $\overline{\left(\dfrac{i + z}{i - z}\right)} = -1$

17. In each part sketch the set of points that satisfy the equation.
  (a) $|z| = 2$ (b) $|z - (1 + i)| = 1$ (c) $|z - i| = |z + i|$ (d) $\text{Im}(\bar{z} + i) = 3$

18. In each part sketch the set of points that satisfy the given condition(s).
  (a) $|z + i| \le 1$ (b) $1 < |z| < 2$ (c) $|2z - 4i| < 1$ (d) $|z| \le |z + i|$

19. Given that $z = x + iy$, find
  (a) $\text{Re}(i\bar{z})$ (b) $\text{Im}(i\bar{z})$ (c) $\text{Re}(i\bar{z})$ (d) $\text{Im}(i\bar{z})$

20. (a) Show that if n is a positive integer, then the only possible values for $(1/i)^n$ are $1, -1,$ i, and $-i$.
  (b) Find $(1/i)^{2509}$. (**Hint.** See Exercise 23(b) of Section 9.1.)

21. Prove:

(a) $\dfrac{1}{2}(z + \bar{z}) = \text{Re}(z)$ (b) $\dfrac{1}{2i}(z - \bar{z}) = \text{Im}(z)$

22. Prove: $z = \bar{z}$ if and only if z is a real number.

23. Given that $z_1 = x_1 + iy_1$ and $z_2 = x_2 + iy_2$, find

(a) $\text{Re}\left(\dfrac{z_1}{z_2}\right)$ (b) $\text{Im}\left(\dfrac{z_1}{z_2}\right)$

24. Prove: If $(\bar{z})^2 = z^2$, then z is either real or pure imaginary.

25. Prove that $|z| = |\bar{z}|$.

26. Prove:

(a) $\overline{z_1 - z_2} = \bar{z}_1 - \bar{z}_2$ (b) $\overline{z_1 z_2} = \bar{z}_1 \bar{z}_2$ (c) $\overline{(z_1/z_2)} = \bar{z}_1/\bar{z}_2$ (d) $\bar{\bar{z}} = z$

27. (a) Prove that $\overline{z^2} = (\bar{z})^2$.

(b) Prove that if n is a positive integer, then $\overline{z^n} = (\bar{z})^n$.

(c) Is the result in (b) true if n is a negative integer? Explain.

In Exercises 28–31 solve the system of linear equations by Cramer's Rule.

28. $ix_1 - ix_2 = -2$
$2x_1 + x_2 = i$

29. $x_1 + x_2 = 2$
$x_1 - x_2 = 2i$

30. $x_1 + x_2 + x_3 = 3$
$x_1 + x_2 - x_3 = 2 + 2i$
$x_1 - x_2 + x_3 = -1$

31. $ix_1 + 3x_2 + (1 + i)x_3 = -i$
$x_1 + ix_2 + 3x_3 = -2i$
$x_1 + x_2 + x_3 = 0$

In Exercises 32 and 33 solve the system of linear equations by Gauss-Jordan elimination.

32. $\begin{bmatrix} -1 & -1-i \\ -1+i & -2 \end{bmatrix}\begin{bmatrix} x_1 \\ x_2 \end{bmatrix} = \begin{bmatrix} 0 \\ 0 \end{bmatrix}$ **33.** $\begin{bmatrix} 2 & -1-i \\ -1+i & 1 \end{bmatrix}\begin{bmatrix} x_1 \\ x_2 \end{bmatrix} = \begin{bmatrix} 0 \\ 0 \end{bmatrix}$

34. Solve the following system of linear equations by Gauss-Jordan elimination.

$$x_1 + ix_2 - ix_3 = 0$$
$$-x_1 + (1 - i)x_2 + 2ix_3 = 0$$
$$2x_1 + (-1 + 2i)x_2 - 3ix_3 = 0$$

35. In each part use the formula given in Example 26 of Section 1.5 to compute the inverse of the matrix and check your result by showing that $AA^{-1} = A^{-1}A = I$.

(a) $A = \begin{bmatrix} i & -2 \\ 1 & i \end{bmatrix}$ (b) $A = \begin{bmatrix} 2 & i \\ 1 & 0 \end{bmatrix}$

36. Let $p(x) = a_0 + a_1 x + a_2 x^2 + \cdots + a_n x^n$ be a polynomial for which the coefficients $a_0, a_1, a_2, \ldots, a_n$ are real. Prove that if z is a solution of the equation $p(x) = 0$, then so is $\bar{z}$.

37. Prove: For any complex number z, $|\text{Re}(z)| \le |z|$ and $|\text{Im}(z)| \le |z|$.

38. Prove that

$$\frac{|\text{Re}(z)| + |\text{Im}(z)|}{\sqrt{2}} \le |z|$$

(*Hint.* Let $z = x + iy$ and use the fact that $(|x| - |y|)^2 \ge 0$.)

39. In each part use the method of Example 31 in Section 1.6 to find A^{-1} and check your result by showing that $AA^{-1} = A^{-1}A = I$.

(a) $A = \begin{bmatrix} 1 & 1+i & 0 \\ 0 & 1 & i \\ -i & 1-2i & 2 \end{bmatrix}$ (b) $A = \begin{bmatrix} i & 0 & -i \\ 0 & 1 & -1-4i \\ 2-i & i & 3 \end{bmatrix}$

9.3 POLAR FORM; DEMOIVRE'S THEOREM

In this section we will discuss some additional basic properties of complex numbers.

If $z = x + iy$ is a nonzero complex number, $r = |z|$, and θ measures the angle from the positive real axis to the vector z, then, as suggested by Figure 9.6,

$$
\begin{aligned}
x &= r \cos \theta \\
y &= r \sin \theta
\end{aligned}
\tag{9.7}
$$

so that $z = x + iy$ can be written as

$$z = r \cos \theta + ir \sin \theta$$

or

$$z = r(\cos \theta + i \sin \theta) \tag{9.8}$$

This is called a *polar form of z.*

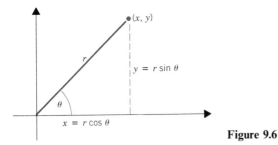

Figure 9.6

The angle θ is called an *argument of z* and is denoted by

$$\theta = \arg z$$

The argument of z is not uniquely determined because we can add or subtract any multiple of 2π from θ to produce another value of the argument. However, there is only one value of the argument in radians that satisfies

$$-\pi < \theta \le \pi$$

This is called the *principal argument of z* and is denoted by

$$\theta = \text{Arg } z$$

Example 11

Express the following complex numbers in polar form using their principal arguments:

(a) $z = 1 + \sqrt{3}i$ (b) $z = -1 - i$

Solution (a). The value of r is

$$r = |z| = \sqrt{1^2 + (\sqrt{3})^2} = \sqrt{4} = 2$$

and since $x = 1$ and $y = \sqrt{3}$, it follows from (9.7) that

$$1 = 2 \cos \theta$$
$$\sqrt{3} = 2 \sin \theta$$

so $\cos \theta = 1/2$ and $\sin \theta = \sqrt{3}/2$. The only value of θ that satisfies these relations and meets the requirement $-\pi < \theta \le \pi$ is $\theta = \pi/3 \ (= 60°)$ (see Figure 9.7a). Thus, a polar form of z is

$$z = 2 \left(\cos \frac{\pi}{3} + i \sin \frac{\pi}{3} \right)$$

Solution (b). The value of r is

$$r = |z| = \sqrt{(-1)^2 + (-1)^2} = \sqrt{2}$$

and since $x = -1$, $y = -1$, it follows from (9.7) that

$$-1 = \sqrt{2} \cos \theta$$
$$-1 = \sqrt{2} \sin \theta$$

so $\cos \theta = -1/\sqrt{2}$ and $\sin \theta = -1/\sqrt{2}$. The only value of θ that satisfies these relations and meets the requirement $-\pi < \theta \le \pi$ is $\theta = -3\pi/4 \ (= -135°)$

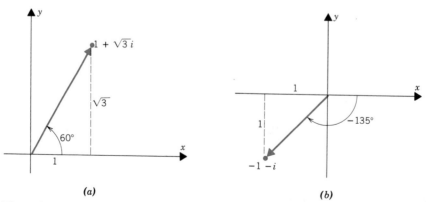

(a)   (b)

Figure 9.7

(Figure 9.7b). Thus, a polar form of z is

$$z = \sqrt{2}\left(\cos\frac{-3\pi}{4} + i\sin\frac{-3\pi}{4}\right)$$

We now show how polar forms can be used to give geometric interpretations of multiplication and division of complex numbers. Let

$$z_1 = r_1(\cos\theta_1 + i\sin\theta_1) \qquad \text{and} \qquad z_2 = r_2(\cos\theta_2 + i\sin\theta_2)$$

Multiplying, we obtain

$$z_1z_2 = r_1r_2[(\cos\theta_1\cos\theta_2 - \sin\theta_1\sin\theta_2) + i(\sin\theta_1\cos\theta_2 + \cos\theta_1\sin\theta_2)]$$

Recalling the trigonometric identities

$$\cos(\theta_1 + \theta_2) = \cos\theta_1\cos\theta_2 - \sin\theta_1\sin\theta_2$$
$$\sin(\theta_1 + \theta_2) = \sin\theta_1\cos\theta_2 + \cos\theta_1\sin\theta_2$$

We obtain

$$z_1z_2 = r_1r_2[\cos(\theta_1 + \theta_2) + i\sin(\theta_1 + \theta_2)] \tag{9.9}$$

which is a polar form of the complex number with modulus r_1r_2 and argument $\theta_1 + \theta_2$. Thus, we have shown that

$$|z_1z_2| = |z_1||z_2| \tag{9.10}$$

and

$$\arg(z_1z_2) = \arg z_1 + \arg z_2$$

(Why?)

In words, *the product of two complex numbers is obtained by multiplying their moduli and adding their arguments* (Figure 9.8).

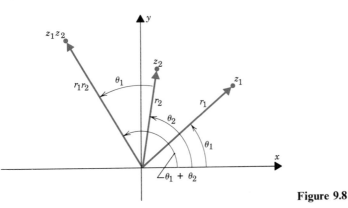

Figure 9.8

We leave it as an exercise to show that if $z_2 \neq 0$, then

$$\frac{z_1}{z_2} = \frac{r_1}{r_2}[\cos(\theta_1 - \theta_2) + i\sin(\theta_1 - \theta_2)] \tag{9.11}$$

from which it follows that

$$\left|\frac{z_1}{z_2}\right| = \frac{|z_1|}{|z_2|} \qquad \text{if } z_2 \neq 0$$

and

$$\arg\left(\frac{z_1}{z_2}\right) = \arg z_1 - \arg z_2$$

In words, *the quotient of two complex numbers is obtained by dividing their moduli and subtracting their arguments (in the appropriate order).*

Example 12

Let

$$z_1 = 1 + \sqrt{3}i \qquad \text{and} \qquad z_2 = \sqrt{3} + i$$

Polar forms of these complex numbers are

$$z_1 = 2\left(\cos\frac{\pi}{3} + i\sin\frac{\pi}{3}\right)$$

$$z_2 = 2\left(\cos\frac{\pi}{6} + i\sin\frac{\pi}{6}\right)$$

(verify) so that from (9.9)

$$z_1 z_2 = 4\left[\cos\left(\frac{\pi}{3} + \frac{\pi}{6}\right) + i\sin\left(\frac{\pi}{3} + \frac{\pi}{6}\right)\right]$$

$$= 4\left[\cos\frac{\pi}{2} + i\sin\frac{\pi}{2}\right]$$

$$= 4[0 + i] = 4i$$

and from (9.11)

$$\frac{z_1}{z_2} = 1 \cdot \left[\cos\left(\frac{\pi}{3} - \frac{\pi}{6}\right) + i\sin\left(\frac{\pi}{3} - \frac{\pi}{6}\right)\right]$$

$$= \cos\frac{\pi}{6} + i\sin\frac{\pi}{6}$$

$$= \frac{\sqrt{3}}{2} + \frac{1}{2}i$$

As a check, we calculate $z_1 z_2$ and z_1/z_2 directly without using polar forms for z_1 and z_2:

$$z_1 z_2 = (1 + \sqrt{3}i)(\sqrt{3} + i) = (\sqrt{3} - \sqrt{3}) + (3 + 1)i = 4i$$

$$\frac{z_1}{z_2} = \frac{1 + \sqrt{3}i}{\sqrt{3} + i} \cdot \frac{\sqrt{3} - i}{\sqrt{3} - i} = \frac{(\sqrt{3} + \sqrt{3}) + (-i + 3i)}{4} = \frac{\sqrt{3}}{2} + \frac{1}{2}i$$

which agrees with our previous results. ▲

The complex number i has a modulus of 1 and an argument of $\pi/2$ $(=90°)$, so the product iz has the same modulus as z, but its argument is $90°$ greater than that of z. In short, *multiplying z by i rotates z counterclockwise by $90°$* (Figure 9.9).

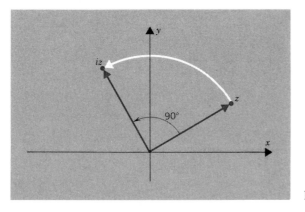

Figure 9.9

If n is a positive integer and $z = r(\cos\theta + i\sin\theta)$, then from Formula (9.9),

$$z^n = \underbrace{z \cdot z \cdot z \cdots z}_{n\text{-factors}} = r^n[\cos(\underbrace{\theta + \theta + \cdots + \theta}_{n\text{-terms}}) + i\sin(\underbrace{\theta + \theta + \cdots + \theta}_{n\text{-terms}})]$$

or

$$z^n = r^n(\cos n\theta + i\sin n\theta) \qquad (9.12)$$

In the special case where $r = 1$, we have $z = \cos\theta + i\sin\theta$, so (9.12) becomes

$$(\cos\theta + i\sin\theta)^n = \cos n\theta + i\sin n\theta \qquad (9.13)$$

which is called **DeMoivre's* Formula**. Although we derived (9.13) assuming n to

---

* *Abraham DeMoivre (1667–1754)* was a French mathematician who made important contributions to probability, statistics, and trigonometry. He developed the concept of statistically independent events, wrote a major influential treatise on probability, and helped transform trigonometry from a branch of geometry into a branch of analysis through his use of complex numbers. In spite of his important work, he barely managed to eke out a living as a tutor and a consultant on gambling and insurance.

be a positive integer, it will be shown in the exercises that this formula is valid for all integers n.

We now show how DeMoivre's Formula can be used to obtain roots of complex numbers. If n is a positive integer, and z is any complex number, then we define an **nth root of** z to be any complex number w that satisfies the equation

$$w^n = z \tag{9.14}$$

We denote an nth root of z by $z^{1/n}$. If $z \neq 0$, then we can derive formulas for the nth roots of z as follows. Let

$$w = \rho(\cos \alpha + i \sin \alpha) \quad \text{and} \quad z = r(\cos \theta + i \sin \theta)$$

If we assume that w satisfies (9.14), then it follows from (9.13) that

$$\rho^n(\cos n\alpha + i \sin n\alpha) = r(\cos \theta + i \sin \theta) \tag{9.15}$$

Comparing the moduli of the two sides, we see that $\rho^n = r$ or

$$\rho = \sqrt[n]{r}$$

where $\sqrt[n]{r}$ denotes the real positive nth root of r. Moreover, in order to have $\cos n\alpha = \cos \theta$ and $\sin n\alpha = \sin \theta$ in (9.15), the angles $n\alpha$ and θ must either be equal or differ by a multiple of 2π. That is,

$$n\alpha = \theta + 2k\pi, \qquad k = 0, \pm 1, \pm 2, \ldots$$

or

$$\alpha = \frac{\theta}{n} + \frac{2k\pi}{n}, \qquad k = 0, \pm 1, \pm 2, \ldots$$

Thus, the values of $w = \rho(\cos \alpha + i \sin \alpha)$ that satisfy (9.14) are given by

$$w = \sqrt[n]{r}\left[\cos\left(\frac{\theta}{n} + \frac{2k\pi}{n}\right) + i \sin\left(\frac{\theta}{n} + \frac{2k\pi}{n}\right)\right] \qquad k = 0, \pm 1, \pm 2, \ldots$$

Although there are infinitely many values of k, it can be shown (Exercise 16) that $k = 0, 1, 2, \ldots, n - 1$ produce distinct values of w satisfying (9.14), but all other choices of k yield duplicates of these. Therefore, there are exactly n different nth roots of $z = r(\cos \theta + i \sin \theta)$, and these are given by

$$z^{1/n} = \sqrt[n]{r}\left[\cos\left(\frac{\theta}{n} + \frac{2k\pi}{n}\right) + i \sin\left(\frac{\theta}{n} + \frac{2k\pi}{n}\right)\right] \qquad k = 0, 1, 2, \ldots, n - 1 \tag{9.16}$$

Example 13

Find all cube roots of -8.

Solution. Since -8 lies on the negative real axis, we can use $\theta = \pi$ as an argument. Moreover, $r = |z| = |-8| = 8$, so a polar form of -8 is

$$-8 = 8(\cos \pi + i \sin \pi)$$

From (9.16) with $n = 3$ it follows that

$$(-8)^{1/3} = \sqrt[3]{8}\left[\cos\left(\frac{\pi}{3} + \frac{2k\pi}{3}\right) + i\sin\left(\frac{\pi}{3} + \frac{2k\pi}{3}\right)\right] \qquad k = 0, 1, 2$$

Thus, the cube roots of -8 are

$$2\left(\cos\frac{\pi}{3} + i\sin\frac{\pi}{3}\right) = 2\left(\frac{1}{2} + \frac{\sqrt{3}}{2}i\right) = 1 + \sqrt{3}i$$

$$2(\cos\pi + i\sin\pi) = 2(-1) = -2$$

$$2\cos\left(\frac{5\pi}{3} + i\sin\frac{5\pi}{3}\right) = 2\left(\frac{1}{2} - \frac{\sqrt{3}}{2}i\right) = 1 - \sqrt{3}i \quad \text{▲}$$

As shown in Figure 9.10, the three cube roots of -8 obtained in Example 13 are equally spaced $\pi/3$ radians ($= 120°$) apart around the circle of radius 2 centered at the origin. This is not accidental. In general, it follows from Formula (9.16) that the nth roots of z lie on the circle of radius $\sqrt[n]{r}$ ($= \sqrt[n]{|z|}$), and are equally spaced $2\pi/n$ radians apart. (Can you see why?) Thus, once one nth root of z is found, the remaining $n - 1$ roots can be generated by rotating this root successively through increments of $2\pi/n$ radians.

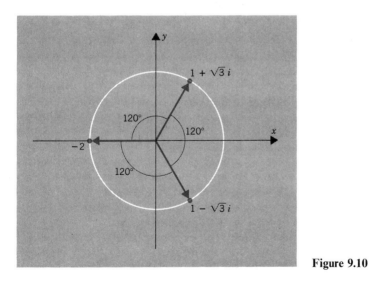

Figure 9.10

Example 14

Find all fourth roots of 1.

Solution. We could apply Formula (9.16). Instead, we observe that $w = 1$ is one fourth root of 1, so that the remaining three roots can be generated by rotating this

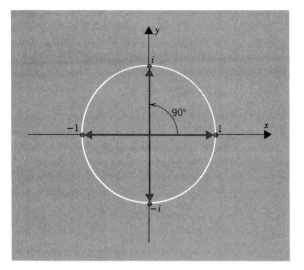

Figure 9.11

root through increments of $2\pi/4 = \pi/2$ radians ($= 90°$). From Figure 9.11 we see that the fourth roots of 1 are

$$1, i, -1, -i$$ ▲

We conclude this section with some comments on notation.

In more detailed studies of complex numbers, complex exponents are defined, and it is proved that

$$\cos \theta + i \sin \theta = e^{i\theta} \tag{9.17}$$

where e is an irrational real number given approximately by $e \approx 2.71828\dots$. (For readers who have studied calculus, a motivation for this result is given in Exercise 18.)

It follows from (9.17) that the polar form

$$z = r(\cos \theta + i \sin \theta)$$

can be written more briefly as

$$z = re^{i\theta} \tag{9.18}$$

Example 15

In Example 11 it was shown that

$$1 + \sqrt{3}i = 2\left(\cos\frac{\pi}{3} + i\sin\frac{\pi}{3}\right)$$

From (9.18) this can also be written as

$$1 + \sqrt{3}i = 2e^{i\pi/3}$$ ▲

It can be proved that complex exponents follow the same laws as real exponents, so that if

$$z_1 = r_1 e^{i\theta_1} \qquad \text{and} \qquad z_2 = r_2 e^{i\theta_2}$$

then

$$z_1 z_2 = r_1 r_2 e^{i\theta_1 + i\theta_2} = r_1 r_2 e^{i(\theta_1 + \theta_2)}$$

$$\frac{z_1}{z_2} = \frac{r_1}{r_2} e^{i\theta_1 - i\theta_2} = \frac{r_1}{r_2} e^{i(\theta_1 - \theta_2)}$$

But these formulas are just Formulas (9.9) and (9.11) in a different notation.

We conclude this section with a useful formula for $\bar{z}$ in polar notation. If

$$z = re^{i\theta} = r(\cos\theta + i\sin\theta)$$

then

$$\bar{z} = r(\cos\theta - i\sin\theta) \qquad\qquad (9.19)$$

Recalling the trigonometric identities

$$\sin(-\theta) = -\sin\theta \qquad \text{and} \qquad \cos(-\theta) = \cos\theta$$

we can rewrite (9.19) as

$$\bar{z} = r[\cos(-\theta) + i\sin(-\theta)] = re^{i(-\theta)}$$

or equivalently

$$\bar{z} = re^{-i\theta} \qquad\qquad (9.20)$$

In the special case where $r = 1$, the polar form of z is $z = e^{i\theta}$, and (9.20) yields the formula

$$\overline{e^{i\theta}} = e^{-i\theta} \qquad\qquad (9.21)$$

EXERCISE SET 9.3

1. In each part find the principal argument of z.
   (a) $z = 1$ (b) $z = i$ (c) $z = -i$
   (d) $z = 1 + i$ (e) $z = -1 + \sqrt{3}i$ (f) $z = 1 - i$

2. In each part find the value of $\theta = \arg(1 - \sqrt{3}i)$ that satisfies the given condition.

   (a) $0 \le \theta < 2\pi$ (b) $-\pi < \theta \le \pi$ (c) $-\dfrac{\pi}{6} \le \theta < \dfrac{11\pi}{6}$

3. In each part express the complex number in polar form using its principal argument.
(a) $2i$ (b) -4 (c) $5 + 5i$
(d) $-6 + 6\sqrt{3}i$ (e) $-3 - 3i$ (f) $2\sqrt{3} - 2i$

4. Given that $z_1 = 2(\cos \pi/4 + i \sin \pi/4)$ and $z_2 = 3(\cos \pi/6 + i \sin \pi/6)$, find a polar form of

(a) $z_1 z_2$ (b) $\dfrac{z_1}{z_2}$ (c) $\dfrac{z_2}{z_1}$ (d) $\dfrac{z_1^5}{z_2^2}$

5. Express $z_1 = i$, $z_2 = 1 - \sqrt{3}i$, and $z_3 = \sqrt{3} + i$ in polar form and use your results to find $z_1 z_2 / z_3$. Check your result by performing the calculation without using polar forms.

6. Use Formula (9.12) to find:

(a) $(1 + i)^{12}$ (b) $\left(\dfrac{1}{\sqrt{2}} - \dfrac{1}{\sqrt{2}} i \right)^{-6}$ (c) $(\sqrt{3} + i)^7$ (d) $(1 - i\sqrt{3})^{-10}$

7. In each part find all the roots and sketch them as vectors in the complex plane.
(a) $(-i)^{1/2}$ (b) $(1 + \sqrt{3}i)^{1/2}$ (c) $(-27)^{1/3}$
(d) $(i)^{1/3}$ (e) $(-1)^{1/4}$ (f) $(-8 + 8\sqrt{3}i)^{1/4}$

8. Use the method of Example 14 to find all cube roots of 1.

9. Use the method of Example 14 to find all sixth roots of 1.

10. Find all square roots of $1 + i$ and express your results in polar form.

11. In each part find all solutions of the equation.
(a) $z^4 - 16 = 0$ (b) $z^{4/3} = -4$

12. Find four solutions of the equation $z^4 + 8 = 0$ and use your results to factor $z^4 + 8$ into two quadratic factors with real coefficients.

13. It was shown in the text that multiplying z by i rotates z counterclockwise by 90°. What is the geometric effect of dividing z by i?

14. In each part use (9.12) to calculate the given power.
(a) $(1 + i)^8$ (b) $(-2\sqrt{3} + 2i)^{-9}$

15. In each part find Re(z) and Im(z).
(a) $z = 3e^{i\pi}$ (b) $z = 3e^{-i\pi}$ (c) $\bar{z} = \sqrt{2}e^{\pi i/2}$ (d) $\bar{z} = -3e^{-2\pi i}$

16. (a) Show that the values of $z^{1/n}$ in Formula (9.16) are all different.
(b) Show that integer values of k other than $k = 0, 1, 2, \ldots, n - 1$ produce values of $z^{1/n}$ that are duplicates of those in Formula (9.16).

17. Show that Formula (9.13) is valid if $n = 0$ or n is a negative integer.

18. **(For readers who have studied calculus.)** To motivate Formula (9.17) recall the Maclaurin series for e^x:

$$e^x = 1 + x + \frac{x^2}{2!} + \cdots + \frac{x^n}{n!} + \cdots$$

(a) By substituting $x = i\theta$ in this series and simplifying, obtain the formula

$$e^{i\theta} = \left(1 - \frac{\theta^2}{2!} + \frac{\theta^4}{4!} - \frac{\theta^6}{6!} + \cdots\right) + i\left(\theta - \frac{\theta^3}{3!} + \frac{\theta^5}{5!} - \frac{\theta^7}{7!} + \cdots\right)$$

(b) Use the result in (a) to obtain (9.17).

19. Derive Formula (9.11).

9.4 COMPLEX VECTOR SPACES

In this section we will develop the basic properties of complex vector spaces and discuss some of the ways in which they differ from real vector spaces. However, before going further the reader should review the definition of a complex vector space given at the beginning of Section 4.2.

In complex vector spaces, linear combinations are defined exactly as in real vector spaces except the scalars are complex. More precisely, a vector **w** is called a *linear combination* of the vectors $v_1, v_2, \ldots, v_r$ if **w** can be expressed in the form

$$\mathbf{w} = k_1 v_1 + k_2 v_2 + \cdots + k_r v_r$$

where $k_1, k_2, \ldots, k_r$ are complex numbers.

The notions of *linear independence, spanning, basis, dimension,* and *subspace* carry over without change to complex vector spaces, and the theorems developed in Sections 4.2–4.6 continue to hold.

Among the real vector spaces the most important one is R^n, the space of n-tuples of real numbers, with addition and scalar multiplication performed coordinatewise. Among the complex vector spaces the most important one is C^n, the space of n-tuples of complex numbers, with addition and scalar multiplication performed coordinatewise. A vector **u** in C^n can be written either in horizontal notation

$$\mathbf{u} = (u_1, u_2, \ldots, u_n)$$

or in matrix notation

$$\mathbf{u} = \begin{bmatrix} u_1 \\ u_2 \\ \vdots \\ u_n \end{bmatrix}$$

where

$$u_1 = a_1 + b_1 i, \quad u_2 = a_2 + b_2 i, \ldots, \quad u_n = a_n + b_n i$$

Example 16

If

$$\mathbf{u} = (i, 1 + i, -2) \quad \text{and} \quad \mathbf{v} = (2 + i, 1 - i, 3 + 2i)$$

then

$$\mathbf{u} + \mathbf{v} = (i, 1 + i, -2) + (2 + i, 1 - i, 3 + 2i) = (2 + 2i, 2, 1 + 2i)$$

and

$$i\mathbf{u} = i(i, 1 + i, -2) = (i^2, i + i^2, -2i) = (-1, -1 + i, -2i)$$

In C^n as in R^n, the vectors

$$\mathbf{e}_1 = (1, 0, 0, \ldots, 0), \quad \mathbf{e}_2 = (0, 1, 0, \ldots, 0), \ldots, \quad \mathbf{e}_n = (0, 0, 0, \ldots, 1)$$

form a basis. It is called the **standard basis** for C^n. Since there are n vectors in this basis, C^n is an n-dimensional vector space.

REMARK. Do not confuse the complex number $i = \sqrt{-1}$ with the vector $\mathbf{i} = (1, 0, 0)$ from the standard basis for R^3 (see Example 14, Section 3.4). The complex number i will always be set in lightface type and the vector $\mathbf{i}$ in boldface.

Example 17

In Example 7 of Section 4.2 we defined the vector space M_{mn} of $m \times n$ matrices with real entries. The complex analog of this space is the vector space of matrices with complex entries and the operations of matrix addition and scalar multiplication. We refer to this space as complex M_{mn}.

Example 18

If $f_1(x)$ and $f_2(x)$ are real-valued functions of the real variable x, then the expression

$$f(x) = f_1(x) + if_2(x)$$

is called a **complex-valued function of the real variable x**. Some examples are:

$$f(x) = 2x + ix^3 \quad \text{and} \quad g(x) = 2\sin x + i\cos x \tag{9.22}$$

Let V be the set of all complex-valued functions defined on the entire line. If $\mathbf{f} = f_1(x) + if_2(x)$ and $\mathbf{g} = g_1(x) + ig_2(x)$ are two such functions and k is any complex number, then we define the **sum** function $\mathbf{f} + \mathbf{g}$ and the **scalar multiple** $k\mathbf{f}$ by

$$(\mathbf{f} + \mathbf{g})(x) = [f_1(x) + g_1(x)] + i[f_2(x) + g_2(x)]$$
$$(k\mathbf{f})(x) = kf_1(x) + ikf_2(x)$$

In words, to form $\mathbf{f} + \mathbf{g}$ add the real parts of $\mathbf{f}$ and $\mathbf{g}$ and add the imaginary parts. To form $k\mathbf{f}$ multiply the real and imaginary parts of $\mathbf{f}$ by k. For example, if $\mathbf{f} = f(x)$ and $\mathbf{g} = g(x)$ are the functions in (9.22), then

$$(\mathbf{f} + \mathbf{g})(x) = (2x + 2\sin x) + i(x^3 + \cos x)$$
$$(i\mathbf{f})(x) = 2xi + i^2x^3 = -x^3 + 2xi$$

It can be shown that V together with the stated operations is a complex vector space. It is the complex analog of the vector space of real-valued functions discussed in Example 8 of Section 4.2.

Example 19 (For readers who have studied calculus.)

If $f(x) = f_1(x) + if_2(x)$ is a complex-valued function of the real variable x, then f is said to be *continuous* if $f_1(x)$ and $f_2(x)$ are continuous. We leave it as an exercise to show that the set of all continuous complex-valued functions of a real variable x is a subspace of the vector space of all complex-valued functions of x. This space is the complex analog of the vector space $C(-\infty, +\infty)$ discussed in Example 14 of Section 4.3 and is called complex $C(-\infty, +\infty)$. A closely related example is complex $C[a, b]$, the vector space of all complex-valued functions that are continuous on the closed interval $[a, b]$. ▲

Recall that in R^n the Euclidean inner product of two vectors

$$\mathbf{u} = (u_1, u_2, \ldots, u_n) \qquad \text{and} \qquad \mathbf{v} = (v_1, v_2, \ldots, v_n)$$

was defined as

$$\mathbf{u} \cdot \mathbf{v} = u_1 v_1 + u_2 v_2 + \cdots + u_n v_n \tag{9.23}$$

and the Euclidean norm (or length) of $\mathbf{u}$ as

$$\|\mathbf{u}\| = (\mathbf{u} \cdot \mathbf{u})^{1/2} = \sqrt{u_1^2 + u_2^2 + \cdots + u_n^2} \tag{9.24}$$

Unfortunately, these definitions are not appropriate for vectors in C^n. For example, if (9.24) were applied to the vector $\mathbf{u} = (i, 1)$ in C^2, we would obtain

$$\|\mathbf{u}\| = \sqrt{i^2 + 1} = \sqrt{0} = 0$$

so $\mathbf{u}$ would be a *nonzero* vector with zero length—a situation that is clearly unsatisfactory.

To extend the notions of norm, distance, and angle to C^n properly, we must modify the inner product slightly.

Definition. If $\mathbf{u} = (u_1, u_2, \ldots, u_n)$ and $\mathbf{v} = (v_1, v_2, \ldots, v_n)$ are vectors in C^n, then their *Euclidean inner product* $\mathbf{u} \cdot \mathbf{v}$ is defined by

$$\mathbf{u} \cdot \mathbf{v} = u_1 \bar{v}_1 + u_2 \bar{v}_2 + \cdots + u_n \bar{v}_n$$

---

where $\bar{v}_1, \bar{v}_2, \ldots, \bar{v}_n$ are the conjugates of $v_1, v_2, \ldots, v_n$.

Example 20

The Euclidean inner product of the vectors

$$\mathbf{u} = (-i, 2, 1 + 3i) \qquad \text{and} \qquad \mathbf{v} = (1 - i, 0, 1 + 3i)$$

is

$$\begin{aligned}
\mathbf{u} \cdot \mathbf{v} &= (-i)\overline{(1 - i)} + (2)(\bar{0}) + (1 + 3i)\overline{(1 + 3i)} \\
&= (-i)(1 + i) + (2)(0) + (1 + 3i)(1 - 3i) \\
&= -i - i^2 + 1 - 9i^2 = 11 - i \quad \text{▲}
\end{aligned}$$

Theorem 2 of Section 4.1 listed the four main properties of the Euclidean inner product on R^n. The following theorem is the corresponding result for the Euclidean inner product on C^n.

Theorem 4. *If* **u**, **v**, *and* **w** *are vectors in* C^n, *and* k *is any complex number, then:*

(a) $\mathbf{u} \cdot \mathbf{v} = \overline{\mathbf{v} \cdot \mathbf{u}}$

(b) $(\mathbf{u} + \mathbf{v}) \cdot \mathbf{w} = \mathbf{u} \cdot \mathbf{w} + \mathbf{v} \cdot \mathbf{w}$

(c) $(k\mathbf{u}) \cdot \mathbf{v} = k(\mathbf{u} \cdot \mathbf{v})$

(d) $\mathbf{v} \cdot \mathbf{v} \geq 0$. *Further,* $\mathbf{v} \cdot \mathbf{v} = 0$ *if and only if* $\mathbf{v} = \mathbf{0}$.

Note the difference between part (a) of this theorem and part (a) of Theorem 2, Section 4.1. We will prove parts (a) and (d) and leave the rest as exercises.

Proof (a). Let $\mathbf{u} = (u_1, u_2, \ldots, u_n)$ and $\mathbf{v} = (v_1, v_2, \ldots, v_n)$. Then

$$\mathbf{u} \cdot \mathbf{v} = u_1 \bar{v}_1 + u_2 \bar{v}_2 + \cdots + u_n \bar{v}_n$$

and

$$\mathbf{v} \cdot \mathbf{u} = v_1 \bar{u}_1 + v_2 \bar{u}_2 + \cdots + v_n \bar{u}_n$$

so

$$\overline{\mathbf{v} \cdot \mathbf{u}} = \overline{v_1 \bar{u}_1 + v_2 \bar{u}_2 + \cdots + v_n \bar{u}_n}$$
$$= \bar{v}_1 \bar{\bar{u}}_1 + \bar{v}_2 \bar{\bar{u}}_2 + \cdots + \bar{v}_n \bar{\bar{u}}_n \qquad \text{(Theorem 3a and 3c, Section 9.2)}$$
$$= \bar{v}_1 u_1 + \bar{v}_2 u_2 + \cdots + \bar{v}_n u_n \qquad \text{(Theorem 3e, Section 9.2)}$$
$$= u_1 \bar{v}_1 + u_2 \bar{v}_2 + \cdots + u_n \bar{v}_n$$
$$= \mathbf{u} \cdot \mathbf{v}$$

Proof (d).

$$\mathbf{v} \cdot \mathbf{v} = v_1 \bar{v}_1 + v_2 \bar{v}_2 + \cdots + v_n \bar{v}_n = |v_1|^2 + |v_2|^2 + \cdots + |v_n|^2 \geq 0$$

Moreover, equality holds if and only if $|v_1| = |v_2| = \cdots = |v_n| = 0$. But this is true if and only if $v_1 = v_2 = \cdots = v_n = 0$, that is, if and only if $\mathbf{v} = \mathbf{0}$.

REMARK. We leave it as an exercise to prove that

$$\mathbf{u} \cdot (k\mathbf{v}) = \bar{k}(\mathbf{u} \cdot \mathbf{v})$$

for vectors in C^n, whereas

$$\mathbf{u} \cdot (k\mathbf{v}) = k(\mathbf{u} \cdot \mathbf{v})$$

for vectors in R^n.

By analogy with (9.24), we define the **Euclidean norm** (or **Euclidean length**) of a vector $\mathbf{u} = (u_1, u_2, \ldots, u_n)$ in C^n by

$$\|\mathbf{u}\| = (\mathbf{u} \cdot \mathbf{u})^{1/2} = \sqrt{|u_1|^2 + |u_2|^2 + \cdots + |u_n|^2}$$

and we define the **Euclidean distance** between the points $\mathbf{u} = (u_1, u_2, \ldots u_n)$ and

$\mathbf{v} = (v_1, v_2, \ldots, v_n)$ by

$$d(\mathbf{u}, \mathbf{v}) = \|\mathbf{u} - \mathbf{v}\| = \sqrt{|u_1 - v_1|^2 + |u_2 - v_2|^2 + \cdots + |u_n - v_n|^2}$$

Example 21

If $\mathbf{u} = (i, 1 + i, 3)$ and $\mathbf{v} = (1 - i, 2, 4i)$, then

$$\|\mathbf{u}\| = \sqrt{|i|^2 + |1 + i|^2 + |3|^2} = \sqrt{1 + 2 + 9} = \sqrt{12} = 2\sqrt{3}$$

and

$$d(\mathbf{u}, \mathbf{v}) = \sqrt{|i - (1 - i)|^2 + |(1 + i) - 2|^2 + |3 - 4i|^2}$$
$$= \sqrt{|-1 + 2i|^2 + |-1 + i|^2 + |3 - 4i|^2} = \sqrt{5 + 2 + 25} = \sqrt{32} = 4\sqrt{2} \quad \blacktriangle$$

The vector space C^n with norm and inner product defined above is called **complex Euclidean n-space**.

EXERCISE SET 9.4

1. Let $\mathbf{u} = (2i, 0, -1, 3)$, $\mathbf{v} = (-i, i, 1 + i, -1)$, and $\mathbf{w} = (1 + i, -i, -1 + 2i, 0)$. Find
   (a) $\mathbf{u} - \mathbf{v}$ (b) $i\mathbf{v} + 2\mathbf{w}$ (c) $-\mathbf{w} + \mathbf{v}$
   (d) $3(\mathbf{u} - (1 + i)\mathbf{v})$ (e) $-i\mathbf{v} + 2i\mathbf{w}$ (f) $2\mathbf{v} - (\mathbf{u} + \mathbf{w})$

2. Let $\mathbf{u}$, $\mathbf{v}$, and $\mathbf{w}$ be the vectors in Exercise 1. Find the vector $\mathbf{x}$ that satisfies
   $\mathbf{u} - \mathbf{v} + i\mathbf{x} = 2i\mathbf{x} + \mathbf{w}$.

3. Let $\mathbf{u}_1 = (1 - i, i, 0)$, $\mathbf{u}_2 = (2i, 1 + i, 1)$, and $\mathbf{u}_3 = (0, 2i, 2 - i)$. Find scalars c_1, c_2, and c_3 such that
   $$c_1\mathbf{u}_1 + c_2\mathbf{u}_2 + c_3\mathbf{u}_3 = (-3 + i, 3 + 2i, 3 - 4i)$$

4. Show that there do not exist scalars c_1, c_2, and c_3 such that
   $$c_1(i, 2 - i, 2 + i) + c_2(1 + i, -2i, 2) + c_3(3, i, 6 + i) = (i, i, i)$$

5. Find the Euclidean norm of $\mathbf{v}$ if
   (a) $\mathbf{v} = (1, i)$ (b) $\mathbf{v} = (1 + i, 3i, 1)$
   (c) $\mathbf{v} = (2i, 0, 2i + 1, -1)$ (d) $\mathbf{v} = (-i, i, i, 3, 3 + 4i)$

6. Let $\mathbf{u} = (3i, 0, -i)$, $\mathbf{v} = (0, 3 + 4i, -2i)$, and $\mathbf{w} = (1 + i, 2i, 0)$. Find
   (a) $\|\mathbf{u} + \mathbf{v}\|$ (b) $\|\mathbf{u}\| + \|\mathbf{v}\|$ (c) $\|-i\mathbf{u}\| + i\|\mathbf{u}\|$

   (d) $\|3\mathbf{u} - 5\mathbf{v} + \mathbf{w}\|$ (e) $\dfrac{1}{\|\mathbf{w}\|}\mathbf{w}$ (f) $\left\|\dfrac{1}{\|\mathbf{w}\|}\mathbf{w}\right\|$

7. Show that if $\mathbf{v}$ is a nonzero vector in C^n, then $(1/\|\mathbf{v}\|)\mathbf{v}$ has Euclidean norm 1.

8. Find all scalars k such that $\|k\mathbf{v}\| = 1$, where $\mathbf{v} = (3i, 4i)$.

9. Find the Euclidean inner product $\mathbf{u} \cdot \mathbf{v}$ if
   (a) $\mathbf{u} = (-i, 3i)$, $\mathbf{v} = (3i, 2i)$
   (b) $\mathbf{u} = (3 - 4i, 2 + i, -6i)$, $\mathbf{v} = (1 + i, 2 - i, 4)$
   (c) $\mathbf{u} = (1 - i, 1 + i, 2i, 3)$, $\mathbf{v} = (4 + 6i, -5i, -1 + i, i)$

In Exercises 10–11 a set of objects is given together with operations of addition and scalar multiplication. Determine which sets are vector spaces under the given operations. For those that are not, list all axioms that fail to hold.

10. The set of all triples of complex numbers (z_1, z_2, z_3) with the operations

$$(z_1, z_2, z_3) + (z_1', z_2', z_3') = (z_1 + z_1', z_2 + z_2', z_3 + z_3')$$

and

$$k(z_1, z_2, z_3) = (\bar{k}z_1, \bar{k}z_2, \bar{k}z_3)$$

11. The set of all complex 2×2 matrices of the form

$$\begin{bmatrix} z & 0 \\ 0 & \bar{z} \end{bmatrix}$$

with the standard matrix operations of addition and scalar multiplication.

12. Is R^n a subspace of C^n? Explain.

13. Use Theorem 4 of Section 4.3 to determine which of the following are subspaces of C^3:
(a) all vectors of the form $(z, 0, 0)$
(b) all vectors of the form (z, i, i)
(c) all vectors of the form (z_1, z_2, z_3), where $z_3 = \bar{z}_1 + \bar{z}_2$
(d) all vectors of the form (z_1, z_2, z_3), where $z_3 = z_1 + z_2 + i$

14. Use Theorem 4 of Section 4.3 to determine which of the following are subspaces of complex M_{22}:
(a) All complex matrices of the form

$$\begin{bmatrix} z_1 & z_2 \\ z_3 & z_4 \end{bmatrix}$$

where z_1 and z_2 are real.
(b) All complex matrices of the form

$$\begin{bmatrix} z_1 & z_2 \\ z_3 & z_4 \end{bmatrix}$$

where $z_1 + z_4 = 0$
(c) All 2×2 complex matrices A such that $(A)^t = A$, where $\bar{A}$ is the matrix whose entries are the conjugates of the corresponding entries in A.

15. Use Theorem 4 of Section 4.3 to determine which of the following are subspaces of the vector space of complex-valued functions of the real variable x:
(a) all f such that $f(1) = 0$
(b) all f such that $f(0) = i$
(c) all f such that $f(-x) = \overline{f(x)}$
(d) all f of the form $k_1 + k_2 e^{ix}$, where k_1 and k_2 are complex numbers.

16. Which of the following are linear combinations of $\mathbf{u} = (i, -i, 3i)$ and $\mathbf{v} = (2i, 4i, 0)$?
(a) $(3i, 3i, 3i)$ (b) $(4i, 2i, 6i)$ (c) $(i, 5i, 6i)$ (d) $(0, 0, 0)$

17. Express the following as linear combinations of $\mathbf{u} = (1, 0, -i)$, $\mathbf{v} = (1 + i, 1, 1 - 2i)$, and $\mathbf{w} = (0, i, 2)$.
(a) $(1, 1, 1)$ (b) $(i, 0, -i)$ (c) $(0, 0, 0)$ (d) $(2 - i, 1, 1 + i)$

18. In each part determine whether the given vectors span C^3.
   (a) $\mathbf{v}_1 = (i, i, i)$, $\mathbf{v}_2 = (2i, 2i, 0)$, $\mathbf{v}_3 = (3i, 0, 0)$
   (b) $\mathbf{v}_1 = (1 + i, 2 - i, 3 + i)$, $\mathbf{v}_2 = (2 + 3i, 0, 1 - i)$
   (c) $\mathbf{v}_1 = (1, 0, -i)$, $\mathbf{v}_2 = (1 + i, 1, 1 - 2i)$, $\mathbf{v}_3 = (0, i, 2)$
   (d) $\mathbf{v}_1 = (1, i, 0)$, $\mathbf{v}_2 = (0, -i, 1)$, $\mathbf{v}_3 = (1, 0, 1)$

19. Determine which of the following lie in the space spanned by

$$\mathbf{f} = e^{ix} \quad \text{and} \quad \mathbf{g} = e^{-ix}$$

   (a) $\cos x$ (b) $\sin x$ (c) $\cos x + 3i \sin x$

20. Explain why the following are linearly dependent sets of vectors. (Solve this problem by inspection.)
   (a) $\mathbf{u}_1 = (1 - i, i)$ and $\mathbf{u}_2 = (1 + i, -1)$ in C^2
   (b) $\mathbf{u}_1 = (1, -i)$, $\mathbf{u}_2 = (2 + i, -1)$, $\mathbf{u}_3 = (4, 0)$ in C^2

   (c) $A = \begin{bmatrix} i & 3i \\ 2i & 0 \end{bmatrix}$ and $B = \begin{bmatrix} 1 & 3 \\ 2 & 0 \end{bmatrix}$ in complex M_{22}.

21. Which of the following sets of vectors in C^3 are linearly independent?
   (a) $\mathbf{u}_1 = (1 - i, 1, 0)$, $\mathbf{u}_2 = (2, 1 + i, 0)$, $\mathbf{u}_3 = (1 + i, i, 0)$
   (b) $\mathbf{u}_1 = (1, 0, -i)$, $\mathbf{u}_2 = (1 + i, 1, 1 - 2i)$, $\mathbf{u}_3 = (0, i, 2)$
   (c) $\mathbf{u}_1 = (i, 0, 2 - i)$, $\mathbf{u}_2 = (0, 1, i)$, $\mathbf{u}_3 = (-i, -1 - 4i, 3)$

22. Let V be the vector space of all complex-valued functions of the real variable x. Show that the following vectors are linearly dependent.

$$\mathbf{f} = 3 + 3i \cos 2x, \quad \mathbf{g} = \sin^2 x + i \cos^2 x, \quad \mathbf{h} = \cos^2 x - i \sin^2 x$$

23. Explain why the following sets of vectors are not bases for the indicated vector spaces. (Solve this problem by inspection.)
   (a) $\mathbf{u}_1 = (i, 2i)$, $\mathbf{u}_2 = (0, 3i)$, $\mathbf{u}_3 = (1, 7i)$ for C^2
   (b) $\mathbf{u}_1 = (-1 + i, 0, 2 - i)$, $\mathbf{u}_2 = (1, -i, 1 + i)$ for C^3

24. Which of the following sets of vectors are bases for C^2?
   (a) $(2i, -i)$, $(4i, 0)$ (b) $(1 + i, 1)$, $(1 + i, i)$
   (c) $(0, 0)$, $(1 + i, 1 - i)$ (d) $(2 - 3i, i)$, $(3 + 2i, -1)$

25. Which of the following sets of vectors are bases for C^3?
   (a) $(i, 0, 0)$, $(i, i, 0)$, (i, i, i)
   (b) $(1, 0, -i)$, $(1 + i, 1, 1 - 2i)$, $(0, i, 2)$
   (c) $(i, 0, 2 - i)$, $(0, 1, i)$, $(-i, -1 - 4i, 3)$
   (d) $(1, 0, i)$, $(2 - i, 1, 2 + i)$, $(0, 3i, 3i)$

In Exercises 26–29 determine the dimension of and a basis for the solution space of the system.

26. $x_1 + (1 + i)x_2 = 0$ 27. $2x_1 - (1 + i)x_2 = 0$
     $(1 - i)x_1 + 2x_2 = 0$ $(-1 + i)x_1 + \quad x_2 = 0$

28. $x_1 + (2 - i)x_2 \qquad = 0$ 29. $x_1 + ix_2 - 2ix_3 + \quad x_4 = 0$
     $\qquad x_2 + 3ix_3 = 0$ $ix_1 + 3x_2 + 4x_3 - 2ix_4 = 0$
     $ix_1 + (2 + 2i)x_2 + 3ix_3 = 0$

30. Prove: If $\mathbf{u}$ and $\mathbf{v}$ are vectors in complex Euclidean n-space, then

$$\mathbf{u} \cdot (k\mathbf{v}) = \bar{k}(\mathbf{u} \cdot \mathbf{v})$$

31. (a) Prove part (b) of Theorem 4.
    (b) Prove part (c) of Theorem 4.

32. (For readers who have studied calculus.) Prove that complex $C(-\infty, +\infty)$ is a subspace of the vector space of complex-valued functions of a real variable.

33. Establish the identity

$$\mathbf{u} \cdot \mathbf{v} = \frac{1}{4} \|\mathbf{u} + \mathbf{v}\|^2 - \frac{1}{4} \|\mathbf{u} - \mathbf{v}\|^2 + \frac{i}{4} \|\mathbf{u} + i\mathbf{v}\|^2 - \frac{i}{4} \|\mathbf{u} - i\mathbf{v}\|^2$$

for vectors in complex Euclidean n-space.

9.5 COMPLEX INNER PRODUCT SPACES

In Section 4.7 we defined the notion of an inner product on a real vector space by using the basic properties of the Euclidean inner product on R^n as axioms. In this section we will define inner products on complex vector spaces by using the properties of the Euclidean inner product on C^n as axioms.

Motivated by Theorem 4 of Section 9.4 we make the following definition.

Definition. An ***inner product on a complex vector space*** V is a function that associates a complex number $\langle \mathbf{u}, \mathbf{v} \rangle$ with each pair of vectors $\mathbf{u}$ and $\mathbf{v}$ in V in such a way that the following axioms are satisfied for all vectors $\mathbf{u}$, $\mathbf{v}$, and $\mathbf{w}$ in V and all scalars k.

(1) $\langle \mathbf{u}, \mathbf{v} \rangle = \overline{\langle \mathbf{v}, \mathbf{u} \rangle}$
(2) $\langle \mathbf{u} + \mathbf{v}, \mathbf{w} \rangle = \langle \mathbf{u}, \mathbf{w} \rangle + \langle \mathbf{v}, \mathbf{w} \rangle$
(3) $\langle k\mathbf{u}, \mathbf{v} \rangle = k \langle \mathbf{u}, \mathbf{v} \rangle$
(4) $\langle \mathbf{v}, \mathbf{v} \rangle \geq 0$ and $\langle \mathbf{v}, \mathbf{v} \rangle = 0$ if and only if $\mathbf{v} = \mathbf{0}$

A complex vector space with an inner product is called a ***complex inner product space*** or a ***unitary space***.

The following additional properties follow immediately from the four inner product axioms:

(i) $\langle \mathbf{0}, \mathbf{v} \rangle = \langle \mathbf{v}, \mathbf{0} \rangle = 0$
(ii) $\langle \mathbf{u}, \mathbf{v} + \mathbf{w} \rangle = \langle \mathbf{u}, \mathbf{v} \rangle + \langle \mathbf{u}, \mathbf{w} \rangle$
(iii) $\langle \mathbf{u}, k\mathbf{v} \rangle = \bar{k} \langle \mathbf{u}, \mathbf{v} \rangle$

Since only (iii) differs from the corresponding results for real inner products, we will prove it and leave the other proofs as exercises.

$$\langle \mathbf{u}, k\mathbf{v} \rangle = \overline{\langle k\mathbf{v}, \mathbf{u} \rangle} \qquad \text{(Axiom (1))}$$
$$= \overline{k\langle \mathbf{v}, \mathbf{u} \rangle} \qquad \text{(Axiom (3))}$$
$$= \bar{k}\overline{\langle \mathbf{v}, \mathbf{u} \rangle} \qquad \text{(Property of conjugates)}$$
$$= \bar{k}\langle \mathbf{u}, \mathbf{v} \rangle \qquad \text{(Axiom (1))}$$

Example 22

Let $\mathbf{u} = (u_1, u_2, \ldots, u_n)$ and $\mathbf{v} = (v_1, v_2, \ldots, v_n)$ be vectors in C^n. The Euclidean inner product $\langle \mathbf{u}, \mathbf{v} \rangle = \mathbf{u} \cdot \mathbf{v} = u_1 \bar{v}_1 + u_2 \bar{v}_2 + \cdots + u_n \bar{v}_n$ satisfies all the inner product axioms by Theorem 4 of Section 9.4.

Example 23

If

$$U = \begin{bmatrix} u_1 & u_2 \\ u_3 & u_4 \end{bmatrix} \qquad \text{and} \qquad V = \begin{bmatrix} v_1 & v_2 \\ v_3 & v_4 \end{bmatrix}$$

are any 2×2 matrices with complex entries, then the following formula defines a complex inner product on complex M_{22} (verify):

$$\langle U, V \rangle = u_1 \bar{v}_1 + u_2 \bar{v}_2 + u_3 \bar{v}_3 + u_4 \bar{v}_4$$

For example, if

$$U = \begin{bmatrix} 0 & i \\ 1 & 1+i \end{bmatrix} \qquad \text{and} \qquad V = \begin{bmatrix} 1 & -i \\ 0 & 2i \end{bmatrix}$$

then

$$\langle U, V \rangle = (0)(\bar{1}) + i(\overline{-i}) + (1)(\bar{0}) + (1+i)(\overline{2i})$$
$$= (0)(1) + i(i) + (1)(0) + (1+i)(-2i)$$
$$= 0 + i^2 + 0 - 2i - 2i^2$$
$$= 1 - 2i$$

Example 24 (For readers who have studied calculus.)

If $f(x) = f_1(x) + if_2(x)$ is a complex-valued function of the real variable x, and if $f_1(x)$ and $f_2(x)$ are continuous on $[a, b]$, then we define

$$\int_a^b f(x)\, dx = \int_a^b [f_1(x) + if_2(x)]\, dx = \int_a^b f_1(x)\, dx + i \int_a^b f_2(x)\, dx$$

In words, *the integral of $f(x)$ is the integral of the real part plus i times the integral of the imaginary part.*

We leave it as an exercise to show that if the functions $\mathbf{f} = f_1(x) + if_2(x)$ and $\mathbf{g} = g_1(x) + ig_2(x)$ are vectors in complex $C[a,b]$, then the following formula defines an inner product on complex $C[a, b]$:

$$\langle \mathbf{f}, \mathbf{g} \rangle = \int_a^b [f_1(x) + if_2(x)]\overline{[g_1(x) + ig_2(x)]}\, dx$$

$$= \int_a^b [f_1(x) + if_2(x)][g_1(x) - ig_2(x)]\, dx$$

$$= \int_a^b [f_1(x)g_1(x) + f_2(x)g_2(x)]\, dx + i \int_a^b [f_2(x)g_1(x) - f_1(x)g_2(x)]\, dx$$

In complex inner product spaces, as in real inner product spaces, the **norm** (or **length**) of a vector $\mathbf{u}$ is defined by

$$\|\mathbf{u}\| = \langle \mathbf{u}, \mathbf{u} \rangle^{1/2}$$

and the **distance** between two vectors $\mathbf{u}$ and $\mathbf{v}$ is defined by

$$d(\mathbf{u}, \mathbf{v}) = \|\mathbf{u} - \mathbf{v}\|$$

It can be shown that with these definitions Theorem 21 of Section 4.8 remains true in complex inner product spaces (Exercise 35).

Example 25

If $\mathbf{u} = (u_1, u_2, \ldots, u_n)$ and $\mathbf{v} = (v_1, v_2, \ldots, v_n)$ are vectors in C^n with the Euclidean inner product, then

$$\|\mathbf{u}\| = \langle \mathbf{u}, \mathbf{u} \rangle^{1/2} = \sqrt{|u_1|^2 + |u_2|^2 + \cdots + |u_n|^2}$$

and

$$d(\mathbf{u}, \mathbf{v}) = \|\mathbf{u} - \mathbf{v}\| = \langle \mathbf{u} - \mathbf{v}, \mathbf{u} - \mathbf{v} \rangle^{1/2}$$

$$= \sqrt{|u_1 - v_1|^2 + |u_2 - v_2|^2 + \cdots + |u_n - v_n|^2}$$

Observe that these are just the formulas for the Euclidean norm and distance discussed in Section 9.4. ▲

Example 26 (For readers who have studied calculus.)

If complex $C[0, 2\pi]$ has the inner product of Example 24, and if $\mathbf{f} = e^{imx}$, where m is any integer, then with the help of Formula (9.21) we obtain

$$\|\mathbf{f}\| = \langle \mathbf{f}, \mathbf{f} \rangle^{1/2} = \left[\int_0^{2\pi} e^{imx}\overline{e^{imx}}\, dx \right]^{1/2}$$

$$= \left[\int_0^{2\pi} e^{imx}e^{-imx}\, dx \right]^{1/2} = \left[\int_0^{2\pi} dx \right]^{1/2} = \sqrt{2\pi} \quad ▲$$

The definitions of such terms as: ***orthogonal vectors, orthogonal set, orthonormal set***, and ***orthonormal basis*** carry over to complex inner product spaces without change. Moreover, Theorems 22–26 in Chapter 4 remain valid in complex inner product spaces, and the Gram-Schmidt process can be used to convert an arbitrary basis for a complex inner product space into an orthonormal basis.

Example 27

The vectors

$$\mathbf{u} = (i, 1) \qquad \text{and} \qquad \mathbf{v} = (1, i)$$

in C^2 are orthogonal with respect to the Euclidean inner product, since

$$\mathbf{u} \cdot \mathbf{v} = (i)(\overline{1}) + (1)(\overline{i}) = (i)(1) + (1)(-i) = 0 \quad \blacktriangle$$

Example 28

Consider the vector space C^3 with the Euclidean inner product. Apply the Gram-Schmidt process to transform the basis $\mathbf{u}_1 = (i, i, i)$, $\mathbf{u}_2 = (0, i, i)$, $\mathbf{u}_3 = (0, 0, i)$ into an orthonormal basis.

Solution.

Step 1. $\mathbf{v}_1 = \dfrac{\mathbf{u}_1}{\|\mathbf{u}_1\|} = \dfrac{(i, i, i)}{\sqrt{3}} = \left(\dfrac{i}{\sqrt{3}}, \dfrac{i}{\sqrt{3}}, \dfrac{i}{\sqrt{3}} \right)$

Step 2. $\mathbf{u}_2 - \text{proj}_{W_1} \mathbf{u}_2 = \mathbf{u}_2 - \langle \mathbf{u}_2, \mathbf{v}_1 \rangle \mathbf{v}_1$

$$= (0, i, i) - \frac{2}{\sqrt{3}} \left(\frac{i}{\sqrt{3}}, \frac{i}{\sqrt{3}}, \frac{i}{\sqrt{3}} \right)$$

$$= \left(-\frac{2i}{3}, \frac{i}{3}, \frac{i}{3} \right)$$

Therefore,

$$\mathbf{v}_2 = \frac{\mathbf{u}_2 - \text{proj}_{W_1} \mathbf{u}_2}{\|\mathbf{u}_2 - \text{proj}_{W_1} \mathbf{u}_2\|} = \frac{3}{\sqrt{6}} \left(-\frac{2i}{3}, \frac{i}{3}, \frac{i}{3} \right) = \left(-\frac{2i}{\sqrt{6}}, \frac{i}{\sqrt{6}}, \frac{i}{\sqrt{6}} \right)$$

Step 3. $\mathbf{u}_3 - \text{proj}_{W_2} \mathbf{u}_3 = \mathbf{u}_3 - \langle \mathbf{u}_3, \mathbf{v}_1 \rangle \mathbf{v}_1 - \langle \mathbf{u}_3, \mathbf{v}_2 \rangle \mathbf{v}_2$

$$= (0, 0, i) - \frac{1}{\sqrt{3}} \left(\frac{i}{\sqrt{3}}, \frac{i}{\sqrt{3}}, \frac{i}{\sqrt{3}} \right) - \frac{1}{\sqrt{6}} \left(-\frac{2i}{\sqrt{6}}, \frac{i}{\sqrt{6}}, \frac{i}{\sqrt{6}} \right)$$

$$= \left(0, -\frac{i}{2}, \frac{i}{2} \right)$$

Therefore,

$$\mathbf{v}_3 = \frac{\mathbf{u}_3 - \text{proj}_{W_2} \mathbf{u}_3}{\|\mathbf{u}_3 - \text{proj}_{W_2} \mathbf{u}_3\|} = \sqrt{2}\left(0, -\frac{i}{2}, \frac{i}{2}\right) = \left(0, -\frac{i}{\sqrt{2}}, \frac{i}{\sqrt{2}}\right)$$

Thus,

$$\mathbf{v}_1 = \left(\frac{i}{\sqrt{3}}, \frac{i}{\sqrt{3}}, \frac{i}{\sqrt{3}}\right), \qquad \mathbf{v}_2 = \left(-\frac{2i}{\sqrt{6}}, \frac{i}{\sqrt{6}}, \frac{i}{\sqrt{6}}\right), \qquad \mathbf{v}_3 = \left(0, -\frac{i}{\sqrt{2}}, \frac{i}{\sqrt{2}}\right)$$

form an orthonormal basis for C^3. ▲

Example 29 (For readers who have studied calculus.)

Let complex $C[0, 2\pi]$ have the inner product of Example 24, and let W be the set of vectors in $C[0, 2\pi]$ of the form

$$e^{im\pi} = \cos m\pi + i \sin m\pi$$

where m is an integer. The set W is orthogonal because if

$$\mathbf{f} = e^{ikx} \qquad \text{and} \qquad \mathbf{g} = e^{ilx}$$

are distinct vectors in W, then

$$\langle \mathbf{f}, \mathbf{g} \rangle = \int_0^{2\pi} e^{ikx}\overline{e^{ilx}}\, dx$$

$$= \int_0^{2\pi} e^{ikx}e^{-ilx}\, dx$$

$$= \int_0^{2\pi} e^{i(k-l)x}\, dx$$

$$= \int_0^{2\pi} \cos(k-l)x\, dx + i\int_0^{2\pi} \sin(k-l)x\, dx$$

$$= \left[\frac{1}{k-l}\sin(k-l)x\right]_0^{2\pi} - i\left[\frac{1}{k-l}\cos(k-l)x\right]_0^{2\pi}$$

$$= (0) - i(0) = 0$$

If we normalize each vector in the orthogonal set W, we obtain an orthonormal set. But in Example 26 we showed that each vector in W has norm $\sqrt{2\pi}$, so the vectors

$$\frac{1}{\sqrt{2\pi}} e^{imx} \qquad m = 0, \pm 1, \pm 2, \ldots$$

form an orthonormal set in complex $C[0, 2\pi]$. ▲

EXERCISE SET 9.5

1. Let $\mathbf{u} = (u_1, u_2)$ and $\mathbf{v} = (v_1, v_2)$. Show that $\langle \mathbf{u}, \mathbf{v} \rangle = 3u_1\bar{v}_1 + 2u_2\bar{v}_2$ defines an inner product on C^2.

2. Compute $\langle \mathbf{u}, \mathbf{v} \rangle$ using the inner product in Exercise 1.
(a) $\mathbf{u} = (2i, -i)$, $\mathbf{v} = (-i, 3i)$
(b) $\mathbf{u} = (0, 0)$, $\mathbf{v} = (1 - i, 7 - 5i)$
(c) $\mathbf{u} = (1 + i, 1 - i)$, $\mathbf{v} = (1 - i, 1 + i)$
(d) $\mathbf{u} = (3i, -1 + 2i)$, $\mathbf{v} = (3i, -1 + 2i)$

3. Let $\mathbf{u} = (u_1, u_2)$ and $\mathbf{v} = (v_1, v_2)$. Show that

$$\langle \mathbf{u}, \mathbf{v} \rangle = u_1\bar{v}_1 + (1 + i)u_1\bar{v}_2 + (1 - i)u_2\bar{v}_1 + 3u_2\bar{v}_2$$

defines an inner product on C^2.

4. Compute $\langle \mathbf{u}, \mathbf{v} \rangle$ using the inner product in Exercise 3.
(a) $\mathbf{u} = (2i, -i)$, $\mathbf{v} = (-i, 3i)$
(b) $\mathbf{u} = (0, 0)$, $\mathbf{v} = (1 - i, 7 - 5i)$
(c) $\mathbf{u} = (1 + i, 1 - i)$, $\mathbf{v} = (1 - i, 1 + i)$
(d) $\mathbf{u} = (3i, -1 + 2i)$, $\mathbf{v} = (3i, -1 + 2i)$

5. Let $\mathbf{u} = (u_1, u_2)$ and $\mathbf{v} = (v_1, v_2)$. Determine which of the following are inner products on C^2. For those that are not, list the axioms that do not hold.
(a) $\langle \mathbf{u}, \mathbf{v} \rangle = u_1\bar{v}_1$
(b) $\langle \mathbf{u}, \mathbf{v} \rangle = u_1\bar{v}_1 - u_2\bar{v}_2$
(c) $\langle \mathbf{u}, \mathbf{v} \rangle = |u_1|^2|v_1|^2 + |u_2|^2|v_2|^2$
(d) $\langle \mathbf{u}, \mathbf{v} \rangle = 2u_1\bar{v}_1 + iu_1\bar{v}_2 + iu_2\bar{v}_1 + 2u_2\bar{v}_2$
(e) $\langle \mathbf{u}, \mathbf{v} \rangle = 2u_1\bar{v}_1 + iu_1\bar{v}_2 - iu_2\bar{v}_1 + 2u_2\bar{v}_2$

6. Use the inner product of Example 23 to find $\langle U, V \rangle$ if

$$U = \begin{bmatrix} -i & 1 + i \\ 1 - i & i \end{bmatrix} \quad \text{and} \quad V = \begin{bmatrix} 3 & -2 - 3i \\ 4i & 1 \end{bmatrix}$$

7. Let $\mathbf{u} = (u_1, u_2, u_3)$ and $\mathbf{v} = (v_1, v_2, v_3)$. Does $\langle \mathbf{u}, \mathbf{v} \rangle = u_1\bar{v}_1 + u_2\bar{v}_2 + u_3\bar{v}_3 - iu_3\bar{v}_1$ define an inner product on C^3? If not, list all axioms that fail to hold.

8. Let V be the vector space of complex-valued functions of the real variable x, and let $\mathbf{f} = f_1(x) + if_2(x)$ and $\mathbf{g} = g_1(x) + ig_2(x)$ be vectors in V. Does

$$\langle \mathbf{f}, \mathbf{g} \rangle = (f_1(0) + if_2(0))\overline{(g_1(0) + ig_2(0))}$$

define an inner product on V? If not, list all axioms that fail to hold.

9. Let C^2 have the inner product of Exercise 1. Find $\|\mathbf{w}\|$ if
(a) $\mathbf{w} = (-i, 3i)$ (b) $\mathbf{w} = (1 - i, 1 + i)$
(c) $\mathbf{w} = (0, 2 - i)$ (d) $\mathbf{w} = (0, 0)$

10. For each vector in Exercise 9, use the Euclidean inner product to find $\|\mathbf{w}\|$.

11. Use the inner product of Exercise 3 to find $\|\mathbf{w}\|$ if
(a) $\mathbf{w} = (1, -i)$ (b) $\mathbf{w} = (1 - i, 1 + i)$ (c) $\mathbf{w} = (3 - 4i, 0)$ (d) $\mathbf{w} = (0, 0)$

12. Use the inner product of Example 23 to find $\|A\|$ if

(a) $A = \begin{bmatrix} -i & 7i \\ 6i & 2i \end{bmatrix}$ (b) $A = \begin{bmatrix} -1 & 1+i \\ 1-i & 3 \end{bmatrix}$

13. Let C^2 have the inner product of Exercise 1. Find $d(\mathbf{x}, \mathbf{y})$ if
  (a) $\mathbf{x} = (1, 1)$, $\mathbf{y} = (i, -i)$ (b) $\mathbf{x} = (1 - i, 3 + 2i)$, $\mathbf{y} = (1 + i, 3)$

14. Repeat the directions of Exercise 13 using the Euclidean inner product on C^2.

15. Repeat the directions of Exercise 13 using the inner product of Exercise 3.

16. Let complex M_{22} have the inner product of Example 23. Find $d(A, B)$ if

(a) $A = \begin{bmatrix} i & 5i \\ 8i & 3i \end{bmatrix}$ and $B = \begin{bmatrix} -5i & 0 \\ 7i & -3i \end{bmatrix}$

(b) $A = \begin{bmatrix} -1 & 1-i \\ 1+i & 2 \end{bmatrix}$ and $B = \begin{bmatrix} 2i & 2-3i \\ i & 1 \end{bmatrix}$

17. Let C^3 have the Euclidean inner product. For which values of k are $\mathbf{u}$ and $\mathbf{v}$ orthogonal?
  (a) $\mathbf{u} = (2i, i, 3i)$, $\mathbf{v} = (i, 6i, k)$ (b) $\mathbf{u} = (k, k, 1 + i)$, $\mathbf{v} = (1, -1, 1 - i)$

18. Let M_{22} have the inner product of Example 23. Determine which of the following are orthogonal to

$$A = \begin{bmatrix} 2i & i \\ -i & 3i \end{bmatrix}$$

(a) $\begin{bmatrix} -3 & 1-i \\ 1-i & 2 \end{bmatrix}$ (b) $\begin{bmatrix} 1 & 1 \\ 0 & -1 \end{bmatrix}$ (c) $\begin{bmatrix} 0 & 0 \\ 0 & 0 \end{bmatrix}$ (d) $\begin{bmatrix} 0 & 1 \\ 3-i & 0 \end{bmatrix}$

19. Let C^3 have the Euclidean inner product. Show that for all values of θ the vector
$$\mathbf{x} = e^{i\theta}\left(\frac{i}{\sqrt{3}}, \frac{1}{\sqrt{3}}, \frac{1}{\sqrt{3}}\right)$$ has norm 1 and is orthogonal to both $(1, i, 0)$ and $(0, i, -i)$.

20. Let C^2 have the Euclidean inner product. Which of the following form orthonormal sets?

(a) $(i, 0)$, $(0, 1 - i)$ (b) $\left(\dfrac{i}{\sqrt{2}}, -\dfrac{i}{\sqrt{2}}\right)$, $\left(\dfrac{i}{\sqrt{2}}, \dfrac{i}{\sqrt{2}}\right)$

(c) $\left(\dfrac{i}{\sqrt{2}}, \dfrac{i}{\sqrt{2}}\right)$, $\left(-\dfrac{i}{\sqrt{2}}, -\dfrac{i}{\sqrt{2}}\right)$ (d) $(i, 0)$, $(0, 0)$

21. Let C^3 have the Euclidean inner product. Which of the following form orthonormal sets?

(a) $\left(\dfrac{i}{\sqrt{2}}, 0, \dfrac{i}{\sqrt{2}}\right)$, $\left(\dfrac{i}{\sqrt{3}}, \dfrac{i}{\sqrt{3}}, -\dfrac{i}{\sqrt{3}}\right)$, $\left(-\dfrac{i}{\sqrt{2}}, 0, \dfrac{i}{\sqrt{2}}\right)$

(b) $\left(\dfrac{2}{3}i, -\dfrac{2}{3}i, \dfrac{1}{3}i\right)$, $\left(\dfrac{2}{3}i, \dfrac{1}{3}i, -\dfrac{2}{3}i\right)$, $\left(\dfrac{1}{3}i, \dfrac{2}{3}i, \dfrac{2}{3}i\right)$

(c) $\left(\dfrac{i}{\sqrt{6}}, \dfrac{i}{\sqrt{6}}, -\dfrac{2i}{\sqrt{6}}\right)$, $\left(\dfrac{i}{\sqrt{2}}, -\dfrac{i}{\sqrt{2}}, 0\right)$

22. Let

$$\mathbf{x} = \left(\dfrac{i}{\sqrt{5}}, -\dfrac{i}{\sqrt{5}}\right) \quad \text{and} \quad \mathbf{y} = \left(\dfrac{2i}{\sqrt{30}}, \dfrac{3i}{\sqrt{30}}\right)$$

Show that $\{\mathbf{x}, \mathbf{y}\}$ is an orthonormal set if C^2 has the inner product

$$\langle \mathbf{u}, \mathbf{v} \rangle = 3u_1\bar{v}_1 + 2u_2\bar{v}_2$$

but is not orthonormal if C^2 has the Euclidean inner product.

23. Show that

$$\begin{aligned}\mathbf{u}_1 &= (i, 0, 0, i) & \mathbf{u}_2 &= (-i, 0, 2i, i)\\ \mathbf{u}_3 &= (2i, 3i, 2i, -2i) & \mathbf{u}_4 &= (-i, 2i, -i, i)\end{aligned}$$

is an orthogonal set in C^4 with the Euclidean inner product. By normalizing each of these vectors, obtain an orthonormal set.

24. Let C^2 have the Euclidean inner product. Use the Gram-Schmidt process to transform the basis $\{\mathbf{u}_1, \mathbf{u}_2\}$ into an orthonormal basis.
(a) $\mathbf{u}_1 = (i, -3i)$, $\mathbf{u}_2 = (2i, 2i)$ (b) $\mathbf{u}_1 = (i, 0)$, $\mathbf{u}_2 = (3i, -5i)$

25. Let C^3 have the Euclidean inner product. Use the Gram-Schmidt process to transform the basis $\{\mathbf{u}_1, \mathbf{u}_2, \mathbf{u}_3\}$ into an orthonormal basis.
(a) $\mathbf{u}_1 = (i, i, i)$ $\mathbf{u}_2 = (-i, i, 0)$ $\mathbf{u}_3 = (i, 2i, i)$
(b) $\mathbf{u}_1 = (i, 0, 0)$ $\mathbf{u}_2 = (3i, 7i, -2i)$ $\mathbf{u}_3 = (0, 4i, i)$

26. Let C^4 have the Euclidean inner product. Use the Gram-Schmidt process to transform the basis $\{\mathbf{u}_1, \mathbf{u}_2, \mathbf{u}_3, \mathbf{u}_4\}$ into an orthonormal basis.

$$\mathbf{u}_1 = (0, 2i, i, 0) \quad \mathbf{u}_2 = (i, -i, 0, 0) \quad \mathbf{u}_3 = (i, 2i, 0, -i) \quad \mathbf{u}_4 = (i, 0, i, i)$$

27. Let C^3 have the Euclidean inner product. Find an orthonormal basis for the subspace spanned by

$$(0, i, 1 - i) \quad \text{and} \quad (-i, 0, 1 + i)$$

28. Let C^4 have the Euclidean inner product. Express $\mathbf{w} = (-i, 2i, 6i, 0)$ in the form $\mathbf{w} = \mathbf{w}_1 + \mathbf{w}_2$, where the vector $\mathbf{w}_1$ is in the space W spanned by $\mathbf{u}_1 = (-i, 0, i, 2i)$ and $\mathbf{u}_2 = (0, i, 0, i)$, and $\mathbf{w}_2$ is orthogonal to W.

29. (a) Prove: If k is a complex number and $\langle \mathbf{u}, \mathbf{v} \rangle$ is an inner product on a complex vector space, then

$$\langle \mathbf{u} - k\mathbf{v}, \mathbf{u} - k\mathbf{v} \rangle = \langle \mathbf{u}, \mathbf{u} \rangle - \bar{k}\langle \mathbf{u}, \mathbf{v} \rangle - k\overline{\langle \mathbf{u}, \mathbf{v} \rangle} + k\bar{k}\langle \mathbf{v}, \mathbf{v} \rangle$$

(b) Use the result in (a) to prove that

$$0 \le \langle \mathbf{u}, \mathbf{u} \rangle - \bar{k}\langle \mathbf{u}, \mathbf{v} \rangle - k\overline{\langle \mathbf{u}, \mathbf{v} \rangle} + k\bar{k}\langle \mathbf{v}, \mathbf{v} \rangle$$

30. Prove that if $\mathbf{u}$ and $\mathbf{v}$ are vectors in a complex inner product space, then

$$|\langle \mathbf{u}, \mathbf{v} \rangle|^2 \le \langle \mathbf{u}, \mathbf{u} \rangle \langle \mathbf{v}, \mathbf{v} \rangle$$

This result, called the *Cauchy-Schwarz inequality for complex inner product spaces*, differs from its real analog (Theorem 20, Section 4.8) in that an absolute value sign must be included in the left side. (*Hint.* Let $k = \langle \mathbf{u}, \mathbf{v} \rangle / \langle \mathbf{v}, \mathbf{v} \rangle$ in the inequality of Exercise 29(b).)

31. Prove: If $\mathbf{u} = (u_1, u_2, \ldots, u_n)$ and $\mathbf{v} = (v_1, v_2, \ldots, v_n)$ are vectors in C^n, then

$$\left| u_1 \bar{v}_1 + u_2 \bar{v}_2 + \cdots + u_n \bar{v}_n \right| \le (|u_1|^2 + |u_2|^2 + \cdots + |u_n|^2)^{1/2}(|v_1|^2 + |v_2|^2 + \cdots + |v_n|^2)^{1/2}$$

This is the complex version of *Cauchy's inequality* discussed in Example 56 of Section 4.8. (*Hint.* Use Exercise 30.)

32. Prove that equality holds in the Cauchy-Schwarz inequality for complex vector spaces if and only if $\mathbf{u}$ and $\mathbf{v}$ are linearly dependent.

33. Prove that if $\langle \mathbf{u}, \mathbf{v} \rangle$ is an inner product on a complex vector space, then

$$\langle \mathbf{0}, \mathbf{v} \rangle = \langle \mathbf{v}, \mathbf{0} \rangle = 0$$

34. Prove that if $\langle \mathbf{u}, \mathbf{v} \rangle$ is an inner product on a complex vector space, then

$$\langle \mathbf{u}, \mathbf{v} + \mathbf{w} \rangle = \langle \mathbf{u}, \mathbf{v} \rangle + \langle \mathbf{u}, \mathbf{w} \rangle$$

35. Theorem 21 of Section 4.8 remains true in complex inner product spaces. In each part prove that this is so.
(a) part L1   (b) part L2   (c) part L3   (d) part L4
(e) part D1   (f) part D2   (g) part D3   (h) part D4

36. In Example 28 it was shown that the vectors

$$\mathbf{v}_1 = \left(\frac{i}{\sqrt{3}}, \frac{i}{\sqrt{3}}, \frac{i}{\sqrt{3}} \right) \qquad \mathbf{v}_2 = \left(-\frac{2i}{\sqrt{6}}, \frac{i}{\sqrt{6}}, \frac{i}{\sqrt{6}} \right) \qquad \mathbf{v}_3 = \left(0, -\frac{i}{\sqrt{2}}, \frac{i}{\sqrt{2}} \right)$$

form an orthonormal basis for C^3. Use Theorem 23 of Section 4.9 to express $\mathbf{u} = (1 - i, 1 + i, 1)$ as a linear combination of these vectors.

37. Prove that if $\mathbf{u}$ and $\mathbf{v}$ are vectors in a complex inner product space, then

$$\langle \mathbf{u}, \mathbf{v} \rangle = \frac{1}{4} \|\mathbf{u} + \mathbf{v}\|^2 - \frac{1}{4} \|\mathbf{u} - \mathbf{v}\|^2 + \frac{i}{4} \|\mathbf{u} + i\mathbf{v}\|^2 - \frac{i}{4} \|\mathbf{u} - i\mathbf{v}\|^2$$

38. Prove: If $\mathbf{v}_1, \mathbf{v}_2, \ldots, \mathbf{v}_n$ is an orthonormal basis for a complex inner product space V, and if $\mathbf{u}$ and $\mathbf{w}$ are any vectors in V, then

$$\langle \mathbf{u}, \mathbf{w} \rangle = \langle \mathbf{u}, \mathbf{v}_1 \rangle \overline{\langle \mathbf{w}, \mathbf{v}_1 \rangle} + \langle \mathbf{u}, \mathbf{v}_2 \rangle \overline{\langle \mathbf{w}, \mathbf{v}_2 \rangle} + \cdots + \langle \mathbf{u}, \mathbf{v}_n \rangle \overline{\langle \mathbf{w}, \mathbf{v}_n \rangle}$$

(*Hint.* Use Theorem 23, Section 4.9, to express $\mathbf{u}$ and $\mathbf{w}$ as linear combinations of the basis vectors.)

39. **(For readers who have studied calculus.)** Prove that if $\mathbf{f} = f_1(x) + i f_2(x)$ and $\mathbf{g} = g_1(x) + i g_2(x)$ are vectors in complex $C[a, b]$, then the formula

$$\langle \mathbf{f}, \mathbf{g} \rangle = \int_a^b [f_1(x) + if_2(x)]\overline{[g_1(x) + ig_2(x)]} \, dx$$

defines a complex inner product on $C[a, b]$.

40. **(For readers who have studied calculus.)**
    (a) Let $\mathbf{f} = f_1(x) + if_2(x)$ and $\mathbf{g} = g_1(x) + ig_2(x)$ be vectors in $C[0, 1]$ and let this space have the inner product

    $$\langle \mathbf{f}, \mathbf{g} \rangle = \int_0^1 [f_1(x) + if_2(x)]\overline{[g_1(x) + ig_2(x)]} \, dx$$

    Show that the vectors

    $$e^{2\pi i m x}, \qquad m = 0, \pm 1, \pm 2, \ldots$$

    form an orthogonal set.
    (b) Obtain an orthonormal set by normalizing the vectors in part (a).

9.6 UNITARY, NORMAL, AND HERMITIAN MATRICES

For matrices with real entries, the orthogonal matrices ($A^{-1} = A^t$) and the symmetric ($A = A^t$) played an important role in the diagonalization problem (Section 6.3). For matrices with complex entries, the orthogonal and symmetric matrices are of relatively little importance; they are superseded by two new classes of matrices, the *unitary* and *Hermitian* matrices, which we will discuss in this section.

If A is a matrix with complex entries, then the **conjugate transpose** of A, denoted by A^*, is defined by

$$A^* = \bar{A}^t$$

where $\bar{A}$ is the matrix whose entries are the complex conjugates of the corresponding entries in A and $\bar{A}^t$ is the transpose of $\bar{A}$.

Example 30

If

$$A = \begin{bmatrix} 1 + i & -i & 0 \\ 2 & 3 - 2i & i \end{bmatrix}$$

then

$$\bar{A} = \begin{bmatrix} 1 - i & i & 0 \\ 2 & 3 + 2i & -i \end{bmatrix}$$

So

$$A^* = \bar{A}^t = \begin{bmatrix} 1 - i & 2 \\ i & 3 + 2i \\ 0 & -i \end{bmatrix}$$

The basic properties of the conjugate transpose operation are similar to those of the tranpose.

Theorem 5. *If A and B are matrices with complex entries and k is any complex number, then*

(a) $(A^*)^* = A$
(b) $(A + B)^* = A^* + B^*$
(c) $(kA)^* = \bar{k}A^*$
(d) $(AB)^* = B^*A^*$

The proofs are left as exercises.

Recall that a matrix with real entries is called *orthogonal* if $A^{-1} = A^t$. The complex analogs of the orthogonal matrices are called *unitary* matrices. They are defined as follows:

Definition. A square matrix A with complex entries is called **unitary** if

$$A^{-1} = A^*$$

The following theorem parallels Theorem 33 of Section 4.10.

Theorem 6. *If A is an n × n matrix with complex entries, then the following are equivalent:*

(a) *A is unitary.*
(b) *The row vectors of A form an orthonormal set in C^n with the Euclidean inner product.*
(c) *The column vectors of A form an orthonormal set in C^n with the Euclidean inner product.*

Example 31

The matrix

$$A = \begin{bmatrix} \dfrac{1+i}{2} & \dfrac{1+i}{2} \\[2mm] \dfrac{1-i}{2} & \dfrac{-1+i}{2} \end{bmatrix} \tag{9.25}$$

has row vectors

$$\mathbf{r}_1 = \left(\frac{1+i}{2}, \frac{1+i}{2}\right) \qquad \mathbf{r}_2 = \left(\frac{1-i}{2}, \frac{-1+i}{2}\right)$$

Relative to the Euclidean inner product on C^n, we have

$$\|\mathbf{r}_1\| = \sqrt{\left|\frac{1+i}{2}\right|^2 + \left|\frac{1+i}{2}\right|^2} = \sqrt{\frac{1}{2} + \frac{1}{2}} = 1$$

$$\|\mathbf{r}_2\| = \sqrt{\left|\frac{1-i}{2}\right|^2 + \left|\frac{-1+i}{2}\right|^2} = \sqrt{\frac{1}{2} + \frac{1}{2}} = 1$$

and

$$\mathbf{r}_1 \cdot \mathbf{r}_2 = \left(\frac{1+i}{2}\right)\overline{\left(\frac{1-i}{2}\right)} + \left(\frac{1+i}{2}\right)\overline{\left(\frac{-1+i}{2}\right)}$$

$$= \left(\frac{1+i}{2}\right)\left(\frac{1+i}{2}\right) + \left(\frac{1+i}{2}\right)\left(\frac{-1-i}{2}\right)$$

$$= i - i = 0$$

so the row vectors form an orthonormal set in C^2. Thus, A is unitary and

$$A^{-1} = A^* = \begin{bmatrix} \dfrac{1-i}{2} & \dfrac{1+i}{2} \\[2mm] \dfrac{1-i}{2} & \dfrac{-1-i}{2} \end{bmatrix} \tag{9.26}$$

The reader should verify that (9.26) is the inverse of (9.25) by showing that $AA^* = A^*A = I$.

Recall that a square matrix A with real entries is called *orthogonally diagonalizable* if there is an orthogonal matrix P such that $P^{-1}AP \ (=P^t AP)$ is diagonal. For complex matrices we have an analogous concept.

Definition. A square matrix A with complex entries is called **unitarily diagonalizable** if there is a unitary P such that $P^{-1}AP(=P^*AP)$ is diagonal; the matrix P is said to **unitarily diagonalize** A.

We have two questions to consider:

1. Which matrices are unitarily diagonalizable?
2. How do we find a unitary matrix P to carry out the diagonalization?

Before pursuing these questions, we note that our earlier definitions of the terms *eigenvector, eigenvalue, eigenspace, characteristic equation,* and *characteristic polynomial* carry over without change to complex vector spaces.

In Section 6.3 we saw that the symmetric matrices played a fundamental role in the problem of orthogonally diagonalizing a matrix with real entries. The most natural complex analogs of the real symmetric matrices are the *Hermitian*[†] matrices, which are defined as follows:

Definition. A square matrix A with complex entries is called **Hermitian**[†] if

$$A = A^*$$

Example 32

If

$$A = \begin{bmatrix} 1 & i & 1+i \\ -i & -5 & 2-i \\ 1-i & 2+i & 3 \end{bmatrix}$$

then

$$\bar{A} = \begin{bmatrix} 1 & -i & 1-i \\ i & -5 & 2+i \\ 1+i & 2-i & 3 \end{bmatrix}$$

so

$$A^* = \bar{A}^t = \begin{bmatrix} 1 & i & 1+i \\ -i & -5 & 2-i \\ 1-i & 2+i & 3 \end{bmatrix} = A$$

which means that A is Hermitian. ▲

It is easy to recognize Hermitian matrices by inspection: the entries on the main diagonal are real numbers (Exercise 17), and the "mirror image" of each entry across the main diagonal is its complex conjugate (Figure 9.12).

[†] *Charles Hermite (1822–1901)* was a French mathematician who made fundamental contributions to algebra, matrix theory, and various branches of analysis. He is noted for giving the first solution of a general fifth-degree polynomial equation and proving that the number *e* (the base for natural logarithms) is not the root of any polynomial equation with rational coefficients.

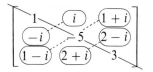

Figure 9.12

Hermitian matrices enjoy many but not all of the properties of real symmetric matrices. For example, just as the real symmetric matrices are orthogonally diagonalizable, so we shall see that the Hermitian matrices are unitarily diagonalizable. However, whereas the real symmetric matrices are the only matrices with real entries that are orthogonally diagonalizable (Theorem 5, Section 6.3), the Hermitian matrices do not constitute the entire class of unitarily diagonalizable matrices; that is, we shall see that there are unitarily diagonalizable matrices that are not Hermitian. To explain why this is so we shall need the following definition:

Definition. A square matrix A with complex entries is called **normal** if

$$AA^* = A^*A$$

Example 33

Every Hermitian matrix A is normal since $AA^* = AA = A^*A$, and every unitary matrix A is normal since $AA^* = I = A^*A$.

The following two theorems are the complex analogs of Theorems 5 and 6 in Section 6.3. The proofs will be omitted.

Theorem 7. *If A is a square matrix with complex entries, then the following are equivalent:*

(a) A is unitarily diagonalizable.
(b) A has an orthonormal set of n eigenvectors.
(c) A is normal.

Theorem 8. *If A is a normal matrix, then eigenvectors from different eigenspaces are orthogonal.*

Theorem 7 tells us that a square matrix A with complex entries is unitarily diagonalizable if and only if it is normal. Theorem 8 will be the key to constructing a matrix that unitarily diagonalizes a normal matrix.

We saw in Section 6.3 that a symmetric matrix A is orthogonally diagonalized by any orthogonal matrix whose column vectors are eigenvectors of A. Similarly, a normal matrix A is diagonalized by any unitary matrix whose column vectors are eigenvectors of A. The procedure for diagonalizing a normal matrix is as follows:

Step 1. Find a basis for each eigenspace of A.

Step 2. Apply the Gram-Schmidt process to each of these bases to obtain an orthonormal basis for each eigenspace.

Step 3. Form the matrix P whose columns are the basis vectors constructed in Step 2. This matrix unitarily diagonalizes A.

The justification of this procedure should be clear. Theorem 8 ensures that eigenvectors from *different* eigenspaces are orthogonal, and the application of the Gram-Schmidt process ensures that the eigenvectors within the *same* eigenspace are orthonormal. Thus, the *entire* set of eigenvectors obtained by this procedure is orthonormal.

Example 34

The matrix

$$A = \begin{bmatrix} 2 & 1+i \\ 1-i & 3 \end{bmatrix}$$

is unitarily diagonalizable because it is Hermitian and therefore normal. Find a matrix P that unitarily diagonalizes A.

Solution. The characteristic polynomial of A is

$$\det(\lambda I - A) = \det \begin{bmatrix} \lambda - 2 & -1-i \\ -1+i & \lambda - 3 \end{bmatrix} = (\lambda - 2)(\lambda - 3) - 2 = \lambda^2 - 5\lambda + 4$$

so the characteristic equation is

$$\lambda^2 - 5\lambda + 4 = (\lambda - 1)(\lambda - 4) = 0$$

and the eigenvalues are $\lambda = 1$ and $\lambda = 4$.

By definition,

$$\mathbf{x} = \begin{bmatrix} x_1 \\ x_2 \end{bmatrix}$$

will be an eigenvector of A corresponding to λ if and only if $\mathbf{x}$ is a nontrivial solution of

$$\begin{bmatrix} \lambda - 2 & -1-i \\ -1+i & \lambda - 3 \end{bmatrix} \begin{bmatrix} x_1 \\ x_2 \end{bmatrix} = \begin{bmatrix} 0 \\ 0 \end{bmatrix} \qquad (9.27)$$

To find the eigenvectors corresponding to $\lambda = 1$, we substitute this value in (9.27):

$$\begin{bmatrix} -1 & -1-i \\ -1+i & -2 \end{bmatrix} \begin{bmatrix} x_1 \\ x_2 \end{bmatrix} = \begin{bmatrix} 0 \\ 0 \end{bmatrix}$$

Solving this system by Gauss-Jordan elimination yields (verify):

$$x_1 = (-1-i)s, \qquad x_2 = s$$

Thus, the eigenvectors of A corresponding to $\lambda = 1$ are the nonzero vectors in C^2 of the form

$$\mathbf{x} = \begin{bmatrix} (-1-i)s \\ s \end{bmatrix} = s \begin{bmatrix} -1-i \\ 1 \end{bmatrix}$$

Thus, this eigenspace is one-dimensional with basis

$$\mathbf{u} = \begin{bmatrix} -1-i \\ 1 \end{bmatrix} \qquad\qquad (9.28)$$

In this case the Gram-Schmidt process involves only one step: normalizing this vector. Since

$$\|\mathbf{u}\| = \sqrt{|-1-i|^2 + |1|^2} = \sqrt{2+1} = \sqrt{3}$$

the vector

$$\mathbf{p}_1 = \frac{\mathbf{u}}{\|\mathbf{u}\|} = \begin{bmatrix} \dfrac{-1-i}{\sqrt{3}} \\ \dfrac{1}{\sqrt{3}} \end{bmatrix}$$

is an orthonormal basis for the eigenspace corresponding to $\lambda = 1$.

To find the eigenvectors corresponding to $\lambda = 4$, we substitute this value in (9.27):

$$\begin{bmatrix} 2 & -1-i \\ -1+i & 1 \end{bmatrix} \begin{bmatrix} x_1 \\ x_2 \end{bmatrix} = \begin{bmatrix} 0 \\ 0 \end{bmatrix}$$

Solving this system by Gauss-Jordan elimination yields (verify)

$$x_1 = \left(\frac{1+i}{2}\right)s, \qquad x_2 = s$$

so the eigenvectors of A corresponding to $\lambda = 4$ are the nonzero vectors in C^2 of the form

$$\mathbf{x} = \begin{bmatrix} \left(\dfrac{1+i}{2}\right)s \\ s \end{bmatrix} = s \begin{bmatrix} \dfrac{1+i}{2} \\ 1 \end{bmatrix}$$

Thus, the eigenspace is one-dimensional with basis

$$\mathbf{u} = \begin{bmatrix} \dfrac{1+i}{2} \\ 1 \end{bmatrix}$$

Applying the Gram-Schmidt process (that is, normalizing this vector) yields

$$\mathbf{p}_2 = \frac{\mathbf{u}}{\|\mathbf{u}\|} = \begin{bmatrix} \dfrac{1+i}{\sqrt{6}} \\ \dfrac{2}{\sqrt{6}} \end{bmatrix}$$

Thus,

$$P = [\mathbf{p}_1 \mid \mathbf{p}_2] = \begin{bmatrix} \dfrac{-1-i}{\sqrt{3}} & \dfrac{1+i}{\sqrt{6}} \\ \dfrac{1}{\sqrt{3}} & \dfrac{2}{\sqrt{6}} \end{bmatrix}$$

diagonalizes A and

$$P^{-1}AP = \begin{bmatrix} 1 & 0 \\ 0 & 4 \end{bmatrix}$$

In Theorem 7 of Section 6.3 it was stated that the characteristic equation of a symmetric matrix with real entries has only real roots, or equivalently, the eigenvalues of a symmetric matrix with real entries are real numbers. This important result is a corollary of the following more general theorem.

Theorem 9. *The eigenvalues of a Hermitian matrix are real numbers.*

Proof. If λ is an eigenvalue and $\mathbf{v}$ a corresponding eigenvector of an $n \times n$ Hermitian matrix A, then

$$A\mathbf{v} = \lambda\mathbf{v}$$

Multiplying both sides of this equation on the left by the conjugate transpose of $\mathbf{v}$ yields

$$\mathbf{v}^*A\mathbf{v} = \mathbf{v}^*(\lambda\mathbf{v}) = \lambda\mathbf{v}^*\mathbf{v} \qquad (9.29)$$

We will show that the 1×1 matrices $\mathbf{v}^*A\mathbf{v}$ and $\mathbf{v}^*\mathbf{v}$ both have real entries, so it will follow from (9.29) that λ must be a real number.

But **v*** *A***v** and **v*****v** are Hermitian, since

$$(\mathbf{v}^* A\mathbf{v})^* = \mathbf{v}^* A^* (\mathbf{v}^*)^* = \mathbf{v}^* A\mathbf{v}$$

and

$$(\mathbf{v}^* \mathbf{v})^* = \mathbf{v}^* (\mathbf{v}^*)^* = \mathbf{v}^* \mathbf{v}$$

Since Hermitian matrices have real entries on the main diagonal, and since **v*** *A***v** and **v*****v** are 1×1, it follows that these matrices have real entries, which completes the proof. ▮

Corollary 10. *The eigenvalues of a symmetric matrix with real entries are real numbers.*

Proof. Let A be a symmetric matrix with real entries. Because the entries in A are real, it follows that

$$\bar{A} = A$$

But this implies that A is Hermitian, since

$$A^* = (\bar{A})^t = A^t = A$$

Thus, A has real eigenvalues by Theorem 9. ▮

EXERCISE SET 9.6

1. In each part find A^*.

(a) $A = \begin{bmatrix} 2i & 1-i \\ 4 & 3+i \\ 5+i & 0 \end{bmatrix}$ (b) $A = \begin{bmatrix} 2i & 1-i & -1+i \\ 4 & 5-7i & -i \\ i & 3 & 1 \end{bmatrix}$

(c) $A = [7i \quad 0 \quad -3i]$ (d) $A = \begin{bmatrix} a_{11} & a_{12} & a_{13} \\ a_{21} & a_{22} & a_{23} \end{bmatrix}$

2. Which of the following are Hermitian matrices?

(a) $\begin{bmatrix} 0 & i \\ i & 2 \end{bmatrix}$ (b) $\begin{bmatrix} 1 & 1+i \\ 1-i & -3 \end{bmatrix}$ (c) $\begin{bmatrix} i & i \\ -i & i \end{bmatrix}$

(d) $\begin{bmatrix} -2 & 1-i & -1+i \\ 1+i & 0 & 3 \\ -1-i & 3 & 5 \end{bmatrix}$ (e) $\begin{bmatrix} 1 & 0 & 0 \\ 0 & 1 & 0 \\ 0 & 0 & 1 \end{bmatrix}$

3. Find k, l, and m to make A a Hermitian matrix.

$$A = \begin{bmatrix} -1 & k & -i \\ 3 - 5i & 0 & m \\ l & 2 + 4i & 2 \end{bmatrix}$$

4. Use Theorem 6 to determine which of the following are unitary matrices.

(a) $\begin{bmatrix} i & 0 \\ 0 & i \end{bmatrix}$

(b) $\begin{bmatrix} \dfrac{i}{\sqrt{2}} & \dfrac{1}{\sqrt{2}} \\ -\dfrac{i}{\sqrt{2}} & \dfrac{1}{\sqrt{2}} \end{bmatrix}$

(c) $\begin{bmatrix} 1 + i & 1 + i \\ 1 - i & -1 + i \end{bmatrix}$

(d) $\begin{bmatrix} -\dfrac{i}{\sqrt{2}} & \dfrac{i}{\sqrt{6}} & \dfrac{i}{\sqrt{3}} \\ 0 & -\dfrac{i}{\sqrt{6}} & \dfrac{i}{\sqrt{3}} \\ \dfrac{i}{\sqrt{2}} & \dfrac{i}{\sqrt{6}} & \dfrac{i}{\sqrt{3}} \end{bmatrix}$

5. In each part verify that the matrix is unitary and find its inverse.

(a) $\begin{bmatrix} \dfrac{3}{5} & \dfrac{4}{5}i \\ -\dfrac{4}{5} & \dfrac{3}{5}i \end{bmatrix}$

(b) $\begin{bmatrix} \dfrac{1}{\sqrt{2}} & \dfrac{1}{\sqrt{2}} \\ -\dfrac{1 + i}{2} & \dfrac{1 + i}{2} \end{bmatrix}$

(c) $\begin{bmatrix} \dfrac{1}{4}(\sqrt{3} + i) & \dfrac{1}{4}(1 - i\sqrt{3}) \\ \dfrac{1}{4}(1 + i\sqrt{3}) & \dfrac{1}{4}(i - \sqrt{3}) \end{bmatrix}$

(d) $\begin{bmatrix} \dfrac{1 + i}{2} & -\dfrac{1}{2} & \dfrac{1}{2} \\ \dfrac{i}{\sqrt{3}} & \dfrac{1}{\sqrt{3}} & -\dfrac{i}{\sqrt{3}} \\ \dfrac{3 + i}{2\sqrt{15}} & \dfrac{4 + 3i}{2\sqrt{15}} & \dfrac{5i}{2\sqrt{15}} \end{bmatrix}$

6. Show that the matrix

$$\frac{1}{\sqrt{2}} \begin{bmatrix} e^{i\theta} & e^{-i\theta} \\ ie^{i\theta} & -ie^{-i\theta} \end{bmatrix}$$

is unitary for every real value of θ.

In Exercises 7–12 find a unitary matrix P that diagonalizes A, and determine $P^{-1} AP$.

7. $A = \begin{bmatrix} 4 & 1 - i \\ 1 + i & 5 \end{bmatrix}$

8. $A = \begin{bmatrix} 3 & -i \\ i & 3 \end{bmatrix}$

9. $A = \begin{bmatrix} 6 & 2 + 2i \\ 2 - 2i & 4 \end{bmatrix}$

10. $A = \begin{bmatrix} 0 & 3 + i \\ 3 - i & -3 \end{bmatrix}$

11. $A = \begin{bmatrix} 5 & 0 & 0 \\ 0 & -1 & -1 + i \\ 0 & -1 - i & 0 \end{bmatrix}$

12. $A = \begin{bmatrix} 2 & \dfrac{i}{\sqrt{2}} & -\dfrac{i}{\sqrt{2}} \\ -\dfrac{i}{\sqrt{2}} & 2 & 0 \\ \dfrac{i}{\sqrt{2}} & 0 & 2 \end{bmatrix}$

13. Show that the eigenvalues of the symmetric matrix

$$A = \begin{bmatrix} 1 & 4i \\ 4i & 3 \end{bmatrix}$$

are not real. Does this violate Corollary 10?

14. Find a 2×2 matrix that is both Hermitian and unitary and whose entries are not all real numbers.

15. Prove: If A is an $n \times n$ matrix with complex entries, then $\det(\bar{A}) = \overline{\det(A)}$. (***Hint.*** First show that the signed elementary products from $\bar{A}$ are the conjugates of the signed elementary products from A.)

16. (a) Use the result of Exercise 15 to prove that if A is an $n \times n$ matrix with complex entries, then $\det(A^*) = \overline{\det(A)}$.
   (b) Prove: If A is Hermitian, then $\overline{\det(A)}$ is real.
   (c) Prove: If A is unitary, then $|\det(A)| = 1$.

17. Prove that the entries on the main diagonal of a Hermitian matrix are real numbers.

18. Let

$$A = \begin{bmatrix} a_{11} & a_{12} & a_{13} \\ a_{21} & a_{22} & a_{23} \\ a_{31} & a_{32} & a_{33} \end{bmatrix} \quad \text{and} \quad B = \begin{bmatrix} b_{11} & b_{12} & b_{13} \\ b_{21} & b_{22} & b_{23} \\ b_{31} & b_{32} & b_{33} \end{bmatrix}$$

be matrices with complex entries. Show that
   (a) $(A^*)^* = A$ (b) $(A + B)^* = A^* + B^*$ (c) $(kA)^* = \bar{k}A^*$ (d) $(AB)^* = B^*A^*$

19. Prove: If A is invertible, then so is A^* in which case $(A^*)^{-1} = (A^{-1})^*$.

20. Show that if A is a unitary matrix, then A^* is also unitary.

21. Prove that an $n \times n$ matrix with complex entries is unitary if and only if its rows form an orthonormal set in C^n with the Euclidean inner product.

22. Use Exercises 20 and 21 to show that an $n \times n$ matrix is unitary if and only if its columns form an orthonormal set in C^n with the Euclidean inner product.

23. Prove: If $A = A^*$, then for every vector $\mathbf{x}$ in C^n, the entry in the 1×1 matrix $\mathbf{x}^*A\mathbf{x}$ is real.

24. Let λ and μ be distinct eigenvalues of a Hermitian matrix A.

(a) Prove that if $\mathbf{x}$ is an eigenvector corresponding to λ and $\mathbf{y}$ an eigenvector corresponding to μ, then

$$\mathbf{x}^*A\mathbf{y} = \lambda\mathbf{x}^*\mathbf{y} \qquad \text{and} \qquad \mathbf{x}^*A\mathbf{y} = \mu\mathbf{x}^*\mathbf{y}$$

(b) Prove Theorem 8. (***Hint.*** Subtract the equations in part (a).)

SUPPLEMENTARY EXERCISES

1. Let $\mathbf{u}=(u_1, u_2, \ldots, u_n)$ and $\mathbf{v}=(v_1, v_2, \ldots, v_n)$ be vectors in C^n, and let $\bar{\mathbf{u}}=(\bar{u}_1, \bar{u}_2, \ldots, \bar{u}_n)$ and $\bar{\mathbf{v}} = (\bar{v}_1, \bar{v}_2, \ldots, \bar{v}_n)$.

(a) Prove: $\overline{\mathbf{u} \cdot \mathbf{v}} = \bar{\mathbf{u}} \cdot \bar{\mathbf{v}}$.

(b) Prove: $\mathbf{u}$ and $\mathbf{v}$ are orthogonal if and only if $\bar{\mathbf{u}}$ and $\bar{\mathbf{v}}$ are orthogonal.

2. Show that if the matrix

$$\begin{bmatrix} a & b \\ -\bar{b} & \bar{a} \end{bmatrix}$$

is nonzero, then it is invertible.

3. Find a basis for the solution space of the system

$$\begin{bmatrix} -1 & -i & 1 \\ -i & 1 & i \\ 1 & i & -1 \end{bmatrix} \begin{bmatrix} x_1 \\ x_2 \\ x_3 \end{bmatrix} = \begin{bmatrix} 0 \\ 0 \\ 0 \end{bmatrix}$$

4. Prove: If a and b are complex numbers such that $|a|^2 + |b|^2 = 1$, and if θ is a real number, then

$$A = \begin{bmatrix} a & b \\ -e^{i\theta}\bar{b} & e^{i\theta}\bar{a} \end{bmatrix}$$

is a unitary matrix.

5. Find the eigenvalues of the matrix

$$\begin{bmatrix} 0 & 0 & 1 \\ 1 & 0 & \omega + 1 + \dfrac{1}{\omega} \\ 0 & 1 & -\omega - 1 - \dfrac{1}{\omega} \end{bmatrix}$$

where $\omega = e^{2\pi i/3}$.

6. (a) Prove that if z is a complex number other than 1, then

$$1 + z + z^2 + \cdots + z^n = \frac{1 - z^{n+1}}{1 - z}$$

(***Hint.*** Let S be the sum on the left side of the equation and consider the quantity $S - zS$.)

(b) Use the result in (a) to prove that if $z^n = 1$ and $z \neq 1$, then

$$1 + z + z^2 + \cdots + z^{n-1} = 0$$

(c) Use the result in (a) to obtain Lagrange's trigonometric identity

$$1 + \cos\theta + \cos 2\theta + \cdots + \cos n\theta = \frac{1}{2} + \frac{\sin[(n + \frac{1}{2})\theta]}{2\sin(\theta/2)}$$

for $0 < \theta < 2\pi$. (**Hint.** Let $z = \cos\theta + i\sin\theta$.)

7. Let $\omega = e^{2\pi i/3}$. Show that the vectors $\mathbf{v}_1 = (1/\sqrt{3})\,(1, 1, 1)$, $\mathbf{v}_2 = (1/\sqrt{3})\,(1, \omega, \omega^2)$, and $\mathbf{v}_3 = (1/\sqrt{3})\,(1, \omega^2, \omega^4)$ form an orthonormal set in C^3. (**Hint.** Use part (b) of Exercise 6.)

8. Show that if U is an $n \times n$ unitary matrix and $|z_1| = |z_2| = \cdots = |z_n| = 1$, then the product

$$U\begin{bmatrix} z_1 & 0 & 0 & \cdots & 0 \\ 0 & z_2 & 0 & \cdots & 0 \\ \vdots & \vdots & \vdots & & \vdots \\ 0 & 0 & 0 & \cdots & z_n \end{bmatrix}$$

is also unitary.

9. Suppose that $A^* = -A$.
   (a) Show that iA is Hermitian.
   (b) Show that A is unitarily diagonalizable and has pure imaginary eigenvalues.

10. (a) Show that the set of complex numbers with the operations $(a + bi) + (c + di) = (a + c) + (b + d)i$ and $k(a + bi) = ka + kbi$, where k is a real number, is a *real* vector space.
    (b) What is the dimension of this space?

Answers
to Exercises

EXERCISE SET 1.1 (page 7)

1. b, d, f

2. (a) $x = \frac{7}{6}t + \frac{1}{2}, y = t$
   (b) $x_1 = -2s + \frac{7}{2}t + 4, x_2 = s, x_3 = t$
   (c) $x_1 = \frac{4}{3}r - \frac{7}{3}s + \frac{8}{3}t - \frac{5}{3}, x_2 = r, x_3 = s, x_4 = t$
   (d) $v = \frac{1}{2}q - \frac{3}{2}r - \frac{1}{2}s + 2t, w = q, x = r, y = s, z = t$

3. (a) $\begin{bmatrix} 1 & -2 & 0 \\ 3 & 4 & -1 \\ 2 & -1 & 3 \end{bmatrix}$
   (b) $\begin{bmatrix} 1 & 0 & 1 & 1 \\ -1 & 2 & -1 & 3 \end{bmatrix}$

   (c) $\begin{bmatrix} 1 & 0 & 1 & 0 & 0 & 1 \\ 0 & 2 & -1 & 0 & 1 & 2 \\ 0 & 0 & 2 & 1 & 0 & 3 \end{bmatrix}$
   (d) $\begin{bmatrix} 1 & 0 & 1 \\ 0 & 1 & 2 \end{bmatrix}$

4. (a) $\begin{aligned} x_1 \quad\;\; - x_3 &= 2 \\ 2x_1 + x_2 + x_3 &= 3 \\ -x_2 + 2x_3 &= 4 \end{aligned}$
   (b) $\begin{aligned} x_1 \qquad\;\; &= 0 \\ x_2 \;\; &= 0 \\ x_1 - x_2 &= 1 \end{aligned}$

   (c) $\begin{aligned} x_1 + 2x_2 + 3x_3 + 4x_4 &= 5 \\ 5x_1 + 4x_2 + 3x_3 + 2x_4 &= 1 \end{aligned}$
   (d) $\begin{aligned} x_1 \qquad\qquad\quad &= 1 \\ x_2 \qquad\quad &= 2 \\ x_3 \quad &= 3 \\ x_4 &= 4 \end{aligned}$

5. $k = 6$ infinitely many solutions
   $k \neq 6$ no solutions

6. (a) The lines have no common point of intersection.
   (b) The lines intersect in exactly one point.
   (c) The three lines coincide.

EXERCISE SET 1.2 (page 17)

1. d, f **2.** b, c, f

3. (a) $x_1 = 4, x_2 = 3, x_3 = 2$ (b) $x_1 = 2 - 3t, x_2 = 4 + t, x_3 = 2 - t, x_4 = t$
(c) $x_1 = -1 - 5s - 5t, x_2 = s, x_3 = 1 - 3t, x_4 = 2 - 4t, x_5 = t$
(d) Inconsistent

4. (a) $x_1 = 4, x_2 = 3, x_3 = 2$ (b) $x_1 = 2 - 3t, x_2 = 4 + t, x_3 = 2 - t, x_4 = t$
(c) $x_1 = -1 - 5s - 5t, x_2 = s, x_3 = 1 - 3t, x_4 = 2 - 4t, x_5 = t$
(d) Inconsistent

5. (a) $x_1 = 3, x_2 = 1, x_3 = 2$ (b) $x_1 = -3t, x_2 = -4t, x_3 = 7t$
(c) $x = s - 1, y = 2r, z = r, w = s$

7. (a) Inconsistent   (b) $x_1 = -4, x_2 = 2, x_3 = 7$ (c) $x_1 = 3 + 2t, x_2 = t$

9. (a) $x_1 = 0, x_2 = -3t, x_3 = t$ (b) Inconsistent

11. (a) $x_1 = \frac{2}{3}a - \frac{1}{9}b, x_2 = -\frac{1}{3}a + \frac{2}{9}b$
(b) $x_1 = a - \frac{1}{3}c, x_2 = a - \frac{1}{2}b, x_3 = -a + \frac{1}{2}b + \frac{1}{3}c$

12. $a = 4$ infinitely many, $a = -4$ none, $a \neq \pm 4$ exactly one.

14. $\begin{bmatrix} 1 & 3 \\ 0 & 1 \end{bmatrix}$ and $\begin{bmatrix} 1 & 0 \\ 0 & 1 \end{bmatrix}$ are possible answers.   **15.** $\alpha = \frac{\pi}{2}, \beta = \pi, \gamma = 0$.

EXERCISE SET 1.3 (page 22)

1. a, c, d **2.** $x_1 = 0, x_2 = 0, x_3 = 0$

3. $x_1 = -\frac{1}{4}s, x_2 = -\frac{1}{4}s - t, x_3 = s, x_4 = t$ **4.** $x_1 = 0, x_2 = 0, x_3 = 0, x_4 = 0$

5. $x = \frac{t}{8}, y = \frac{5t}{16}, z = t$ **6.** $\lambda = 4, \lambda = 2$

11. One possible answer is $x + y + z = 0$
$$x + y + z = 1$$

EXERCISE SET 1.4 (page 30)

1. (a) Undefined   (b) 4×2 (c) Undefined   (d) Undefined
(e) 5×5 (f) 5×2 (g) Undefined   (h) 5×2

4. (a) $\begin{bmatrix} 12 & -3 \\ -4 & 5 \\ 4 & 1 \end{bmatrix}$ (b) $\begin{bmatrix} 7 & 6 & 5 \\ -2 & 1 & 3 \\ 7 & 3 & 7 \end{bmatrix}$ (c) $\begin{bmatrix} -5 & 4 & -1 \\ 0 & -1 & -1 \\ -1 & 1 & 1 \end{bmatrix}$

(d) $\begin{bmatrix} 9 & 8 & 19 \\ -2 & 0 & 0 \\ 32 & 9 & 25 \end{bmatrix}$ (e) $\begin{bmatrix} 14 & 36 & 25 \\ 4 & -1 & 7 \\ 12 & 26 & 21 \end{bmatrix}$ (f) $\begin{bmatrix} -28 & 7 \\ 0 & -14 \end{bmatrix}$

5. (a) Undefined (b) $\begin{bmatrix} 42 & 108 & 75 \\ 12 & -3 & 21 \\ 36 & 78 & 63 \end{bmatrix}$ (c) $\begin{bmatrix} 3 & 45 & 9 \\ 11 & -11 & 17 \\ 7 & 17 & 13 \end{bmatrix}$

(d) $\begin{bmatrix} 3 & 45 & 9 \\ 11 & -11 & 17 \\ 7 & 17 & 13 \end{bmatrix}$ (e) Undefined (f) $\begin{bmatrix} 48 & 15 & 31 \\ 0 & 2 & 6 \\ 38 & 10 & 27 \end{bmatrix}$

6. (a) $\begin{bmatrix} 3 & -1 & 1 \\ 0 & 2 & 1 \end{bmatrix}$ (b) $\begin{bmatrix} 1 & -1 & 3 \\ 5 & 0 & 2 \\ 2 & 1 & 4 \end{bmatrix}$ (c) $\begin{bmatrix} 12 & 0 \\ -3 & 6 \end{bmatrix}$

(d) Not possible (e) $\begin{bmatrix} 11 & 66 & 24 \\ 10 & -4 & 16 \end{bmatrix}$ (f) $\begin{bmatrix} 0 & 0 & 0 \\ 0 & 0 & 0 \\ 0 & 0 & 0 \end{bmatrix}$

7. (a) $[67 \quad 41 \quad 41]$ (b) $[63 \quad 67 \quad 57]$ (c) $\begin{bmatrix} 41 \\ 21 \\ 67 \end{bmatrix}$

(d) $\begin{bmatrix} 6 \\ 6 \\ 63 \end{bmatrix}$ (e) $[24 \quad 56 \quad 97]$ (f) $\begin{bmatrix} 76 \\ 98 \\ 97 \end{bmatrix}$

8. 182

EXERCISE SET 1.5 (page 40)

4. $A^{-1} = \begin{bmatrix} 2 & -1 \\ -5 & 3 \end{bmatrix}$ $B^{-1} = \begin{bmatrix} \frac{1}{5} & \frac{3}{20} \\ -\frac{1}{5} & \frac{1}{10} \end{bmatrix}$ $C^{-1} = \begin{bmatrix} \frac{1}{2} & 0 \\ 0 & \frac{1}{3} \end{bmatrix}$

6. No **7.** $\begin{bmatrix} -3 & 2 \\ \frac{5}{2} & -\frac{3}{2} \end{bmatrix}$ **8.** $\begin{bmatrix} 1 & \frac{2}{7} \\ \frac{4}{7} & \frac{1}{7} \end{bmatrix}$

9. $A^3 = \begin{bmatrix} 1 & 0 \\ 26 & 27 \end{bmatrix}$ $A^{-3} = \begin{bmatrix} 1 & 0 \\ -\frac{26}{27} & \frac{1}{27} \end{bmatrix}$ $A^2 - 2A + I = \begin{bmatrix} 0 & 0 \\ 4 & 4 \end{bmatrix}$

10. $A^{-1} = \begin{bmatrix} \frac{1}{2} & -\frac{1}{2} & \frac{1}{2} \\ \frac{1}{2} & \frac{1}{2} & -\frac{1}{2} \\ -\frac{1}{2} & \frac{1}{2} & \frac{1}{2} \end{bmatrix}$ **11.** $\begin{bmatrix} \cos\theta & -\sin\theta \\ \sin\theta & \cos\theta \end{bmatrix}$

12. (c) $(A + B)^2 = A^2 + AB + BA + B^2$ **13.** $A^{-1} = \begin{bmatrix} \dfrac{1}{a_{11}} & 0 & \cdots & 0 \\ 0 & \dfrac{1}{a_{22}} & \cdots & 0 \\ \vdots & \vdots & & \vdots \\ 0 & 0 & \cdots & \dfrac{1}{a_{nn}} \end{bmatrix}$

18. $0A$ and $A0$ may not have the same size.   **19.** $\begin{bmatrix} \pm 1 & 0 & 0 \\ 0 & \pm 1 & 0 \\ 0 & 0 & \pm 1 \end{bmatrix}$

EXERCISE SET 1.6 (page 49)

1. a, b, d, f, g **2.** (a) Add -5 times the first row to the second.
(b) Interchange the first and third rows.
(c) Multiply the second row by $\frac{1}{8}$.

3. (a) $E_1 = \begin{bmatrix} 0 & 0 & 1 \\ 0 & 1 & 0 \\ 1 & 0 & 0 \end{bmatrix}$ (b) $E_2 = \begin{bmatrix} 0 & 0 & 1 \\ 0 & 1 & 0 \\ 1 & 0 & 0 \end{bmatrix}$

(c) $E_3 = \begin{bmatrix} 1 & 0 & 0 \\ 0 & 1 & 0 \\ 2 & 0 & 1 \end{bmatrix}$ (d) $E_4 = \begin{bmatrix} 1 & 0 & 0 \\ 0 & 1 & 0 \\ -2 & 0 & 1 \end{bmatrix}$

4. No, since C cannot be obtained by performing a single row operation on B.

5. (a) $\begin{bmatrix} -5 & 2 \\ 3 & -1 \end{bmatrix}$ (b) $\begin{bmatrix} -5 & -3 \\ -3 & -2 \end{bmatrix}$ (c) Not invertible

6. (a) $\begin{bmatrix} \frac{3}{2} & -\frac{11}{10} & -\frac{6}{5} \\ -1 & 1 & 1 \\ -\frac{1}{2} & \frac{7}{10} & \frac{2}{5} \end{bmatrix}$ (b) Not invertible   (c) $\begin{bmatrix} \frac{1}{2} & -\frac{1}{2} & \frac{1}{2} \\ -\frac{1}{2} & \frac{1}{2} & \frac{1}{2} \\ \frac{1}{2} & \frac{1}{2} & -\frac{1}{2} \end{bmatrix}$

(d) $\begin{bmatrix} \frac{7}{2} & 0 & -3 \\ -1 & 1 & 0 \\ 0 & -1 & 1 \end{bmatrix}$ (e) $\begin{bmatrix} \frac{1}{2} & -\frac{1}{2} & \frac{1}{2} \\ 0 & 0 & 1 \\ \frac{1}{2} & \frac{1}{2} & -\frac{1}{2} \end{bmatrix}$ (f) $\begin{bmatrix} 1 & 0 & -2 \\ 3 & 1 & 2 \\ 1 & -1 & 0 \end{bmatrix}$

7. (a) $\begin{bmatrix} \frac{1}{2}\sqrt{2} & -\frac{1}{2}\sqrt{2} & 0 \\ \frac{1}{2}\sqrt{2} & \frac{1}{2}\sqrt{2} & 0 \\ 0 & 0 & 1 \end{bmatrix}$ (b) $\begin{bmatrix} 1 & 0 & 0 & 0 \\ -\frac{1}{2} & \frac{1}{2} & 0 & 0 \\ 0 & -\frac{1}{4} & \frac{1}{4} & 0 \\ 0 & 0 & -\frac{1}{8} & \frac{1}{8} \end{bmatrix}$ (c) Not invertible

8. $A^{-1} = \begin{bmatrix} \cos \theta & -\sin \theta & 0 \\ \sin \theta & \cos \theta & 0 \\ 0 & 0 & 1 \end{bmatrix}$

9. (a) $E_1 = \begin{bmatrix} 1 & 0 \\ -3 & 1 \end{bmatrix}$, $E_2 = \begin{bmatrix} 1 & 0 \\ 0 & \frac{1}{4} \end{bmatrix}$ (b) $A^{-1} = \begin{bmatrix} 1 & 0 \\ 0 & \frac{1}{4} \end{bmatrix}\begin{bmatrix} 1 & 0 \\ -3 & 1 \end{bmatrix}$

(c) $A = \begin{bmatrix} 1 & 0 \\ 3 & 1 \end{bmatrix}\begin{bmatrix} 1 & 0 \\ 0 & 4 \end{bmatrix}$

11. $A = \begin{bmatrix} 1 & 0 & 0 \\ -2 & 1 & 0 \\ 0 & 0 & 1 \end{bmatrix} \begin{bmatrix} 1 & 0 & 0 \\ 0 & 1 & 0 \\ 0 & 1 & 1 \end{bmatrix} \begin{bmatrix} 1 & 3 & 3 & 8 \\ 0 & 1 & 7 & 8 \\ 0 & 0 & 0 & 0 \end{bmatrix}$

13. (a) $\begin{bmatrix} \frac{1}{k_1} & 0 & 0 & 0 \\ 0 & \frac{1}{k_2} & 0 & 0 \\ 0 & 0 & \frac{1}{k_3} & 0 \\ 0 & 0 & 0 & \frac{1}{k_4} \end{bmatrix}$ (b) $\begin{bmatrix} 0 & 0 & 0 & \frac{1}{k_4} \\ 0 & 0 & \frac{1}{k_3} & 0 \\ 0 & \frac{1}{k_2} & 0 & 0 \\ \frac{1}{k_1} & 0 & 0 & 0 \end{bmatrix}$ (c) $\begin{bmatrix} \frac{1}{k} & 0 & 0 & 0 \\ -\frac{1}{k^2} & \frac{1}{k} & 0 & 0 \\ \frac{1}{k^3} & -\frac{1}{k^2} & \frac{1}{k} & 0 \\ -\frac{1}{k^4} & \frac{1}{k^3} & -\frac{1}{k^2} & \frac{1}{k} \end{bmatrix}$

EXERCISE SET 1.7 (page 58)

1. $x_1 = 41, x_2 = -17$ **2.** $x_1 = \frac{46}{27}, x_2 = -\frac{13}{27}$ **3.** $x_1 = -7, x_2 = 4, x_3 = -1$

4. $x_1 = 1, x_2 = -11, x_3 = 16$ **5.** $x = 1, y = 5, z = -1$

6. $w = 1, x = -6, y = 10, z = -7$ **7.** $x_1 = 2b_1 - 5b_2, x_2 = -b_1 + 3b_2$

8. $x_1 = -\frac{15}{2}b_1 + \frac{1}{2}b_2 + \frac{5}{2}b_3$ **9.** (a) $x_1 = \frac{16}{3}, x_2 = -\frac{4}{3}, x_3 = -\frac{11}{3}$
$\quad\ x_2 = \frac{1}{2}b_1 + \frac{1}{2}b_2 - \frac{1}{2}b_3$ $\qquad$ (b) $x_1 = -\frac{5}{3}, x_2 = \frac{5}{3}, x_3 = \frac{10}{3}$
$\quad\ x_3 = \frac{5}{2}b_1 - \frac{1}{2}b_2 - \frac{1}{2}b_3$ $\qquad$ (c) $x_1 = 3, x_2 = 0, x_3 = -4$

11. (a) $x_1 = 1, x_2 = 0$ (b) $x_1 = \frac{19}{10}, x_2 = \frac{13}{10}$

12. (a) $x_1 = 14, x_2 = 1, x_3 = 11$ (b) $x_1 = -35, x_2 = -2, x_3 = -26$

13. (a) $x_1 = \frac{5}{11}, x_2 = \frac{2}{11}$ (b) $x_1 = \frac{18}{11}, x_2 = \frac{16}{11}$
$\quad\ $ (c) $x_1 = \frac{12}{11}, x_2 = \frac{7}{11}$ (d) $x_1 = -\frac{10}{11}, x_2 = \frac{7}{11}$

14. (a) $x_1 = -21, x_2 = 10, x_3 = -2$ **15.** (a) $x_1 = -12 - 3t, x_2 = -5 - t$
$\quad\ $ (b) $x_1 = -2, x_2 = 1, x_3 = 0$ $\qquad$ (b) $x_1 = 7 - 3t, x_2 = 3 - t$
$\quad\ $ (c) $x_1 = 23, x_2 = -11, x_3 = 2$

16. $b_2 = \frac{1}{2}b_1$ **17.** $b_2 = 3b_1, b_3 = -2b_1$

18. No conditions; system is consistent for all values of $b_1, b_2,$ and b_3.

19. $b_3 = b_2 - b_1, b_4 = 2b_1 - b_2$ **20.** (a) $X = \begin{bmatrix} 0 \\ 0 \\ 0 \end{bmatrix}$ (b) $X = \begin{bmatrix} 4t \\ 5t \\ 2 \\ t \end{bmatrix}$

21. (a) Invertible (b) Not invertible

CHAPTER 1 SUPPLEMENTARY EXERCISES (page 60)

1. $x' = \frac{3}{5}x + \frac{4}{5}y,\ y' = -\frac{4}{5}x + \frac{3}{5}y$

2. $x' = x \cos\theta + y \sin\theta,\ y' = -x \sin\theta + y \cos\theta$

3. One possible answer is
$$x_1 - 2x_2 - x_3 - \quad x_4 = 0$$
$$x_1 + 5x_2 \qquad + 2x_4 = 0$$

4. 3 pennies, 4 nickels, 6 dimes

5. $x = 4,\ y = 2,\ z = 3$

6. Infinitely many if $a = 2$ or $a = -\frac{3}{2}$; none otherwise.

7. (a) $a \neq 0, b \neq 2$ (b) $a \neq 0, b = 2$
(c) $a = 0, b = 2$ (d) $a = 0, b \neq 2$

8. $K = \begin{bmatrix} 0 & 2 \\ 1 & 1 \end{bmatrix}$

9. $a = 2,\ b = -1,\ c = 1$

10. (a) $X = \begin{bmatrix} -1 & 3 & -1 \\ 6 & 0 & 1 \end{bmatrix}$ (b) $X = \begin{bmatrix} 1 & -2 \\ 3 & 1 \end{bmatrix}$

(c) $X = \begin{bmatrix} -\frac{113}{37} & -\frac{160}{37} \\ -\frac{20}{37} & -\frac{46}{37} \end{bmatrix}$

11. (a) $Z = \begin{bmatrix} -1 & -7 & 11 \\ 14 & 10 & -26 \end{bmatrix} X$ (b) $\begin{aligned} z_1 &= -x_1 - 7x_2 + 11x_3 \\ z_2 &= 14x_1 + 10x_2 - 26x_3 \end{aligned}$

12. *mpn* multiplications and $mp(n-1)$ additions

14. $a = 1,\ b = -2,\ c = 3$ **15.** $a = 1,\ b = -4,\ c = -5$

25. $A = -\frac{7}{5},\ B = \frac{4}{5},\ C = \frac{3}{5}$

EXERCISE SET 2.1 (page 71)

1. (a) 5 (b) 7 (c) 10 (d) 0 (e) 4 (f) 5

2. (a) Odd (b) Odd (c) Even (d) Even (e) Even (f) Odd

3. 5   **4.** 0   **5.** 59   **6.** $k^2 - 4k - 5$ **7.** 0   **8.** 425   **9.** 104

10. $-k^4 - k^3 + 18k^2 + 9k - 21$ **11.** (a) $\lambda = 3, \lambda = 2$ (b) $\lambda = 2, \lambda = 6$

14. 275   **15.** (a) 120 (b) -120

EXERCISE SET 2.2 (page 76)

1. (a) 6 (b) -16 (c) 0 (d) 0   **2.** -21 **3.** -5 **4.** -7

5. 18   **6.** -21 **7.** 6   **8.** $\frac{1}{6}$ **9.** -2 **10.** (a) 5 (b) 10
(c) 5 (d) 10

EXERCISE SET 2.3 (page 82)

1. (a) $\begin{bmatrix} 2 & -3 & 0 \\ 1 & 1 & 2 \end{bmatrix}$ (b) $\begin{bmatrix} 6 & -8 & 0 \\ 1 & 4 & 1 \\ 1 & 3 & 3 \end{bmatrix}$ (c) $\begin{bmatrix} 7 \\ 0 \\ 2 \end{bmatrix}$ (d) $\begin{bmatrix} a_{11} & a_{21} \\ a_{12} & a_{22} \\ a_{13} & a_{23} \end{bmatrix}$

4. (a) Invertible (b) Not invertible **5.** (a) 135 (b) $\frac{8}{5}$
(c) Not invertible (d) Not invertible (c) $\frac{1}{40}$ (d) -5

6. If $x = 0$, the first and third rows are proportional.
If $x = 2$, the first and second rows are proportional.

12. (a) $k = \frac{1}{2}(5 + \sqrt{17})$, $k = \frac{1}{2}(5 - \sqrt{17})$ (b) $k = -1$

EXERCISE SET 2.4 (page 93)

1. (a) $M_{11} = 29$, $M_{12} = -11$, $M_{13} = -19$, $M_{21} = 21$, $M_{22} = 13$, $M_{23} = -19$
$M_{31} = 27$, $M_{32} = -5$, $M_{33} = 19$
 (b) $C_{11} = 29$, $C_{12} = 11$, $C_{13} = -19$, $C_{21} = -21$, $C_{22} = 13$
$C_{23} = 19$, $C_{31} = 27$, $C_{32} = 5$, $C_{33} = 19$

2. (a) $M_{13} = 36$, $C_{13} = 36$ (b) $M_{23} = 24$, $C_{23} = -24$
 (c) $M_{22} = -48$, $C_{22} = -48$ (d) $M_{21} = -108$, $C_{21} = 108$

3. 152 **4.** (a) $\begin{bmatrix} 29 & -21 & 27 \\ 11 & 13 & 5 \\ -19 & 19 & 19 \end{bmatrix}$ (b) $\left(\frac{1}{152}\right)\begin{bmatrix} 29 & -21 & 27 \\ 11 & 13 & 5 \\ -19 & 19 & 19 \end{bmatrix}$

5. 48 **6.** -66 **7.** 0 **8.** $k^3 - 8k^2 - 10k + 95$ **9.** -120 **10.** 0

11. $A^{-1} = \begin{bmatrix} 3 & 8 & -5 \\ -3 & -6 & 4 \\ 2 & 3 & -2 \end{bmatrix}$ **12.** $A^{-1} = \begin{bmatrix} 2 & 0 & -1 \\ \frac{2}{3} & \frac{1}{3} & -\frac{1}{3} \\ -1 & 0 & 1 \end{bmatrix}$

13. $A^{-1} = \begin{bmatrix} \frac{1}{2} & 2 & -\frac{1}{2} \\ 0 & 1 & \frac{1}{2} \\ 0 & 0 & \frac{1}{2} \end{bmatrix}$ **14.** $A^{-1} = \begin{bmatrix} \frac{1}{3} & 0 & 0 \\ -3 & 1 & 0 \\ \frac{11}{6} & -\frac{1}{2} & \frac{1}{4} \end{bmatrix}$

15. $A^{-1} = \begin{bmatrix} -4 & 3 & 0 & -1 \\ 2 & -1 & 0 & 0 \\ -7 & 0 & -1 & 8 \\ 6 & 0 & 1 & -7 \end{bmatrix}$ **16.** $x_1 = 1$, $x_2 = 2$

17. $x = \frac{3}{11}$, $y = \frac{2}{11}$, $z = -\frac{1}{11}$ **18.** $x = \frac{26}{21}$, $y = \frac{25}{21}$, $z = \frac{5}{7}$

19. $x_1 = -\frac{30}{11}$, $x_2 = -\frac{38}{11}$, $x_3 = -\frac{40}{11}$ **20.** $x_1 = 3$, $x_2 = 5$, $x_3 = -1$, $x_4 = 8$

21. Cramer's rule does not apply. **22.** $z = 2$

23. (a) $x = 1$, $y = 0$, $z = 2$, $w = 0$

CHAPTER 2 SUPPLEMENTARY EXERCISES (page 95)

1. $x' = \frac{3}{5}x + \frac{4}{5}y, \qquad y' = -\frac{4}{5}x + \frac{3}{5}y$

2. $x' = x\cos\theta + y\sin\theta,\ y' = -x\sin\theta + y\cos\theta$ **4.** 2

5. $\cos\beta = \dfrac{c^2 + a^2 - b^2}{2ac}, \ \cos\gamma = \dfrac{a^2 + b^2 - c^2}{2ab}$ **10.** (b) $\frac{19}{2}$

12. $\det(B) = -1^{m(m-1)/2}\det(A)$

13. (a) The ith and jth columns will be interchanged.
    (b) The ith column will be divided by c.
    (c) $-c$ times the jth column will be added to the ith column.

EXERCISE SET 3.1 (page 108)

3. (a) $\overrightarrow{P_1P_2} = (-1, 3)$ (b) $\overrightarrow{P_1P_2} = (-7, 2)$
    (c) $\overrightarrow{P_1P_2} = (2, -12, -11)$ (d) $\overrightarrow{P_1P_2} = (-8, 7, 4)$

4. $\overrightarrow{PQ}$, where $Q = (9, 5, 1)$ is one possible answer.

5. $\overrightarrow{PQ}$, where $P = (0, 4, -8)$ is one possible answer.

6. (a) $(-2, 0, 4)$ (b) $(23, -15, 4)$ **7.** $\mathbf{x} = (-\frac{1}{2}, \frac{5}{6}, 1)$
    (c) $(-1, -5, 2)$ (d) $(-39, 69, -12)$
    (e) $(-30, -7, 5)$ (f) $(0, -10, 0)$

8. $c_1 = 1, c_2 = -2, c_3 = 3$

10. $c_1 = -t, c_2 = -t, c_3 = t$ (where t is arbitrary)

11. (a) $(\frac{9}{2}, -\frac{1}{2}, -\frac{1}{2})$ (b) $(\frac{23}{4}, -\frac{9}{4}, \frac{1}{4})$

12. (a) $x' = 5, y' = 8$ (b) $x = -1, y = 3$

EXERCISE SET 3.2 (page 111)

1. (a) 5   (b) $5\sqrt{2}$ (c) 3   (d) $\sqrt{3}$ (e) $\sqrt{129}$ (f) 9

2. (a) $\sqrt{13}$ (b) $2\sqrt{26}$ (c) $\sqrt{209}$ (d) $\sqrt{93}$

3. (a) $2\sqrt{3}$ (b) $\sqrt{14} + \sqrt{2}$ (c) $4\sqrt{14}$ (d) $2\sqrt{37}$

    (e) $(1/\sqrt{6}, 1/\sqrt{6}, -2/\sqrt{6})$ (f) 1   **4.** $k = \pm\dfrac{3}{\sqrt{21}}$

7. $(1/\sqrt{3}, 1/\sqrt{3}, 1/\sqrt{3})$ **8.** A sphere of radius 1 centered at (x_0, y_0, z_0)

EXERCISE SET 3.3 (page 121)

1. (a) -10 (b) -3 (c) 0   (d) -20

2. (a) $-\dfrac{1}{\sqrt{5}}$ (b) $-\dfrac{3}{\sqrt{58}}$ (c) 0   (d) $-\dfrac{20}{3\sqrt{70}}$

3. (a) Obtuse   (b) Acute   (c) Obtuse   (d) Orthogonal

4. (a) $(\frac{12}{13}, -\frac{8}{13})$ (b) $(0, 0)$ (c) $(-\frac{80}{13}, 0, -\frac{16}{13})$ (d) $(\frac{32}{89}, \frac{12}{89}, \frac{16}{89})$

5. (a) $(\frac{14}{13}, \frac{21}{13})$ (b) $(2, 6)$ (c) $(-\frac{11}{13}, 1, \frac{55}{13})$ (d) $(-\frac{32}{89}, -\frac{12}{89}, \frac{73}{89})$

6. (a) $\dfrac{2}{5}$ (b) $\dfrac{6}{\sqrt{5}}$ (c) 2   (d) $\dfrac{37}{7}$ **8.** $\pm\left(\dfrac{2}{\sqrt{13}}, \dfrac{3}{\sqrt{13}}\right)$

9. (a) 6   (b) 36   (c) $24\sqrt{5}$ (d) $24\sqrt{5}$

11. $\cos\theta_1 = 0$, $\cos\theta_2 = \dfrac{3}{\sqrt{10}}$, $\cos\theta_3 = \dfrac{1}{\sqrt{10}}$ **12.** The right angle is at B.

13. No. The equation will hold if **b** and **c** are perpendicular to **a** and $\mathbf{b} \neq \mathbf{c}$.

14. (a) $k = -\frac{3}{4}$ (b) $k = \frac{1}{7}$ (c) $k = \dfrac{48 \pm 25\sqrt{3}}{11}$ (d) $k = \frac{4}{3}$

15. (a) $\frac{2}{5}$ (b) $\dfrac{2}{\sqrt{5}}$ (c) $\dfrac{2}{\sqrt{5}}$ **18.** $\theta = \cos^{-1}\left(\dfrac{2}{\sqrt{6}}\right)$

19. $\cos\beta = \dfrac{b}{\sqrt{a^2 + b^2 + c^2}}$, $\cos\gamma = \dfrac{c}{\sqrt{a^2 + b^2 + c^2}}$

EXERCISE SET 3.4 (page 129)

1. (a) $(-23, 7, -1)$ (b) $(-20, -67, -9)$ (c) $(-78, 52, -26)$
   (d) $(0, -56, -392)$ (e) $(24, 0, -16)$ (f) $(-12, -22, -8)$

2. (a) $(12, 30, -6)$ (b) $(-2, 0, 2)$ **3.** (a) $\frac{1}{2}\sqrt{374}$ (b) $9\sqrt{13}$

7. $x = (\frac{1}{2} - \frac{1}{2}t, -\frac{1}{2} + \frac{3}{2}t, t)$, where t is arbitrary   **9.** 227

10. (a) $\mathbf{u} = (0, 1, 0)$ and $\mathbf{v} = (1, 0, 0)$
   (b) $(-1, 0, 0)$
   (c) $(0, 0, -1)$

EXERCISE SET 3.5 (page 139)

1. (a) $(x - 2) + 4(y - 6) + 2(z - 1) = 0$
   (b) $-(x + 1) + 7(y + 1) + 6(z - 2) = 0$

(c) $z = 0$
(d) $2x + 3y + 4z = 0$

2. (a) $x + 4y + 2z - 28 = 0$ (b) $-x + 7y + 6z - 6 = 0$
(c) $z = 0$ (d) $2x + 3y + 4z = 0$

3. (a) $(5, 0, 0)$ is a point in the plane and $\mathbf{n} = (2, -3, 7)$ is a normal vector so that $2(x - 5) - 3y + 7z = 0$ is a point-normal form; other points and normals yield other correct answers.
(b) $x + 3z = 0$ is one possible answer.

4. (a) $2y - z - 1 = 0$ (b) $x + 9y - 5z - 16 = 0$

5. (a) No, because $(3, -2, 1)$ and $(6, -4, 3)$ are not parallel.
(b) Yes, because $(2, -8, -6)$ and $(-1, 4, 3)$ are parallel.
(c) Yes, because $(4, -1, -2)$ and $(1, -1/4, -1/2)$ are parallel.

6. (a) Yes, because $(2, -1, -4)$ and $(3, 2, 1)$ are perpendicular.
(b) No, because $(1, 2, 3)$ and $(1, -1, 2)$ are not perpendicular.

7. (a) No, because $(1, -1, 3)$ and $(2, 0, 1)$ are not perpendicular.
(b) Yes, because $(3, -2, 1)$ and $(4, 5, -2)$ are perpendicular.

8. (a) Yes, because $(2, 1, -1)$ and $(4, 2, -2)$ are parallel.
(b) No, because $(-1, 1, -3)$ and $(2, 2, 0)$ are not parallel.

9. (a) $x = 2 + t, y = 4 + 2t, z = 6 + 5t$
(b) $x = -3 + 5t, y = 2 - 7t, z = -4 - 3t$
(c) $x = 1, y = 1, z = 5 + t$
(d) $x = t, y = t, z = t$

10. (a) $x - 2 = \dfrac{y - 4}{2} = \dfrac{z - 6}{5}$ (b) $\dfrac{x + 3}{5} = \dfrac{y - 2}{-7} = \dfrac{z + 4}{-3}$

11. (a) $x = 6 + t, y = -1 + 3t, z = 5 - 9t$ or $x = 7 + t, y = 2 + 3t,$
$z = -4 - 9t$ are possible answers.
(b) $x = -t, y = -t, z = -t$ or $x = -1 - t, y = -1 - t,$
$z = -1 - t$ are possible answers.

12. (a) $x = -\frac{11}{7} + \frac{23}{7}t, y = -\frac{12}{7} - \frac{1}{7}t, z = t$ (b) $x = \frac{5}{3}t, y = t, z = 0$

13. (a) $x - 2y - 17 = 0, y + 2z - 5 = 0$ (b) $3x - 5y = 0, 2y - z = 0$

15. $x + 2y - z = 10$ 16. (a) $z = 0$ (b) $y = 0$ (c) $x = 0$

17. (a) $z = z_0$ (b) $x = x_0$ (c) $y = y_0$ 18. $4x - 2y + 7z = 0$

19. $5x - 2y + z - 30 = 0$ 20. $\left(-\frac{222}{7}, -\frac{64}{7}, \frac{78}{7}\right)$ 21. $y + 2z = 10$

22. $4x - 13y + 21z = -14$ 24. $x = 5 - 2t, y = 5t, z = -2 + 11t$

25. $x + 5y + 3z = -6$ 26. $x + y - 3z = 6$ 27. $4x + 13y - z = 1$

28. $7x + y + 9z = 25$ 29. $7x - y - 3z = 5$ 30. $x + 2y + 4z = \frac{29}{2}$

31. $3x - 5y - 2z = -5$ **33.** $(-17, -1, 1)$ **34.** $x - 4y + 4z + 9 = 0$

35. (a) $x = -\frac{11}{7} - 23t$, $y = -\frac{12}{7} + t$, $z = -7t$ (b) $x = -5t$, $y = -3t$, $z = 0$

37. (a) $\frac{5}{3}$ (b) $\frac{9}{7}$ (c) $\frac{137}{21}$ **38.** (a) $\dfrac{11}{\sqrt{116}}$ (b) $\dfrac{5}{\sqrt{54}}$ (c) $\dfrac{2}{\sqrt{3}}$

EXERCISE SET 4.1 (page 148)

1. (a) $(-3, -4, -8, 4)$ (b) $(53, 34, 49, 20)$
(c) $(-1, 2, 7, -10)$ (d) $(-99, -84, -150, 30)$
(e) $(-63, -28, -21, -69)$ (f) $(2, 6, 15, -14)$

2. $(-\frac{7}{6}, -1, -\frac{3}{2}, -\frac{1}{3})$ **3.** $c_1 = 1, c_2 = 1, c_3 = -1, c_4 = 1$

5. (a) 5   (b) $\sqrt{11}$ (c) $\sqrt{14}$ (d) $\sqrt{48}$

6. (a) $\sqrt{73}$ (b) $\sqrt{14} + 3\sqrt{7}$ (c) $4\sqrt{14}$ (d) $\sqrt{1801}$

(e) $\left(\dfrac{2}{\sqrt{6}}, 0, \dfrac{1}{\sqrt{6}}, \dfrac{1}{\sqrt{6}} \right)$ (f) 1

8. $k = \pm \dfrac{3}{\sqrt{14}}$ **9.** (a) -1 (b) -1 (c) 0   (d) 27

10. (a) $\left(\dfrac{2}{\sqrt{5}}, \dfrac{1}{\sqrt{5}} \right)$ and $\left(-\dfrac{2}{\sqrt{5}}, -\dfrac{1}{\sqrt{5}} \right)$

11. (a) $\sqrt{10}$ (b) $3\sqrt{3}$ (c) $\sqrt{59}$ (d) 10

EXERCISE SET 4.2 (page 154)

1. Not a vector space. Axiom 8 fails.

2. Not a vector space. Axiom 10 fails.

3. Not a vector space. Axioms 9 and 10 fail.

4. The set is a vector space under the given operations.

5. The set is a vector space under the given operations.

6. Not a vector space. Axioms 5 and 6 fail.

7. The set is a vector space under the given operations.

8. Not a vector space. Axioms 7 and 8 fail.

9. The set is a vector space under the given operations.

10. Not a vector space. Axioms 1, 4, 5, and 6 fail.

11. The set is a vector space under the given operations.

12. The set is a vector space under the given operations.

13. The set is a vector space under the given operations.

14. The set is a vector space under the given operations.

EXERCISE SET 4.3 (page 162)

1. a, c **2.** b, c **3.** a, b, d **4.** b, d, e **5.** a, b, d

6. (a) $(5, 9, 5) = 3\mathbf{u} - 4\mathbf{v} + \mathbf{w}$ (b) $(2, 0, 6) = 4\mathbf{u} - 2\mathbf{w}$
(c) $(0, 0, 0) = 0\mathbf{u} + 0\mathbf{v} + 0\mathbf{w}$ (d) $(2, 2, 3) = \frac{1}{2}\mathbf{u} - \frac{1}{2}\mathbf{v} + \frac{1}{2}\mathbf{w}$

7. (a) $5 + 9x + 5x^2 = 3\mathbf{p}_1 - 4\mathbf{p}_2 + \mathbf{p}_3$
(b) $2 + 6x^2 = 4\mathbf{p}_1 - 2\mathbf{p}_3$
(c) $0 = 0\mathbf{p}_1 + 0\mathbf{p}_2 + 0\mathbf{p}_3$
(d) $2 + 2x + 3x^2 = \frac{1}{2}\mathbf{p}_1 - \frac{1}{2}\mathbf{p}_2 + \frac{1}{2}\mathbf{p}_3$

8. a, c, d

9. (a) The vectors span.   (b) The vectors do not span.
(c) The vectors do not span.   (d) The vectors span.

10. a, c **11.** The polynomials do not span P_2.   **12.** a, b, d

13. $8x - 7y + z = 0$ **14.** $x = 2t,\ y = 7t,\ z = -t$, where $-\infty < t < +\infty$

EXERCISE SET 4.4 (page 169)

1. (a) $\mathbf{u}_2$ is a scalar multiple of $\mathbf{u}_1$.
(b) The vectors are linearly dependent by Theorem 7.
(c) $\mathbf{p}_2$ is a scalar multiple of $\mathbf{p}_1$.
(d) B is a scalar multiple of A.

2. (a) Independent   (b) Independent   (c) Independent   (d) Dependent

3. (a) Independent   (b) Independent   (c) Independent   (d) Independent

4. (a) Independent   (b) Independent   (c) Independent   (d) Dependent

5. (a) Dependent   (b) Independent   (c) Independent   (d) Dependent
(e) Dependent   (f) Dependent

6. (a) They do not lie in a plane.   **7.** (a) They do not lie on the same line.
(b) They do lie in a plane.   (b) They do not lie on the same line.
(c) They do lie on the same line.

8. $\lambda = -\frac{1}{2}, \lambda = 1$

9. (b) $v_1 = \frac{2}{7}v_2 - \frac{3}{7}v_3$, $v_2 = \frac{7}{2}v_1 + \frac{3}{2}v_3$, $v_3 = -\frac{7}{3}v_1 + \frac{2}{3}v_2$

20. If and only if the vector is not zero.

EXERCISE SET 4.5 (page 177)

1. (a) A basis for R^2 has two vectors.
(b) A basis for R^3 has three vectors.
(c) A basis for P_2 has three vectors.
(d) A basis for M_{22} has four vectors.

2. *a, b*   **3.** *a, b*   **4.** *c, d*

6. Any two of the vectors v_1, v_2, v_3 **7.** Basis: $(1, 0, 1)$; dimension $= 1$

8. Basis: $(-\frac{1}{4}, -\frac{1}{4}, 1, 0)$, $(0, -1, 0, 1)$; dimension $= 2$

9. Basis: $(4, 1, 0, 0)$, $(-3, 0, 1, 0)$, $(1, 0, 0, 1)$; dimension $= 3$

10. Basis: $(3, 1, 0)$, $(-1, 0, 1)$; dimension $= 2$ **11.** No basis; dimension $= 0$

12. Basis: $(4, -5, 1)$; dimension $= 1$

13. (a) $(\frac{2}{3}, 1, 0)$, $(-\frac{5}{3}, 0, 1)$ (b) $(1, 1, 0)$, $(0, 0, 1)$
(c) $(2, -1, 4)$ (d) $(1, 1, 0)$, $(0, 1, 1)$

14. (a) 3-dimensional   (b) 2-dimensional   (c) 1-dimensional

15. 3-dimensional

EXERCISE SET 4.6 (page 189)

1. $r_1 = (2, -1, 0, 1)$ $c_1 = \begin{bmatrix} 2 \\ 3 \\ 1 \end{bmatrix}$ $c_2 = \begin{bmatrix} -1 \\ 5 \\ 4 \end{bmatrix}$
$r_2 = (3, 5, 7, -1)$
$r_3 = (1, 4, 2, 7)$ $c_3 = \begin{bmatrix} 0 \\ 7 \\ 2 \end{bmatrix}$ $c_4 = \begin{bmatrix} 1 \\ -1 \\ 7 \end{bmatrix}$

2. (a) $(1, -3)$ (b) $\begin{bmatrix} 1 \\ 2 \end{bmatrix}$ (c) 1

3. (a) $(1, 2, 0)$, $(0, 0, 1)$ (b) $\begin{bmatrix} 1 \\ 2 \\ 0 \end{bmatrix}$, $\begin{bmatrix} 0 \\ 1 \\ -1 \end{bmatrix}$ (c) 2

4. (a) $(1, 0, 1, 2), (0, 1, 1, 0), (0, 0, 0, 1)$ (b) $\begin{bmatrix} 1 \\ 0 \\ 0 \end{bmatrix}, \begin{bmatrix} 0 \\ 1 \\ 0 \end{bmatrix}, \begin{bmatrix} 0 \\ 0 \\ 1 \end{bmatrix}$ (c) 3

5. (a) $(1, 0, 5, 2, 0), (0, 1, 0, 0, 0), (0, 0, -3, 0, 1)$ (b) $\begin{bmatrix} 0 \\ 0 \\ 1 \\ 1 \\ 2 \end{bmatrix}, \begin{bmatrix} 1 \\ 0 \\ 0 \\ 1 \\ 1 \end{bmatrix}, \begin{bmatrix} 0 \\ -1 \\ 1 \\ 0 \\ 0 \end{bmatrix}$ (c) 3

6. (a) $(1, 1, -4, -3), (0, 1, -5, -2), (0, 0, 1, -\frac{1}{2})$
(b) $(1, -1, 2, 0), (0, 1, 0, 0), (0, 0, 1, -\frac{1}{6})$
(c) $(1, 1, 0, 0), (0, 1, 1, 1), (0, 0, 1, 1), (0, 0, 0, 1)$

8. (a) $\{\mathbf{v}_1, \mathbf{v}_2, \mathbf{v}_4\}$ (b) $\mathbf{v}_3 = 2\mathbf{v}_1 - \mathbf{v}_2, \mathbf{v}_5 = -\mathbf{v}_1 + 3\mathbf{v}_2 + 2\mathbf{v}_4$

9. (a) $\{\mathbf{v}_1, \mathbf{v}_2\}$ (b) $\mathbf{v}_3 = 2\mathbf{v}_1 + \mathbf{v}_2, \mathbf{v}_4 = -2\mathbf{v}_1 + \mathbf{v}_2$

10. (a) First and third rows, or second and third rows; any pair of columns except the pair consisting of the first and fourth columns.
(b) Any two of the first three rows, together with the last row; any three columns.

11. (a) 3   (b) The minimum of m and n.   **12.** (a) $\mathbf{b} = \begin{bmatrix} 1 \\ 4 \end{bmatrix} - \begin{bmatrix} 3 \\ -6 \end{bmatrix}$

(b) $\mathbf{b}$ is not in the column space of A.
(c) $\mathbf{b} = \mathbf{c}_1 + (r - 1)\mathbf{c}_2 + r\mathbf{c}_3$ (r arbitrary)

13. (a) 0   (b) Inconsistent   (c) 2
(d) 7   (e) Inconsistent   (f) 4   (g) 4

EXERCISE SET 4.7 (page 196)

1. (a) -12 (b) 0   (c) 0   (d) 120

2. (a) -5 (b) 0   (c) 3   (d) 52

3. (a) 16   (b) 56   **4.** (a) -6 (b) 0

5. (b) 0   **6.** (b) -39 **7.** (a) $\begin{bmatrix} 2 & 0 \\ 0 & 4 \end{bmatrix}$ (b) $\begin{bmatrix} \sqrt{2} & 0 \\ 0 & \sqrt{7} \end{bmatrix}$

9. (a) Not an inner product, axiom 4 fails.   **10.** No. Axiom 4 fails.
(b) Not an inner product, axioms 2, 3 fail.
(c) $\langle \mathbf{u}, \mathbf{v} \rangle$ is an inner product.
(d) Not an inner product, axiom 4 fails.

15. (a) $-\frac{28}{15}$ (b) 0   **16.** (a) 0   (b) 1   (c) $-\frac{4}{\pi} \ln\left(\frac{1}{\sqrt{2}}\right)$ or $\frac{2}{\pi} \ln 2$

EXERCISE SET 4.8 (page 205)

1. (a) $\sqrt{10}$ (b) $\sqrt{21}$ (c) $5\sqrt{5}$ **2.** (a) $\sqrt{18}$ (b) $\sqrt{45}$ (c) $3\sqrt{13}$

3. (a) $\sqrt{6}$ (b) 5 **4.** (a) $\sqrt{90}$ (b) 0 **5.** $\sqrt{18}$ **6.** (a) $\sqrt{98}$ (b) 0

7. (a) Yes (b) No (c) Yes (d) Yes (e) No (f) Yes **8.** No

9. (a) $\dfrac{-1}{\sqrt{2}}$ (b) $\dfrac{-3}{\sqrt{73}}$ (c) 0 (d) $\dfrac{-20}{9\sqrt{10}}$ (e) $\dfrac{-1}{\sqrt{2}}$ (f) $\dfrac{2}{\sqrt{55}}$

10. (a) 0 (b) 0 **11.** (a) $\dfrac{19}{10\sqrt{7}}$ (b) 0

12. (a) $k = -3$ (b) $k = -2, k = -3$ **14.** a, b, c

15. $\pm\dfrac{1}{57}(-34, 44, -6, 11)$

32. (a) $\|1\| = \sqrt{2}, \|x\| = \sqrt{\tfrac{2}{3}}, \|x^2\| = \sqrt{\tfrac{2}{5}},$ (b) $2\sqrt{\tfrac{2}{3}}$

EXERCISE SET 4.9 (page 218)

1. b **2.** b, d **3.** a **4.** a

6. (b) $\dfrac{2}{\sqrt{2}}\mathbf{v}_1 + \dfrac{2}{\sqrt{6}}\mathbf{v}_2 + \dfrac{5}{\sqrt{21}}\mathbf{v}_3 + \dfrac{1}{\sqrt{7}}\mathbf{v}_4$

7. (a) $\left(\dfrac{1}{\sqrt{10}}, -\dfrac{3}{\sqrt{10}}\right), \left(\dfrac{3}{\sqrt{10}}, \dfrac{1}{\sqrt{10}}\right)$ (b) $(1, 0), (0, -1)$

8. (a) $\left(\dfrac{1}{\sqrt{3}}, \dfrac{1}{\sqrt{3}}, \dfrac{1}{\sqrt{3}}\right), \left(-\dfrac{1}{\sqrt{2}}, \dfrac{1}{\sqrt{2}}, 0\right), \left(\dfrac{1}{\sqrt{6}}, \dfrac{1}{\sqrt{6}}, -\dfrac{2}{\sqrt{6}}\right)$

(b) $(1, 0, 0), \left(0, \dfrac{7}{\sqrt{53}}, -\dfrac{2}{\sqrt{53}}\right), \left(0, \dfrac{2}{\sqrt{53}}, \dfrac{7}{\sqrt{53}}\right)$

9. $\left(0, \dfrac{2}{\sqrt{5}}, \dfrac{1}{\sqrt{5}}, 0\right), \left(\dfrac{5}{\sqrt{30}}, -\dfrac{1}{\sqrt{30}}, \dfrac{2}{\sqrt{30}}, 0\right), \left(\dfrac{1}{\sqrt{10}}, \dfrac{1}{\sqrt{10}}, -\dfrac{2}{\sqrt{10}}, -\dfrac{2}{\sqrt{10}}\right),$

$\left(\dfrac{1}{\sqrt{15}}, \dfrac{1}{\sqrt{15}}, -\dfrac{2}{\sqrt{15}}, \dfrac{3}{\sqrt{15}}\right)$

10. $\left(0, \dfrac{1}{\sqrt{5}}, \dfrac{2}{\sqrt{5}}\right), \left(-\dfrac{\sqrt{5}}{\sqrt{6}}, -\dfrac{2}{\sqrt{30}}, \dfrac{1}{\sqrt{30}}\right)$

11. $\left(\dfrac{1}{\sqrt{6}}, \dfrac{1}{\sqrt{6}}, \dfrac{1}{\sqrt{6}}\right), \left(\dfrac{1}{\sqrt{6}}, \dfrac{1}{\sqrt{6}}, -\dfrac{1}{\sqrt{6}}\right), \left(\dfrac{2}{\sqrt{6}}, -\dfrac{1}{\sqrt{6}}, 0\right)$

12. $\mathbf{w}_1 = (-\frac{4}{5}, 2, \frac{3}{5})$, $\mathbf{w}_2 = (\frac{9}{5}, 0, \frac{12}{5})$ **13.** $\mathbf{w}_1 = (\frac{39}{42}, \frac{93}{42}, \frac{120}{42})$, $\mathbf{w}_2 = (\frac{3}{42}, -\frac{9}{42}, \frac{6}{42})$

14. $\mathbf{w}_1 = (-\frac{5}{4}, -\frac{1}{4}, \frac{5}{4}, \frac{9}{4})$, $\mathbf{w}_2 = (\frac{1}{4}, \frac{9}{4}, \frac{19}{4}, -\frac{9}{4})$

19. $\mathbf{v}_1 = \dfrac{1}{\sqrt{2}}$, $\mathbf{v}_2 = \sqrt{\dfrac{3}{2}}\, x$, $\mathbf{v}_3 = \dfrac{\sqrt{5}}{2\sqrt{2}}(3x^2 - 1)$

20. (a) $1 + x + 4x^2 = \frac{7}{3}\sqrt{2}\mathbf{v}_1 + \frac{1}{3}\sqrt{6}\mathbf{v}_2 + \frac{8}{15}\sqrt{5}\mathbf{v}_3$

   (b) $2 - 7x^2 = -\dfrac{\sqrt{2}}{3}\mathbf{v}_1 - \dfrac{28}{15}\sqrt{\dfrac{5}{2}}\mathbf{v}_3$ (c) $4 + 3x = 4\sqrt{2}\mathbf{v}_1 + \sqrt{6}\mathbf{v}_2$

21. $\mathbf{v}_1 = 1$, $\mathbf{v}_2 = \sqrt{3}(2x - 1)$, $\mathbf{v}_3 = \sqrt{5}(6x^2 - 6x + 1)$ **22.** $Q(-\frac{8}{7}, -\frac{5}{7}, \frac{25}{7})$; $\frac{3}{7}\sqrt{35}$

23. $Q(-\frac{8}{7}, \frac{4}{7}, -\frac{16}{7})$

EXERCISE SET 4.10 (page 237)

1. (a) $(\mathbf{w})_S = (3, -7)$, $[\mathbf{w}]_S = \begin{bmatrix} 3 \\ -7 \end{bmatrix}$ (b) $(\mathbf{w})_S = (\frac{5}{28}, \frac{3}{14})$, $[\mathbf{w}]_S = \begin{bmatrix} \frac{5}{28} \\ \frac{3}{14} \end{bmatrix}$

   (c) $(\mathbf{w})_S = \left(a, \dfrac{b-a}{2}\right)$, $[\mathbf{w}]_S = \begin{bmatrix} a \\ \dfrac{b-a}{2} \end{bmatrix}$

2. (a) $(\mathbf{v})_S = (3, -2, 1)$, $[\mathbf{v}]_S = \begin{bmatrix} 3 \\ -2 \\ 1 \end{bmatrix}$ (b) $(\mathbf{v})_S = (-2, 0, 1)$, $[\mathbf{v}]_S = \begin{bmatrix} -2 \\ 0 \\ 1 \end{bmatrix}$

3. (a) $(\mathbf{p})_S = (4, -3, 1)$, $[\mathbf{p}]_S = \begin{bmatrix} 4 \\ -3 \\ 1 \end{bmatrix}$ (b) $(\mathbf{p})_S = (0, 2, -1)$, $[\mathbf{p}]_S = \begin{bmatrix} 0 \\ 2 \\ -1 \end{bmatrix}$

4. $(A)_S = (-1, 1, -1, 3)$, $[A]_S = \begin{bmatrix} -1 \\ 1 \\ -1 \\ 3 \end{bmatrix}$

5. (a) $(\mathbf{w})_S = (-2\sqrt{2}, 5\sqrt{2})$, $[\mathbf{w}]_S = \begin{bmatrix} -2\sqrt{2} \\ 5\sqrt{2} \end{bmatrix}$

   (b) $(\mathbf{w})_S = (0, -2, 1)$, $[\mathbf{w}]_S = \begin{bmatrix} 0 \\ -2 \\ 1 \end{bmatrix}$

6. (a) $\mathbf{w} = (16, 10, 12)$ (b) $\mathbf{q} = 3 + 4x^2$ (c) $B = \begin{bmatrix} 15 & -1 \\ 6 & 3 \end{bmatrix}$

7. (a) $\|\mathbf{u}\| = \sqrt{2}$, $d(\mathbf{u}, \mathbf{v}) = \sqrt{13}$, $\langle \mathbf{u}, \mathbf{v} \rangle = 3$

8. (a) $\begin{bmatrix} 2 & -3 \\ 1 & 4 \end{bmatrix}$ (b) $\begin{bmatrix} \frac{4}{11} & \frac{3}{11} \\ -\frac{1}{11} & \frac{2}{11} \end{bmatrix}$ (c) $[\mathbf{w}]_B = \begin{bmatrix} 3 \\ -5 \end{bmatrix}$, $[\mathbf{w}]_{B'} = \begin{bmatrix} -\frac{3}{11} \\ -\frac{13}{11} \end{bmatrix}$

9. (a) $\begin{bmatrix} \frac{13}{10} & -\frac{1}{2} \\ -\frac{2}{5} & 0 \end{bmatrix}$ (b) $\begin{bmatrix} 0 & -\frac{5}{2} \\ -2 & -\frac{13}{2} \end{bmatrix}$

(c) $[\mathbf{w}]_B = \begin{bmatrix} -\frac{17}{10} \\ \frac{8}{5} \end{bmatrix}$, $[\mathbf{w}]_{B'} = \begin{bmatrix} -4 \\ -7 \end{bmatrix}$

10. (a) $\begin{bmatrix} \frac{3}{4} & \frac{3}{4} & \frac{1}{12} \\ -\frac{3}{4} & -\frac{17}{12} & -\frac{17}{12} \\ 0 & \frac{2}{3} & \frac{2}{3} \end{bmatrix}$ (b) $\begin{bmatrix} \frac{19}{12} \\ -\frac{43}{12} \\ \frac{4}{3} \end{bmatrix}$

11. (a) $\begin{bmatrix} 3 & 2 & \frac{5}{2} \\ -2 & -3 & -\frac{1}{2} \\ 5 & 1 & 6 \end{bmatrix}$ (b) $\begin{bmatrix} -\frac{7}{2} \\ \frac{23}{2} \\ 6 \end{bmatrix}$

12. (a) $\begin{bmatrix} -\frac{2}{9} & \frac{7}{9} \\ \frac{1}{3} & -\frac{1}{6} \end{bmatrix}$ (b) $\begin{bmatrix} \frac{3}{4} & \frac{7}{2} \\ \frac{3}{2} & 1 \end{bmatrix}$ (c) $[\mathbf{p}]_B = \begin{bmatrix} 1 \\ -1 \end{bmatrix}$

(d) $[\mathbf{p}]_{B'} = \begin{bmatrix} -\frac{11}{4} \\ \frac{1}{2} \end{bmatrix}$

13. (b) $\begin{bmatrix} 2 & 0 \\ 1 & 3 \end{bmatrix}$ (c) $\begin{bmatrix} \frac{1}{2} & 0 \\ -\frac{1}{6} & \frac{1}{3} \end{bmatrix}$ (d) $[\mathbf{h}]_B = \begin{bmatrix} 2 \\ -5 \end{bmatrix}$, $[\mathbf{h}]_{B'} = \begin{bmatrix} 1 \\ -2 \end{bmatrix}$

14. (a) $(4\sqrt{2}, -2\sqrt{2})$ (b) $(-\frac{7}{2}\sqrt{2}, \frac{3}{2}\sqrt{2})$

15. (a) $(-1 + 3\sqrt{3}, 3 + \sqrt{3})$ (b) $(\frac{5}{2} - \sqrt{3}, \frac{5}{2}\sqrt{3} + 1)$

16. (a) $(\frac{1}{2}\sqrt{2}, \frac{3}{2}\sqrt{2}, 5)$ (b) $(-\frac{5}{2}\sqrt{2}, \frac{7}{2}\sqrt{2}, -3)$

17. (a) $(-\frac{1}{2} - \frac{5}{2}\sqrt{3}, 2, \frac{5}{2} - \frac{1}{2}\sqrt{3})$ (b) $(\frac{1}{2} - \frac{3}{2}\sqrt{3}, 6, -\frac{3}{2} - \frac{1}{2}\sqrt{3})$

18. (a) $(-1, \frac{3}{2}\sqrt{2}, -\frac{7}{2}\sqrt{2})$ (b) $(1, -\frac{3}{2}\sqrt{2}, \frac{9}{2}\sqrt{3})$ **19.** a, b, d, e

20. (a) $\begin{bmatrix} 1 & 0 \\ 0 & 1 \end{bmatrix}$ (b) $\begin{bmatrix} \frac{1}{\sqrt{2}} & \frac{1}{\sqrt{2}} \\ -\frac{1}{\sqrt{2}} & \frac{1}{\sqrt{2}} \end{bmatrix}$

(d) $\begin{bmatrix} -\frac{1}{\sqrt{2}} & 0 & \frac{1}{\sqrt{2}} \\ \frac{1}{\sqrt{6}} & -\frac{2}{\sqrt{6}} & \frac{1}{\sqrt{6}} \\ \frac{1}{\sqrt{3}} & \frac{1}{\sqrt{3}} & \frac{1}{\sqrt{3}} \end{bmatrix}$ (e) $\begin{bmatrix} \frac{1}{2} & \frac{1}{2} & \frac{1}{2} & \frac{1}{2} \\ \frac{1}{2} & -\frac{5}{6} & \frac{1}{6} & \frac{1}{6} \\ \frac{1}{2} & \frac{1}{6} & \frac{1}{6} & -\frac{5}{6} \\ \frac{1}{2} & \frac{1}{6} & -\frac{5}{6} & \frac{1}{6} \end{bmatrix}$

22. (a) $\begin{bmatrix} \cos\theta & \sin\theta \\ -\sin\theta & \cos\theta \end{bmatrix}$ **(b)** $\begin{bmatrix} \cos\theta & \sin\theta & 0 \\ -\sin\theta & \cos\theta & 0 \\ 0 & 0 & 1 \end{bmatrix}$

23. (a) $(-2, -1)$ **(b)** $(-\frac{4}{5}, -\frac{22}{5})$ **(c)** $(-\frac{11}{5}, \frac{52}{5})$ **(d)** $(0, 0)$ **25.** *a, c*

27. (a) $(\frac{12}{5}, -\frac{9}{5}, -7)$ **(b)** $(2, 1, 6)$ **(c)** $(-\frac{42}{5}, \frac{19}{5}, -3)$ **(d)** $(0, 0, 0)$ **29.** *b*

31. (a) $A = \begin{bmatrix} \cos\theta & 0 & -\sin\theta \\ 0 & 1 & 0 \\ \sin\theta & 0 & \cos\theta \end{bmatrix}$ **(b)** $A = \begin{bmatrix} 1 & 0 & 0 \\ 0 & \cos\theta & \sin\theta \\ 0 & -\sin\theta & \cos\theta \end{bmatrix}$

32. $\begin{bmatrix} \dfrac{\sqrt{2}}{4} & \dfrac{\sqrt{6}}{4} & -\dfrac{\sqrt{2}}{2} \\[2mm] -\dfrac{\sqrt{3}}{2} & \dfrac{1}{2} & 0 \\[2mm] \dfrac{\sqrt{2}}{4} & \dfrac{\sqrt{6}}{4} & \dfrac{\sqrt{2}}{2} \end{bmatrix}$

CHAPTER 4 SUPPLEMENTARY EXERCISES (page 241)

1. (a) All of R^3 **(b)** Plane: $2x - 3y + z = 0$
(c) Line: $x = 2t, y = t, z = 0$ **(d)** The origin: $(0, 0, 0)$

2. (a) $(0, a, a, 0)$ with $a \neq 0$ **(b)** $\pm\left(0, \dfrac{2}{\sqrt{5}}, \dfrac{1}{\sqrt{5}}, 0\right)$

7. (a) $a(4, 1, 1) + b(0, -1, 2)$
(b) $(a + c)(3, -1, 2) + b(1, 4, 1)$
(c) $a(2, 3, 0) + b(-1, 0, 4) + c(4, -1, 1)$

8. (a) $\mathbf{v} = (-1 + r)\mathbf{v}_1 + (\frac{2}{3} - r)\mathbf{v}_2 + r\mathbf{v}_3$; r arbitrary   **10.** $\pm\left(\dfrac{1}{\sqrt{2}}, 0, \dfrac{1}{\sqrt{2}}\right)$

11. $w_k = \dfrac{1}{k}$, $k = 1, 2, \ldots, n$ **12.** No   **20.** *A* must be invertible.

21. No   **22. (a)** 2   **(b)** 1   **(c)** 2   **(d)** 3   **23.** 0, 1, or 2

EXERCISE SET 5.1 (page 252)

1. Linear   **2.** Nonlinear   **3.** Linear   **4.** Linear

5. Nonlinear   **6.** Linear   **7.** Linear   **8.** Nonlinear

9. Linear   **10.** Linear   **11.** Nonlinear   **12.** Linear

13. Linear   **14.** Nonlinear   **15.** Linear   **16.** Nonlinear

17. Linear   **18.** Linear   **19.** Linear   **20.** Nonlinear

21. $F(x, y) = (-x, y)$

23. (a) $\begin{bmatrix} 1 & 3 & 4 \\ 1 & 0 & -7 \end{bmatrix}$ (b) $\begin{bmatrix} 42 \\ -55 \end{bmatrix}$ (c) $\begin{bmatrix} x + 3y + 4z \\ x - 7z \end{bmatrix}$

24. (a) $T(x, y, z) = (x, 0, z)$ (b) $(2, 0, -1)$

25. (a) $T(x, y, z) = \frac{1}{3}(2x - y - z, -x + 2y - z, -x - y + 2z)$
(b) $T(3, 8, 4) = (-2, 3, -1)$

26. (a) $T(-1, 2) = \left(-\frac{3}{\sqrt{2}}, \frac{1}{\sqrt{2}}\right)$; $T(x, y) = \left(\frac{x}{\sqrt{2}} - \frac{y}{\sqrt{2}}, \frac{x}{\sqrt{2}} + \frac{y}{\sqrt{2}}\right)$

(b) $T(-1, 2) = (1, -2)$; $T(x, y) = (-x, -y)$

(c) $T(-1, 2) = \left(-\frac{\sqrt{3}}{2} - 1, \sqrt{3} - \frac{1}{2}\right)$; $T(x, y) = \left(\frac{\sqrt{3}}{2}x - \frac{1}{2}y, \frac{1}{2}x + \frac{\sqrt{3}}{2}y\right)$

(d) $T(-1, 2) = \left(-\frac{1}{2} + \sqrt{3}, \frac{\sqrt{3}}{2} + 1\right)$; $T(x, y) = \left(\frac{1}{2}x + \frac{\sqrt{3}}{2}y, -\frac{\sqrt{3}}{2}x + \frac{1}{2}y\right)$

EXERCISE SET 5.2 (page 260)

1. a, c **2.** a **3.** a, b, c **4.** a **5.** b **6.** a

7. $\ker(T) = \{0\}$; $R(T) = V$

8. Rank $(T) = 1$, nullity $(T) = 1$ **9.** Rank $(T) = 3$, nullity $(T) = 0$

10. (a) Rank $(T) = n$, nullity $(T) = 0$
(b) Rank $(T) = 0$, nullity $(T) = n$
(c) Rank $(T) = n$, nullity $(T) = 0$

11. $T(x, y, z) = (30x - 10y - 3z, -9x + 3y + z)$, $T(1, 1, 1) = (17, -5)$

12. $T(a_0 + a_1 x + a_2 x^2) = (a_0 + 3a_1 + 4a_2) + (a_0 + 2a_2)x + (-a_1 - 3a_2)x^2$
$T(2 - 2x + 3x^2) = 8 + 8x - 7x^2$

13. (a) Nullity $(T) = 2$ (b) Nullity $(T) = 4$
(c) Nullity $(T) = 3$ (d) Nullity $(T) = 1$

14. Nullity $(T) = 0$, rank $(T) = 6$

15. (a) Dimension = nullity $(T) = 3$
(b) No. In order for $A\mathbf{x} = \mathbf{b}$ to be consistent for all $\mathbf{b}$ in R^5, we must have $R(T) = R^5$. But $R(T) \neq R^5$, since rank $(T) = \dim R(T) = 4$.

16. (a) $\begin{bmatrix} 1 \\ 5 \\ 7 \end{bmatrix}, \begin{bmatrix} 0 \\ 1 \\ 1 \end{bmatrix}$ (b) $\begin{bmatrix} -\frac{14}{11} \\ \frac{19}{11} \\ 1 \end{bmatrix}$ (c) Rank $(T) = 2$, nullity $(T) = 1$

17. (a) $\begin{bmatrix} 1 \\ 2 \\ 0 \end{bmatrix}$ (b) $\begin{bmatrix} \frac{1}{2} \\ 0 \\ 1 \end{bmatrix}, \begin{bmatrix} 0 \\ 1 \\ 0 \end{bmatrix}$ (c) Rank $(T) = 1$, nullity $(T) = 2$

18. (a) $\begin{bmatrix} 1 \\ \frac{1}{4} \end{bmatrix}, \begin{bmatrix} 0 \\ 1 \end{bmatrix}$ (b) $\begin{bmatrix} -1 \\ -1 \\ 1 \\ 0 \end{bmatrix}, \begin{bmatrix} -\frac{4}{7} \\ \frac{2}{7} \\ 0 \\ 1 \end{bmatrix}$ (c) Rank $(T) = 2$, nullity $(T) = 2$

19. (a) $\begin{bmatrix} 1 \\ 3 \\ -1 \\ 2 \end{bmatrix}, \begin{bmatrix} 0 \\ 1 \\ -\frac{2}{7} \\ \frac{5}{14} \end{bmatrix}, \begin{bmatrix} 0 \\ 0 \\ 0 \\ 1 \end{bmatrix}$ (c) $\begin{bmatrix} -1 \\ -1 \\ 1 \\ 0 \\ 0 \end{bmatrix}, \begin{bmatrix} -1 \\ -2 \\ 0 \\ 0 \\ 1 \end{bmatrix}$

(c) Rank $(T) = 3$, nullity $(T) = 2$

22. (a) $x = -t, y = -t, z = t, -\infty < t < +\infty$ (b) $14x - 8y - 5z = 0$

26. ker(D) consists of all constant polynomials.

27. ker(J) consists of all polynomials of the form kx.

EXERCISE SET 5.3 (page 277)

1. (a) $\begin{bmatrix} 2 & -1 \\ 1 & 1 \end{bmatrix}$ (b) $\begin{bmatrix} 1 & 0 \\ 0 & 1 \end{bmatrix}$ (c) $\begin{bmatrix} 1 & 2 & 1 \\ 1 & 5 & 0 \\ 0 & 0 & 1 \end{bmatrix}$ (d) $\begin{bmatrix} 4 & 0 & 0 \\ 0 & 7 & 0 \\ 0 & 0 & -8 \end{bmatrix}$

2. (a) $\begin{bmatrix} 0 & 1 \\ -1 & 0 \\ 1 & 3 \\ 1 & -1 \end{bmatrix}$ (b) $\begin{bmatrix} 7 & 2 & -1 & 1 \\ 0 & 1 & 1 & 0 \\ -1 & 0 & 0 & 0 \end{bmatrix}$ (c) $\begin{bmatrix} 0 & 0 & 0 \\ 0 & 0 & 0 \\ 0 & 0 & 0 \\ 0 & 0 & 0 \\ 0 & 0 & 0 \end{bmatrix}$

(d) $\begin{bmatrix} 0 & 0 & 0 & 1 \\ 1 & 0 & 0 & 0 \\ 0 & 0 & 1 & 0 \\ 0 & 1 & 0 & 0 \\ 1 & 0 & -1 & 0 \end{bmatrix}$

3. (a) $\begin{bmatrix} 0 & -1 \\ -1 & 0 \end{bmatrix}$ (b) $\begin{bmatrix} -1 & 0 \\ 0 & -1 \end{bmatrix}$ (c) $\begin{bmatrix} 1 & 0 \\ 0 & 0 \end{bmatrix}$ (d) $\begin{bmatrix} 0 & 0 \\ 0 & 1 \end{bmatrix}$

4. (a) $(-1, -2)$ (b) $(-2, -1)$ (c) $(2, 0)$ (d) $(0, 1)$

8. (a) Rectangle with vertices at $(0, 0)$, $(1, 0)$, $(1, -2)$, $(0, -2)$
(b) Rectangle with vertices at $(0, 0)$, $(-1, 0)$, $(-1, 2)$, $(0, 2)$
(c) Rectangle with vertices at $(0, 0)$, $(1, 0)$, $(1, \frac{1}{2})$, $(0, \frac{1}{2})$
(d) Square with vertices at $(0, 0)$, $(2, 0)$, $(2, 2)$, $(0, 2)$
(e) Parallelogram with vertices at $(0, 0)$, $(1, 0)$, $(7, 2)$, $(6, 2)$
(f) Parallelogram with vertices at $(0, 0)$, $(1, -2)$, $(1, 0)$, $(0, 2)$

9. Rectangle with vertices at $(0, 0)$, $(-3, 0)$, $(0, 1)$, $(-3, 1)$

10. (a) $\begin{bmatrix} \dfrac{1}{\sqrt{2}} & -\dfrac{1}{\sqrt{2}} \\ \dfrac{1}{\sqrt{2}} & \dfrac{1}{\sqrt{2}} \end{bmatrix}$ (b) $\begin{bmatrix} 0 & -1 \\ 1 & 0 \end{bmatrix}$ (c) $\begin{bmatrix} -1 & 0 \\ 0 & -1 \end{bmatrix}$

(d) $\begin{bmatrix} 0 & 1 \\ -1 & 0 \end{bmatrix}$ (e) $\begin{bmatrix} \dfrac{\sqrt{3}}{2} & \dfrac{1}{2} \\ -\dfrac{1}{2} & \dfrac{\sqrt{3}}{2} \end{bmatrix}$

11. (a) $\begin{bmatrix} 1 & 0 \\ 4 & 1 \end{bmatrix}$ (b) $\begin{bmatrix} 1 & -2 \\ 0 & 1 \end{bmatrix}$ **12.** (a) $\begin{bmatrix} 1 & 0 \\ 0 & \frac{1}{3} \end{bmatrix}$ (b) $\begin{bmatrix} 6 & 0 \\ 0 & 1 \end{bmatrix}$

13. (a) Expansion by a factor of 3 in the x-direction.
(b) Expansion by a factor of -5 in the y-direction.
(c) Shear by a factor of 4 in the x-direction.

14. (a) $\begin{bmatrix} 2 & 0 \\ 0 & 1 \end{bmatrix}\begin{bmatrix} 3 & 0 \\ 0 & 1 \end{bmatrix}$; expansion in the y-direction by a factor of 3, then
expansion in the x-direction by a factor of 2.

(b) $\begin{bmatrix} 1 & 0 \\ 2 & 1 \end{bmatrix}\begin{bmatrix} 1 & 4 \\ 0 & 1 \end{bmatrix}$; shear in the x-direction by a factor of 4, then shear in
the y-direction by a factor of 2.

(c) $\begin{bmatrix} 0 & 1 \\ 1 & 0 \end{bmatrix}\begin{bmatrix} 4 & 0 \\ 0 & 1 \end{bmatrix}\begin{bmatrix} 1 & 0 \\ 0 & -2 \end{bmatrix}$; expansion in the y-direction by a factor of
-2, then expansion in the x-direction by a factor of 4, then reflection
about $y = x$.

(d) $\begin{bmatrix} 1 & 0 \\ 4 & 1 \end{bmatrix}\begin{bmatrix} 1 & 0 \\ 1 & 18 \end{bmatrix}\begin{bmatrix} 1 & -3 \\ 0 & 1 \end{bmatrix}$; shear in the x-direction by a factor of -3,
then expansion in the y-direction by a factor of 18, then shear in the
y-direction by a factor of 4.

15. (a) $\begin{bmatrix} \frac{1}{2} & 0 \\ 0 & 5 \end{bmatrix}$ (b) $\begin{bmatrix} 1 & 0 \\ 2 & 5 \end{bmatrix}$ (c) $\begin{bmatrix} 0 & -1 \\ -1 & 0 \end{bmatrix}$

16. (a) $\begin{bmatrix} 0 & 1 \\ -5 & 0 \end{bmatrix}$ **(b)** $\dfrac{1}{2}\begin{bmatrix} \sqrt{3} & -1 \\ -6\sqrt{3}+3 & 6+3\sqrt{3} \end{bmatrix}$

18. $16y - 11x - 3 = 0$

19. (a) $y = \frac{2}{7}x$ **(b)** $y = x$ **(c)** $y = \frac{1}{2}x$ **(d)** $y = -2x$

21. $\begin{bmatrix} 1 & -2 \\ 0 & 1 \end{bmatrix}$ **22. (b)** No; A is not invertible.

EXERCISE SET 5.4 (page 288)

1. $\begin{bmatrix} 1 & 1 & 0 \\ 0 & -2 & -3 \end{bmatrix}$ **2. (a)** $\begin{bmatrix} 0 & 0 \\ -\frac{1}{2} & 1 \\ \frac{8}{3} & \frac{4}{3} \end{bmatrix}$ **(b)** $\begin{bmatrix} 14 \\ -8 \\ 0 \end{bmatrix}$

3. (a) $\begin{bmatrix} 1 & -\frac{3}{2} & \frac{1}{2} \\ -1 & \frac{1}{2} & \frac{1}{2} \\ 0 & \frac{1}{2} & -\frac{1}{2} \end{bmatrix}$ **(b)** $\begin{bmatrix} 2 \\ -2 \\ 2 \end{bmatrix}$

4. (a) $\begin{bmatrix} 0 & 0 & 0 \\ 0 & 0 & 0 \\ 1 & 1 & 4 \\ 0 & 2 & 5 \\ 1 & 3 & 1 \end{bmatrix}$ **(b)** $-3x^2 + 5x^3 - 2x^4$

5. (a) $[T(\mathbf{v}_1)]_B = \begin{bmatrix} 1 \\ -2 \end{bmatrix}, \; [T(\mathbf{v}_2)]_B = \begin{bmatrix} 3 \\ 5 \end{bmatrix}$

(b) $T(\mathbf{v}_1) = \begin{bmatrix} 3 \\ -5 \end{bmatrix}, \; T(\mathbf{v}_2) = \begin{bmatrix} -2 \\ 29 \end{bmatrix}$

(c) $T\left(\begin{bmatrix} x_1 \\ x_2 \end{bmatrix}\right) = \begin{bmatrix} \frac{18}{7} & \frac{1}{7} \\ -\frac{107}{7} & \frac{24}{7} \end{bmatrix}\begin{bmatrix} x_1 \\ x_2 \end{bmatrix}$ **(d)** $\begin{bmatrix} \frac{19}{7} \\ -\frac{83}{7} \end{bmatrix}$

6. (a) $[T(\mathbf{v}_1)]_{B'} = \begin{bmatrix} 3 \\ 1 \\ -3 \end{bmatrix}, \; [T(\mathbf{v}_2)]_{B'} = \begin{bmatrix} -2 \\ 6 \\ 0 \end{bmatrix}, \; [T(\mathbf{v}_3)]_{B'} = \begin{bmatrix} 1 \\ 2 \\ 7 \end{bmatrix}, \; [T(\mathbf{v}_4)]_{B'} = \begin{bmatrix} 0 \\ 1 \\ 1 \end{bmatrix}$

(b) $T(\mathbf{v}_1) = \begin{bmatrix} 11 \\ 5 \\ 22 \end{bmatrix}, \; T(\mathbf{v}_2) = \begin{bmatrix} -42 \\ 32 \\ -10 \end{bmatrix}, \; T(\mathbf{v}_3) = \begin{bmatrix} -56 \\ 87 \\ 17 \end{bmatrix}, \; T(\mathbf{v}_4) = \begin{bmatrix} -13 \\ 17 \\ 2 \end{bmatrix}$

(c) $T\left(\begin{bmatrix} x_1 \\ x_2 \\ x_3 \\ x_4 \end{bmatrix}\right) = \begin{bmatrix} -\frac{253}{10} & \frac{49}{5} & \frac{241}{10} & -\frac{229}{10} \\ \frac{115}{2} & -39 & -\frac{65}{2} & \frac{153}{2} \\ 66 & -60 & -9 & 91 \end{bmatrix}\begin{bmatrix} x_1 \\ x_2 \\ x_3 \\ x_4 \end{bmatrix}$ **(d)** $\begin{bmatrix} -31 \\ 37 \\ 12 \end{bmatrix}$

7. (a) $[T(\mathbf{v}_1)]_B = \begin{bmatrix} 1 \\ 2 \\ 6 \end{bmatrix}$, $[T(\mathbf{v}_2)]_B = \begin{bmatrix} 3 \\ 0 \\ -2 \end{bmatrix}$, $[T(\mathbf{v}_3)]_B = \begin{bmatrix} -1 \\ 5 \\ 4 \end{bmatrix}$

(b) $T(\mathbf{v}_1) = 16 + 51x + 19x^2$, $T(\mathbf{v}_2) = -6 - 5x + 5x^2$, $T(\mathbf{v}_3) = 7 + 40x + 15x^2$

(c) $T(a_0 + a_1x + a_2x^2) = \dfrac{239a_0 - 161a_1 + 289a_2}{24} + \dfrac{201a_0 - 111a_1 + 247a_2}{8}x$
$+ \dfrac{61a_0 - 31a_1 + 107a_2}{12}x^2$

(d) $T(Hx^2) = 22 + 56x + 14x^2$

10. $\begin{bmatrix} 0 & 0 & 0 & 1 \\ 1 & 0 & 0 & 0 \\ 0 & 1 & 0 & 0 \\ 0 & 0 & 1 & 0 \end{bmatrix}$

11. (a) $\begin{bmatrix} 0 & 1 & 0 \\ 0 & 0 & 2 \\ 0 & 0 & 0 \end{bmatrix}$ (b) $\begin{bmatrix} 0 & -\frac{3}{2} & \frac{23}{6} \\ 0 & 0 & -\frac{16}{3} \\ 0 & 0 & 0 \end{bmatrix}$ (c) $-6 + 48x$

12. (a) $\begin{bmatrix} 0 & 0 & 0 \\ 0 & 0 & -1 \\ 0 & 1 & 0 \end{bmatrix}$ (b) $\begin{bmatrix} 0 & 0 & 0 \\ 0 & 1 & 0 \\ 0 & 0 & 2 \end{bmatrix}$ (c) $\begin{bmatrix} 2 & 1 & 0 \\ 0 & 2 & 2 \\ 0 & 0 & 2 \end{bmatrix}$

EXERCISE SET 5.5 (page 295)

1. $[T]_B = \begin{bmatrix} 1 & -2 \\ 0 & -1 \end{bmatrix}$ $[T]_{B'} = \begin{bmatrix} -\frac{3}{11} & -\frac{56}{11} \\ -\frac{2}{11} & \frac{3}{11} \end{bmatrix}$

2. $[T]_B = \begin{bmatrix} \frac{4}{5} & \frac{61}{10} \\ \frac{18}{5} & -\frac{19}{5} \end{bmatrix}$ $[T]_{B'} = \begin{bmatrix} -\frac{31}{2} & \frac{9}{2} \\ -\frac{75}{2} & \frac{25}{2} \end{bmatrix}$

3. $[T]_B = \begin{bmatrix} \frac{1}{\sqrt{2}} & -\frac{1}{\sqrt{2}} \\ \frac{1}{\sqrt{2}} & \frac{1}{\sqrt{2}} \end{bmatrix}$ $[T]_{B'} = \begin{bmatrix} \frac{13}{11\sqrt{2}} & -\frac{25}{11\sqrt{2}} \\ \frac{5}{11\sqrt{2}} & \frac{9}{11\sqrt{2}} \end{bmatrix}$

4. $[T]_B = \begin{bmatrix} 1 & 2 & -1 \\ 0 & -1 & 0 \\ 1 & 0 & 7 \end{bmatrix}$ $[T]_{B'} = \begin{bmatrix} 1 & 4 & 3 \\ -1 & -2 & -9 \\ 1 & 1 & 8 \end{bmatrix}$

5. $[T]_B = \begin{bmatrix} 1 & 0 & 0 \\ 0 & 1 & 0 \\ 0 & 0 & 0 \end{bmatrix}$ $[T]_{B'} = \begin{bmatrix} 1 & 0 & 0 \\ 0 & 1 & 1 \\ 0 & 0 & 0 \end{bmatrix}$

6. $[T]_B = \begin{bmatrix} 5 & 0 \\ 0 & 5 \end{bmatrix}$ $[T]_{B'} = \begin{bmatrix} 5 & 0 \\ 0 & 5 \end{bmatrix}$

7. $[T]_B = \begin{bmatrix} \frac{2}{3} & -\frac{2}{9} \\ \frac{1}{2} & \frac{4}{3} \end{bmatrix}$ $[T]_{B'} = \begin{bmatrix} 1 & 1 \\ 0 & 1 \end{bmatrix}$

CHAPTER 5 SUPPLEMENTARY EXERCISES (page 296)

1. No; $T(\mathbf{x}_1 + \mathbf{x}_2) = A(\mathbf{x}_1 + \mathbf{x}_2) + B \neq (A\mathbf{x}_1 + B) + (A\mathbf{x}_2 + B) = T(\mathbf{x}_1) + T(\mathbf{x}_2)$, and if $c \neq 1$, then $T(c\mathbf{x}) = cA\mathbf{x} + B \neq c(A\mathbf{x} + B) = cT(\mathbf{x})$.

2. (b) $A^n = \begin{bmatrix} \cos n\theta & -\sin n\theta \\ \sin n\theta & \cos n\theta \end{bmatrix}$

5. (a) $T(\mathbf{e}_3)$ and any two of $T(\mathbf{e}_1)$, $T(\mathbf{e}_2)$, and $T(\mathbf{e}_4)$ form bases for the range; $(-1, 1, 0, 1)$ is a basis for the kernel.
(b) Rank $= 3$, nullity $= 1$

6. (a) 2   (b) 0   **7.** (b) $\left(\dfrac{3 + 5\sqrt{3}}{4}, \dfrac{\sqrt{3} + 5}{4} \right)$

11. Rank $= 3$, nullity $= 1$ **13.** $\begin{bmatrix} 1 & 0 & 0 & 0 \\ 0 & 0 & 1 & 0 \\ 0 & 1 & 0 & 0 \\ 0 & 0 & 0 & 1 \end{bmatrix}$

14. (a) $\mathbf{v}_1 = 2\mathbf{u}_1 + \mathbf{u}_2$, $\mathbf{v}_2 = -\mathbf{u}_1 + \mathbf{u}_2 + \mathbf{u}_3$, $\mathbf{v}_3 = 3\mathbf{u}_1 + 4\mathbf{u}_2 + 2\mathbf{u}_3$
(b) $\mathbf{u}_1 = -2\mathbf{v}_1 - 2\mathbf{v}_2 + \mathbf{v}_3$, $\mathbf{u}_2 = 5\mathbf{v}_1 + 4\mathbf{v}_2 - 2\mathbf{v}_3$, $\mathbf{u}_3 = -7\mathbf{v}_1 - 5\mathbf{v}_2 + 3\mathbf{v}_3$

15. $[T]_{B'} = \begin{bmatrix} -4 & 0 & 9 \\ 1 & 0 & -2 \\ 0 & 1 & 1 \end{bmatrix}$ **17.** $[T]_B = \begin{bmatrix} 1 & -1 & 1 \\ 0 & 1 & 0 \\ 1 & 0 & -1 \end{bmatrix}$

18. (a) $\begin{bmatrix} 1 & 0 & 0 \\ 0 & 0 & 1 \\ 0 & 1 & 0 \end{bmatrix}$ (b) $\begin{bmatrix} 0 & 0 & 1 \\ 0 & 1 & 0 \\ 1 & 0 & 0 \end{bmatrix}$ (c) $\begin{bmatrix} 0 & 1 & 0 \\ 1 & 0 & 0 \\ 0 & 0 & 1 \end{bmatrix}$

19. (a) $\begin{bmatrix} 1 & 0 & k \\ 0 & 1 & k \\ 0 & 0 & 1 \end{bmatrix}$ (b) xz-direction: $\begin{bmatrix} 1 & k & 0 \\ 0 & 1 & 0 \\ 0 & k & 1 \end{bmatrix}$ yz-direction: $\begin{bmatrix} 1 & 0 & 0 \\ k & 1 & 0 \\ k & 0 & 1 \end{bmatrix}$

21. $\begin{bmatrix} 0 & 1 & 0 & 0 & \cdots & 0 \\ 0 & 0 & 1 & 0 & \cdots & 0 \\ 0 & 0 & 0 & 1 & \cdots & 0 \\ \vdots & \vdots & \vdots & \vdots & & \vdots \\ 0 & 0 & 0 & 0 & \cdots & 1 \\ 0 & 0 & 0 & 0 & \cdots & 0 \end{bmatrix}$ **22.** $\begin{bmatrix} 0 & 0 & 0 & \cdots & 0 \\ 1 & 0 & 0 & \cdots & 0 \\ 0 & \frac{1}{2} & 0 & \cdots & 0 \\ 0 & 0 & \frac{1}{3} & \cdots & 0 \\ \vdots & \vdots & \vdots & & \vdots \\ 0 & 0 & 0 & \cdots & \frac{1}{n+1} \end{bmatrix}$

EXERCISE SET 6.1 (page 308)

1. (a) $\lambda^2 - 2\lambda - 3 = 0$ (b) $\lambda^2 - 8\lambda + 16 = 0$ (c) $\lambda^2 - 12 = 0$
   (d) $\lambda^2 + 3 = 0$ (e) $\lambda^2 = 0$ (f) $\lambda^2 - 2\lambda + 1 = 0$

2. (a) $\lambda = 3, \lambda = -1$ (b) $\lambda = 4$ (c) $\lambda = \sqrt{12}, \lambda = -\sqrt{12}$
   (d) No real eigenvalues   (e) $\lambda = 0$ (f) $\lambda = 1$

3. (a) Basis for eigenspace corresponding to $\lambda = 3$: $\begin{bmatrix} \frac{1}{2} \\ 1 \end{bmatrix}$

   basis for eigenspace corresponding to $\lambda = -1$: $\begin{bmatrix} 0 \\ 1 \end{bmatrix}$

   (b) Basis for eigenspace corresponding to $\lambda = 4$: $\begin{bmatrix} \frac{3}{2} \\ 1 \end{bmatrix}$

   (c) Basis for eigenspace corresponding to $\lambda = \sqrt{12}$: $\begin{bmatrix} \dfrac{3}{\sqrt{12}} \\ 1 \end{bmatrix}$

   basis for eigenspace corresponding to $\lambda = -\sqrt{12}$: $\begin{bmatrix} -\dfrac{3}{\sqrt{12}} \\ 1 \end{bmatrix}$

   (d) There are no eigenspaces.

   (e) Basis for eigenspace corresponding to $\lambda = 0$: $\begin{bmatrix} 1 \\ 0 \end{bmatrix}, \begin{bmatrix} 0 \\ 1 \end{bmatrix}$

   (f) Basis for eigenspace corresponding to $\lambda = 1$: $\begin{bmatrix} 1 \\ 0 \end{bmatrix}, \begin{bmatrix} 0 \\ 1 \end{bmatrix}$

5. (a) $\lambda^3 - 6\lambda^2 + 11\lambda - 6 = 0$ (b) $\lambda^3 - 2\lambda = 0$
   (c) $\lambda^3 + 8\lambda^2 + \lambda + 8 = 0$ (d) $\lambda^3 - \lambda^2 - \lambda - 2 = 0$
   (e) $\lambda^3 - 6\lambda^2 + 12\lambda - 8 = 0$ (f) $\lambda^3 - 2\lambda^2 - 15\lambda + 36 = 0$

6. (a) $\lambda = 1, \lambda = 2, \lambda = 3$ (b) $\lambda = 0, \lambda = \sqrt{2}, \lambda = -\sqrt{2}$ (c) $\lambda = -8$
   (d) $\lambda = 2$ (e) $\lambda = 2$ (f) $\lambda = -4, \lambda = 3$

7. (a) $\lambda = 1$: basis $\begin{bmatrix} 0 \\ 1 \\ 0 \end{bmatrix}$, $\lambda = 2$: basis $\begin{bmatrix} -\frac{1}{2} \\ 1 \\ 1 \end{bmatrix}$, $\lambda = 3$: basis $\begin{bmatrix} -1 \\ 1 \\ 1 \end{bmatrix}$

   (b) $\lambda = 0$: basis $\begin{bmatrix} \frac{5}{3} \\ \frac{1}{3} \\ 1 \end{bmatrix}$, $\lambda = \sqrt{2}$: basis $\begin{bmatrix} \frac{1}{7}(15 + 5\sqrt{2}) \\ \frac{1}{7}(-1 + 2\sqrt{2}) \\ 1 \end{bmatrix}$

   $\lambda = -\sqrt{2}$: basis $\begin{bmatrix} \frac{1}{7}(15 - 5\sqrt{2}) \\ \frac{1}{7}(-1 - 2\sqrt{2}) \\ 1 \end{bmatrix}$

(c) $\lambda = -8$: basis $\begin{bmatrix} -\frac{1}{6} \\ -\frac{1}{6} \\ 1 \end{bmatrix}$ (d) $\lambda = 2$: basis $\begin{bmatrix} \frac{1}{3} \\ \frac{1}{3} \\ 1 \end{bmatrix}$

(e) $\lambda = 2$: basis $\begin{bmatrix} -\frac{1}{3} \\ -\frac{1}{3} \\ 1 \end{bmatrix}$ (f) $\lambda = -4$: basis $\begin{bmatrix} -2 \\ \frac{8}{3} \\ 1 \end{bmatrix}$, $\lambda = 3$: basis $\begin{bmatrix} 5 \\ -2 \\ 1 \end{bmatrix}$

8. (a) $(\lambda - 1)^2(\lambda + 2)(\lambda + 1) = 0$ (b) $(\lambda - 4)^2(\lambda^2 + 3) = 0$

9. (a) $\lambda = 1, \lambda = -2, \lambda = -1$ (b) $\lambda = 4$

10. (a) $\lambda = 1$: basis $\begin{bmatrix} 0 \\ 0 \\ 0 \\ 1 \end{bmatrix}$ and $\begin{bmatrix} 2 \\ 3 \\ 1 \\ 0 \end{bmatrix}$, $\lambda = -2$: basis $\begin{bmatrix} -1 \\ 0 \\ 1 \\ 0 \end{bmatrix}$, $\lambda = -1$: basis $\begin{bmatrix} -2 \\ 1 \\ 1 \\ 0 \end{bmatrix}$

(b) $\lambda = 4$: basis $\begin{bmatrix} \frac{3}{2} \\ 1 \\ 0 \\ 0 \end{bmatrix}$

11. (a) $\lambda = -4, \lambda = 3$
(b) Basis for eigenspace corresponding to $\lambda = -4$: $-2 + \frac{8}{3}x + x^2$
basis for eigenspace corresponding to $\lambda = 3$: $5 - 2x + x^2$

12. (a) $\lambda = 1, \lambda = -2, \lambda = -1$

(b) Basis for eigenspace corresponding to $\lambda = 1$: $\begin{bmatrix} 0 & 0 \\ 0 & 1 \end{bmatrix}$ and $\begin{bmatrix} 2 & 3 \\ 1 & 0 \end{bmatrix}$;

basis for eigenspace corresponding to $\lambda = -2$: $\begin{bmatrix} -1 & 0 \\ 1 & 0 \end{bmatrix}$;

basis for eigenspace corresponding to $\lambda = -1$: $\begin{bmatrix} -2 & 1 \\ 1 & 0 \end{bmatrix}$

19. $1, -1, -2^9, 2^9$.

EXERCISE SET 6.2 (page 317)

5. $P = \begin{bmatrix} \frac{4}{5} & \frac{3}{4} \\ 1 & 1 \end{bmatrix}$ $P^{-1}AP = \begin{bmatrix} 1 & 0 \\ 0 & 2 \end{bmatrix}$

6. $P = \begin{bmatrix} \frac{1}{3} & 0 \\ 1 & 1 \end{bmatrix}$ $P^{-1}AP = \begin{bmatrix} 1 & 0 \\ 0 & -1 \end{bmatrix}$

7. $P = \begin{bmatrix} 0 & 1 & 0 \\ 1 & 0 & 1 \\ -1 & 0 & 1 \end{bmatrix}$ $\quad P^{-1}AP = \begin{bmatrix} 0 & 0 & 0 \\ 0 & 1 & 0 \\ 0 & 0 & 2 \end{bmatrix}$

8. $P = \begin{bmatrix} -2 & 0 & 1 \\ 0 & 1 & 0 \\ 1 & 0 & 0 \end{bmatrix}$ $\quad P^{-1}AP = \begin{bmatrix} 3 & 0 & 0 \\ 0 & 3 & 0 \\ 0 & 0 & 2 \end{bmatrix}$

9. Not diagonalizable   **10.** $P = \begin{bmatrix} 1 & 2 & 1 \\ 1 & 3 & 3 \\ 1 & 3 & 4 \end{bmatrix}$ $\quad P^{-1}AP = \begin{bmatrix} 1 & 0 & 0 \\ 0 & 2 & 0 \\ 0 & 0 & 3 \end{bmatrix}$

11. Not diagonalizable   **12.** $P = \begin{bmatrix} -\frac{1}{3} & 0 & 0 \\ 0 & 1 & 0 \\ 1 & 0 & 1 \end{bmatrix}$ $\quad P^{-1}AP = \begin{bmatrix} 0 & 0 & 0 \\ 0 & 0 & 0 \\ 0 & 0 & 1 \end{bmatrix}$

13. Not diagonalizable

14. $P = \begin{bmatrix} 1 & 1 & 0 & 0 \\ 0 & 1 & 1 & 0 \\ 0 & 0 & 1 & 1 \\ 0 & 0 & 0 & 1 \end{bmatrix}$ $\quad P^{-1}AP = \begin{bmatrix} -2 & 0 & 0 & 0 \\ 0 & -2 & 0 & 0 \\ 0 & 0 & 3 & 0 \\ 0 & 0 & 0 & 3 \end{bmatrix}$

15. $\begin{bmatrix} 2 \\ 1 \end{bmatrix}, \begin{bmatrix} 1 \\ -1 \end{bmatrix}$ **16.** $\begin{bmatrix} 1 \\ 1 \\ -1 \end{bmatrix}, \begin{bmatrix} 1 \\ 0 \\ 1 \end{bmatrix}, \begin{bmatrix} 1 \\ 1 \\ 0 \end{bmatrix}$

17. $\mathbf{p}_1 = \frac{1}{3} + x, \mathbf{p}_2 = x$ **19.** $\begin{bmatrix} 1 & 0 \\ -1023 & 1024 \end{bmatrix}$

EXERCISE SET 6.3 (page 324)

1. (a) $\lambda = 0$: 1-dimensional, $\lambda = 2$: 1-dimensional
    (b) $\lambda = 6$: 1-dimensional, $\lambda = -3$: 2-dimensional
    (c) $\lambda = 3$: 1-dimensional, $\lambda = 0$: 2-dimensional
    (d) $\lambda = 0$: 1-dimensional, $\lambda = 6$: 2-dimensional
    (e) $\lambda = 0$: 3-dimensional, $\lambda = 8$: 1-dimensional
    (f) $\lambda = -2$: 3-dimensional, $\lambda = 4$: 1-dimensional

2. $P = \begin{bmatrix} \dfrac{1}{\sqrt{2}} & -\dfrac{1}{\sqrt{2}} \\ \dfrac{1}{\sqrt{2}} & \dfrac{1}{\sqrt{2}} \end{bmatrix}$ $\quad P^{-1}AP = \begin{bmatrix} 4 & 0 \\ 0 & 2 \end{bmatrix}$

3. $P = \begin{bmatrix} \dfrac{\sqrt{3}}{2} & -\dfrac{1}{2} \\ \dfrac{1}{2} & \dfrac{\sqrt{3}}{2} \end{bmatrix}$ $\quad P^{-1}AP = \begin{bmatrix} 8 & 0 \\ 0 & -4 \end{bmatrix}$

4. $P = \begin{bmatrix} \frac{3}{5} & -\frac{4}{5} \\ \frac{4}{5} & \frac{3}{5} \end{bmatrix}$ $P^{-1}AP = \begin{bmatrix} 25 & 0 \\ 0 & -25 \end{bmatrix}$

5. $P = \begin{bmatrix} -\frac{4}{5} & 0 & \frac{3}{5} \\ 0 & 1 & 0 \\ \frac{3}{5} & 0 & \frac{4}{5} \end{bmatrix}$ $P^{-1}AP = \begin{bmatrix} 25 & 0 & 0 \\ 0 & -3 & 0 \\ 0 & 0 & -50 \end{bmatrix}$ **6.** $\begin{bmatrix} \frac{1}{\sqrt{2}} & \frac{1}{\sqrt{2}} & 0 \\ \frac{1}{\sqrt{2}} & -\frac{1}{\sqrt{2}} & 0 \\ 0 & 0 & 1 \end{bmatrix}$

7. $\begin{bmatrix} \frac{1}{\sqrt{3}} & \frac{1}{\sqrt{6}} & \frac{1}{\sqrt{2}} \\ \frac{1}{\sqrt{3}} & -\frac{2}{\sqrt{6}} & 0 \\ \frac{1}{\sqrt{3}} & \frac{1}{\sqrt{6}} & -\frac{1}{\sqrt{2}} \end{bmatrix}$ **8.** $\begin{bmatrix} 0 & 0 & \frac{1}{\sqrt{2}} & \frac{1}{\sqrt{2}} \\ 0 & 0 & \frac{1}{\sqrt{2}} & -\frac{1}{\sqrt{2}} \\ 1 & 0 & 0 & 0 \\ 0 & 1 & 0 & 0 \end{bmatrix}$

9. $\begin{bmatrix} \frac{1}{\sqrt{5}} & 0 & -\frac{2}{\sqrt{5}} & 0 \\ \frac{2}{\sqrt{5}} & 0 & \frac{1}{\sqrt{5}} & 0 \\ 0 & \frac{1}{\sqrt{5}} & 0 & -\frac{2}{\sqrt{5}} \\ 0 & \frac{2}{\sqrt{5}} & 0 & \frac{1}{\sqrt{5}} \end{bmatrix}$

CHAPTER 6 SUPPLEMENTARY EXERCISES (page 325)

1. (b) The transformation rotates vectors through the angle θ; therefore, if $0 < \theta < \pi$, then no nonzero vector is transformed into a vector in the same or opposite direction.

2. $\lambda = k$ with multiplicity 3. **3.** (c) $\begin{bmatrix} 1 & 1 & 0 \\ 0 & 2 & 1 \\ 0 & 0 & 3 \end{bmatrix}$

9. $A^2 = \begin{bmatrix} 15 & 30 \\ 5 & 10 \end{bmatrix}$, $A^3 = \begin{bmatrix} 75 & 150 \\ 25 & 50 \end{bmatrix}$, $A^4 = \begin{bmatrix} 375 & 750 \\ 125 & 250 \end{bmatrix}$,

$A^5 = \begin{bmatrix} 1875 & 3750 \\ 625 & 1250 \end{bmatrix}$

10. $A^3 = \begin{bmatrix} 1 & -3 & 3 \\ 3 & -8 & 6 \\ 6 & -15 & 10 \end{bmatrix}$, $A^4 = \begin{bmatrix} 3 & -8 & 6 \\ 6 & -15 & 10 \\ 10 & -24 & 15 \end{bmatrix}$

12. (b) $\begin{bmatrix} 0 & 0 & 0 & -1 \\ 1 & 0 & 0 & 2 \\ 0 & 1 & 0 & -1 \\ 0 & 0 & 1 & -3 \end{bmatrix}$

17. (a) $\lambda_1 = 1: \begin{bmatrix} 1 \\ 0 \\ 1 \end{bmatrix}, \lambda_2 = \frac{1}{2}: \begin{bmatrix} 1 \\ \frac{1}{2} \\ 0 \end{bmatrix}, \lambda_3 = \frac{1}{3}: \begin{bmatrix} 1 \\ 1 \\ 1 \end{bmatrix}$

(b) $\lambda_1 = -2: \begin{bmatrix} 1 \\ 0 \\ 1 \end{bmatrix}, \lambda_2 = -1: \begin{bmatrix} 1 \\ \frac{1}{2} \\ 0 \end{bmatrix}, \lambda_3 = 0: \begin{bmatrix} 1 \\ 1 \\ 1 \end{bmatrix}$

(c) $\lambda_1 = 3: \begin{bmatrix} 1 \\ 0 \\ 1 \end{bmatrix}, \lambda_2 = 4: \begin{bmatrix} 1 \\ \frac{1}{2} \\ 0 \end{bmatrix}, \lambda_3 = 5: \begin{bmatrix} 1 \\ 1 \\ 1 \end{bmatrix}$

18. (a) $\lambda_1 = 1: \begin{bmatrix} 1 \\ 0 \end{bmatrix}, \lambda_2 = -1: \begin{bmatrix} 0 \\ 1 \end{bmatrix}$ (b) $\lambda_1 = 1: \begin{bmatrix} 0 \\ 1 \end{bmatrix}, \lambda_2 = -1: \begin{bmatrix} 1 \\ 0 \end{bmatrix}$

(c) $\lambda_1 = 1: \begin{bmatrix} 1 \\ 1 \end{bmatrix}, \lambda_2 = -1: \begin{bmatrix} -1 \\ 1 \end{bmatrix}$ (d) $\lambda = 1: \begin{bmatrix} 1 \\ 0 \end{bmatrix}$

(e) $\lambda = 1: \begin{bmatrix} 0 \\ 1 \end{bmatrix}$

(f) θ an odd integer multiple of π: $\lambda = -1: \begin{bmatrix} 1 \\ 0 \end{bmatrix}$

θ an even integer with multiple of π: $\lambda = 1: \begin{bmatrix} 1 \\ 0 \end{bmatrix}, \begin{bmatrix} 0 \\ 1 \end{bmatrix}$

θ not an integer multiple of π: no real eigenvalues.

EXERCISE SET 7.1 (page 334)

1. (a) $y_1 = c_1 e^{5x} - 2c_2 e^{-x}$
   $y_2 = c_1 e^{5x} + c_2 e^{-x}$

   (b) $y_1 = 0$
   $y_2 = 0$

2. (a) $y_1 = c_1 e^{7x} - 3c_2 e^{-x}$
   $y_2 = 2c_1 e^{7x} + 2c_2 e^{-x}$

   (b) $y_1 = -\frac{1}{40} e^{7x} + \frac{81}{40} e^{-x}$
   $y_2 = -\frac{1}{20} e^{7x} - \frac{27}{20} e^{-x}$

3. (a) $y_1 = -c_2 e^{2x} + c_3 e^{3x}$
   $y_2 = c_1 e^x + 2c_2 e^{2x} - c_3 e^{3x}$
   $y_3 = 2c_2 e^{2x} - c_3 e^{3x}$

   (b) $y_1 = e^{2x} - 2e^{3x}$
   $y_2 = e^x - 2e^{2x} + 2e^{3x}$
   $y_3 = -2e^{2x} + 2e^{3x}$

4. $y_1 = (c_1 + c_2)e^{2x} + c_3 e^{8x}$
   $y_2 = -c_2 e^{2x} + c_3 e^{8x}$
   $y_3 = -c_1 e^{2x} + c_3 e^{8x}$

5. $y = c_1 e^{3x} + c_2 e^{-2x}$ 6. $y = c_1 e^x + c_2 e^{2x} + c_3 e^{3x}$

EXERCISE SET 7.2 (page 341)

1. (a) $(1 + \pi) - 2 \sin x - \sin 2x$

(b) $(1 + \pi) - 2 \left[\sin x + \dfrac{\sin 2x}{2} + \dfrac{\sin 3x}{3} + \cdots + \dfrac{\sin nx}{n} \right]$

2. (a) $\frac{4}{3}\pi^2 + 4 \cos x + \cos 2x + \frac{4}{9} \cos 3x - 4\pi \sin x - 2\pi \sin 2x - \dfrac{4\pi}{3} \sin 3x$

(b) $\frac{4}{3}\pi^2 + 4 \sum\limits_{k=1}^{n} \dfrac{\cos kx}{k^2} - 4\pi \sum\limits_{k=1}^{n} \dfrac{\sin kx}{k}$

3. (a) $-\dfrac{1}{2} + \dfrac{1}{e-1} e^x$ (b) $\dfrac{1}{12} - \dfrac{3-e}{2e-2}$

4. (a) $(4e - 10) + (18 - 6e)x$ (b) $\dfrac{(3-e)(7e-19)}{2}$

5. (a) $\dfrac{3}{\pi} x$ (b) $1 - \dfrac{6}{\pi^2}$ **8.** $\sum\limits_{k=1}^{\infty} \dfrac{2}{k} \sin(kx)$

EXERCISE SET 7.3 (page 349)

1. (a), (b), (c), (e), (g), (h)

2. (a) $A = \begin{bmatrix} 3 & 0 \\ 0 & 7 \end{bmatrix}$ (b) $A = \begin{bmatrix} 4 & -3 \\ -3 & -9 \end{bmatrix}$ (c) $A = \begin{bmatrix} 5 & \frac{5}{2} \\ \frac{5}{2} & 0 \end{bmatrix}$

(d) $A = \begin{bmatrix} 0 & -\frac{7}{2} \\ -\frac{7}{2} & 0 \end{bmatrix}$

3. (a) $A = \begin{bmatrix} 9 & 3 & -4 \\ 3 & -1 & \frac{1}{2} \\ -4 & \frac{1}{2} & 4 \end{bmatrix}$ (b) $A = \begin{bmatrix} 1 & -\frac{5}{2} & \frac{9}{2} \\ -\frac{5}{2} & 1 & 0 \\ \frac{9}{2} & 0 & -3 \end{bmatrix}$

(c) $A = \begin{bmatrix} 0 & \frac{1}{2} & \frac{1}{2} \\ \frac{1}{2} & 0 & \frac{1}{2} \\ \frac{1}{2} & \frac{1}{2} & 0 \end{bmatrix}$ (d) $A = \begin{bmatrix} \sqrt{2} & \sqrt{2} & -4\sqrt{3} \\ \sqrt{2} & 0 & 0 \\ -4\sqrt{3} & 0 & -\sqrt{3} \end{bmatrix}$

(e) $A = \begin{bmatrix} 1 & 1 & 0 & -5 \\ 1 & 1 & 0 & 0 \\ 0 & 0 & -1 & 2 \\ -5 & 0 & 2 & -1 \end{bmatrix}$

4. (a) $2x^2 + 5y^2 - 6xy$ (b) $7x_1^2 + 5x_1 x_2$ (c) $x^2 - 3y^2 + 5z^2$
(d) $-2x_1^2 + 3x_3^2 + 7x_1 x_2 + x_1 x_3 + 12 x_2 x_3$
(e) $2x_1 x_2 + 2x_1 x_3 + 2x_1 x_4 + 2x_2 x_3 + 2x_2 x_4 + 2x_3 x_4$

5. (a) max value $= 5$ at $\pm(1, 0)$
min value $= -1$ at $\pm(0, 1)$

(b) max value $= \dfrac{11 + \sqrt{10}}{2}$ at $\pm\left(\dfrac{-1}{\sqrt{20 + 6\sqrt{10}}}, \dfrac{1}{\sqrt{20 + 6\sqrt{10}}}\right)$

min value $= \dfrac{11 - \sqrt{10}}{2}$ at $\pm\left(\dfrac{-1}{\sqrt{20 + 6\sqrt{10}}}, \dfrac{1}{\sqrt{20 + 6\sqrt{10}}}\right)$

(c) max value $= \dfrac{7 + \sqrt{10}}{2}$ at $\pm\left(\dfrac{3 + \sqrt{10}}{\sqrt{20 + 6\sqrt{10}}}, \dfrac{-1}{\sqrt{20 + 6\sqrt{10}}}\right)$

min value $= \dfrac{7 - \sqrt{10}}{2}$ at $\pm\left(\dfrac{1}{\sqrt{20 + 6\sqrt{10}}}, \dfrac{3 + \sqrt{10}}{\sqrt{20 + 6\sqrt{10}}}\right)$

(d) max value $= \dfrac{7}{2}$ at $\pm\left(\dfrac{1}{\sqrt{2}}, \dfrac{1}{\sqrt{2}}\right)$

min value $= \dfrac{1}{2}$ at $\pm\left(\dfrac{1}{\sqrt{2}}, -\dfrac{1}{\sqrt{2}}\right)$

6. (a) max value $= 4$ at $\pm\left(\dfrac{1}{\sqrt{5}}, \dfrac{1}{\sqrt{5}}, \dfrac{2}{\sqrt{5}}\right)$

min value $= -2$ at $\pm\left(-\dfrac{1}{\sqrt{3}}, -\dfrac{1}{\sqrt{3}}, \dfrac{1}{\sqrt{3}}\right)$

(b) max value $= 3$ at $\pm\left(\dfrac{2}{\sqrt{6}}, \dfrac{1}{\sqrt{6}}, \dfrac{1}{\sqrt{6}}\right)$

min value $= 0$ at $\pm\left(\dfrac{1}{\sqrt{3}}, -\dfrac{1}{\sqrt{3}}, \dfrac{-1}{\sqrt{3}}\right)$

(c) max value $= 4$ at $\pm\left(\dfrac{1}{\sqrt{2}}, 0, \dfrac{1}{\sqrt{2}}\right)$

min value $= 2$ at $\pm\left(\dfrac{1}{\sqrt{2}}, 0, -\dfrac{1}{\sqrt{2}}\right)$ and $\pm(0, 1, 0)$

7. (b)   **9.** (a)

11. (a) Positive definite   (b) Negative definite   (c) Positive semidefinite
(d) Negative semidefinite   (e) Indefinite   (f) Indefinite

12. (a) Indefinite   (b) Indefinite   (c) Positive definite   (d) Indefinite
(e) Positive and negative semidefinite   (f) Positive definite

13. (c) No; $T(k\mathbf{x}) \neq kT(\mathbf{x})$, unless $k = 1$.

14. (a) $k > 4$ (b) $k > 2$ (c) $-\frac{1}{3}\sqrt{15} < k < \frac{1}{3}\sqrt{15}$

15. $A = \begin{bmatrix} c_1^2 & c_1 c_2 & c_1 c_3 & \cdots & c_1 c_n \\ c_1 c_2 & c_2^2 & c_2 c_3 & \cdots & c_2 c_n \\ \vdots & \vdots & \vdots & & \vdots \\ c_1 c_n & c_2 c_n & c_3 c_n & \cdots & c_n^2 \end{bmatrix}$

16. (a) $A = \begin{bmatrix} \dfrac{1}{n} & \dfrac{-1}{n(n-1)} & \dfrac{-1}{n(n-1)} & \cdots & \dfrac{-1}{n(n-1)} \\ \dfrac{-1}{n(n-1)} & \dfrac{1}{n} & \dfrac{-1}{n(n-1)} & \cdots & \dfrac{-1}{n(n-1)} \\ \vdots & \vdots & \vdots & & \vdots \\ \dfrac{-1}{n(n-1)} & \dfrac{-1}{n(n-1)} & \dfrac{-1}{n(n-1)} & \cdots & \dfrac{1}{n} \end{bmatrix}$

(b) Positive semidefinite

EXERCISE SET 7.4 (page 362)

1. (a) $\begin{bmatrix} x_1 \\ x_2 \end{bmatrix} = \begin{bmatrix} \dfrac{1}{\sqrt{2}} & \dfrac{1}{\sqrt{2}} \\ \dfrac{1}{\sqrt{2}} & -\dfrac{1}{\sqrt{2}} \end{bmatrix} \begin{bmatrix} y_1 \\ y_2 \end{bmatrix}$; $y_1^2 + 3y_2^2$

(b) $\begin{bmatrix} x_1 \\ x_2 \end{bmatrix} = \begin{bmatrix} \dfrac{1}{\sqrt{5}} & \dfrac{2}{\sqrt{5}} \\ -\dfrac{2}{\sqrt{5}} & \dfrac{1}{\sqrt{5}} \end{bmatrix} \begin{bmatrix} y_1 \\ y_2 \end{bmatrix}$; $y_1^2 + 6y_2^2$

(c) $\begin{bmatrix} x_1 \\ x_2 \end{bmatrix} = \begin{bmatrix} \dfrac{1}{\sqrt{2}} & \dfrac{1}{\sqrt{2}} \\ \dfrac{1}{\sqrt{2}} & -\dfrac{1}{\sqrt{2}} \end{bmatrix} \begin{bmatrix} y_1 \\ y_2 \end{bmatrix}$; $y_1^2 - y_2^2$

(d) $\begin{bmatrix} x_1 \\ x_2 \end{bmatrix} = \begin{bmatrix} \dfrac{\sqrt{17}-4}{\sqrt{34-8\sqrt{17}}} & \dfrac{\sqrt{17}-4}{\sqrt{34+8\sqrt{17}}} \\ \dfrac{1}{\sqrt{34-8\sqrt{17}}} & \dfrac{-1}{\sqrt{34+8\sqrt{17}}} \end{bmatrix} \begin{bmatrix} y_1 \\ y_2 \end{bmatrix}$; $(1+\sqrt{17})y_1^2 + (1-\sqrt{17})y_2^2$

2. (a) $\begin{bmatrix} x_1 \\ x_2 \\ x_3 \end{bmatrix} = \begin{bmatrix} \dfrac{1}{\sqrt{2}} & \dfrac{-4}{\sqrt{66+2\sqrt{33}}} & \dfrac{-4}{\sqrt{66-2\sqrt{33}}} \\[4mm] 0 & \dfrac{-1-\sqrt{33}}{\sqrt{66+2\sqrt{33}}} & \dfrac{1-\sqrt{33}}{\sqrt{66-2\sqrt{33}}} \\[4mm] \dfrac{1}{\sqrt{2}} & \dfrac{4}{\sqrt{66+2\sqrt{33}}} & \dfrac{4}{\sqrt{66-2\sqrt{33}}} \end{bmatrix} \begin{bmatrix} y_1 \\ y_2 \\ y_3 \end{bmatrix};$

$3y_1^2 + \dfrac{7\sqrt{33}}{2}y_2^2 + \dfrac{7-\sqrt{33}}{2}y_3^2$

(b) $\begin{bmatrix} x_1 \\ x_2 \\ x_3 \end{bmatrix} = \begin{bmatrix} \frac{1}{3} & \frac{2}{3} & \frac{2}{3} \\[1mm] \frac{2}{3} & \frac{1}{3} & -\frac{2}{3} \\[1mm] -\frac{2}{3} & \frac{2}{3} & -\frac{1}{3} \end{bmatrix} \begin{bmatrix} y_1 \\ y_2 \\ y_3 \end{bmatrix};\ 7y_1^2 + 4y_2^2 + y_3^2$

(c) $\begin{bmatrix} x_1 \\ x_2 \\ x_3 \end{bmatrix} = \begin{bmatrix} \dfrac{1}{\sqrt{14}} & \dfrac{1}{\sqrt{6}} & -\dfrac{4}{\sqrt{21}} \\[3mm] -\dfrac{2}{\sqrt{14}} & \dfrac{2}{\sqrt{6}} & \dfrac{1}{\sqrt{21}} \\[3mm] \dfrac{3}{\sqrt{14}} & \dfrac{1}{\sqrt{6}} & \dfrac{2}{\sqrt{21}} \end{bmatrix} \begin{bmatrix} y_1 \\ y_2 \\ y_3 \end{bmatrix};\ 2y_2^2 - 7y_3^2$

(d) $\begin{bmatrix} x_1 \\ x_2 \\ x_3 \end{bmatrix} = \begin{bmatrix} \dfrac{3}{\sqrt{10}} & \dfrac{1}{\sqrt{20}} & \dfrac{1}{\sqrt{20}} \\[3mm] -\dfrac{1}{\sqrt{10}} & \dfrac{3}{\sqrt{20}} & \dfrac{3}{\sqrt{20}} \\[3mm] 0 & \dfrac{1}{\sqrt{2}} & -\dfrac{1}{\sqrt{2}} \end{bmatrix} \begin{bmatrix} y_1 \\ y_2 \\ y_3 \end{bmatrix};\ \sqrt{10}y_2^2 - \sqrt{10}y_3^2$

3. (a) $2x^2 - 3xy + 4y^2$ (b) $x^2 - xy$ (c) $5xy$ (d) $4x^2 - 2y^2$ (e) y^2

4. (a) $\begin{bmatrix} 2 & -\frac{3}{2} \\ -\frac{3}{2} & 4 \end{bmatrix}$ (b) $\begin{bmatrix} 1 & -\frac{1}{2} \\ -\frac{1}{2} & 0 \end{bmatrix}$ (c) $\begin{bmatrix} 0 & \frac{5}{2} \\ \frac{5}{2} & 0 \end{bmatrix}$ (d) $\begin{bmatrix} 4 & 0 \\ 0 & -2 \end{bmatrix}$ (e) $\begin{bmatrix} 0 & 0 \\ 0 & 1 \end{bmatrix}$

5. (a) $[x\ \ y]\begin{bmatrix} 2 & -\frac{3}{2} \\ -\frac{3}{2} & 4 \end{bmatrix}\begin{bmatrix} x \\ y \end{bmatrix} + [-7\ \ 2]\begin{bmatrix} x \\ y \end{bmatrix} + 7 = 0$

(b) $[x\ \ y]\begin{bmatrix} 1 & -\frac{1}{2} \\ -\frac{1}{2} & 0 \end{bmatrix}\begin{bmatrix} x \\ y \end{bmatrix} + [5\ \ 8]\begin{bmatrix} x \\ y \end{bmatrix} - 3 = 0$

(c) $[x\ \ y]\begin{bmatrix} 0 & \frac{5}{2} \\ \frac{5}{2} & 0 \end{bmatrix}\begin{bmatrix} x \\ y \end{bmatrix} - 8 = 0$ (d) $[x\ \ y]\begin{bmatrix} 4 & 0 \\ 0 & -2 \end{bmatrix}\begin{bmatrix} x \\ y \end{bmatrix} - 7 = 0$

(e) $[x\ \ y]\begin{bmatrix} 0 & 0 \\ 0 & 1 \end{bmatrix}\begin{bmatrix} x \\ y \end{bmatrix} + [7\ \ -8]\begin{bmatrix} x \\ y \end{bmatrix} - 5 = 0$

6. (a) Ellipse (b) Ellipse (c) Hyperbola (d) Hyperbola
(e) Circle (f) Parabola (g) Parabola (h) Parabola
(i) Parabola (j) Circle

7. (a) $9x'^2 + 4y'^2 = 36$, ellipse (b) $x'^2 - 16y'^2 = 16$, hyperbola
(c) $y'^2 = 8x'$, parabola (d) $x'^2 + y'^2 = 16$, circle
(e) $18y'^2 - 12x'^2 = 419$, hyperbola (f) $y' = -\frac{1}{7}x'^2$, parabola

8. (a) Hyperbola; possible equations are:
$$3x'^2 - 2y'^2 + 8 = 0, \quad -2x'^2 + 3y'^2 + 8 = 0$$
(b) Parabola; possible equations are:
$$2\sqrt{2}x'^2 + 9x' - 7y' = 0, \quad 2\sqrt{2}y'^2 + 7x' + 9y' = 0$$
$$2\sqrt{2}y'^2 - 7x' - 9y' = 0, \quad 2\sqrt{2}x'^2 - 9x' + 7y' = 0$$
(c) Ellipse; possible equations are:
$$7x'^2 + 3y'^2 = 9, \quad 3x'^2 + 7y'^2 = 9$$
(d) Hyperbola; possible equations are:
$$4x'^2 - y'^2 = 3, \quad 4y'^2 - x'^2 = 3$$

9. $2x''^2 + y''^2 = 6$, ellipse   10. $13y''^2 - 4x''^2 = 81$, hyperbola

11. $2x''^2 - 3y''^2 = 24$, hyperbola   12. $6x''^2 + 11y''^2 = 66$, ellipse

13. $4y''^2 - x''^2 = 0$, hyperbola   14. $\sqrt{29}x'^2 - 3y'^2 = 0$, parabola

15. (a) Two intersecting lines, $y = x$ and $y = -x$.
(b) No graph.
(c) The graph is the single point $(0, 0)$.
(d) The graph is the line $y = x$.
(e) The graph consists of two parallel lines $\dfrac{3}{\sqrt{13}}x + \dfrac{2}{\sqrt{13}}y = \pm 2$.
(f) The graph is the single point $(1, 2)$.

EXERCISE SET 7.5 (page 369)

1. (a) $x^2 + 2y^2 - z^2 + 4xy - 5yz$ (b) $3x^2 + 7z^2 + 2xy - 3xz + 4yz$
(c) $xy + xz + yz$ (d) $x^2 + y^2 - z^2$
(e) $3z^2 + 3xz$ (f) $2z^2 + 2xz + y^2$

2. (a) $\begin{bmatrix} 1 & 2 & 0 \\ 2 & 2 & -\frac{5}{2} \\ 0 & -\frac{5}{2} & -1 \end{bmatrix}$ (b) $\begin{bmatrix} 3 & 1 & -\frac{3}{2} \\ 1 & 0 & 2 \\ -\frac{3}{2} & 2 & 7 \end{bmatrix}$ (c) $\begin{bmatrix} 0 & \frac{1}{2} & \frac{1}{2} \\ \frac{1}{2} & 0 & \frac{1}{2} \\ \frac{1}{2} & \frac{1}{2} & 0 \end{bmatrix}$

(d) $\begin{bmatrix} 1 & 0 & 0 \\ 0 & 1 & 0 \\ 0 & 0 & -1 \end{bmatrix}$ (e) $\begin{bmatrix} 0 & 0 & \frac{3}{2} \\ 0 & 0 & 0 \\ \frac{3}{2} & 0 & 3 \end{bmatrix}$ (f) $\begin{bmatrix} 0 & 0 & 1 \\ 0 & 1 & 0 \\ 1 & 0 & 2 \end{bmatrix}$

3. (a) $\begin{bmatrix} x & y & z \end{bmatrix} \begin{bmatrix} 1 & 2 & 0 \\ 2 & 2 & -\frac{5}{2} \\ 0 & -\frac{5}{2} & -1 \end{bmatrix} \begin{bmatrix} x \\ y \\ z \end{bmatrix} + \begin{bmatrix} 7 & 0 & 2 \end{bmatrix} \begin{bmatrix} x \\ y \\ z \end{bmatrix} - 3 = 0$

(b) $[x \quad y \quad z] \begin{bmatrix} 3 & 1 & -\frac{3}{2} \\ 1 & 0 & 2 \\ -\frac{3}{2} & 2 & 7 \end{bmatrix} \begin{bmatrix} x \\ y \\ z \end{bmatrix} + [-3 \quad 0 \quad 0] \begin{bmatrix} x \\ y \\ z \end{bmatrix} - 4 = 0$

(c) $[x \quad y \quad z] \begin{bmatrix} 0 & \frac{1}{2} & \frac{1}{2} \\ \frac{1}{2} & 0 & \frac{1}{2} \\ \frac{1}{2} & \frac{1}{2} & 0 \end{bmatrix} \begin{bmatrix} x \\ y \\ z \end{bmatrix} - 1 = 0$

(d) $[x \quad y \quad z] \begin{bmatrix} 1 & 0 & 0 \\ 0 & 1 & 0 \\ 0 & 0 & -1 \end{bmatrix} \begin{bmatrix} x \\ y \\ z \end{bmatrix} - 7 = 0$

(e) $[x \quad y \quad z] \begin{bmatrix} 0 & 0 & \frac{3}{2} \\ 0 & 0 & 0 \\ \frac{3}{2} & 0 & 3 \end{bmatrix} \begin{bmatrix} x \\ y \\ z \end{bmatrix} + [0 \quad -14 \quad 0] \begin{bmatrix} x \\ y \\ z \end{bmatrix} + 9 = 0$

(f) $[x \quad y \quad z] \begin{bmatrix} 0 & 0 & 1 \\ 0 & 1 & 0 \\ 1 & 0 & 2 \end{bmatrix} \begin{bmatrix} x \\ y \\ z \end{bmatrix} + [2 \quad -1 \quad 3] \begin{bmatrix} x \\ y \\ z \end{bmatrix} = 0$

4. (a) Ellipsoid   (b) Hyperboloid of one sheet
  (c) Hyperboloid of two sheets   (d) Elliptic cone   (e) Elliptic paraboloid
  (f) Hyperbolic paraboloid   (g) Sphere

5. (a) $9x'^2 + 36y'^2 + 4z'^2 = 36$, ellipsoid
  (b) $6x'^2 + 3y'^2 - 2z'^2 = 18$, hyperboloid of one sheet
  (c) $3x'^2 - 3y'^2 - z'^2 = 3$, hyperboloid of two sheets
  (d) $4x'^2 + 9y'^2 - z'^2 = 0$, elliptic cone
  (e) $x'^2 + 16y'^2 - 16z' = 0$, elliptic paraboloid
  (f) $7x'^2 - 3y'^2 + z' = 0$, hyperbolic paraboloid
  (g) $x'^2 + y'^2 + z'^2 = 25$, sphere

6. (a) $25x'^2 - 3y'^2 - 50z'^2 - 150 = 0$, hyperboloid of two sheets
  (b) $2x'^2 + 2y'^2 + 8z'^2 - 5 = 0$, ellipsoid
  (c) $9x'^2 + 4y'^2 - 36z = 0$, elliptic paraboloid
  (d) $x'^2 - y'^2 + z' = 0$, hyperbolic paraboloid

7. $x''^2 + y''^2 - 2z''^2 = -1$, hyperboloid of two sheets

8. $x''^2 + y''^2 + 2z''^2 = 4$, ellipsoid

9. $x''^2 - y''^2 + z'' = 0$, hyperbolic paraboloid

10. $6x''^2 + 3y''^2 - 8\sqrt{2}z'' = 0$, elliptic paraboloid

EXERCISE SET 8.1 (page 371)

1. Multiplications: mpn
  additions: $mp(n - 1)$

2. Multiplications: $(k - 1)n^3$
  additions: $(k - 1)(n^3 - n^2)$

3.

	$n = 5$	$n = 10$	$n = 100$	$n = 1000$
Solve $Ax = b$ by Gauss-Jordan elimination	+: 50 ×: 65	+: 375 ×: 430	+: 383,250 ×: 343,300	+: 333,832,500 ×: 334,333,000
Solve $Ax = b$ by Gaussian elimination	+: 50 ×: 65	+: 375 ×: 430	+: 383,250 ×: 343,300	+: 333,832,500 ×: 334,333,000
Find A^{-1} by reducing $[A\|I]$ to $[I\|A^{-1}]$	+: 80 ×: 125	+: 810 ×: 1000	+: 980,100 ×: 1,000,000	+: 998,001,000 ×: 1,000,000,000
Solve $Ax = b$ as $x = A^{-1}b$	+: 100 ×: 150	+: 900 ×: 1100	+: 990,000 ×: 1,010,000	+: 999,000,000 ×: 1,001,000,000
Find det(A) by row reduction	+: 30 ×: 44	+: 285 ×: 339	+: 328,350 ×: 333,399	+: 332,833,500 ×: 333,333,999
Solve $Ax = b$ by Cramer's rule	+: 180 ×: 264	+: 3135 ×: 3729	+: 33,163,350 ×: 33,673,299	+: $33,299,933 \times 10^4$ ×: $33,366,733 \times 10^4$

4.

	$n = 5$ Execution Time (sec)	$n = 10$ Execution Time (sec)	$n = 100$ Execution Time (sec)	$n = 1000$ Execution Time (sec)
Solve $Ax = b$ by Gauss-Jordan elimination	1.55×10^{-4}	1.05×10^{-3}	.878	836
Solve $Ax = b$ by Gaussian elimination	1.55×10^{-4}	1.05×10^{-3}	.878	836
Find A^{-1} by reducing $[A\|I]$ to $[I\|A^{-1}]$	2.84×10^{-4}	2.41×10^{-3}	2.49	2499
Solve $Ax = b$ as $x = A^{-1}b$	3.50×10^{-4}	2.65×10^{-3}	2.52	2502
Find det(A) by row reduction	1.03×10^{-4}	8.21×10^{-4}	.831	833
Solve $Ax = b$ by Cramer's rule	6.18×10^{-4}	90.3×10^{-4}	83.9	834×10^3

EXERCISE SET 8.2 (page 389)

1. $x_1 = 2, x_2 = 1$

2. $x_1 = -2, x_2 = 1, x_3 = -3$

3. $x_1 = 3, x_2 = -1$

4. $x_1 = 4, x_2 = -1$

5. $x_1 = -1, x_2 = 1, x_3 = 0$

6. $x_1 = 1, x_2 = -2, x_3 = 1$

7. $x_1 = -1, x_2 = 1, x_3 = 0$

8. $x_1 = -1, x_2 = 1, x_3 = 1$

9. $x_1 = -3, x_2 = 1, x_3 = 2, x_4 = 1$

10. $x_1 = 2, x_2 = -1, x_3 = 0, x_4 = 0$

11. (a) $A = Lu = \begin{bmatrix} 2 & 0 & 0 \\ -2 & 1 & 0 \\ 2 & 1 & 1 \end{bmatrix} \begin{bmatrix} 1 & \frac{1}{2} & -\frac{1}{2} \\ 0 & 0 & 1 \\ 0 & 0 & 0 \end{bmatrix}$

(b) $A = L_1 DU = \begin{bmatrix} 1 & 0 & 0 \\ -1 & 1 & 0 \\ 1 & 1 & 1 \end{bmatrix} \begin{bmatrix} 2 & 0 & 0 \\ 0 & 1 & 0 \\ 0 & 0 & 1 \end{bmatrix} \begin{bmatrix} 1 & \frac{1}{2} & -\frac{1}{2} \\ 0 & 0 & 1 \\ 0 & 0 & 0 \end{bmatrix}$

(c) $A = L_2 U_2 = \begin{bmatrix} 1 & 0 & 0 \\ -1 & 1 & 0 \\ 1 & 1 & 1 \end{bmatrix} \begin{bmatrix} 2 & 1 & -1 \\ 0 & 0 & 1 \\ 0 & 0 & 0 \end{bmatrix}$

13. (b) $\begin{bmatrix} a & b \\ b & d \end{bmatrix} = \begin{bmatrix} 1 & 0 \\ \dfrac{c}{a} & 1 \end{bmatrix} \begin{bmatrix} a & b \\ 0 & \dfrac{ad - bc}{a} \end{bmatrix}$

14. Additions: $\dfrac{n^3}{3} + \dfrac{n^2}{2} - \dfrac{5n}{6}$; multiplications: $\dfrac{n^3}{3} + n^2 - \dfrac{n}{3}$

18. $A = PLU = \begin{bmatrix} 1 & 0 & 0 \\ 0 & 0 & 1 \\ 0 & 1 & 0 \end{bmatrix} \begin{bmatrix} 3 & 0 & 0 \\ 0 & 2 & 0 \\ 3 & 0 & 1 \end{bmatrix} \begin{bmatrix} 1 & -\frac{1}{3} & 0 \\ 0 & 1 & \frac{1}{2} \\ 0 & 0 & 1 \end{bmatrix}$

EXERCISE SET 8.3 (page 397)

1. $x_1 \approx 2.81, x_2 \approx .940$; exact solution is $x_1 = 3, x_2 = 1$

2. $x_1 \approx .954, x_2 \approx -1.90$; exact solution is $x_1 = 1, x_2 = -2$

3. $x_1 \approx -2.99, x_2 \approx -.999$; exact solution is $x_1 = -3, x_2 = -1$

4. $x_1 \approx 0.00, x_2 \approx 2.00$; exact solution is $x_1 = 0, x_2 = 2$

5. $x_1 \approx 3.03, x_2 \approx 1.02$; exact solution is $x_1 = 3, x_2 = 1$

6. $x_1 \approx 1.03, x_2 \approx -2.02$; exact solution is $x_1 = 1, x_2 = -2$

7. $x_1 \approx -3.00, x_2 \approx -1.00$; exact solution is $x_1 = -3, x_2 = -1$

8. $x_1 \approx .005, x_2 \approx 2.00$; exact solution is $x_1 = 0, x_2 = 2$

9. $x_1 \approx .492, x_2 \approx .006, x_3 \approx -.996$; exact solution is $x_1 = \frac{1}{2}, x_2 = 0, x_3 = -1$

10. $x_1 \approx 1.00$, $x_2 \approx .998$, $x_3 = 1.00$; exact solution is $x_1 = 1$, $x_2 = 1$, $x_3 = 1$

11. $x_1 \approx .499$, $x_2 \approx .0004$, $x_3 \approx -1.00$; exact solution is $x_1 = \frac{1}{2}$, $x_2 = 0$, $x_3 = -1$

12. $x_1 \approx 1.00$, $x_2 \approx 1.00$, $x_3 \approx 1.00$; exact solution is $x_1 = 1$, $x_2 = 1$, $x_3 = 1$

13. *a, d, e*

EXERCISE SET 8.4 (page 403)

1. (a) $.28 \times 10^1$ (b) $.3452 \times 10^4$ (c) $.3879 \times 10^{-5}$
    (d) $-.135 \times 10^0$ (e) $.17921 \times 10^2$ (f) $-.863 \times 10^{-1}$

2. (a) $.280 \times 10^1$ (b) $.345 \times 10^4$ (c) $.388 \times 10^{-5}$
    (d) $-.135 \times 10^0$ (e) $.179 \times 10^2$ (f) $-.863 \times 10^{-1}$

3. (a) $.28 \times 10^1$ (b) $.35 \times 10^4$ (c) $.39 \times 10^{-5}$
    (d) $-.14 \times 10^0$ (e) $.18 \times 10^2$ (f) $-.86 \times 10^{-1}$

4. $x_1 = -3$, $x_2 = 7$ **5.** $x_1 = \frac{13}{8}$, $x_2 = \frac{14}{8}$, $x_3 = \frac{21}{8}$

6. $x_1 = 1$, $x_2 = 2$, $x_3 = 3$ **7.** $x_1 = 0$, $x_2 = 0$, $x_3 = 1$, $x_4 = -1$

8. $x_1 = .997$, $x_2 = 1.00$ **9.** $x_1 = -2$, $x_2 = 0$, $x_3 = 1$

10. $x_1 = 0$, $x_2 = 1$ (without pivoting); $x_1 = 1$, $x_2 = 1$ (with pivoting)

EXERCISE SET 8.5 (page 412)

1. (a) $\lambda = -3$ (b) No dominant eigenvalue   (c) $\lambda = 6$ (d) $\lambda = 3$

2. (a) $\begin{bmatrix} 1.00 \\ .503 \end{bmatrix}$ (b) 5.02

(c) The dominant eigenvector is $\begin{bmatrix} 1 \\ \frac{1}{2} \end{bmatrix}$; the dominant eigenvalue is 5

(d) The percentage error is .4%

3. (a) $\begin{bmatrix} 1.00 \\ .750 \end{bmatrix}$ (b) 8.01

(c) The dominant eigenvector is $\begin{bmatrix} 1 \\ \frac{3}{4} \end{bmatrix}$; the dominant eigenvalue is 8

(d) The percentage error is .125%

4. (a) $\begin{bmatrix} 1.00 \\ -.560 \end{bmatrix}$ (b) -4.00

(c) The dominant eigenvector is $\begin{bmatrix} 1 \\ -\frac{1}{2} \end{bmatrix}$; the dominant eigenvalue is -4

(d) The percentage error is 0%

5. (a) At the end of two iterations the dominant eigenvalue and eigenvector are approximately $\lambda_1 \approx 20.1$ and $x \approx \begin{bmatrix} 1 \\ .119 \end{bmatrix}$

(b) The exact values of the dominant eigenvalue and eigenvector are $\lambda_1 = 20$ and $x = \begin{bmatrix} 1 \\ \frac{2}{17} \end{bmatrix}$

6. (a) At the end of three iterations, the dominant eigenvalue and eigenvector are approximately $\lambda_1 \approx -9.95$ and $x \approx \begin{bmatrix} -.978 \\ 1 \end{bmatrix}$

(b) The exact values of the dominant eigenvalue and eigenvector are $\lambda_1 = -10$ and $x = \begin{bmatrix} -1 \\ 1 \end{bmatrix}$

7. (a) $\begin{bmatrix} .027 \\ .027 \\ 1 \end{bmatrix}$ (b) 10.0

(c) The dominant eigenvector is $\begin{bmatrix} 0 \\ 0 \\ 1 \end{bmatrix}$; the dominant eigenvalue is 10

EXERCISE SET 8.6 (page 418)

1. (a) $\begin{bmatrix} 1 \\ .509 \end{bmatrix}$ (b) 7.00 (c) $\lambda_2 \approx 2.00$, $v_2 \approx \begin{bmatrix} -.51 \\ 1 \end{bmatrix}$

(d) Exact eigenvalues 7, 2;
exact eigenvectors $v_1 = \begin{bmatrix} 1 \\ \frac{1}{2} \end{bmatrix}$, $v_2 = \begin{bmatrix} -\frac{1}{2} \\ 1 \end{bmatrix}$

2. (a) $\begin{bmatrix} 1 \\ .503 \end{bmatrix}$ (b) 12.0 (c) $\lambda_2 \approx 2.02$, $v_2 \approx \begin{bmatrix} -.532 \\ 1 \end{bmatrix}$

(d) Exact eigenvalues 12, 2;
exact eigenvectors $v_1 = \begin{bmatrix} 1 \\ \frac{1}{2} \end{bmatrix}$, $v_2 = \begin{bmatrix} -\frac{1}{2} \\ 1 \end{bmatrix}$

3. $\lambda_{max} \approx -1.37$; $v_{max} \approx \begin{bmatrix} -.448 \\ 1.00 \end{bmatrix}$ $\lambda_{min} \approx \dfrac{1}{.482} \approx 2.07$; $v_{min} \approx \begin{bmatrix} .125 \\ 1.00 \end{bmatrix}$

4. $\lambda_{max} \approx -3.99$; $v_{max} \approx \begin{bmatrix} -1.00 \\ .560 \end{bmatrix}$ $\lambda_{min} \approx \dfrac{1}{1.00} = 1.00$; $v_{min} \approx \begin{bmatrix} .494 \\ 1.00 \end{bmatrix}$

EXERCISE SET 9.1 (page 427)

1. (a–d)

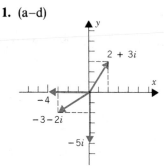

2. (a) $(2, 3)$ (b) $(-4, 0)$ (c) $(-3, -2)$ (d) $(0, -5)$

3. (a) $x = -2, y = -3$ (b) $x = 2, y = 1$

4. (a) $5 + 3i$ (b) $-3 - 7i$ (c) $4 - 8i$
   (d) $-4 - 5i$ (e) $19 + 14i$ (f) $-\frac{11}{2} - \frac{17}{2}i$

5. (a) $2 + 3i$ (b) $-1 - 2i$ (c) $-2 + 9i$

6. (a)

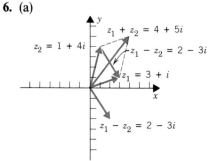

(b)

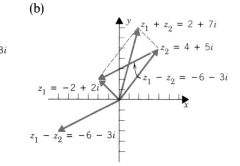

7. (a)

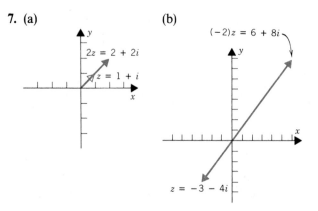

(b)

(c)

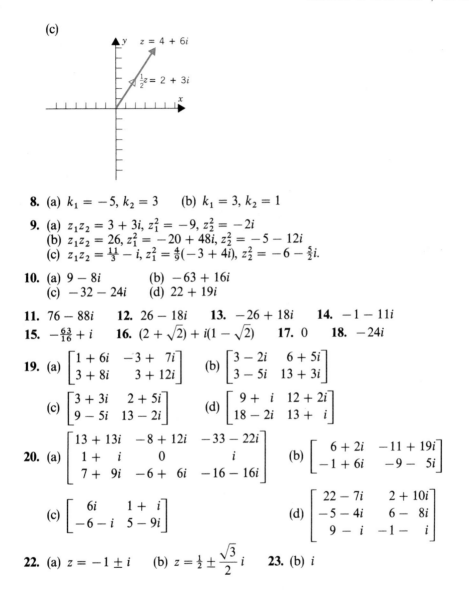

8. (a) $k_1 = -5, k_2 = 3$ (b) $k_1 = 3, k_2 = 1$

9. (a) $z_1z_2 = 3 + 3i, z_1^2 = -9, z_2^2 = -2i$
   (b) $z_1z_2 = 26, z_1^2 = -20 + 48i, z_2^2 = -5 - 12i$
   (c) $z_1z_2 = \frac{11}{3} - i, z_1^2 = \frac{4}{9}(-3 + 4i), z_2^2 = -6 - \frac{5}{2}i.$

10. (a) $9 - 8i$ (b) $-63 + 16i$
    (c) $-32 - 24i$ (d) $22 + 19i$

11. $76 - 88i$ **12.** $26 - 18i$ **13.** $-26 + 18i$ **14.** $-1 - 11i$
15. $-\frac{63}{16} + i$ **16.** $(2 + \sqrt{2}) + i(1 - \sqrt{2})$ **17.** 0   **18.** $-24i$

19. (a) $\begin{bmatrix} 1 + 6i & -3 + 7i \\ 3 + 8i & 3 + 12i \end{bmatrix}$ (b) $\begin{bmatrix} 3 - 2i & 6 + 5i \\ 3 - 5i & 13 + 3i \end{bmatrix}$

   (c) $\begin{bmatrix} 3 + 3i & 2 + 5i \\ 9 - 5i & 13 - 2i \end{bmatrix}$ (d) $\begin{bmatrix} 9 + i & 12 + 2i \\ 18 - 2i & 13 + i \end{bmatrix}$

20. (a) $\begin{bmatrix} 13 + 13i & -8 + 12i & -33 - 22i \\ 1 + i & 0 & i \\ 7 + 9i & -6 + 6i & -16 - 16i \end{bmatrix}$ (b) $\begin{bmatrix} 6 + 2i & -11 + 19i \\ -1 + 6i & -9 - 5i \end{bmatrix}$

   (c) $\begin{bmatrix} 6i & 1 + i \\ -6 - i & 5 - 9i \end{bmatrix}$ (d) $\begin{bmatrix} 22 - 7i & 2 + 10i \\ -5 - 4i & 6 - 8i \\ 9 - i & -1 - i \end{bmatrix}$

22. (a) $z = -1 \pm i$ (b) $z = \frac{1}{2} \pm \frac{\sqrt{3}}{2}i$ **23.** (b) i

EXERCISE SET 9.2 (page 434)

1. (a) $2 - 7i$ (b) $-3 + 5i$ (c) $-5i$
   (d) i (e) -9 (f) 0

2. (a) 1   (b) 7   (c) 5   (d) $\sqrt{2}$ (e) 8   (f) 0

4. (a) $-\frac{17}{25} - \frac{19}{25}i$ (b) $\frac{23}{25} + \frac{11}{25}i$ (c) $\frac{23}{25} - \frac{11}{25}i$ (d) $-\frac{17}{25} + \frac{19}{25}i$

(e) $\frac{1}{5} - i$ (f) $\dfrac{\sqrt{26}}{5}$

5. (a) $-i$ (b) $\frac{1}{26} + \frac{5}{26}i$ (c) $7i$

6. (a) $\frac{6}{5} + \frac{2}{5}i$ (b) $-\frac{2}{5} + \frac{1}{5}i$ (c) $\frac{3}{5} + \frac{11}{5}i$ (d) $\frac{3}{5} + \frac{1}{5}i$

7. $\frac{1}{2} + \frac{1}{2}i$ **8.** $\frac{2}{5} + \frac{1}{5}i$ **9.** $-\frac{7}{625} - \frac{24}{625}i$ **10.** $-\frac{11}{25} + \frac{2}{25}i$

11. $\dfrac{1 - \sqrt{3}}{4} + \dfrac{1 + \sqrt{3}}{4}i$ **12.** $-\frac{1}{26} - \frac{5}{26}i$ **13.** $-\frac{1}{10} + \frac{1}{10}i$ **14.** $-\frac{2}{5}$

15. (a) $-1 - 2i$ (b) $-\frac{3}{25} - \frac{4}{25}i$

17. (a)
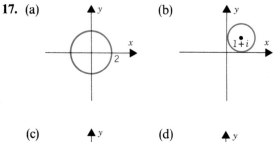

(b)

(c)
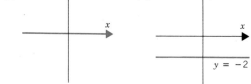

(d)

18. (a)
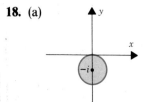

(b)
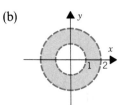

(c)
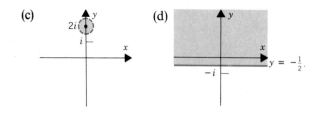

(d)

19. (a) $-y$ (b) $-x$ (c) y (d) x **20.** (b) $-i$

23. (a) $\dfrac{x_1 x_2 + y_1 y_2}{x_2^2 + y_2^2}$ (b) $\dfrac{x_2 y_1 - x_1 y_2}{x_2^2 + y_2^2}$ **27.** (c) Yes, if $z \neq 0$.

28. $x_1 = i,\ x_2 = -i$

29. $x_1 = 1 + i,\ x_2 = 1 - i$

30. $x_1 = \frac{1}{2} + i,\ x_2 = 2,\ x_3 = \frac{1}{2} - i$

31. $x_1 = i,\ x_2 = 0,\ x_3 = -i$.

32. $x_1 = -(1 + i)t,\ x_2 = t$

33. $x_1 = (1 + i)t,\ x_2 = 2t$

34. $x_1 = -(1 - i)t,\ x_2 = -it,\ x_3 = t$

35. (a) $\begin{bmatrix} i & 2 \\ -1 & i \end{bmatrix}$ (b) $\begin{bmatrix} 0 & 1 \\ -i & 2i \end{bmatrix}$

38. (a) $\begin{bmatrix} -i & -2-2i & -1+i \\ 1 & 2 & -i \\ i & i & 1 \end{bmatrix}$ (b) $\begin{bmatrix} 1+i & -i & 1 \\ -7+6i & 5-i & 1+4i \\ 1+2i & -i & 1 \end{bmatrix}$

EXERCISE SET 9.3 (page 445)

1. (a) 0 (b) $\pi/2$ (c) $-\pi/2$ (d) $\pi/4$ (e) $2\pi/3$ (f) $-\pi/4$

2. (a) $5\pi/3$ (b) $-\pi/3$ (c) $5\pi/3$

3. (a) $2\left[\cos\left(\dfrac{\pi}{2}\right) + i \sin\left(\dfrac{\pi}{2}\right)\right]$ (b) $4[\cos \pi + i \sin \pi]$

(c) $5\sqrt{2}\left[\cos\left(\dfrac{\pi}{4}\right) + i \sin\left(\dfrac{\pi}{4}\right)\right]$ (d) $12\left[\cos\left(\dfrac{2\pi}{3}\right) + i \sin\left(\dfrac{2\pi}{3}\right)\right]$

(e) $3\sqrt{2}\left[\cos\left(-\dfrac{3\pi}{4}\right) + i \sin\left(-\dfrac{3\pi}{4}\right)\right]$ (f) $4\left[\cos\left(-\dfrac{\pi}{6}\right) + i \sin\left(-\dfrac{\pi}{6}\right)\right]$

4. (a) $6\left[\cos\left(\dfrac{5\pi}{12}\right) + i \sin\left(\dfrac{5\pi}{12}\right)\right]$ (b) $\dfrac{2}{3}\left[\cos\left(\dfrac{\pi}{12}\right) + i \sin\left(\dfrac{\pi}{12}\right)\right]$

(c) $\dfrac{3}{2}\left[\cos\left(-\dfrac{\pi}{12}\right) + i \sin\left(-\dfrac{\pi}{12}\right)\right]$ (d) $\dfrac{32}{9}\left[\cos\left(\dfrac{11\pi}{12}\right) + i \sin\left(\dfrac{11\pi}{12}\right)\right]$

5. 1 **6.** (a) -64 (b) $-i$ (c) $-64\sqrt{3} - 64i$ (d) $-\dfrac{1 + \sqrt{3}i}{2048}$

7. (a)

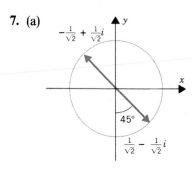

(b)

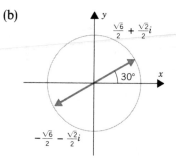

(c)

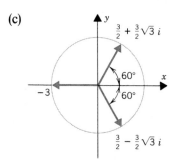

(d)

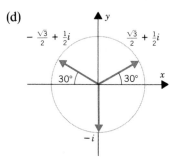

(e)

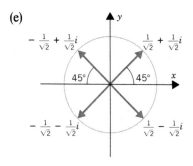

(f)

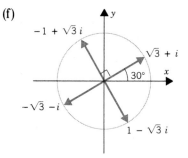

8.

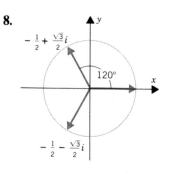

9.

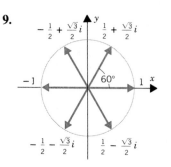

10. $\sqrt[4]{2}\left[\cos\left(\dfrac{\pi}{8}\right) + i\sin\left(\dfrac{\pi}{8}\right)\right], \sqrt[4]{2}\left[\cos\left(\dfrac{9\pi}{8}\right) + i\sin\left(\dfrac{9\pi}{8}\right)\right]$

11. (a) $\pm 2, \pm 2i$ (b) $\pm(2 + 2i), \pm(2 - 2i)$

12. The roots are: $\pm(2^{1/4} + 2^{1/4}i), \pm(2^{1/4} - 2^{1/4}i)$ and the factorization is:
$z^4 + 8 = (z^2 - 2^{5/4}z + 2^{3/2})(z^2 + 2^{5/4}z + 2^{3/2})$

13. Rotates z clockwise by $90°$.   14. (a) 16   (b) $\dfrac{i}{4^9}$

15. (a) $\text{Re}(z) = -3, \text{Im}(z) = 0$ (b) $\text{Re}(z) = -3, \text{Im}(z) = 0$
    (c) $\text{Re}(z) = 0, \text{Im}(z) = -\sqrt{2}$ (d) $\text{Re}(z) = -3, \text{Im}(z) = 0$

EXERCISE SET 9.4 (page 451)

1. (a) $(3i, -i, -2 - i, 4)$ (b) $(3 + 2i, -1 - 2i, -3 + 5i, -i)$
    (c) $(-1 - 2i, 2i, 2 - i, -1)$ (d) $(-3 + 9i, 3 - 3i, -3 - 6i, 12 + 3i)$
    (e) $(-3 + 2i, 3, -3 - 3i, i)$ (f) $(-1 - 5i, 3i, 4, -5)$

2. $(2 + i, 0, -3 + i, -4i)$ 3. $c_1 = -2 - i, c_2 = 0, c_3 = 2 - i$

5. (a) $\sqrt{2}$ (b) $2\sqrt{3}$ (c) $\sqrt{10}$ (d) $\sqrt{37}$

6. (a) $\sqrt{43}$ (b) $\sqrt{10} + \sqrt{29}$ (c) $\sqrt{10} + \sqrt{10}i$

    (d) $\sqrt{699}$ (e) $\left(\dfrac{1 + i}{\sqrt{6}}, \dfrac{2i}{\sqrt{6}}, 0\right)$ (f) 1

8. All k such that $|k| = \frac{1}{5}$

9. (a) 3   (b) $2 - 27i$ (c) $-5 - 10i$

10. The set is a vector space under the given operations.

11. Not a vector space. Axiom 6 fails, that is, the set is not closed under scalar multiplication. (Multiply by i, for example.)

12. No, R^n is not closed under scalar multiplication. (Multiply a nonzero vector in R^n by i.)

13. (a)   14. (b)   15. (a) and (d)   16. (a), (b), and (d)

17. (a) $(3 - 2i)\mathbf{u} + (3 - i)\mathbf{v} + (1 + 2i)\mathbf{w}$ (b) $(2 + i)\mathbf{u} + (-1 + i)\mathbf{v} + (-1 - i)\mathbf{w}$
    (c) $0\mathbf{u} + 0\mathbf{v} + 0\mathbf{w}$ (d) $(-5 - 4i)\mathbf{u} + (5 - 2i)\mathbf{v} + (2 + 4i)\mathbf{w}$

18. (a) Yes   (b) No   (c) Yes   (d) No   19. (a), (b), (c)

20. (a) $\mathbf{u}_2 = i\mathbf{u}_1$,   (b) Three vectors in a two-dimensional space.
    (c) A is a scalar multiple of B.

21. (b) and (c)   22. $\mathbf{f} - 3\mathbf{g} - 3\mathbf{h} = \mathbf{0}$

23. (a) Three vectors in a two-dimensional space.
(b) Two vectors in a three-dimensional space.

24. (a) and (b)   **25.** (a), (b), (c), (d)   **26.** $(-1 - i, 1)$; dimension $= 1$

27. $(1, 1 - i)$; dimension $= 1$ **28.** $(3 + 6i, -3i, 1)$; dimension $= 1$

29. $(\frac{5}{2}i, -\frac{1}{2}, 1, 0), (-\frac{1}{4}, \frac{3}{4}i, 0, 1)$; dimension $= 2$

EXERCISE SET 9.5 (page 459)

2. (a) -12 (b) 0   (c) $2i$ (d) 37

4. (a) $-4 + 5i$ (b) 0   (c) $4 - 4i$ (d) 42

5. (a) Axiom 4 fails.   (b) Axiom 4 fails.   (c) Axioms 2 and 3 fail.
(d) Axioms 1 and 4 fail.   (e) This is an inner product.

6. $-9 - 5i$ **7.** No. Axioms 1 and 4 fail.

9. (a) $\sqrt{21}$ (b) $\sqrt{10}$ (c) $\sqrt{10}$ (d) 0

10. (a) $\sqrt{10}$ (b) 2   (c) $\sqrt{5}$ (d) 0

11. (a) $\sqrt{2}$ (b) $2\sqrt{3}$ (c) 5   (d) 0

12. (a) $3\sqrt{10}$ (b) $\sqrt{14}$ **13.** (a) $\sqrt{10}$ (b) $2\sqrt{5}$ **14.** (a) 2   (b) $2\sqrt{2}$

15. (a) $2\sqrt{3}$ (b) $2\sqrt{2}$ **16.** (a) $7\sqrt{2}$ (b) $2\sqrt{3}$

17. (a) $-\frac{8}{3}i$ (b) None   **18.** (a), (b), (c)   **20.** (b)   **21.** (b), (c)

23. $\left(\dfrac{i}{\sqrt{2}}, 0, 0, \dfrac{i}{\sqrt{2}}\right), \left(-\dfrac{i}{\sqrt{6}}, 0, \dfrac{2i}{\sqrt{6}}, \dfrac{i}{\sqrt{6}}\right), \left(\dfrac{2i}{\sqrt{21}}, \dfrac{3i}{\sqrt{21}}, \dfrac{2i}{\sqrt{21}}, \dfrac{-2i}{\sqrt{21}}\right),$

$\left(-\dfrac{i}{\sqrt{7}}, \dfrac{2i}{\sqrt{7}}, -\dfrac{i}{\sqrt{7}}, \dfrac{i}{\sqrt{7}}\right)$

24. (a) $\mathbf{v}_1 = \left(\dfrac{i}{\sqrt{10}}, -\dfrac{3i}{\sqrt{10}}\right), \mathbf{v}_2 = \left(\dfrac{3i}{\sqrt{10}}, \dfrac{i}{\sqrt{10}}\right)$ (b) $\mathbf{v}_1 = (i, 0), \mathbf{v}_2 = (0, -i)$

25. (a) $\mathbf{v}_1 = \left(\dfrac{i}{\sqrt{3}}, \dfrac{i}{\sqrt{3}}, \dfrac{i}{\sqrt{3}}\right), \mathbf{v}_2 = \left(-\dfrac{i}{\sqrt{2}}, \dfrac{i}{\sqrt{2}}, 0\right), \mathbf{v}_3 = \left(\dfrac{i}{\sqrt{6}}, \dfrac{i}{\sqrt{6}}, -\dfrac{2i}{\sqrt{6}}\right)$

(b) $\mathbf{v}_1 = (i, 0, 0), \mathbf{v}_2 = \left(0, \dfrac{7i}{\sqrt{53}}, \dfrac{-2i}{\sqrt{53}}\right), \mathbf{v}_3 = \left(0, \dfrac{2i}{\sqrt{53}}, \dfrac{7i}{\sqrt{53}}\right)$

26. $\left(0, \dfrac{2i}{\sqrt{5}}, \dfrac{i}{\sqrt{5}}, 0\right), \left(\dfrac{5i}{\sqrt{30}}, -\dfrac{i}{\sqrt{30}}, \dfrac{2i}{\sqrt{30}}, 0\right),$

$\left(\dfrac{i}{\sqrt{10}}, \dfrac{i}{\sqrt{10}}, -\dfrac{2i}{\sqrt{10}}, \dfrac{2i}{\sqrt{10}}\right), \left(\dfrac{i}{\sqrt{15}}, \dfrac{i}{\sqrt{15}}, -\dfrac{2i}{\sqrt{15}}, \dfrac{3i}{\sqrt{15}}\right)$

27. $\mathbf{v}_1 = \left(0, \dfrac{i}{\sqrt{3}}, \dfrac{1-i}{\sqrt{3}}\right), \mathbf{v}_2 = \left(-\dfrac{3i}{\sqrt{15}}, \dfrac{2}{\sqrt{15}}, \dfrac{1+i}{\sqrt{15}}\right)$

28. $\mathbf{w}_1 = \left(-\dfrac{5i}{4}, -\dfrac{i}{4}, \dfrac{5i}{4}, \dfrac{9i}{4}\right), \mathbf{w}_2 = \left(\dfrac{i}{4}, \dfrac{9i}{4}, \dfrac{19i}{4}, -\dfrac{9i}{4}\right)$

36. $\mathbf{u} = -\sqrt{3}i\mathbf{v}_1 + \dfrac{3}{\sqrt{6}}\mathbf{v}_2 - \dfrac{1}{\sqrt{2}}\mathbf{v}_3$

EXERCISE SET 9.6 (page 451)

1. (a) $\begin{bmatrix} -2i & 4 & 5-i \\ 1+i & 3-i & 0 \end{bmatrix}$ (b) $\begin{bmatrix} -2i & 4 & -i \\ 1+i & 5+7i & 3 \\ -1-i & i & 1 \end{bmatrix}$

(c) $\begin{bmatrix} -7i \\ 0 \\ 3i \end{bmatrix}$ (d) $\begin{bmatrix} \bar{a}_{11} & \bar{a}_{21} \\ \bar{a}_{12} & \bar{a}_{22} \\ \bar{a}_{13} & \bar{a}_{23} \end{bmatrix}$

2. (b), (d), (e) **3.** $k = 3 + 5i, l = i, m = 2 - 4i$ **4.** (a), (b)

5. (a) $A^{-1} = \begin{bmatrix} \dfrac{3}{5} & -\dfrac{4}{5} \\ -\dfrac{4}{5}i & -\dfrac{3}{5}i \end{bmatrix}$ (b) $A^{-1} = \begin{bmatrix} \dfrac{1}{\sqrt{2}} & \dfrac{-1+i}{2} \\ \dfrac{1}{\sqrt{2}} & \dfrac{1-i}{2} \end{bmatrix}$

(c) $A^{-1} = \begin{bmatrix} \frac{1}{4}(\sqrt{3} - i) & \frac{1}{4}(1 - i\sqrt{3}) \\ \frac{1}{4}(1 + \sqrt{3}i) & \frac{1}{4}(1 + \sqrt{3}i) \end{bmatrix}$

(d) $A^{-1} = \begin{bmatrix} \dfrac{1-i}{2} & -\dfrac{i}{\sqrt{3}} & \dfrac{3-i}{2\sqrt{15}} \\ -\dfrac{1}{2} & \dfrac{1}{\sqrt{3}} & \dfrac{4-3i}{2\sqrt{15}} \\ \dfrac{1}{2} & \dfrac{i}{\sqrt{3}} & -\dfrac{5i}{2\sqrt{15}} \end{bmatrix}$

7. $P = \begin{bmatrix} \dfrac{-1+i}{\sqrt{3}} & \dfrac{1-i}{\sqrt{6}} \\ \dfrac{1}{\sqrt{3}} & \dfrac{2}{\sqrt{6}} \end{bmatrix}$; $P^{-1}AP = \begin{bmatrix} 3 & 0 \\ 0 & 6 \end{bmatrix}$

8. $P = \begin{bmatrix} -i/\sqrt{2} & i/\sqrt{2} \\ 1/\sqrt{2} & 1/\sqrt{2} \end{bmatrix}$; $P^{-1}AP = \begin{bmatrix} 4 & 0 \\ 0 & 2 \end{bmatrix}$

9. $P = \begin{bmatrix} -\dfrac{1+i}{\sqrt{6}} & \dfrac{1+i}{\sqrt{3}} \\[2mm] \dfrac{2}{\sqrt{6}} & \dfrac{1}{\sqrt{3}} \end{bmatrix}$; $P^{-1}AP = \begin{bmatrix} 2 & 0 \\ 0 & 8 \end{bmatrix}$

10. $P = \begin{bmatrix} -\dfrac{2}{\sqrt{14}} & \dfrac{5}{\sqrt{35}} \\[2mm] \dfrac{3-i}{\sqrt{14}} & \dfrac{3-i}{\sqrt{35}} \end{bmatrix}$; $P^{-1}AP = \begin{bmatrix} -5 & 0 \\ 0 & 2 \end{bmatrix}$

11. $P = \begin{bmatrix} 0 & 1 & 0 \\[1mm] -\dfrac{1-i}{\sqrt{6}} & 0 & \dfrac{1-i}{\sqrt{3}} \\[2mm] \dfrac{2}{\sqrt{6}} & 0 & \dfrac{1}{\sqrt{3}} \end{bmatrix}$; $P^{-1}AP = \begin{bmatrix} 1 & 0 & 0 \\ 0 & 5 & 0 \\ 0 & 0 & -2 \end{bmatrix}$

12. $P = \begin{bmatrix} \dfrac{i}{\sqrt{2}} & 0 & -\dfrac{i}{\sqrt{2}} \\[2mm] -\dfrac{1}{2} & \dfrac{1}{\sqrt{2}} & -\dfrac{1}{2} \\[2mm] \dfrac{1}{2} & \dfrac{1}{\sqrt{2}} & \dfrac{1}{2} \end{bmatrix}$; $P^{-1}AP = \begin{bmatrix} 1 & 0 & 0 \\ 0 & 2 & 0 \\ 0 & 0 & 3 \end{bmatrix}$

13. $\lambda = 2 \pm i\sqrt{15}$; no, since A has complex entries.

14. $\begin{bmatrix} 0 & i \\ -i & 0 \end{bmatrix}$ is one possibility.

SUPPLEMENTARY EXERCISES (page 474)

3. $\begin{bmatrix} -i \\ 1 \\ 0 \end{bmatrix}, \begin{bmatrix} 1 \\ 0 \\ 1 \end{bmatrix}$ is one possibility.   **5.** $\lambda = 1, \omega, \omega^2(=\bar{\omega})$

10. (b) Dimension $= 2$

Index

NOTES

NOTES

NOTES

NOTES

NOTES

NOTES

NOTES

NOTES

NOTES

NOTES

NOTES

NOTES

NOTES